D0609441

ENGINEERING ANALYSIS OF FLIGHT VEHICLES

HOLT ASHLEY
Stanford University

ADDISON-WESLEY PUBLISHING COMPANY
READING, MASSACHUSETTS
MENLO PARK, CALIFORNIA • LONDON • DON MILLS, ONTARIO

This book is in the
ADDISON-WESLEY AEROSPACE SERIES

Consulting Editor
DANIEL BERSHADER

Copyright © 1974 by Addison-Wesley Publishing Company, Inc. Philippines copyright 1974 by Addison-Wesley Publishing Company, Inc.

All rights reserved. No part of this publication may be reproduced, stored in a retrieval system, or transmitted, in any form or by any means, electronic, mechanical, photocopying, recording, or otherwise, without the prior written permission of the publisher. Printed in the United States of America. Published simultaneously in Canada. Library of Congress Catalog Card No. 73-8390.

ISBN 0-201-00306-6
ABCDEFGHIJ-MA-787654

PREFACE

Alfred J. Lotka (1956) has said, "The preface is that part of a book which is written last, placed first, and read least." When one sets out to produce a textbook which is also a pedagogic experiment, the preface provides his only opportunity to explain what he is up to. I hope, therefore, that the third part of Lotka's description is not predictive for the paragraphs which follow.

At the Stanford Department of Aeronautics and Astronautics I have been engaged for five years in evolving a course whose title coincides with this book's and whose first aim is to introduce entering graduate students to the applications of some important engineering sciences, especially as they affect the complex process of designing atmospheric flight vehicles. Most of these students come from backgrounds other than four-year aerospace programs. There are among them mechanical engineers, civil engineers, mathematicians, physicists, and representatives of a dozen other fields, but they all share what I consider the most important prerequisite for the course: admission as advanced degree candidates in the School of Engineering. Well-motivated seniors from our undergraduate aeronautical option also have no difficulty with the material. At the other end of a potential population distribution, a number of experienced aerospace engineers, who returned to Stanford for further study, have told me that they benefited from the review offered by this course.

As mentioned in the opening paragraphs of Section 1.1, people who first join the aeronautics and astronautics curriculum at the graduate level tend to be impatient with the traditional approach, wherein dynamics, aerodynamics, propulsion, structures, stability, control, and guidance are taught as separate disciplines before they are integrated into the process of flight-vehicle design. Now I would be the last to claim for myself the very precious qualities of the designer. As conceived of here, these are skills attained by only a few, and then usually after the better part of a lifetime of hard work in a place in which vehicle systems (not research and Ph.D.'s) are the principal product. What I do know is something of the analytical tools that enter the armamentarium of a design team. Experience has shown me that one way of describing them is through the vehicle's equations of motion as a unifying framework. In the process of examining the terms that appear in these equations and the manner in which a vehicle's capabilities are predicted, I find it natural to explore several of the most widely used tools and refer to many more.

The program of this book agrees approximately with the order in which the one-year course is presented. Chapter 1—to the surprise of some students and colleagues—talks about why aircraft, boosters, entry gliders, and the like look the way they do. This is attempted through discussion of a series of examples, whose photographs illustrate the text. The equations for rigid-vehicle motion are developed in Chapter 2, along with some essential geometrical considerations, derivation of gravity forces, and brief allusions to certain phenomena that I must neglect in the interests of brevity. In Chapter 3 the conventional scheme of presenting aerodynamic data is described, and the longitudinal equations serve as a basis for listing important problems of performance, stability, and response. Chapters 4 and 10 deal, respectively, with the aerodynamic theory of wings and bodies. Some will disagree with this program, but delaying the subject of flow over bodies to the point at which the treatment of boost and entry vehicles creates a special need for it seems to cause my students little inconvenience.

The propulsive terms are the topic of Chapter 5; one-dimensional cycle analysis is applied to turbojets, ramjets, rockets, and other systems of the same family. The words in this chapter are mine, but the ideas are a legacy from Professor Jack Kerrebrock and the "Caltech school." Chapters 6 and 7 comprise a standard but telegraphic development of small-perturbation dynamic stability and response, with the B-70 airplane used for typical calculations. Having introduced the characteristics of the short-period mode to justify the requirement of static stability, I go more systematically into this topic in Chapter 8. Also included are some simple questions of trim, control, and static performance.

The remainder of the book focuses on the vital subject which I choose to call "dynamic performance." As examples, analyses of boost trajectories, entry, and optimal trajectories make up the material of Chapters 9, 11, and 12, respectively.

I have discovered that this totality is perhaps ten percent too large for a three-quarter academic year taught at my rather leisurely pace. Others may wish to pick and choose or to cover the entire text, for that is up to the individual instructor. I regret that my own practice to date has been to short-change the discussion of entry and nearly to omit the fascinating subject of optimization. Perhaps the availability of the printed word will help me to reform.

I should also mention that the manner in which our graduate program is arranged causes some students to elect only the first quarter. By covering Chapters 1 through 4, with perhaps some brief reference to propulsion, I find myself offering them a fairly self-contained course on applied aerodynamics.

One plea must be made to experts in the particular disciplines which the book tries to synthesize. I beg their tolerance of the omissions and simplifications that are forced on me as I strive for succinctness. I ask these people to judge the book on what it is trying to achieve and to question only whether no harm has been done to the essence of their specialties. Although not as fully informed as they, I am more knowledgeable about their subjects than may appear here. With malice aforethought, some chapters contain mere hints and allusions to the further depth to which the student may wish to explore a given issue. I also attempt to furnish a selective guide to readily available literature, in the hope that readers may be stimulated to follow leads, to flesh out specific topics by the familiar process of searching libraries and tracing back references. When it is impossible to be comprehensive, one is wise to seek all the help he can get.

There is one sincere regret. It is that I have not been able to devote as much attention as they deserve to two indispensible disciplines: structural analysis, and the complex of tools which my M.I.T. friends refer to as guidance, automatic control, communication, and navigation. Some flavor of these is introduced in problems and qualitative discussions.

If ever a course employed problem assignments as an essential complement to lectures, it is the one behind this book. Among the suggested homework at the ends of Chapters 1 through 11, a few questions are of the traditional variety; that is, the student will know when he has reached the "answer" and can neatly box it for submission. The majority of the problems, however, are open-ended. Their aim is to encourage the expansion of textual material, the study of a new idea, or even, in some cases, the exploration of a new area. I find that this scheme is so novel that it must be carefully explained, lest students become either angry with me for overloading them or so captivated by a particular question that they neglect other work. Although a teacher can hardly be disappointed by the latter reaction, he is well advised to mention the problem philosophy in advance and even to suggest rough time limits on certain assignments. The teacher should also realize that what I offer here is only typical and that, if he approves the open-ended format, there are unlimited opportunities for building additional problems around his own interests.

Most graduate programs have ready access these days to computational facilities. It is largely through the homework that I attempt to bring in both digital and analog computers. This is done in such a way that it should be easily adaptable to the machinery available at the instructor's institution.

The reference list at the end of the book is not intended to be overwhelming. Rather I have followed two general guidelines in choosing citations: that they should be on the shelves of all aerospace libraries and most comprehensive engineering libraries; and that they should have proved useful and enjoyable in my own studies. There are only a few exceptions to these rules. At the same time, another teacher may wish to make quite a different selection. I have intentionally concentrated on American literature, notably on the output of two or three very productive agencies, such as the National Aeronautics and Space Administration (NASA), and its predecessor, the National Advisory Committee for Aeronautics (NACA).

The degree of standardization of such things as physical and mathematical symbols, units, and abbreviations is the result of my modest effort to conform to whatever strong consensuses seem to exist. Any minor failures do not disturb me, because soon enough the student will be immersed in the notational chaos which is engineering practice. A little exposure in advance will prepare him better. I stoutly resisted any temptation to "go metric." That transition is a "consummation devoutly to be wished," and I hope America achieves it during my lifetime. Engineers should find the personal adjustment relatively easy, however, after dealing routinely with speeds in ft/sec, mph, knots, kilometers per hour, and millimeters per microsecond.

With joy in their remembrance, I recognize the generous assistance of numerous friends, students, and colleagues. From the last group, I am especially indebted to three: Professor Bernard Etkin, University of Toronto, for inspiration received from his two splendid books on flight dynamics; Professor Jack Kerrebrock, Massachusetts Institute of Technology, for permission to adapt from his notes the Chapter 5 approach to propulsion systems; and Dr. Benjamin O. Lange, formerly of Stanford, for his notes on boost and entry which underly Chapters 9 and 11. Encouragement and advice have also been freely supplied by Professors Daniel Bershader, S. C. McIntosh, Jr., Terrence A. Weisshaar, C. Stark Draper, N. J. Hoff, and A. E. Bryson, Jr., the last three my former and present department heads. I owe a great deal to three institutions: to Stanford for creating the stimulating but tolerant environment in which a manuscript can be forged; to the University of Maryland for the visiting appointment in Aerospace Engineering during 1971–72, which yielded me precious time to get the project under way; and to the Air Force Office of Scientific Research, whose understanding support for me and my students contributed indirectly toward its completion.

Too many students gave me aid and assistance to permit my naming them all, but I hope they take pleasure from the knowledge of what they did and from the product. I shall acknowledge the substantial effort on numerical examples by James M. Summa, K. V. Krishna Rao, and Herbert Basik. My teaching assistants in the various years the course was given surely deserve particular thanks: Susann N. Shaw, James I. Lerner, Mr. Summa, Walter Apley, Ajoy Kundu, Andrew Thompson, and Michael Hirtle. Typing and reproduction of the manuscript were skillfully handled by Mrs. Mary Goodno Weisshaar, and the latter portions were completed by Miss Andrea Dinsmore.

Every person mentioned above shares in any credit which this book earns, but I must add the customary stipulation that errors and imperfections are my responsibility alone. I shall deeply appreciate having corrections and improvements called to my attention.

One postscript: my conscience tells me that in 1972 there are much more urgent human concerns than the engineering analysis of flight vehicles. Higher on my own list of priorities are that the United States shall disengage completely in Southeast Asia, that world population shall be stabilized, and that conservation of energy and material resources shall become a universal practice. Nevertheless, there are innumerable people better qualified than I to discourse on any of these matters. This book speaks only of what I love and also know.

Stanford, California H. A.
October 1972

CONTENTS

Chapter 1 The Morphology of Flight Vehicles

1.1 Introduction 1
1.2 Key factors affecting vehicle configuration 2
1.3 Some representative flight vehicles 5
1.4 Problems 25

Chapter 2 Equations of Motion for Rigid Flight Vehicles

2.1 Definitions; vector and scalar realizations of Newton's second law . . 27
2.2 The tensor of inertia 30
2.3 Choice of vehicle axes 32
2.4 Orientation of the vehicle relative to the ground; flight-path determination 34
2.5 Gravitational terms in the equations of motion 38
2.6 The state vector 39
2.7 Three significant phenomena that have been neglected 41
2.8 Problems 47

Chapter 3 Introduction to Vehicle Aerodynamics

3.1 Aerodynamic contributions to X, Z, and M_p; dimensionless coefficients defined 51
3.2 Equations of perturbed longitudinal motion; categories of problems in flight dynamics 58
3.3 Problems 62

Chapter 4 Aerodynamic Terms for Equations of Motion; Airloads on Thin Wings

4.1 Inviscid fluid motion past streamlined shapes 65
4.2 Aerodynamic theory for small disturbances from a uniform flow . . . 67
4.3 Thin airfoil in two-dimensional, incompressible flow 74
4.4 Compressibility correction at subsonic and supersonic speeds 84
4.5 Thin airfoil in two-dimensional supersonic flow 89
4.6 Shock-expansion methods for thicker supersonic airfoils 96
4.7 Three-dimensional wings in subsonic flight 102
4.8 Three-dimensional wings in supersonic flight 117
4.9 Problems 124

Chapter 5 Propulsive Terms—One-Dimensional Analysis of Jet and Rocket Propulsion

5.1 Turbojet operation; steady-flow energy equation; partition of thrust and drag 133
5.2 One-dimensional analysis of ideal ramjet and turbojet 140
5.3 Rocket propulsion 152
5.4 One-dimensional analysis of turbojet with realistic component efficiencies. 156
5.5 Actual turbojet performance 163
5.6 Problems 169

Chapter 6 Small-Perturbation Response and Dynamic Stability of Flight Vehicles

6.1 Equations of motion; aerodynamic approximations; stability derivatives . 173
6.2 Dimensionless equations of motion 180
6.3 Estimation of stability derivatives: longitudinal 186
6.4 Estimation of stability derivatives: lateral 197
6.5 Problems 203

Chapter 7 Solution of the Small-Perturbation Equations of Motion

7.1 A simplified look at a lateral response 207
7.2 Lateral-directional normal modes; comparison with the complete lateral equations of motion; roots-locus technique 219
7.3 Longitudinal stability and response; exact and approximate properties of the normal modes 230
7.4 Handling qualities 238
7.5 Problems 241

Chapter 8 Static Stability, Trim, Static Performance and Related Subjects

8.1 Impact of stability requirements on design and longitudinal control . . 247
8.2 Static performance 256
8.3 Problems 260

Chapter 9 Dynamic Performance: Boost from Nonrotating and Rotating Planets; Numerical Integration of Ordinary Differential Equations

9.1 Introduction 265
9.2 Numerical integration of ordinary differential equations 268
9.3 Simplified treatment of boost from a nonrotating planet 276
9.4 An elementary look at staging 281
9.5 Equations of boost from a rotating planet 286
9.6 Problems 300

Chapter 10 Aerodynamic Terms for Equations of Motion; Slender and Blunt Bodies

10.1 Some information on booster airloads 303
10.2 Aerodynamic theory for slender, pointed bodies 312
10.3 Improved methods for supersonic bodies of revolution: hypersonic flow . 326
10.4 Newtonian theory and heat transfer 332
10.5 Problems 334

Chapter 11 Dynamic Performance: Atmospheric Entry

11.1 Introduction; equations of motion 337
11.2 Approximate analysis of gliding entry into a planetary atmosphere . . 338
11.3 Problems 347

Chapter 12 The Uses of Optimization

12.1 Introduction 349
12.2 The mathematics of optimization 350
12.3 Two elementary examples 357
12.4 Optimal trajectories 363
12.5 Problems 369

References 371

Index 379

CHAPTER 1

THE MORPHOLOGY OF FLIGHT VEHICLES

1.1 INTRODUCTION

For someone who is fascinated with the phenomenon of flight and the aircraft that achieve it, there are several paths to their quantitative understanding. The traditional approach in schools of engineering proceeds from the mathematical, physical, and chemical fundamentals through a series of engineering sciences focused on solid and fluid mechanics to disciplinary studies with such names as structures, aerodynamics, propulsion, stability, control, and guidance. Only when all these foundations and pillars are erected for the "temple of aerospace engineering" is the student usually invited to synthesize his knowledge in a design project or in analyses of the overall configurations of flight vehicles.

Particularly for those who enter the field at an advanced or postgraduate level, we are concerned that this traditional path may call for unwarranted patience. It demands faith that, in the long run, all the pieces will fall together. Still building on the engineering-science fundamentals and trying to preserve much that is attractive about the disciplinary route, we offer in this book a partial alternative. We have found, in the vehicle's equations of motion, a unifying framework for studying many important questions that affect its design and arise during the extensive analytical investigation which contributes to that design. In Chapter 2 we shall present those equations and begin our discussion of their various terms. By way of introduction and motivation, however, we begin here with some qualitative descriptions of flight vehicles and the reasons for their general arrangement.

Much of what we say is concentrated on powered aircraft, that class of vehicles which is capable of sustained cruising flight within the lower atmosphere. Clearly, our point of view can be extended to vehicles which transport spacecraft to and from their extra-atmospheric environment. Even spacecraft themselves, when their structures are sufficiently rigid, fall within our scope, and some attention will be paid to all of these. The static and dynamic operation of submersible ships can also be analyzed, in common with balloons and airships, by rather standard aeronautical methods. Only when vehicles contact the interface between two terrestrial media, as in the case of surface ships and wheeled landcraft, do some of the force systems which act on them fall beyond the range which we hope to cover.

1.2 KEY FACTORS AFFECTING VEHICLE CONFIGURATION

Whole treatises have been written on why airplanes look the way they do. Stinton (1966) is a splendid example; the reader who would enjoy a discursive and less concentrated treatment of the subject is invited to browse in Stinton's book and others like it. Since our aim here is introductory, we must be brief and therefore skim over numerous important details. The several photographs and sketches accompanying this chapter will illustrate our remarks.

Every vehicle serves some transportation function—for people, cargo, armament, measuring instruments, or even just its own equipment and fuel. The enclosed volume which carries and shields this payload is the fuselage, although we observe that other vehicle elements usually assist in this function, an extreme case being the flying wing. The most efficient shape for simply containing material would be a sphere, whose minimum surface area for a given volume would need the least weight of structure for protection and support. This is a good shape for spacecraft and has often been adopted. When rapidly moved through an atmosphere, however, the sphere experiences too high an aerodynamic force (drag D) opposite to its direction of motion. Early designers, looking at birds and fish, found the remedy in streamlining—making the fuselage length in the flight direction quite large compared to its cross-sectional dimensions, and inclining most of its surface area at small angles to the relative wind.

Drag is unavoidable in practice. Hence, except for such transitory operations as gliding and atmospheric entry, a propulsive force is required, and it is furnished by one or more engines or rocket motors. Whether this system involves a reciprocating engine and propeller, turboprop, turbojet, turbofan, ramjet, liquid rocket, or solid rocket, it works by the principle of reaction propulsion. Backward momentum is imparted to the atmospheric gases (propeller), to material stored within the vehicle (rocket), or to a combination of these two, including products of combustion (most other devices). Thrust—which is the reactive force thereby applied to the vehicle—may simply counterbalance drag or may produce longitudinal acceleration, increased altitude, or, for a spacecraft, changes in the parameters of an inertial orbit.

The propulsion system is often housed in a distinct element of the vehicle, such as a nacelle or jet-engine pod. Alternatively it may be internal, with only an air inlet and exhaust nozzle visible from the outside. On some extreme configurations, such as the hypersonic ramjet, large portions of the aircraft's surface participate in slowing down the air (compression) before it enters the combustion chamber, and then reexpanding the combustion products into a high-speed exhaust slipstream.

Another force which dominates the performance on all but interstellar craft is the weight W exerted by gravitational fields. In level cruising flight, weight is counterbalanced by an aerodynamic force (lift L) normal to the flight direction. Some lift is usually contributed by the fuselage, but a more efficient device for its production is the wing—a flattened, often cambered or twisted surface which intersects the fuselage, but usually has its longest dimension (span) normal to the airspeed vector. In rectilinear cruising flight, the wing lies close to a plane parallel to the local horizontal. A well-designed wing is a marvelously effective device for lift generation, to the degree that the ratio L/D may approach 20 on powered aircraft flying below the speed of sound and exceed 40 on a high-performance sailplane.

The next most prominent feature of atmospheric aircraft is a group of lifting surfaces known as the tail or empennage. The most common arrangement has its location at the rear of the fuselage and consists of one portion (horizontal stabilizer) roughly parallel to the wing plane and a second (vertical stabilizer or fin) which is perpendicular, lying in the vehicle's central plane of symmetry. The immobile parts of these surfaces play a stabilizing role similar to a weathervane. Thus, when the vehicle sideslips, acquiring a component of relative wind normal to its symmetry plane, the vertical stabilizer gives rise to a *yawing moment* about a vertical axis through the center of mass (CM) and tries to rotate the fuselage axis toward the relative wind. As we shall discuss later, one historical feature of design evolution has been the tendency of the fin's area to grow progressively larger compared to the wing's.

The horizontal stabilizer applies *pitching moments*, which work to fix the inclination of the relative wind to the wing plane (angle of attack). It also assists with the "trimming" process of canceling pitching moments about the CM due to the wing lift, fuselage, etc.

The need for *controls*—movable portions of the various surfaces in the form of rotatable flaps along the trailing edges—is most clearly understood in connection with trimming. The wing lift depends on both angle of attack and airspeed, so that this angle must be readily adjustable to ensure that the weight can be supported in various flight conditions. Clearly, equilibrium demands that the resultant pitching moment about the CM be zero, but the contribution of wing lift to this moment often varies rapidly with angle of attack. The most efficient way to make the required pitching-moment adjustments has usually proved to be by controlling the tail lift with a trailing-edge "elevator." On vehicles which are to fly near or above the speed of sound, however, the tail must normally possess so much power that the whole horizontal stabilizer is rotated at the pilot's command. For instance, the Boeing 707 and many other transports are furnished with both a screw-jack arrangement that sets the stabilizer incidence for gradual trim changes and a rapidly actuated elevator for maneuvering, pulling up the nose during landing, and the like. Occasionally the horizontal stabilizer is placed far forward on the fuselage; in such cases it is called a *canard* (like a duck's bill).

For altering flight direction, banking the wings relative to the horizon plane, and performing other maneuvers, control surfaces must also be provided which can affect either the yawing moment or the moment about a longitudinal axis through the CM (*rolling moment*). Yawing control is supplied by the *rudder*, a flap acting at the trailing edge of the vertical stabilizer. The rudder has a trimming function in such situations as a steady turn or multi-engine flight when one engine is inoperable. Rolling is accomplished by *ailerons* and/or *spoilers*, placed near each wing tip and deflected in an antisymmetrical manner. At high speeds, rolling moment may be exerted simply by differential rotation of two all-movable horizontal stabilizers. The wing flaps resemble control surfaces, but they are actuated slowly and only at low speeds, where they augment the lift to facilitate landing or takeoff.

Devices available to the pilot for moving the various control and trimming surfaces, cockpit layout, landing gear, and other internal details of the vehicle fall somewhat out of this chapter's scope. The book by Langewiesche (1944) has an excellent discussion of how the airplane flies and the proper use of controls. It is worth mentioning here that many large controls on high-performance vehicles are actuated by hydraulic or electric devices, since their direct mechanical operation by cables may require excessive effort from the pilot. Both power-boosted and manual controls may also be provided with automatic gadgetry which assists the pilot: *autopilots* are employed to help him maintain the direction, speed, and altitude of flight; elaborate *stability augmentation systems* (SAS's) modify the apparent dynamic behavior so as to improve controllability and render the "handling qualities" more acceptable to him.

Section 1.3 describes a historical series of flight vehicles, suggesting some of the reasons for their configurations in the light of the foregoing generalities. We must initially mention a few more observations which, we believe, apply quite broadly. The first is that any significant feature of a design can best be understood as the result of compromise among the recommendations of several engineering specialists who view it from different perspectives. Such compromises are known as "tradeoffs." An example might be determining the proportions of a low-speed wing. The aerodynamicist concerned with drag and cruising efficiency wants the wing area to be small (to minimize skin friction) and the span to be very large (to reduce the "induced-drag" penalty for producing lift, as we shall discuss further below). He is opposed by another specialist, seeking safe takeoffs and landings on short runways, who insists on high area to give plenty of lift at low speed. The structures man argues for reduced wingspan, so that the internal bending loads due to lift can be sustained with the least amount of material and the weight thereby reduced. Several others may simultaneously insert their requirements into the wing-design process. The role of skillful engineering leadership is then to arrive at an optimum configuration—a tradeoff which, in some sense, represents the best resolution of the conflict.

Another key concept is that of the *design point*. For many aircraft, considerable flying time is spent near a certain combination of standard atmospheric altitude and *Mach number*,* and there is a desire to ensure especially efficient

* The *Mach number* M, or the ratio of airspeed to speed of sound in the surrounding gas, is often used in place of airspeed. Flow fields have distinctly different characteristics in different ranges of M, such as the *subsonic* ($M < 1$), *transonic* (M near 1), and *supersonic* ($M > 1$).

operation at this point in the envelope of possible flight conditions. The various parts are therefore optimized with this condition in view, although constraints such as a limit on takeoff run cannot be completely overlooked. Some vehicles give rise to extreme difficulties because of having two distinct and incompatible design points, one example being the penetration bomber which must cruise supersonically at high altitude and also fly down over the "nap of the earth" just below $M = 1$. There are also air-superiority fighters and multipurpose aircraft whose design is dominated by considerations of dynamic maneuverability or by the need to do several different things with equal efficiency.

As you know, flight vehicles come in a bewildering multiplicity of shapes, sizes, and arrangements of components. As we attempt to clarify the rationale for these differences, it is useful to ask questions such as "What was the design point?" and "What were the technological resources available at the time of development?" These are not always easy to answer, but one is surprised, after having reached some depth of understanding, at how often the actual configuration seems to lie pretty close to an optimum under the given circumstances. Aerospace engineers tend to set for themselves very demanding objectives and requirements. Someone has remarked that, given the technology when they were first built, most successful aircraft are just barely able to do what is asked of them. Whether one is studying a Wright biplane, a supersonic transport, or a man-rated booster, he discovers much truth in this "theorem of the barely possible."

1.3 SOME REPRESENTATIVE FLIGHT VEHICLES

Since pictures are a poor substitute for the real thing when the latter has the dimensions of, say, the B-70, we urge you to seek opportunities for first-hand inspection of machines like those reviewed in this section. The Smithsonian Institution, in Washington, D.C., exhibits an outstanding collection of historical aircraft, engines, and spacecraft, as does England's National Science Museum in the Kensington area of London. At the Wright-Patterson Air Force Base near Dayton, Ohio, there is free admission to the Air Force Museum, whose unique series of transport, military, and experimental aircraft covers the period from 1912 through to the present.

Limited both by space and the aims of this book, we shall point to only a few highlights of a dozen or so designs. Much more extensive, for instance, is the treatment of Stinton (1966); the forthcoming treatise by Hoff (1974) promises to be a compendium of engineering information and critical judgments on the world's aircraft. Each year, McGraw-Hill publishes *Jane's All The World's Aircraft*, an illustrated summary of performance and other features of current types (see Taylor, Ed.).

1. Strut and Wire-Braced Biplane of World War I Era (Curtiss R-4)

During the first decade of powered flight, 1903–1913, configuration was going through rapid, somewhat haphazard evolution. There were therefore many

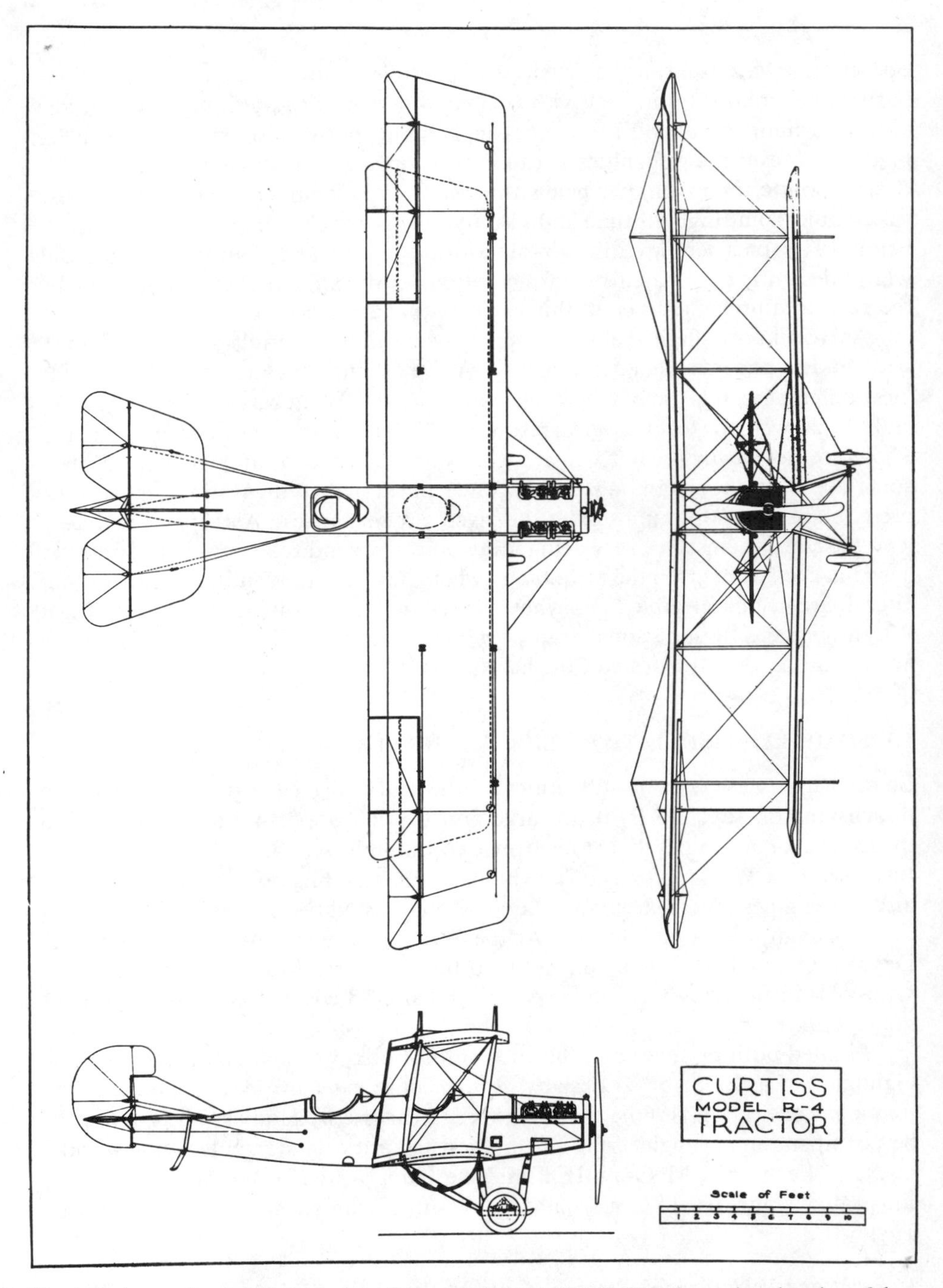

Fig. 1.1. A strut-braced biplane aircraft flown in World War I. This three-view drawing of the Curtiss R-4 is reproduced courtesy of Mr. Arnold Kossar, Curtiss-Wright Corporation.

deviations from the prototypic arrangement outlined in Section 1.2, e.g., the Wright Flyers had canard stabilizers and employed antisymmetric warping of the wing structure for roll control. By the time of the R-4 (Fig. 1.1), however, morphology had evolved until it was close to that of more recent aircraft.

A modern designer would naturally reject the biplane wing because the numerous compression struts and bracing wires required for strength and stiffness cost a huge penalty in drag. Several factors forced the choice of biplanes and even occasional triplanes. The tradition persisted of adopting very thin airfoil sections, with maximum thickness only a few percent of wing chord, and thus providing little depth for the structural beam which sustained the lift loads. A certain minimum span of wing was essential to ensure that thrust exceeded drag just after takeoff and in slow flight generally, as can be seen from the formula relating *induced drag* (due to lift) D_i, wing span b, weight W, and dynamic pressure* q:

$$D_i = \frac{W^2}{\pi q b^2} \tag{1.1}$$

D_i constitutes the bulk of drag at low v_c, with the contribution of skin friction, which is roughly proportional to v_c^2, predominating near top speed. Finally, the best available materials were high-strength woods, coated fabric for skins, and steel. The steel was so heavy that its application had to be limited to engine parts, wires, and certain truss structures. With all these constraints, the designer had to go to a biplane in order to obtain a structural "box" adequate to the bending moments, shears, and torques which it had to withstand.

Another vital limiting factor was thrust, inasmuch as reciprocating engines of the early twentieth century were far heavier than the 1 or 2 lb per horsepower (HP) that characterize current models. The thrust available, without the aid of a variable-pitch propeller, tended to be inversely proportional to v_c (i.e., constant "thrust HP"). Working against the v_c^2-proportional drag term, such a propulsion system could achieve only modest peak performance. At the same time, minimum safe flying speeds could not be reduced, since the elaborate, variable-geometry devices now used to generate high lift at low v_c were not available. There resulted both a narrow performance envelope between lowest and highest possible speeds and a modest "ceiling" (maximum altitude of continuous flight). The ability to turn and carry out other maneuvers which rely on power was similarly restricted. Looking at the rather small fixed and movable control surfaces which are typical of this aircraft generation, we suggest that, in most cases, they were adequate for trim and acceleration during the maneuvers of which it was otherwise capable.

---

* The quantity $q \equiv (\rho_\infty/2)v_c^2$, where v_c is the speed of flight and ρ_∞ ambient atmospheric density, appears in many aerodynamic formulas. It is approximately equal to the pressure rise felt near the center of a plane surface held normal to wind v_c. The derivation of (1.1) will be given in Chapter 4.

2. Cantilever Monoplane of All-Metal Construction (Douglas DC-3)

Although certainly not the first, the DC-3 (Fig. 1.2) is the most widely recognized of an important family of vehicles which began to emerge in the 1920's. Among features which distinguish them from most predecessors are enclosed cockpits and cabins, engines blended into wing or fuselage outlines, cowlings to assist with

Fig. 1.2. The DC-3, a transport with straight, cantilever monoplane wing and all-metal construction (reproduced courtesy of Douglas Aircraft Company).

air-cooling of cylinders, and retractable landing gear. Perhaps their most prominent innovation, however, was a combination of all-metal primary structure with thick, cantilever, monoplane wings.

The availability of strong, light, and relatively inexpensive aluminum alloy offered a very efficient alternative to wood and steel. Equally important was the realization, stimulated by aerodynamic research in several European countries, that high L/D ratios were attainable with airfoil thicknesses in the range of 20–25% of local wing chord. The phenomenon of *stalling*—breaking away of flow from the upper surface at high angle of attack—was rendered less violent by these thick wings, while simultaneously the designer was given much greater "box" depth. With no external bracing or, at most, a faired diagonal strut from a wing spar down to a fuselage frame, a light structure could be engineered out of thin sheets of aluminum, carrying spanwise and chordwise stiffeners. This arrangement could support a lift equal to several times the gross vehicle weight.

For such configurations, a parameter which measures both the aerodynamic efficiency and the challenge to the structural designer is *aspect ratio* $\mathcal{R} \equiv b^2/S$ (b = wingspan or tip-to-tip distance, as before; S = total area of wing, seen in plan view, and including area within the fuselage intercepted by inboard projections of the leading and trailing edges). The optimum value of Æ for a cantilever wing seems to depend on available structural technology and on the *limit load factor* n_z (ratio of the maximum value of L, expected to be encountered during flight operations, to W) required when determining the structural strength. For the DC-3, Æ $= 9.14$ was associated with an $n_z =$ slightly over four assumed in the original design.

Fig. 1.3. Diamant 18 Sailplane (reproduced courtesy of the Soaring Society of America).

An extreme instance of the effect of the structural–aerodynamic tradeoff in establishing aspect ratio is furnished by high-performance sailplanes. Thus the modern Diamant 18 (Fig. 1.3) uses Æ $= 22.7$ to achieve a claimed peak L/D of 45 in gliding flight. Of wooden construction, this remarkable vehicle has a maximum all-up weight of 970 lb and a wingspan of 59 ft.

3. Subsonic, Swept-Wing Jet Aircraft, Optimized for Cruising at a Single Design Point (Boeing B-52)

In the late 1940's appeared the first of a long series of transport and bombardment aircraft that are consistently marked by quite high-aspect-ratio wings and by

Fig. 1.4. Cruising aircraft with swept-back, cantilever wing and tail surfaces, designed for flight at high subsonic M. The B-52H is reproduced courtesy of the Boeing Company.

swept-back lifting surfaces on wing and empennage. The B-52H (Fig. 1.4) is representative of these, except that its Æ = 8.55 is higher than average and associated with an n_z slightly less than 2 (based on the absolute maximum weight at which the fully loaded vehicle is permitted to fly).

This type was made possible, first of all, by the development of efficient turbojet engines. These simple and light propulsive devices combined substantial thrust with fairly low fuel consumption in the stratosphere at M = 0.7 to 0.9+, where the vehicle design points generally fall. Operation being restricted to the subsonic régime, the wings must still have both high aspect ratio *and* span. The former ensures favorable cruising L/D, whereas the latter is connected with safe runway liftoff, followed by a lively angle of climbout.

The novel feature is sweep—a contribution from aerodynamic research during World War II. It was discovered that undesirable drag increases and flow unsteadiness, which arise on an airfoil as flight M approaches the transonic range, could be delayed by inclining the wing axis away from normal to the plane of symmetry. Indeed, as we shall see later, the appearance of transonic effects on a large-aspect-ratio wing is approximately governed by the parameter M cos Λ, where Λ is the angle which some representative line from root to tip forms with this normal. A wing with $\Lambda = 35°$, like that of the Boeing 707, therefore experiences flow equivalent to Mach number 0.69 at its design point (about $M = 0.84$); L/D of nearly 20 can be maintained.

Unfortunately, the structural engineer was confronted with a beam whose slenderness is proportional to $b/\cos \Lambda$. His task has remained tolerable only because of the appearance of aluminum alloys with improved ratios of strength to density. Advances in manufacturing technique, such as huge milling machines and controlled chemical removal of metal, have led to thick, tapered wing skins with integral stiffeners which reduce the weight of bolts and rivets. Interstices of the primary wing structure have been used to contain fuel; together with wing-mounted engines, such wing tanks help to lower design stresses by exerting weight and inertia loads that partially counterbalance those associated with lift.

Some of the structural material in these wings is there because of *aeroelastic* phenomena—static and dynamic deformations due to interaction with the aerodynamic forces. It is found, for instance, that the rolling moments generated by an aileron at high q are diminished both by the twisting and the bending produced when it is rotated. Its power can be increased by stiffening the structure, but sometimes the designer chooses alternative means for high-speed roll control (inboard ailerons, spoilers, or differential rotation of the horizontal stabilizer). A very dangerous aeroelastic problem is *flutter*—the onset, beyond some speed–altitude combinations, of unstable and destructive vibration of a lifting surface in an airstream. Although it often demands greater stiffness, flutter can sometimes be avoided on this class of aircraft by "balancing" through favorable chordwise and spanwise location of the engine masses.

The success of the high-aspect-ratio swept-back wing, especially for transport vehicles, is attested by its incorporation into numerous designs over a period of some 25 years. Prominent examples in the United States include the 707/720 series, DC-8, Convair 880/990, C-141, 727, 737, DC-9, C-5A, 747, DC-10, and L-1011. Their evolution is marked by the availability of progressively better engines, notably turbofans with ever-increasing *bypass ratio*,* and by a growing knowledge of how to shape wing contours so as to minimize unwanted transonic-flow effects. Structural progress has made possible the use of airfoils with smaller thickness ratio, down to 10% on the 747 with its cruising M near 0.9. The aerodynamicist has learned how to achieve more favorable interference between the fuselage and wing flow fields, one consequence being the larger body diameters and greater useful volumes which characterize the current generation.

Swept-wing commercial transports are expected to be developed during the

---

* Chapter 5 contains a quantitative discussion of these propulsion systems. The turbofan combines a turbojet "core" with an annular ducted fan, which acts like a small propeller imparting momentum directly to the airstream. The mass flow rate through the fan, divided by that through the central turbojet, is the bypass ratio.

1970's which will benefit, in speed and efficiency, from research on refined transonic airfoils conducted by Whitcomb and collaborators at NASA Langley Research Center (see e.g., Braslow and Ayers 1971). Exactly what value of M will mark the cruising design point is still under discussion, but it will occur near and slightly below unity.

4. Subsonic Jet Fighter (North American F-86)

The turbojet engine made its operational debut on pursuit and interceptor aircraft (P-80, HE-262, etc.) near the end of World War II. One of its main advantages was to overcome the speed limitations associated with transonic effects on conventional propellers, which appear at distinctly subsonic flight M because of the added relative velocity due to rotation. It was therefore natural that this higher-speed propulsion should soon be wedded to the technology of the swept lifting surface. Illustrated in Fig. 1.5, the F-86F is an outstanding representative of the American fighters which emerged in this category.

This airplane does not have a single design point, but is intended for close air-to-air engagement against hostile fighters at intermediate to high altitudes. Its many configurational dissimilarities from the larger cruising jet aircraft may

Fig. 1.5. Subsonic, swept-winged fighter aircraft, the F-86F (reproduced courtesy of North American Rockwell).

be explained in terms of the resulting need for rapid speed changes, high maneuverability, and small turning radius. Much more powerful propulsion was required, such that the F-86's ratio of peak thrust available at sea level to maximum weight at takeoff was about 0.39 for the clean airplane. Although swept back to nearly $\Lambda = 45°$, the wing obviously has much lower aerodynamic and structural aspect ratios than the B-52/707 class—one consequence of a very different structural tradeoff. To attain the desired turn rates, this wing was expected to develop lift greater than seven times the weight at limit conditions. In actuality, n_z's above eight were occasionally observed, and high loads were applied repeatedly, so that fatigue had an important influence on the structural "life."

Also clear from Fig. 1.5 is the large air inlet at the fuselage nose. For such single-engine fighters, this arrangement proves convenient unless the fuselage becomes so extended that friction with the walls of a long internal duct may deprive the air of too much momentum. On more recent designs, however, another reason for turning to inlets at the fuselage side or some other aft position is to reserve the nose for a radar antenna, needed in the detection of enemy aircraft, turbulent clouds, and other targets.

5. Supersonic Delta-Winged Interceptor (Convair F-102)

Among the products of wartime research whose values were less immediately recognized was the triangular or delta lifting-surface planform. Its advantages are especially apparent when a mission calls for a supersonic "dash" as well as acceptable subsonic performance and/or relatively short takeoff and landing runs. The F-102 (Fig. 1.6) exemplifies these circumstances and also permits us to note certain additional aerodynamic features.

We have already referred to the key function of the elevator and horizontal stabilizer as balancers of pitching moments about the CM. Not only must the ordinary changes (due to angle of attack) in the wing and fuselage contributions to these moments be counterbalanced, but a further burden is put on the tail due to the transition from subsonic through transonic to supersonic flight. For a large-aspect-ratio wing with zero to moderate sweep, this effect of passing through the sonic barrier may amount to a rearward shift of nearly 25% of the wing chord in the apparent center of action of wing lift.

By contrast, a delta with leading-edge Λ higher than about 60° experiences only slight Mach-number variations in its moment characteristics. Recognizing this insensitivity, together with the substantial pitching moment arm which the large chord length of the delta offers to trailing-edge controls, the designers of such vehicles as F-102, B-58, and the British "V-bombers" discovered that they could dispense entirely with the horizontal stabilizer. The F-102 controls are called *elevons*, since their symmetrical displacement accomplishes the pitching function of the elevator, whereas an antisymmetrical rotation produces rolling.

The slender fuselage and 60°-delta of the F-102 are optimized for supersonic drag performance. We shall see later that a large proportion of D at $M > 1$

Fig. 1.6. Supersonic, delta-winged interceptor aircraft, the F-102 (reproduced courtesy of General Dynamics).

comes from "wave-making" in the atmosphere and that the part of this due to wing thickness is approximately proportional to the square of the ratio of maximum wing depth to chord. The small structural aspect ratio and large streamwise dimensions of wings like the F-102 permit thickness ratios in the range of 3–5% to be used without adversely affecting the structure or weight tradeoffs.

Both transonic and supersonic drag are also favorably influenced by the fuselage "waisting" which is evident in Fig. 1.6. Indeed, this vehicle represents the first practical adaptation of the "area rule" idea (Hayes 1947, Whitcomb 1956), according to which the cross-sectional area distribution of a streamlined wing–body combination can be tailored so as to ameliorate the penetration of the sonic barrier. Pursuant to other dictates of aerodynamic theory, the wing is carefully cambered and twisted in such ways that supersonic drag due to lift is minimized around the design point.

A few other features of the delta planform are worth mentioning. Its *wing loading* W/S turns out lower than that of other designs to similar requirements (compare $W/S = 53\ \text{lb/ft}^2$, based on the gross weight of the F-102, with $148\ \text{lb/ft}^2$ for the F-104G). Not only is maneuverability thereby enhanced, but the speeds for liftoff and landing touchdown, which tend to be inversely dependent on

$(W/S)^{1/2}$, are reduced. As its angle of attack is increased, the delta does not exhibit the direct proportionality of lift to angle α followed by sharp stalling which characterizes straighter wings. Rather, a stable aerodynamic "vortex" begins to separate from each of the leading edges at α's as low as 4–5°, its intensity growing with α, its influence on the L-versus-α curve giving rise first to a parabolic nonlinearity and ultimately to a very smooth peak. Because flight at high angles is possible, however, the pilot's visibility may be restricted, and awkwardly long landing gear may be needed.

Study of other supersonic interceptors and air-superiority fighters of the past two decades shows us that many designers have not regarded the delta as the best choice. Individual ingenuity and differing requirements have led to such diverse solutions as the following:

- The F-100 series, with 45°-swept-back wing and conventional empennage. (Several USSR MiG fighters are similarly arranged.)
- The F-104, with rather low-aspect-ratio straight wing and T-tail.
- The F-4, which combines a highly tapered, moderately swept wing with a horizontal stabilizer whose *anhedral** improves stability and control.
- The F-111 and F-14, whose variable sweep is discussed in Subsection 7 below.
- The Swedish Viggen, on which the delta canard trimming surface is located so as to achieve favorable vortex interaction with the main wing.
- The F-15, whose requirements for sustained energy and controllability at high turn rates result in lower-than-average wing loading plus a ratio of thrust to weight (at fully loaded takeoff) in excess of 1.2.

It is worth mentioning that nearly every nondelta supersonic fighter needs an all-movable horizontal for adequate trim throughout its wide performance envelope. All of them employ one or two jet engines, mounted within the aft fuselage. There is a remarkable variety of single and dual air inlet arrangements.

6. Aircraft Designed for Long-Range Supersonic Cruise (North American B-70)

As we shall see later when we discuss the subject of cruising performance, the distance an airplane can fly with a given fraction of its weight in the form of fuel is proportional to the *range factor* $M(L/D)$. Since this product falls in the vicinity of 15–16 for the 707/DC-8 family, any transportation vehicle designed for continuous flight at supersonic speeds must equal or exceed $M(L/D) = 16$ before it can compete either economically or in range of operation. In the late 1950's it was realized that a combination of high gross weight with inlets, engines, and

---

* This adjective identifies a lifting surface whose plane is inclined downward at a solid angle to the airplane's horizontal plane. The opposite, upward inclination is called *dihedral*, and is often seen on unswept wings in the low position (cf. the DC-3, Fig. 1.2).

Fig. 1.7. Long-range, supersonic cruising aircraft, the North American B-70 (reproduced courtesy of North American Rockwell).

airframe carefully optimized for a design point between M = 2 and 3 could meet the stated goal.

The 500,000-lb B-70 (Fig. 1.7) was the first vehicle in this class to fly. It attained its intended cruise M near 3.0 in late 1963. The greatest challenge to the B-70's success lay in coping with the temperatures generated by motion through the atmosphere at these high speeds, since nearly 600°F was experienced on certain portions of the outer surface. Advanced steel alloys had to be used for structure in place of the traditional aluminum, which loses strength due to softening at such temperatures. Only two B-70's were completed, and their principal role turned out to be research on various aspects of supersonic flight. Cost of development and difficulties with fabrication and assembly of the untried structural materials prevented further construction and operation. Yet the engines, inlets, controls, and many other elements of the design functioned very well.

An extraordinary achievement, the B-70 was clearly ahead of its time. Although the lift-to-drag ratio was never accurately measured, its maximum value, with account taken of drag contributed by the canard trimming surface, was believed to be between 6 and 7. The corresponding range factor is approximately $M(L/D) = 20$.

In common with the second generation of supersonic cruisers (Concorde, Tu-144, and the Boeing SST—a development which was suspended by the U.S.

Congress in March, 1971), B-70 uses the delta main lifting surface, its leading-edge Λ being about 70°. In connection with the comments on delta-wing aerodynamics in Subsection 5 above, we remark that Fig. 1.7, with the airplane in the touchdown condition, illustrates a dramatic mechanism for visualizing the leading-edge vortices at large α. The air near their "cores" has a high angular velocity about the axes of these vortices, which are lines originating near the roots, then extending downstream and above each wing. The swirling motion reduces the pressure in these cores (cf. the surface depression near the center of a whirlpool). Moisture in the atmosphere—already at high relative humidity in this case—is caused to condense so that the droplets reveal the vortex path. With close scrutiny, we are also able to discern a smaller vortex that trails backward from the tip of the right canard.

The B-70's canard consists of a trailing-edge elevator and a stabilizer which can be rotated slowly for trimming; these surfaces are required to reduce trimming contributions to drag and for the maneuvers of takeoff and landing at satisfactorily low speeds.

The B-70 embodies many other aerodynamic innovations. Each elevon is split into six separately actuated "fingers." For yawing control there is a large-area rudder on each of the twin vertical stabilizers. The wingtips can be folded downward, to an anhedral angle of 25° for intermediate cruising conditions and to 65° for M = 3 at high altitude. The reason for this use of variable geometry relates to lateral stability in the low-density gas of the upper atmosphere. As we shall see in Chapter 7, the critical damping ratio for lateral oscillations of a fixed aerodynamic configuration varies as $\rho^{1/2}$. Greater area of vertical stabilizing surface can compensate for this loss of damping, however, and folding the B-70 wingtips did the job effectively.

Not visible in Fig. 1.7 are large under-wing inlets on either side of the centerline, each capturing air for three of the six J-93 turbojet engines. These inlets have "ramps," in the form of segmented plane and curved surfaces that are rotatable about vertical axes so as to produce favorable patterns of oblique "shock waves" (cf. Chapter 4) for slowing down the air from supersonic speeds. The B-70 inlet arrangement not only achieved efficient compression, which assisted the engine performance, but also augmented the surface pressures over the central portion of the lower wing. The resulting "shock lift" helped generate high L/D at reduced α.

The general outlines of the B-70 are seen to be reflected in the Concorde and Tu-144 supersonic transports, with the addition of numerous refinements contributed by aerodynamic research during the intervening years. These are, however, basically aluminum airplanes. Their range factors are similar to the B-70, but the cruise Mach numbers are restricted structurally below about 2.3. The final 1970 version of the U.S. SST design (Boeing Model 2707-300) differs in several important respects. Notably, its wider delta wing (leading-edge Λ = 45°) and all-movable horizontal stabilizer were chosen so as to combine maximum range factor at the M = 2.7 primary design point with comparably efficient high subsonic operation. The latter alternative would permit flight over populous land areas without the sonic boom from shock waves, unavoidably associated with the lift and the volume displacement at $M > 1$. Most of the SST structure is of titanium alloy. This material tends to be lighter than steel for a given load-carrying capacity, and its high-temperature properties allow even for some increase in the supersonic cruise speed. $M(L/D)$ for this vehicle will be in the 23–24 range, giving it an advantage over the subsonic competition with the vital proviso that adequate payload and low enough cost of operation can be realized.

7. The Variable-Sweep Wing (General Dynamics F-111)

Figure 1.8 shows the first operational aircraft which availed itself of research done in the 1950's, mainly by NASA and its predecessor NACA, on the performance gains derivable from varying Λ on a relatively large-aspect-ratio wing. By pivoting the halves of this surface about vertical axes just outboard of the fuselage, one creates a vehicle which may be said to have an infinity of design points. On the F-111, Λ of the quarter-chordline can be altered continuously from 20° to

Fig. 1.8. Supersonic fighter with variable-sweep wing, the F-111, shown with wings in the forward position (reproduced courtesy of General Dynamics).

over 70°. A similar variation of sweep angles characterizes the larger, longer-range B-1 bomber. At one time, U.S. SST designers contemplated such a wing, with the additional feature that, at the full-aft position, it would mate with the stabilizer into a continuous, delta-like surface. This scheme was abandoned because the mechanical complexity and extra weight overcame its advantages for a transport.

We shall briefly describe a few of the operating conditions in which F-111, FB-111, and B-1 performance is enhanced by variable geometry. Landing and takeoff are, of course, normally accomplished with the wings fully forward. Not only are the associated speeds reduced by the greater lifting capacity of this relatively unswept configuration, but the large aerodynamic span facilitates brisk climbout after liftoff. When it is necessary to "ferry" the airplane a long distance, without any constraint on the time of flight, one also finds that the lowest sweep maximizes range through a combination of large L/D, favorable fuel consumption, and the ability to carry auxiliary fuel supplies in tanks suspended from pylons on the outer wing.

There are several attractive cruising options involving higher speeds and altitudes. One, at an intermediate value of Λ, resembles the cruise of a subsonic jet transport. There is also the possibility, however, of fairly efficient operation with full sweepback at M well in excess of 2. Here the performance benefits from an extreme instance of the influence of sweep in delaying "compressibility": Even at $M = 2.5$ the component of airspeed normal to a surface at $\Lambda = 70°$ is subsonic, the consequence being substantially reduced wave-making drag.

Perhaps the most interesting design point involves "nap of the earth" flying at the lowest possible altitude and high subsonic M. Great demands are placed on the structure, the pilot, and the control system, but the problem of running low-bypass-ratio turbofan engines at reasonable rates of fuel consumption imposes the severest limits. Near the ground one often encounters atmospheric turbulence; in its presence and at the dynamic pressure q in excess of half an atmosphere (standard sea-level atmospheric pressure equals 2116 lb/ft^2), violent vibratory airloads can be experienced. It is therefore fortunate that the highly swept wing acts as a gust alleviator, since any increase or decrease of lift causes an α-change due to bending that almost instantaneously reduces the magnitude of the lift increment.

Finally, it is worth noting that this same aeroelastic effect of wing bending ameliorates the task of trimming a variable-sweep wing. Thus, when greater Λ is applied, the center of action of the lift on a completely rigid wing would be moved aft by a distance proportional to the increase in $\sin \Lambda$, and the stabilizer would be required to counterbalance a large pitching-moment change. The aforementioned modification, due to bending, in the distribution of α along the wingspan fortunately tends to move the center of lift inboard and forward. Although not entirely canceled on the F-111, the influence of sweep on trim is thereby kept within tolerable bounds.

8. Vehicle Optimized for High-Altitude Supersonic Cruising (Lockheed SR-71 and YF-12)

A uniquely different supersonic configuration is the SR-71 type, illustrated in Fig. 1.9. Intended for long-range reconnaissance and high-speed interception missions, these aircraft have a design point slightly above M = 3 in the 100,000-ft altitude range. The photograph shows clearly the wide inlets needed to capture enough low-density air and their conical centerbodies which generate efficient patterns of compression shocks. Much of the lift at supersonic conditions is carried on the fuselage. The strakes or "chine strips" extending forward from the wing leading edge are an example of wing-body "blending," in that they assist with this redistribution of lift and help to increase L/D.

Initially flown in the early 1960's, the SR-71 was another very innovative vehicle. Among many features that deserve mention, it was the first to make use of titanium throughout the primary structure and over the external, aerodynamically heated surface.

Fig. 1.9. High-altitude, supersonic cruising aircraft, the SR-71 (reproduced courtesy of Lockheed-California Company).

9. Ballisti ype of Boost Vehicle (Saturn V)

Durin e first 30 years of flight into and back from the environment of space, the of boost and entry vehicles can be understood—albeit simplistically—as ev ension of the ballistician's desire to hurl projectiles over longer and longer ges. Rocket motors, burning propellants stored internally in the liquid or lid phase, have so far proved the best and lightest means of supplying the necessary high thrust without reliance on the atmosphere for an oxidizer. There is no question that ballistic operation ensures the absolute minimum in structural weight and complexity for a given payload. But these advantages must be traded off, at every stage of the evolution, against the high cost of using each vehicle for only a single flight.

Of the many boosters that might have been illustrated, we choose Saturn V (Fig. 1.10) as the culmination, in size and design refinement, of those developed or modified for NASA's space program. Like its predecessors, Saturn V is designed to a single, tightly defined mission, when we think of it as a device for injecting a

ehicle, the Saturn V with Apollo spacecraft (reproduced e Administration).

certain mass into orbit around the earth. In the absence of winds
through the lower atmosphere at essentially zero angle of attack, and ae[illegible]ould fly
effects become negligible above, say, 200,000 ft altitude. In the unlikely ev[illegible]amic
α exceeded 15°, due to control-system malfunction or an extremely rapid cha[illegible]at
wind with altitude, the flight would be immediately aborted.

The principal structural loads are therefore due to the following:

- Rocket thrust, which acts in a nearly axial direction, exceeds the weight by about 25% at launch, and produces increasing acceleration along the trajectory as propellant is depleted.
- Inertia forces due to this acceleration.
- Weight, whose component normal to the trajectory is employed to cause a gradual curving toward the horizontal at orbit injection.
- Drag and lift (or side force), the latter usually remaining small both because of the limited α and the inefficiency of an elongated body of revolution as a generator of side force.

Since this load system is mild compared to the design conditions of a typical airplane and since metal fatigue due to repeated flights is not a consideration, booster structure can be remarkably light. For instance, less than 10% of the gross weight at takeoff of each of the first two Saturn V stages is devoted to load-carrying metal. It is not surprising that boosters have been described as "flying gas tanks."

We note that all boosters to date have used multiple stages, even for low earth orbits. The number ranges from four on early Explorer vehicles to two for Atlas and others; on Saturn V, the third stage is involved in initial boost, but its main purpose is later to transfer the payload into a higher-energy orbit toward the moon. The need for staging is connected with the inability of current propellant–rocket combinations to bring the full empty weight of a single vehicle all the way up to orbital speed. We shall treat this subject further in Chapters 5 and 8.

When Saturn V leaves the pad, its 6×10^6 lb weight is supported and accelerated by five motors of about 1.5×10^6 lb thrust each. The functions of control and of trimming in pitch and yaw are performed by swiveling the four outer rockets in such a way that the thrust components normal to the vehicle's long axis generate the required accelerations or balancing moments. Unlike an airpl[illegible]
the booster is "statically unstable" in the sense that a change in α [illegible]
aerodynamic moment about the CM which tries to increase this [illegible]
matic devices, which sense the unwanted angular motions a[illegible]
so as to compensate, furnish the vehicle with "clos[illegible]
tribute to control-system engineering that Mer[illegible]
Saturn V–Apollo have all been qualified fo[illegible]
automatic stabilization.

10. Atmospheric Entry Vehicle (Apollo Command Module and Others)

The other element of the ballistic approach to space transportation is exemplified by "capsules" like the manned Mercury, Gemini, and Apollo command module. The familiar conical shape of Apollo is discernible near the top of the stack in Fig. 1.10, with the "launch escape tower" truss mounted above it. These and most similar vehicles developed to date are also designed to a single mission. High drag is employed for deceleration from orbital to subsonic speeds in preparation for parachute descent to the earth's surface, while "ablative" coating attached to a steel shell absorbs the extreme atmospheric heating due to entry and protects the payload within.

Research in the 1950's (e.g., Allen and Eggers 1957) demonstrated that the blunt-nosed shape has an important aerodynamic advantage, in that a large fraction of the heat generated by deceleration is dissipated in the air rather than absorbed at the surface. This configuration is also the simplest structurally, requiring minimum total vehicle weight for a given payload. We must, however, point to certain undesirable features. Ballistic vehicles of the sort used to date impose high forces on their contents; thus during normal Apollo operation the astronauts are exposed to a peak deceleration greater than 10 times earth's gravity. Moreover, there is little opportunity to employ aerodynamic forces for modifying the trajectory or point of impact once the initial conditions of entry have been established.

For almost 15 years, proposals have been made to develop vehicles whose hypersonic lift and drag characteristics would make possible trajectory control and return to a designated site (airbase or land recovery zone). Reference has been made to high-, moderate-, and low-L/D entry, the corresponding values of this ratio (at hypersonic M) being three or above, the vicinity of two, and one or below, respectively. Blunt-nosed gliders with delta planform, flattened lower surfaces, and blended fuselages have been studied. It generally turns out that the fraction of weight available for payload decreases as L/D increases, but that maneuverability is enhanced and the problem of thermal protection is ameliorated to the point that reusable, radiating material may be adopted for most of the outer surface. Materials research has concentrated on nickel-based "superalloys" and coated metals for these radiative surfaces, as well as on ceramic oxides like Thoria for blunted nose regions in which the heating load is greatest. A vehicle constructed along these lines has a real possibility of being recycled for several missions.

Apollo itself represents a primitive example of trajectory control by lift. Although there are no surfaces to perform the trimming function of an elevator, the capsule's CM is intentionally located off the axis of symmetry. Hence, during exposure to high aerodynamic forces, a small angle of attack is "locked in," and a usable L/D of about 0.25 is achieved. The rocket "reaction control" motors are then able to roll the vehicle about its velocity vector, in such a way that the lift is pointed in the desired direction to produce flight-path modifications. If the entry is on-course and only a drag force is needed, Apollo can be rolled steadily so that the lift effects average to zero.

The proposed entry gliders not only reach or exceed $L/D = 1$, but they are furnished with horizontal and vertical stabilizers and flapped elevators. Changes in lift and banking maneuvers like those of a cruising aircraft permit down-range and lateral adjustments to the landing point of the order ± 1–2000 miles. Unmanned gliders, in the low- to moderate-L/D class, have already been flown in from earth orbit. These include the ASSET structural test vehicle and the X-24.

11. Reusable Space Transportation System (the NASA Shuttle)

NASA and several industrial organizations have extensively studied candidate two-stage systems, with one or both stages recoverable and reusable, for operation to and from near-earth orbits. Figure 1.11 depicts the final configuration which evolved during 1973 as the first-generation Shuttle; it is now under development.

The original concept of the Shuttle was that two mated vehicles were to be launched vertically from the same sort of pad as a ballistic booster. When separated, each was to be capable of atmospheric flight and landing like a conventional aircraft. In order to reduce costs of development, the recoverable first

Fig. 1.11. Full booster configuration of the first-generation Shuttle, shortly after lift-off. Overall length, including the main fuel tank, is 55.2 meters. (Reproduced courtesy of NASA.)

stage has now been replaced by a pair of solid-propellant rockets, attached to a large tank which externally carries the liquid hydrogen and liquid oxygen propellants required for the three engines of the second, or Orbiter, stage. These engines are to be burned throughout the entire boost, with the solid rockets dropping off when their fuel is exhausted. It is also expected that the solids' cases can be recovered from the ocean and refurbished for ten or more launches.

Because liquid hydrogen has a specific gravity of 0.07 and is stored at a temperature of around $-423°$F, much volume in the Orbiter is saved by the decision to employ a heavily insulated external tank. The Orbiter will have internal provision for payloads in the 25–65,000-lb range. With great flexibility, it will be able to carry to, or return from, space the majority of manned and unmanned satellites that might be realistically conceived of for the early 1980's when it becomes operational. Its most challenging design problem relates to the protection of crew and contents from entry heating with surface materials which, with minor maintenance, can survive 100 or more missions.

The greatest dynamic pressure encountered during Shuttle flight will be experienced in the early phase of boost (about 600 lb/ft^2 near $M = 1$). This is because the planned entry of each stage is to occur at extremely high α and low W/S. Most of the deceleration will therefore take place gradually in the upper atmosphere, so that the peak inertia loads and heating rates will stay far below those characteristic of ballistic entry. In one version of the entry profile, α is maintained at some 60° (a "deep stall") until the stage slows down to near $M = 0.5$, whereupon the transition is made to a cruising condition and flight to base occurs in the manner of a subsonic airliner.

Because so many decisions will be required and so much development effort will have to follow before the Shuttle becomes a reality, we confine ourselves here to the foregoing brief description. We emphasize that this system's merit lies in rendering operations into space routine and much less expensive than they are now. If we may assume that the world's requirements for satellite launchings and recoveries remain near present levels, and that the Shuttle attains its goal of 100 or so flights per vehicle, the cost of placing a given weight in a given orbit can ultimately be reduced by a factor of 100 or more.

1.4 PROBLEMS

1A For values of flight speed v_c ranging from low subsonic to orbital, compare numerically the potential energy of altitude h with the kinetic energy. Vehicle mass is m, but some may find it interesting to work with "specific energies" or energies per unit mass. Potential energy is zero at sea level, and when getting into the orbital speed range you will want to work with radius r and earth radius r_e. Estimate the ranges of speed, and reasonable corresponding altitude ranges, where:

a) Most of the fuel is expended in gaining altitude and overcoming drag.
b) Potential and kinetic energies are of the same orders of magnitude, so that "dynamic performance" methods are needed to study maneuvers, ceilings, and the like.
c) Most of the fuel goes into attaining high speed.

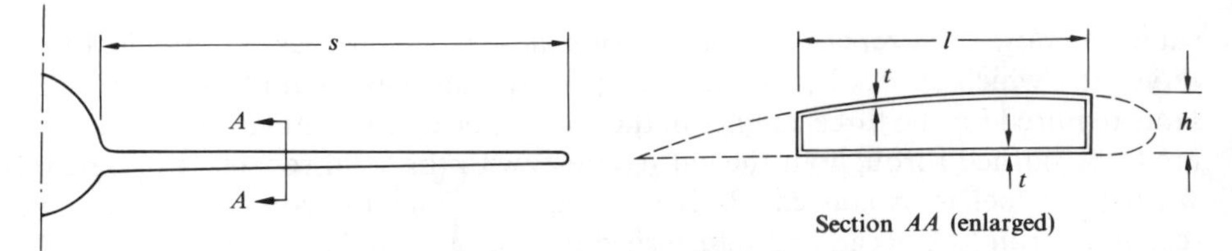

Section AA (enlarged)

1B An airplane of fixed weight W is to be provided with an unswept wing having structural semispan s. Recall that the wing lift is roughly proportional to W. The principal load-carrying structure in the wing consists of a nearly rectangular "box beam" with depth h, chordwise length l, and skin thickness t (see the figure).

The structural weight W_s of the wing can be estimated in an obvious way from t and the other dimensions. At any cross section, t is determined by the requirement that the bending stress σ cannot exceed some allowable value, which is a material property. σ can be estimated from

$$\sigma = (\text{BM})\frac{y}{I} \cong (\text{BM})\frac{h}{2I},$$

where I is the area moment of inertia of the cross section about its horizontal neutral axis, which can easily be expressed when t is small compared with h. The maximum bending moment (BM) is proportional to the product of lift and wingspan.

For subsonic flight, the aerodynamic designer wants to make the aspect ratio (proportional to s/l) as large as possible for a fixed wing area. He also won't let the thickness ratio (proportional to h/l) get too large. Study the tradeoff between structures and aerodynamics by finding the effects of aspect-ratio change on W_s. What happens to W_s when the allowable stress is doubled or the permissible airfoil thickness ratio is doubled?

1C Given that the imposed load on the wing is proportional to gross weight W, explain why and how the stress in a family of geometrically similar airplanes increases with increasing linear dimension. Does this imply that skin thickness must grow faster than in direct proportion to the linear dimension, so that structural weight may become a higher *percentage* of W on large airplanes? After studying this simple case, look at the more realistic scaling-up in which the external shape is held fixed but the bending stress and the wing loading (W/S, on which landing and takeoff runs depend) are constants. Again you should find that larger size implies larger structural-weight fraction. For a discussion of how the designer has coped with this problem, see Cleveland (1970, Section III).

1D Try to extend the approach of the preceding problem to the effects of scale on jet engines. Other scale-independent quantities being held constant, how would you expect the thrust and weight to vary with linear size? Justify the fact that, in a family of airplanes to perform similar missions, thrust at any given flight condition should be proportional to gross weight W. What does this say about weight of the propulsion system as a fraction of W?

1E In preparation for later study of vehicle dynamics, derive the equation of motion of a weathervane in a wind. Treat it as a single-degree-of-freedom linear system, with damping less than critical. The aerodynamic surface of the vane acts like a wing, with "lift" force proportional to angle of attack over some limited range of the latter. Use $\psi(t)$ for angular displacement about the vertical.

Following some disturbance (e.g., a 20° change in wind direction), calculate the response and describe physically what the vane does.

CHAPTER 2

EQUATIONS OF MOTION FOR RIGID FLIGHT VEHICLES

2.1 DEFINITIONS; VECTOR AND SCALAR REALIZATIONS OF NEWTON'S SECOND LAW

Having devoted a chapter to qualitative description of flight vehicles and the rationale for their configuration, we now turn to the quantitative tools which are the foundations for their design analysis. Out of respect for brevity and the reader's prior education, we omit derivations of principles which are fully discussed in the basic texts of engineering science. Newton's laws, as they apply to an isolated and unconstrained rigid body, are certainly an example. We personally have been enlightened by the developments of vector and scalar forms to be found in Goldstein (1950) and Halfman (Vol. I, 1962).

Although we must obviously later qualify the assumption that a typical vehicle structure is so rigid that the distance between any pair of mass points remains constant during any maneuvers it may perform, for many purposes this approximation proves quite serviceable. Such a vehicle is illustrated in Fig. 2.1. The angular velocity vector $\boldsymbol{\omega}$ and the linear velocity vector $\mathbf{v}_c$ of the CM are measured relative to an inertial reference frame. Until we come to discuss boost and entry, this frame will be taken either as the earth's surface or a portion of the atmosphere with constant wind velocity relative to earth. (A useful estimate of the errors involved in regarding earth frames as inertial appears on pages 57–60 of Halfman 1962.) An atmosphere reference at instantaneous flight altitude is the most natural choice for studying airplane motions, since the magnitude v_c then becomes "true airspeed."

Body mass m is composed of fixed infinitesimal elements dm, whose position vectors relative to the CM are denoted $\mathbf{r}$. Unlike $\mathbf{v}_c$ and external force and moment vectors, $\mathbf{r}$ is usually measured in a frame attached to the vehicle with its origin at the CM. The rigid-body approximation requires that $\mathbf{r}$ be constant in direction and magnitude. We can write

$$m = \int_{\text{Body}} dm \tag{2.1}$$

$$0 = \int_{\text{Body}} \mathbf{r}\, dm \qquad \text{(the CM condition)} \tag{2.2}$$

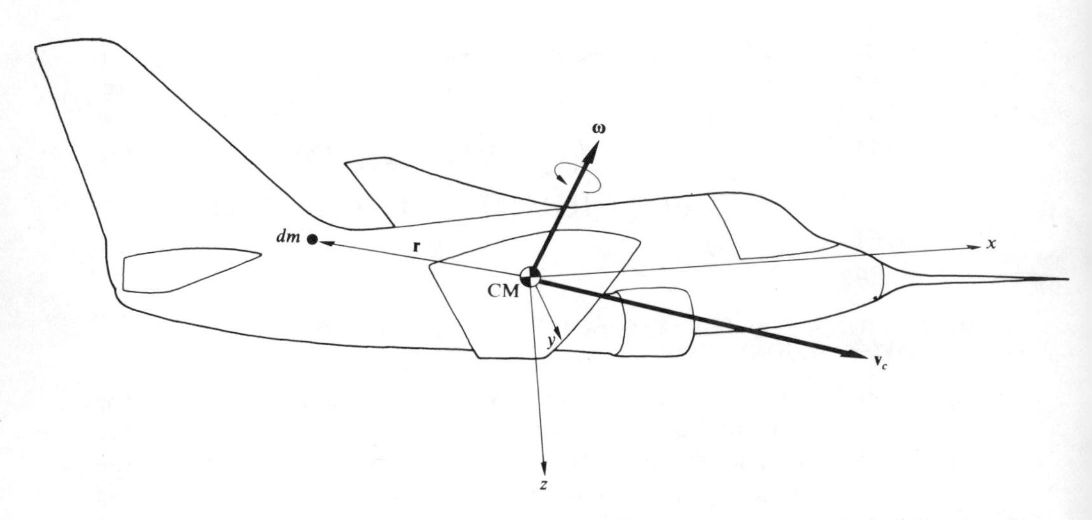

Fig. 2.1. Lockheed S-3A, illustrating Cartesian axes with origin at center of mass, vectors of linear and angular velocity relative to an inertial frame, and position vector for typical mass element dm.

Let the body be exposed to the following:

$\mathbf{F}(t)$ = time-dependent resultant of all external forces,
$\mathbf{G}(t)$ = time-dependent resultant of all external moments, with their axes so transferred that the axis of $\mathbf{G}$ always passes through the vehicle CM.

The equations of motion governing the time histories of $\mathbf{v}_c$ and $\boldsymbol{\omega}$ may then be summarized as

$$\mathbf{F} = m\frac{d\mathbf{v}_c}{dt} \tag{2.3}$$

$$\mathbf{G} = \frac{d\mathbf{h}}{dt} \tag{2.4}$$

In (2.4), $\mathbf{h}$ is the (inertial) angular momentum of the body. An important result of mechanics is that, when the CM is picked as the origin for moments, (2.4) retains its simple form even though that origin is accelerating,

$$\mathbf{h} = \int_{\text{Body}} \mathbf{r} \times [\mathbf{v}_c + \boldsymbol{\omega} \times \mathbf{r}]\, dm \tag{2.5}$$

Derivative $d(\quad)/dt$ refers to the rate of change of a vector, as seen by an observer fixed in the inertial frame. The term containing $\mathbf{v}_c$ in the brackets of (2.5) can be dropped, since $\mathbf{v}_c(t)$ may be factored outside the integral, whereupon (2.2) shows that this term always vanishes.

For purposes of analysis and computation, we need to take components in (2.3)–(2.4) and introduce scalar variables that relate naturally to motion through the air. Let us associate with the CM origin a rectangular coordinate system, with directions attached to the vehicle (Fig. 2.1) and such that $\mathbf{r}$ may be written

$$\mathbf{r} = x\mathbf{i} + y\mathbf{j} + z\mathbf{k} \tag{2.6}$$

Evidently $\mathbf{i}, \mathbf{j}, \mathbf{k}$ are the triad of unit vectors in the respective coordinate directions. Let the components of other vectors in (2.3)–(2.5) be defined as follows:

$$\mathbf{v}_c = U\mathbf{i} + V\mathbf{j} + W\mathbf{k} \tag{2.7}$$

$$\boldsymbol{\omega} = P\mathbf{i} + Q\mathbf{j} + R\mathbf{k} \tag{2.8}$$

$$\mathbf{h} = h_x\mathbf{i} + h_y\mathbf{j} + h_z\mathbf{k} \tag{2.9}$$

We also need the Coriolis Law (e.g., Halfman 1962, pages 46–47), which connects the rates of change of a vector as seen from nonrotating and rotating references. For any $\mathbf{A}$,

$$\frac{d\mathbf{A}}{dt} = \frac{\delta\mathbf{A}}{\delta t} + \boldsymbol{\omega} \times \mathbf{A} \tag{2.10}$$

Here $\delta(\ )/\delta t$ means the rate determined in a frame like x, y, z which has instantaneous angular velocity $\boldsymbol{\omega}$ relative to the frame for $d(\ )/dt$. It is vital to understand that $\mathbf{A}$ may itself be a vector ($\mathbf{v}_c$, $\boldsymbol{\omega}$, $\mathbf{h}$, etc.) measured with respect to inertial space, in which case the observer who reports $\delta(\ )/\delta t$ must be thought of as partaking of the angular velocity but not the translational velocity of the second frame. Equation (2.10) enables us to rewrite the laws of motion

$$\mathbf{F} = m\left[\frac{\delta\mathbf{v}_c}{\delta t} + \boldsymbol{\omega} \times \mathbf{v}_c\right] \tag{2.11}$$

$$\mathbf{G} = \frac{\delta\mathbf{h}}{\delta t} + \boldsymbol{\omega} \times \mathbf{h} \tag{2.12}$$

When components of a quantity like $\delta\mathbf{v}_c/\delta t$ are taken, the rotating observer sees $\mathbf{i}, \mathbf{j}$, and $\mathbf{k}$ as fixed in magnitude and direction. The alteration of $\mathbf{v}_c$ is therefore due entirely to variations in the magnitudes of U, V, W in (2.7), and we may write

$$\frac{\delta\mathbf{v}_c}{\delta t} = \dot{U}\mathbf{i} + \dot{V}\mathbf{j} + \dot{W}\mathbf{k} \tag{2.13}$$

In what follows, we shall generally reserve the superimposed dot $(\dot{\ })$ for rates of change of scalars. When such a scalar happens to be a vector component, uniquely defined relative to some frame, there is no uncertainty regarding what "kind" of time derivative is meant.

For breaking out components* of a vector cross product, let us remind

* Except in cases in which it is convenient or customary to choose non-subscripted components for a vector, the subscripts x, y, z on the italicized equivalent of the boldface vector will identify its respective Cartesian components.

ourselves that

$$\mathbf{A} \times \mathbf{B} = \begin{vmatrix} \mathbf{i} & \mathbf{j} & \mathbf{k} \\ A_x & A_y & A_z \\ B_x & B_y & B_z \end{vmatrix} = \mathbf{i}(A_yB_z - A_zB_y) + \mathbf{j}(A_zB_x - A_xB_z) + \mathbf{k}(A_xB_y - A_yB_x) \tag{2.14}$$

Bringing together (2.7), (2.8), (2.13), and (2.14), we calculate for the x-component of (2.11)

$$F_x = m[\dot{U} + QW - RV] \tag{2.15}$$

We recall (e.g., Rauscher 1953) that the three terms in brackets here may be visualized as due to the "stretching" of U along x and the x-contributions from the "swinging" of W and V at angular rates Q and R, respectively, about the y- and z-directions. It is easily shown that the other two components of (2.11) may be found from (2.15) by *cyclic permutation*, a process whereby each quantity associated with x is permuted into the corresponding y-quantity, y into z, and z into x.

Recognizing the convention, we choose at once to call the components of $\mathbf{G}$ rolling moment L_r, pitching moment M_p, and yawing moment N. (Note that the only reason for carrying subscripts—which will sometimes be omitted below—on the first two is to approximate the conventional aeronautical symbolism while avoiding confusion with lift and Mach number.) The six scalar equivalents of (2.11) and (2.12) are therefore

$$F_x = m[\dot{U} + QW - RV] \tag{2.16a}$$
$$F_y = m[\dot{V} + RU - PW] \tag{2.16b}$$
$$F_z = m[\dot{W} + PV - QU] \tag{2.16c}$$
$$L_r = \dot{h}_x + Qh_z - Rh_y \tag{2.16d}$$
$$M_p = \dot{h}_y + Rh_x - Ph_z \tag{2.16e}$$
$$N = \dot{h}_z + Ph_y - Qh_x \tag{2.16f}$$

At this point we again acknowledge our debt to Etkin (1959); these are his Eqs. (4.4,3) and (4.4,4). Those portions of our book which relate to static stability, dynamic stability, and response of aircraft rely heavily on his approach, as developed in both the original and revised (Etkin 1972) texts. All his chapters will be found to contain useful parallel reading, and several go much more deeply than we into their subject matter.

2.2 THE TENSOR OF INERTIA

In order to recast (2.5) into its more familiar scalar form, we introduce the vector identity

$$\mathbf{A} \times (\mathbf{B} \times \mathbf{C}) \equiv \mathbf{B}(\mathbf{A} \cdot \mathbf{C}) - \mathbf{C}(\mathbf{A} \cdot \mathbf{B}) \tag{2.17}$$

which makes it possible to write the triple product as

$$\begin{aligned}\mathbf{r} \times (\boldsymbol{\omega} \times \mathbf{r}) \equiv \boldsymbol{\omega} r^2 - \mathbf{r}(\boldsymbol{\omega} \cdot \mathbf{r}) = [P\mathbf{i} + Q\mathbf{j} + R\mathbf{k}][x^2 + y^2 + z^2] \\ - \mathbf{i}[Px^2 + Qxy + Rxz] - \mathbf{j}[Pyx + Qy^2 + Ryz] \\ - \mathbf{k}[Pzx + Qzy + Rz^2]\end{aligned} \tag{2.18}$$

Dropping the zero $\mathbf{v}_c$-term, we find for the components of (2.5)

$$h_x = I_{xx}P - I_{xy}Q - I_{xz}R \tag{2.19a}$$

$$h_y = -I_{yx}P + I_{yy}Q - I_{yz}R \tag{2.19b}$$

$$h_z = -I_{zx}P - I_{zy}Q + I_{zz}R \tag{2.19c}$$

In (2.19) there appear the following nine properties, the moments and products of inertia of the body:

$$I_{xx} = \int_{\text{Body}} (y^2 + z^2)\, dm \tag{2.20a}$$

$$I_{xy} = \int_{\text{Body}} xy\, dm = I_{yx} \tag{2.20b}$$

$$\vdots$$

It is convenient for some purposes to realize that the nine $I_{ij}\,(i, j = x, y, z)$ happen to be Cartesian components of a tensor of second rank (cf. Halfman 1962, page 206). Although we do not use the idea extensively in this book, you should be aware of tensors, as a generalization of vectors, and of the tensor nature of many field variables that arise in mechanics, electromagnetics, and elsewhere. The definition of a tensor rests on the manner in which its components change (or "transform") when the reference frame of observation is subjected to a rotational displacement about its origin. Two ways in which this concept proves valuable are the following.

- As soon as a tensor is identified, we know a great deal about it; for instance, that any second-rank tensor with symmetry ($I_{ij} = I_{ji}$) has a set of "principal directions" for Cartesian axes in which all its components vanish except the three with $i = j$. (An interesting corollary is that I_{ij} is determined by only three independent scalars, since all I_{ij} for any orientation of axes can be calculated from the three "principal moments of inertia.")
- The writing and manipulation of equations can be enormously simplified.

An example of the latter is to abbreviate the tensor of inertia as follows:

$$I_{ij} \equiv \begin{pmatrix} I_{xx} & (-I_{xy}) & (-I_{xz}) \\ (-I_{yx}) & I_{yy} & (-I_{yz}) \\ (-I_{zx}) & (-I_{zy}) & I_{zz} \end{pmatrix} \tag{2.21}$$

We adopt the Einstein convention that any subscript which appears twice in a product is to be *summed on*—that is, that i, j (or whatever) is to be equated successively to x, y, and z, and the resulting three terms are to be added. We can then replace (2.19) by

$$h_i = i_{ij}\omega_j \tag{2.22}$$

where ω_x, ω_y, ω_z represent components of $\boldsymbol{\omega}$. The three equations are recovered by specifying that i may stand for x, y, or z.

Once axes have been attached to our vehicle, I_{xx}, I_{xy}, . . . are fixed numbers. The derivatives of the angular-momentum components in (2.16d, e, f) may therefore be written

$$\dot{h}_x = I_{xx}\dot{P} - I_{xy}\dot{Q} - I_{xz}\dot{R} \tag{2.23a}$$

$$\dot{h}_y = -I_{yx}\dot{P} + I_{yy}\dot{Q} - I_{yz}\dot{R} \tag{2.23b}$$

$$\dot{h}_z = -I_{zx}\dot{P} - I_{zy}\dot{Q} + I_{zz}\dot{R} \tag{2.23c}$$

Imagining momentarily that the external forces and moments are known functions of time, we recognize (2.16) and (2.23) as reducible to six ordinary differential equations governing the six unknowns $U(t)$, $V(t)$, $W(t)$, $P(t)$, $Q(t)$, and $R(t)$.

2.3 CHOICE OF VEHICLE AXES

In principle, any CM axes may be employed to study motion. But, in common practice, only two or three sets appear. These all take advantage of the observation that almost every flight vehicle has at least one plane of symmetry. During rectilinear level flight, climbing or diving, the "longitudinal" plane of symmetry is vertical; x and z are located in this plane, with y to the right in the direction that would be outward along an unswept right wing.

Two consequences of symmetry are that the CM lies in the xz-plane and that each mass element dm at (x, y, z) is balanced by an equal dm at $(x, -y, z)$. It follows at once from the definitions (2.20) that

$$I_{xy} = 0 = I_{yz} \tag{2.24}$$

and y coincides with the CM principal axis $\bar{y}$.

Specification of x will now complete the attachment of axes to body. Let us discuss three ways in which this is done.

1. Principal Axes

This selection also makes $I_{xz} = 0$. With superimposed bars to emphasize "principality," we note that the equations of motion become simpler because $h_{\bar{x}} = I_{\bar{x}\bar{x}}P$, etc. For dynamic analysis of vehicles, $\bar{x}$ is given a "forward" direction, whereupon $\bar{z}$ is seen to point "downward" in level flight. On a conventional airplane during cruise, climb, etc., $\bar{x}$ is usually found to be a few degrees above the velocity vector. On a vehicle with two symmetry planes (cruciform missile) or with an axis of symmetry (rifle bullet, ballistic booster), it coincides with the longitudinal axis.

2. Stability Axes

For convenience in handling the aerodynamic terms, x is made parallel to the airspeed vector $\mathbf{v}_{c0}$ associated with some condition of steady, rectilinear flight. As long as the motion remains undisturbed, this means that $V = 0 = W$, and (2.7) reduces to

$$\mathbf{v}_{c0} = v_{c0}\mathbf{i} \tag{2.25}$$

We emphasize that, after the vehicle begins a maneuver or its straight-line path is otherwise disturbed, the projections of the airspeed vector onto (x, y, z) become functions of time. Moreover, the manner in which x is attached may differ from one portion of a flight to another, as the trimmed angle of attack varies with speed, altitude, and weight.

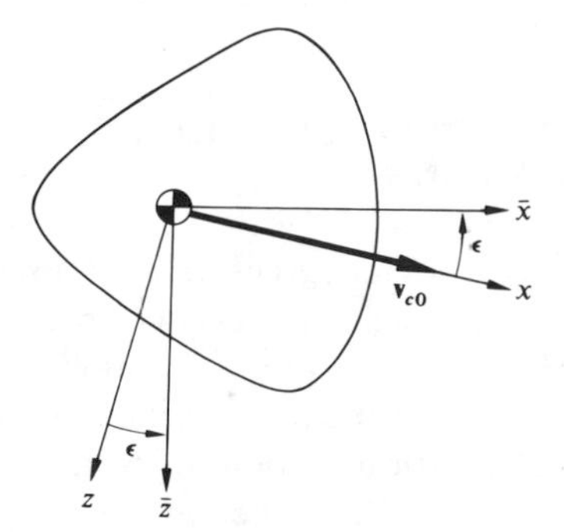

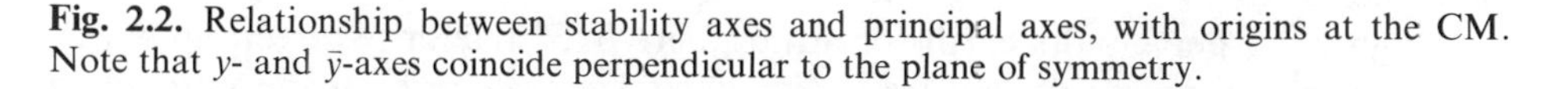
Fig. 2.2. Relationship between stability axes and principal axes, with origins at the CM. Note that y- and $\bar{y}$-axes coincide perpendicular to the plane of symmetry.

Figure 2.2 illustrates one particular set of stability axes. Also shown is their orientation with respect to the principal directions, which tend more nearly to be a fixed property of a given vehicle. In terms of the angle ϵ, which is positive when a right-handed rotation about $y \equiv \bar{y}$ carries x into $\bar{x}$, the two sets of inertias are connected by the *Mohr circle relations:*

$$I_{xx} = I_{\bar{x}\bar{x}} \cos^2 \epsilon + I_{\bar{z}\bar{z}} \sin^2 \epsilon \tag{2.26a}$$

$$I_{zz} = I_{\bar{x}\bar{x}} \sin^2 \epsilon + I_{\bar{z}\bar{z}} \cos^2 \epsilon \tag{2.26b}$$

$$I_{xz} = [I_{\bar{x}\bar{x}} - I_{\bar{z}\bar{z}}] \frac{\sin 2\epsilon}{2} \tag{2.26c}$$

[You may find it instructive to confirm the correctness of (2.26), making use of (2.20) and rotational transformations like $x = \bar{x} \cos \epsilon + \bar{z} \sin \epsilon$.]

3. Aerodynamic Axes

For the purpose of theoretical aerodynamic developments like those in Chapter 4 we are forced to adopt the almost universal convention of orienting x parallel to, rather than opposite to, the relative wind blowing past the vehicle; since y is unchanged, z then takes an upward direction to ensure right-handedness. We do this with some reluctance, since the practice is inconsistent with those of vehicle dynamics. It will, however, facilitate reading the extensive aerodynamic literature.

The fluid mechanicist conceives of an aircraft as a set of boundary conditions for the flow field which is his main interest. His viewpoint resembles that of the experimental aerodynamicist, who also ties his "wind-tunnel axes" to the fixed direction of the oncoming flow and often measures forces and moments with balances arranged parallel to these axes (internal balances in "sting mounts" are a modern exception). The vehicle angle of attack and sideslip are changed relative to these axes, and a rotational transformation must be used to transfer measured airload components from them to the "stability" frame. You will note, incidentally, that lift and drag are defined according to the aerodynamicist's conventions.

2.4 ORIENTATION OF THE VEHICLE RELATIVE TO THE GROUND; FLIGHT-PATH DETERMINATION

It is said that the rigid body has six degrees of freedom, meaning that three linear position coordinates and three angles are needed to fix uniquely its location and orientation relative to the inertial frame. Complete time histories of the six quantities $U(t)$ through $R(t)$, plus the specification of an initial position at, say, $t = 0$, provide enough information to determine these coordinates and angles. Here we discuss one scheme by which they are often defined and show how to calculate them.

Figure 2.3 depicts the earth- or atmosphere-fixed reference directions x', y', z'. Note that z' points downward, so that this system can be made parallel to stability axes when the vehicle flies horizontally; an equally common convention is to take z' positive upward. When analyzing problems of navigation or performance, it is enough to know $x'(t)$, $y'(t)$, and $z'(t)$ for the CM. For the study of maneuvers and dynamic response, however, we must also determine the three angles. In either case, all six quantities must be found if we are starting from components of $\mathbf{v}_c(t)$ and $\boldsymbol{\omega}(t)$ referred to vehicle axes.

There are two alternative ways of specifying the orientation of x, y, z relative to x', y', z': by means of nine *direction cosines;* or by prescribing three angular rotations taken in a prescribed order about three coordinate directions, known as *Euler angles.* Although there is no single agreed convention, the latter approach seems to be favored in engineering applications, since the minimum amount of information is involved. The following four ordered steps define the Euler angles shown in Fig. 2.3 (all rotations are right-handed).

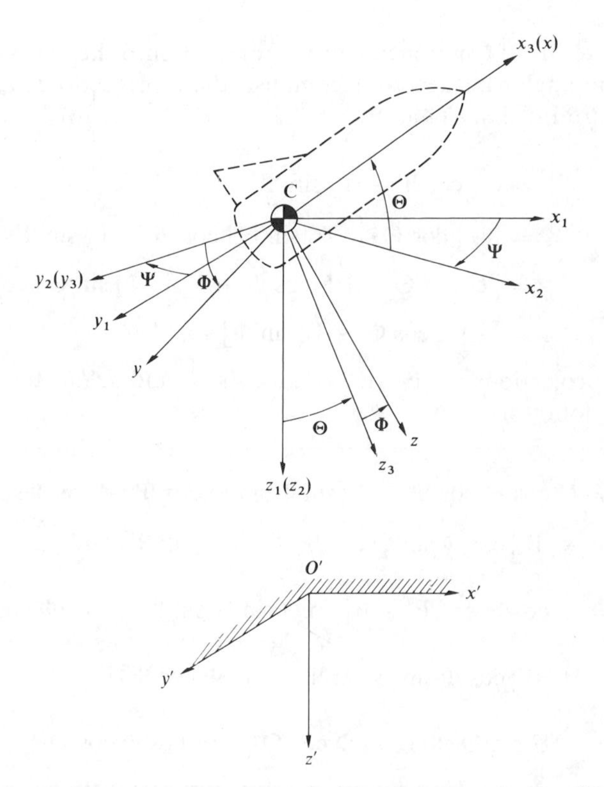

Fig. 2.3. The three Euler-angle rotations required to carry axes $x_1y_1z_1$, which are parallel to the inertial axes, into vehicle axes xyz. Rotations are taken in the order Ψ, Θ, Φ.

- Place axes $Cx_1y_1z_1$ with origin at the instantaneous CM and directions parallel to x', y', and z'.
- Rotate the system through angle Ψ about the vertical direction Cz_1 until the x-axis is in the vertical plane containing vehicle axis Cx. The new axes are called $Cx_2y_2z_2$; z_2 and z_1 coincide.
- Rotate the system through angle Θ about the horizontal direction Cy_2 until the x-axis coincides with Cx. The new axes are called $Cx_3y_3z_3$; y_3 and y_2 coincide.
- Rotate the system through angle Φ about $Cx_3 \equiv Cx$ until the y- and z-axes coincide with Cy and Cz.

As an illustration of how these angles are used, suppose we are given the components of $\mathbf{v}_c$ and $\boldsymbol{\omega}$ and are asked to compute position vector

$$\mathbf{r}'(t) = x'\mathbf{i}' + y'\mathbf{j}' + z'\mathbf{k}' \tag{2.27}$$

for the CM. Let us begin with the inertial rate of change dx'/dt, expressing it successively in terms of components of $\mathbf{v}_c$ projected onto the axis systems 1, 2, 3, and $Cxyz$. The angles needed to accomplish these projections, as can be seen from Fig. 2.3, are Euler's.

$$\begin{aligned}\frac{dx'}{dt} \equiv U_1 &= U_2 \cos\Psi - V_2 \sin\Psi \\ &= [U_3 \cos\Theta + W_3 \sin\Theta]\cos\Psi - V_3 \sin\Psi \\ &= [U\cos\Theta + (W\cos\Phi + V\sin\Phi)\sin\Theta]\cos\Psi \\ &\quad - [V\cos\Phi - W\sin\Phi]\sin\Psi \end{aligned} \tag{2.28}$$

When similar projections are carried out for dy'/dt and dz'/dt, the results can be reorganized as follows.

$$\begin{aligned}\frac{dx'}{dt} &= U\cos\Theta\cos\Psi + V[\sin\Phi\sin\Theta\cos\Psi - \cos\Phi\sin\Psi] \\ &\quad + W[\cos\Phi\sin\Theta\cos\Psi + \sin\Phi\sin\Psi] \end{aligned} \tag{2.29a}$$

$$\begin{aligned}\frac{dy'}{dt} &= U\cos\Theta\sin\Psi + V[\sin\Phi\sin\Theta\sin\Psi + \cos\Phi\cos\Psi] \\ &\quad + W[\cos\Phi\sin\Theta\sin\Psi - \sin\Phi\cos\Psi] \end{aligned} \tag{2.29b}$$

$$\frac{dz'}{dt} = -U\sin\Theta + V\sin\Phi\cos\Theta + W\cos\Phi\cos\Theta \tag{2.29c}$$

We note that the use of matrix notation* and the idea of a "rotation matrix" can simplify component projections like the foregoing. For instance, the transformation of $\mathbf{v}_c$ components from the 2 to the 1 axis system may be abbreviated, in terms of pre-multiplication of a column matrix by a 3 × 3 square matrix,

$$\begin{Bmatrix} \dfrac{dx'}{dt} \\ \dfrac{dy'}{dt} \\ \dfrac{dz'}{dt} \end{Bmatrix} = [(\theta_{\text{Rot}})_{12}] \begin{Bmatrix} U_2 \\ V_2 \\ W_2 \end{Bmatrix} \tag{2.30}$$

Here

$$[(\theta_{\text{Rot}})_{12}] \equiv \begin{bmatrix} \cos\Psi & (-\sin\Psi) & 0 \\ \sin\Psi & \cos\Psi & 0 \\ 0 & 0 & 1 \end{bmatrix} \tag{2.31}$$

---

* For general information on matrices and their manipulation, refer to Frazer *et al.* (1938). Halfman (1962) contains both a useful Appendix in Volume I and detailed discussion of rotations on pages 23–28 and 185–191.

is a rotation matrix which summarizes the effect of the single right-handed angular displacement Ψ about the axis $Cz_1 = Cz_2$. In terms of three easily constructed individual rotations, the complete transformation (2.29) is expressible as

$$\begin{Bmatrix} \dfrac{dx'}{dt} \\ \dfrac{dy'}{dt} \\ \dfrac{dz'}{dt} \end{Bmatrix} = [(\theta_{\mathrm{Rot}})_{12}] \quad [(\theta_{\mathrm{Rot}})_{23}] \quad [(\theta_{\mathrm{Rot}})_{3Cxyz}] \quad \begin{Bmatrix} U \\ V \\ W \end{Bmatrix} \tag{2.32}$$

You will find it instructive to write out the three matrices in (2.32) and verify that their product yields (2.29). Such transformations can be applied to Cartesian components of any vector. It is precisely this manner of transforming under coordinate rotations which identifies a vector as a first-rank tensor.

The next step toward computing the components in (2.27) is to find the Euler angles. For this purpose, we introduce appropriately subscripted unit vectors for each axis system in Fig. 2.3, and observe that the angular velocity of the vehicle can be componentized in two alternative ways.

$$\begin{aligned} \boldsymbol{\omega} &= \dot{\Phi}\mathbf{i} + \dot{\Theta}\mathbf{j}_3 + \dot{\Psi}\mathbf{k}_2 \\ &= P\mathbf{i} + Q\mathbf{j} + R\mathbf{k} \end{aligned} \tag{2.33}$$

[Although the three terms in the middle member of (2.33) are not mutually perpendicular, they are nevertheless three "axial" vectors which instantaneously add up to $\boldsymbol{\omega}$. This fact follows from the vector character of the three infinitesimal rotations $d\Phi, d\Theta, d\Psi$ that occur during any dt; cf. Halfman (1962, pages 191–192).] Our rotation matrices can also be applied to unit vectors, and they lead us to the relations

$$\mathbf{j}_3 = \mathbf{j}\cos\Phi - \mathbf{k}\sin\Phi \tag{2.34}$$

$$\begin{aligned} \mathbf{k}_2 &= \mathbf{k}_3\cos\Theta - \mathbf{i}_3\sin\Theta \\ &= [\mathbf{k}\cos\Phi + \mathbf{j}\sin\Phi]\cos\Theta - \mathbf{i}\sin\Theta \end{aligned} \tag{2.35}$$

When we substitute (2.34)–(2.35) into the middle member of (2.33), we find, by equating x-, y-, z-components, that

$$P = \dot{\Phi} - \dot{\Psi}\sin\Theta \tag{2.36a}$$

$$Q = \dot{\Theta}\cos\Phi + \dot{\Psi}\sin\Phi\cos\Theta \tag{2.36b}$$

$$R = -\dot{\Theta}\sin\Phi + \dot{\Psi}\cos\Phi\cos\Theta \tag{2.36c}$$

Equations (2.36) are a system of linear simultaneous equations, which may be solved by determinants or matrix inversion for the relations sought:

$$\dot{\Phi} = P + Q \sin \Phi \tan \Theta + R \cos \Phi \tan \Theta \tag{2.37a}$$

$$\dot{\Theta} = Q \cos \Phi - R \sin \Phi \tag{2.37b}$$

$$\dot{\Psi} = Q \sin \Phi \sec \Theta + R \cos \Phi \sec \Theta \tag{2.37c}$$

Equations (2.29) and (2.37) embody our objective of developing a nonlinear set of first-order, ordinary differential equations which contain explicitly the first derivatives of the six desired quantities x', y', z', Φ, Θ, Ψ. Given suitable initial conditions and time histories of U through R, we can solve this set by numerical integration over successive short time intervals. Chapter 9 discusses more fully the treatment of such systems, in connection with dynamic performance. We observe, in (2.37a) and (2.37c), one weakness of our choice of Euler angles: Special care must be taken when the vehicle's x-axis passes through the vertical ($\Theta = \pm\pi/2$) because $\tan \Theta$ and $\sec \Theta$ formally take on infinite values. This singularity is, however, integrable.

The foregoing approach is particularly useful whenever a vehicle carries equipment capable of observing and recording $U(t)$ through $R(t)$. The angular rates are most commonly measured by a set of single-axis gyroscopic instruments strapped down near the CM so as to read about x, y, and z. U, V, W are found, relative to the atmosphere, from a Pitot tube which measures the magnitude of $\mathbf{v}_c$ and a set of angle-of-attack and yaw vanes whose readings permit $\mathbf{v}_c$ to be projected onto x, y, z. Alternatively, three linear accelerometers may be installed alongside the gyros; by appropriate integration and interpretation, one can process their signals to yield velocity components.

Some "inertial navigation systems" now carry an instrument platform which is rotationally stabilized to hold fixed directions in inertial space. From the history of its orientation relative to the carrier vehicle, one may calculate the Euler angles directly; signals from accelerometers mounted on this platform might also be processed to furnish dx'/dt, dy'/dt, dz'/dt. It might then prove convenient to invert (2.29) and solve for U, V, W. Equations (2.36) are already suitable for determining angular velocities in roll, pitch, and yaw.

2.5 GRAVITATIONAL TERMS IN THE EQUATIONS OF MOTION

The forces and moments on the left-hand sides of (2.16) can usually be separated into terms of aerodynamic, propulsive, and gravitational origin. When flight takes place close enough to the earth and at low enough speeds that we may assume a uniform, parallel, and vertical gravity field, the last of these are very simple. The total effect is a downward-force vector, acting through the CM and having magnitude mg, where g is the local gravitational acceleration. Obviously there are no moments about the CM. By reference to Fig. 2.3, we see that the components in vehicle axes are

$$F_{xg} = -mg \sin \Theta \tag{2.38a}$$

$$F_{yg} = mg \cos \Theta \sin \Phi \tag{2.38b}$$

$$F_{zg} = mg \cos \Theta \cos \Phi \tag{2.38c}$$

In order to arrive at a second working form of the equations, let us adopt the symbols X, Y, Z for the aerodynamic plus propulsive parts of F_x, F_y, F_z. Into (2.16), we then substitute (2.23), (2.38) and the plane-of-symmetry condition (2.24), obtaining

$$X - mg \sin \Theta = m[\dot{U} + QW - RV] \tag{2.39a}$$

$$Y + mg \cos \Theta \sin \Phi = m[\dot{V} + RU - PW] \tag{2.39b}$$

$$Z + mg \cos \Theta \cos \Phi = m[\dot{W} + PV - QU] \tag{2.39c}$$

$$L_r = I_{xx}\dot{P} - I_{xz}\dot{R} + [I_{zz} - I_{yy}]QR - I_{xz}PQ \tag{2.39d}$$

$$M_p = I_{yy}\dot{Q} + [I_{xx} - I_{zz}]RP + I_{xz}[P^2 - R^2] \tag{2.39e}$$

$$N = -I_{xz}\dot{P} + I_{zz}\dot{R} + [I_{yy} - I_{xx}]PQ + I_{xz}QR \tag{2.39f}$$

We set down here an important special case of (2.39). It describes longitudinal motion, with the wings level ($\Phi = 0$), in the vehicle's own vertical plane of symmetry. When this symmetry is perfect, such motion cannot give rise to any side force, rolling moment, or yawing moment. If P, R, and V are initially zero, we therefore conclude from (2.39b, d, f) that they must remain zero throughout the maneuver. Without any restriction on the sizes of U, W, and Q, Eqs. (2.39) reduce to

$$\boxed{\begin{aligned} X - mg \sin \Theta &= m[\dot{U} + QW] \quad (2.40a)\\ Z + mg \cos \Theta &= m[\dot{W} - QU] \quad (2.40b)\\ M_p &= I_{yy}\dot{Q} \quad (2.40c)\end{aligned}}$$

2.6 THE STATE VECTOR

We have already characterized the rigid flight vehicle as having six degrees of freedom. An alternative characterization, now in common use, is arrived at by solving the equations of motion explicitly for the first derivative of each of the dependent variables. They may then be written in the symbolic matrix form

$$\{\dot{\mathscr{X}}\} = \{f(\{\mathscr{X}\}; t)\} \tag{2.41}$$

Here $\{\mathscr{X}\}$ is an $n \times 1$ column matrix or n-dimensional vector.

As a simple example of (2.41), we make the assumption* in (2.40) that X, Z,

* This approximation would be valid for slow maneuvering at nearly constant altitude, so that ρ_∞, a_∞ are fixed and $v_c = \sqrt{U^2 + W^2}$. For larger-altitude excursions, $\dot{z}' = -U \sin \Theta + W \cos \Theta$ would have to be added to the state vector equations; ambient air properties would be specified as functions of z'.

and M_p can be expressed wholly in terms of U, W, and Q. The system can be recast as follows:

$$\dot{U} = -QW + \frac{X}{m} - g \sin \Theta \tag{2.42a}$$

$$\dot{W} = QU + \frac{Z}{m} + g \cos \Theta \tag{2.42b}$$

$$\dot{Q} = \frac{M_p}{I_{yy}} \tag{2.42c}$$

$$\dot{\Theta} = Q \tag{2.42d}$$

Clearly, the state vector is

$$\{\mathscr{X}\} = \begin{Bmatrix} U \\ W \\ Q \\ \Theta \end{Bmatrix} \tag{2.43}$$

and the four functions f on the right of (2.41) may be identified by substituting on the right of (2.42) the known relations between the aerodynamic terms and the velocity components. (Note that Θ does not appear aerodynamically because a change in flight-path angle *only* does not alter the speed or angle of attack.)

The vehicle moving in its vertical plane of symmetry is said to be a "system of state four" because its $\{\mathscr{X}\}$ has $n = 4$ dimensions. It is instructive to examine (2.37) and (2.39), under similar aerodynamic assumptions, and identify the state vector for more general motions. This process reveals that (2.39) can be solved explicitly for the first derivatives of U, V, W, P, Q, R. Equations (2.37a, b) must be added to provide information on Φ and Θ. The *azimuth angle* Ψ does not appear, however, since its only effect is to orient the vertical plane containing the instantaneous flight path and this orientation influences neither gravity nor the airloads. Thus we find a vehicle with state eight, having

$$\{\mathscr{X}\} = \begin{Bmatrix} U \\ V \\ W \\ P \\ Q \\ R \\ \Phi \\ \Theta \end{Bmatrix} \tag{2.44}$$

A ninth equation for $\dot{z}'(t)$ must be included if the altitude changes significantly.

When manual or automatic control is being analyzed, an $m \times 1$ "control vector" $\{\mathscr{U}\}$ is added on the right of the first-order equations (2.41).

$$\{\dot{\mathscr{X}}\} = \{f(\{\mathscr{X}\}, \{\mathscr{U}\}; t)\} \tag{2.45}$$

The components of $\mathscr{U}$ are physical quantities, regarded as functions of time or "inputs" specified through the action of a human or automatic pilot. For longitudinal motion, $\mathscr{U}$ might consist of a single scalar $\delta_e(t)$, the *elevator angle*. Rotational displacements of ailerons, rudder, elevator, spoilers, and possibly flaps or dive brakes might all appear in $\mathscr{U}$ for general maneuvers. The thrust or engine throttle setting is often added, this being the most common input for controlling rate of climb or descent.

The concept of state vector arises from the physicist's approach to the dynamics of groups of particles. Each mass point is characterized by the components of its displacement and velocity—or displacement and momentum—so that first-order equations of motion are obtained. A single particle in three dimensions, acted on by "spring" forces dependent on its displacements, would thus have state six. The special significance of forms (2.41) and (2.45) is that, if an initial state $\{\mathscr{X}(t_0)\}$ is given at some convenient instant, one can determine the subsequent motion completely from them. This is possible, in principle, for all $t \geq t_0$ and, in fact, for all times until disturbing effects not comprehended by these equations begin to cause appreciable deviations from the unique "trajectory in state space."

Equations of motion are usually nonlinear, despite the deceivingly "linear" appearance of $f = ma$. This fact highlights another key feature of the state vector formulation, namely that the most efficient schemes for numerical integration of ordinary differential systems are based on the form (2.45). You will find that every high-speed digital computing facility possesses a selection of routines for this sort of integration. We shall discuss the subject further in Chapter 9.

2.7 THREE SIGNIFICANT PHENOMENA THAT HAVE BEEN NEGLECTED

1. Angular Momentum of Rotating Machinery

As discussed by Etkin (1959, Section 4.6), some aircraft have components, such as large propellers or the rotating stages of gas turbines and turbofan engines, whose angular momentum is so great that it cannot be neglected by comparison with other terms in (2.19). Designated

$$\mathbf{h}' = h'_x\mathbf{i} + h'_y\mathbf{j} + h'_z\mathbf{k} \tag{2.46}$$

these momenta tend to be nearly fixed in vehicle axes ($\delta\mathbf{h}'/\delta t \cong 0$), or at most they vary slowly relative to the time constants of most maneuvers. The inclusion of (2.46) in the equations of motion is then straightforward. It is of interest that a significant h'_x (engine, propeller) or h'_z ("lift fan") couples the longitudinal and lateral degrees of freedom. In this case the reasoning that leads to (2.40) would clearly lose its validity.

The helicopter rotor is another major contributor to z-angular momentum, but it must often be treated as a source of additional degrees of freedom because of the possibility of deformations and angular displacements relative to the fuselage.

We leave the subject to specialized books on rotary-wing aircraft. For instance, Seckel (1964, Chapter XX) discusses rotor inertial effects.

2. Control-System Degrees of Freedom

Ailerons, elevators, rudders, and tabs are typical parts of a flight vehicle that can move relative to the main airframe. We shall go into their aerodynamic contributions to the equations of motion in subsequent chapters, but some words are in order here about their inertial effects. A much more thorough treatment of the latter appears in Etkin (1959, Sections 4.7–4.10).

Control surfaces fall into two broad categories. Most aerodynamic controls on the modern vehicles mentioned in Chapter 1 are "irreversible." The forces that would be required for the pilot to displace them directly, over some regions of the flight envelope, far exceed human capability. They are therefore provided with "power boost," in the form of hydraulic actuators, electrically driven screw jacks, and the like. "Feel" devices are furnished to simulate, at the stick or control column, the sensations of being directly connected. The term "irreversibility" refers to the fact that the stick cannot be moved by applying a moment about the control-surface hingeline, as it can on a reversible control which is attached by cables or push rods. So stiff are the back restraints in the irreversible case that the natural frequency of rotational vibration about the hingeline ranges—on most primary controls, including all-movable stabilizers—between 10 and 50 Hz. Since dynamic phenomena, other than vibratory response or flutter involving structural oscillations, have time constants of one second or longer, the inertial coupling between irreversible controls and vehicle rigid-body degrees of freedom is nearly always negligible; hence our lack of emphasis on such effects in this book.

Reversible controls are the rule, however, on light and low-speed aircraft. We therefore wish to present at least one example of how they are handled. It also gives us the opportunity to illustrate Lagrange's technique (Halfman 1962, Chapter 11) as a systematic, foolproof way of developing equations of motion for complex systems. We choose the case of an elevator, connected by elastic cables to a fixed control column. Since the principal inertia coupling of the elevator is with the longitudinal freedoms, we assume that only U, W, and Q differ from zero. Figure 2.4 pictures the vehicle and the various motion coordinates. We make the further approximation that the cables act like a linear elastic restraint, without inertia or friction, which opposes rotation $\delta_e(t)$ with a torsional spring constant K_δ (typical units: foot-pounds per radian).

Lagrange's equations require the identification of just enough independent generalized coordinates $q_i(t)$ to specify completely the position of the system. External generalized forces Q_i are associated with these coordinates by giving each of them an infinitesimal virtual displacement, and writing the virtual work associated with these displacements as

$$\delta W = \sum_{i=1} Q_i \, \delta q_i \tag{2.47}$$

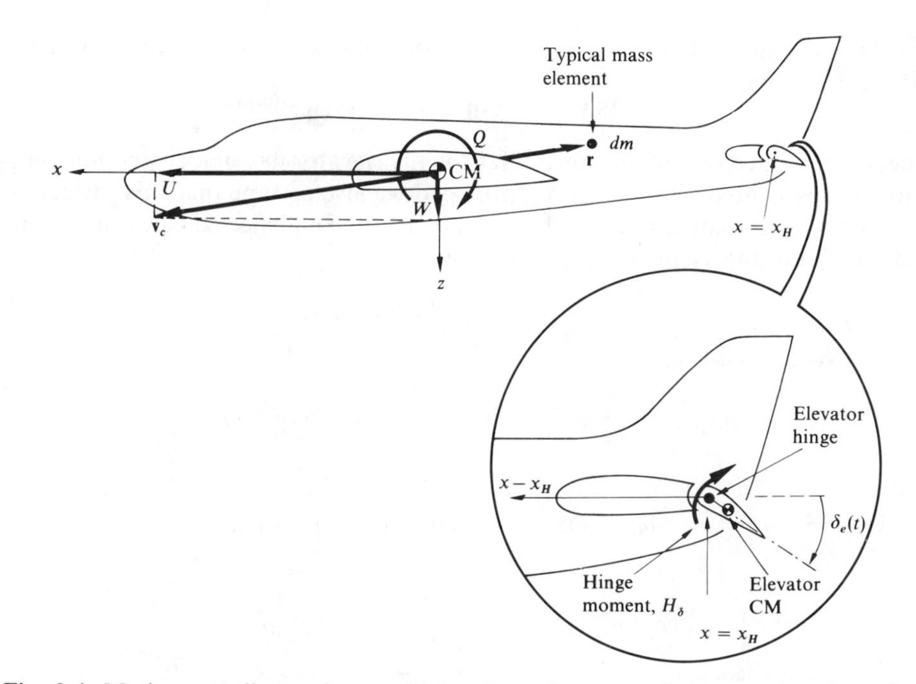

Fig. 2.4. Motion coordinates for a vehicle whose elevator undergoes a time-dependent displacement $\delta_e(t)$.

The kinetic energy τ and potential energy U (of any conservative forces not included in Q_i) are then written in terms of the q_i. Finally, the equations of motion are developed from

$$\frac{d}{dt}\left[\frac{\partial(\tau - U)}{\partial \dot{q}_i}\right] - \frac{\partial(\tau - U)}{\partial q_i} = Q_i, \qquad i = 1, 2, \ldots \tag{2.48}$$

Looking at Fig. 2.4, we temporarily introduce an inertial frame xyz that coincides with the vehicle's stability axes at a typical instant t. We choose the four coordinates $q_i = x_c, z_c, \Theta, \delta_e$, where the first three have the significance $\dot{x}_c = U$, $\dot{z}_c = W, \dot{\Theta} = Q$. We include weight effects in the Q_i and find that three of the four generalized forces are identical with those introduced in (2.16a, c, e):

$$\delta W = F_x\, \delta x_c + F_z\, \delta z_c + M_p\, \delta\Theta + H_\delta\, \delta(\delta_e) \tag{2.49}$$

H_δ is the "hinge moment" exerted about the elevator hinge line by the gravity and aerodynamic forces that act on that movable surface. It is positive in the same sense as δ_e.

The only potential energy not already encompassed in (2.49) is that associated with elastic straining of the control cables, which can be expressed as

$$U = \tfrac{1}{2}K_\delta\, \delta_e^2 \tag{2.50}$$

To calculate the kinetic energy, we note that the vector velocity of any typical mass element dm is

$$\mathbf{v} = (\dot{x}_c + \dot{\Theta}z)\mathbf{i} + (\dot{z}_c - \dot{\Theta}x)\mathbf{j} \tag{2.51}$$

unless it is located on the elevator. If it is on the elevator its z velocity component must be augmented by $\dot{\delta}_e[x_H - x]$, provided we also assume that this surface is flat, that δ_e is a small angle, and that the plane of the undeflected elevator is near and parallel to the xy-plane. It follows that

$$\begin{aligned}
\tau = {} & \tfrac{1}{2} \int_{\text{(Vehicle less elevator)}} [(\dot{x}_c + \dot{\Theta}z)^2 + (\dot{z}_c - \dot{\Theta}x)^2]\, dm \\
& + \tfrac{1}{2} \int_{\text{Elevator}} \{(\dot{x}_c + \dot{\Theta}z)^2 + [\dot{z}_c - \dot{\Theta}x + \dot{\delta}_e(x_H - x)]^2\}\, dm \\
= {} & \tfrac{1}{2}[\dot{x}_c^2 + \dot{z}_c^2] \int_{\text{Vehicle}} dm + \dot{\Theta}\dot{x}_c \int_{\text{Vehicle}} z\, dm - \dot{\Theta}\dot{z}_c \int_{\text{Vehicle}} x\, dm \\
& + \tfrac{1}{2}\dot{\Theta}^2 \int_{\text{Vehicle}} [x^2 + z^2]\, dm + \dot{z}_c\, \dot{\delta}_e \int_{\text{Elevator}} [x_H - x]\, dm \\
& - \dot{\Theta}\dot{\delta}_e \int_{\text{Elevator}} x[x_H - x]\, dm + \tfrac{1}{2}\, \dot{\delta}_e^2 \int_{\text{Elevator}} [x_H - x]^2\, dm \\
= {} & \tfrac{1}{2}m[\dot{x}_c^2 + \dot{z}_c^2] + \tfrac{1}{2}\dot{\Theta}^2 I_{yy} + \dot{z}_c\, \dot{\delta}_e S_{y\delta} - \dot{\Theta}\, \dot{\delta}_e \int_{\text{Elevator}} x[x_H - x]\, dm + \tfrac{1}{2}\, \dot{\delta}_e^2 I_{y\delta}
\end{aligned} \tag{2.52}$$

The second and third integrals in the second member of (2.52) vanish because of the CM condition. The two new quantities introduced in the final line are

$$I_{y\delta} \equiv \int_{\text{Elevator}} [x_H - x]^2\, dm = \text{Moment of inertia about the hingeline}$$

$$S_{y\delta} \equiv \int_{\text{Elevator}} [x_H - x]\, dm = \text{``static unbalance'' about the hingeline}$$

The other integral in (2.52) can be manipulated as follows, in recognition that $x_H < 0$ on a conventional tail:

$$\begin{aligned}
- \int_{\text{Elevator}} x[x_H - x]\, dm &= -x_H \int_{\text{Elevator}} [x_H - x]\, dm + \int_{\text{Elevator}} [x_H - x]^2\, dm \\
&= -x_H S_{y\delta} + I_{y\delta}
\end{aligned} \tag{2.53}$$

Into (2.48), we now substitute (2.49), (2.50), (2.52), and (2.53), applying the equation successively for $q_i = x_c, z_c, \Theta, \delta_e$, and replacing $\dot{x}_c$, $\dot{z}_c$, and $\dot{\Theta}$ with the corresponding motion coordinates. Since d/dt in (2.48) represents an inertial rate of change, we also observe that the Coriolis law requires us to add the terms QW and $(-QU)$ to $\dot{U}$ and $\dot{W}$, respectively.

$$F_x = m[\dot{U} + QW] \tag{2.54a}$$

$$F_z = m[\dot{W} - QU] + S_{y\delta}\,\ddot{\delta}_e \tag{2.54b}$$

$$M_p = I_{yy}\dot{Q} + [-x_H S_{y\delta} + I_{y\delta}]\,\ddot{\delta}_e \tag{2.54c}$$

$$H_\delta = S_{y\delta}[\dot{W} - QU] + [-x_H S_{y\delta} + I_{y\delta}]\dot{Q} + I_{y\delta}\,\ddot{\delta}_e + K_\delta\,\delta_e \tag{2.54d}$$

We note that $S_{y\delta}$, which is positive when the CM of the elevator mass lies behind the hingeline, produces "inertia coupling" with vertical motion of the vehicle, in the sense that a z-acceleration tends to rotate the control, and conversely. Often, to prevent flutter, the control is mass-balanced so as to make $S_{y\delta} = 0$. In this event, there is still seen to be some inertia coupling with pitch through the $I_{y\delta}$ terms in (2.54c, d).

3. Aeroelasticity

When one is analyzing the dynamic response of high-performance vehicles, it is almost never acceptable to overlook entirely the influence of structural deformations. Their importance was once vividly brought home to a colleague of mine when he visited the designers of the B-36. In one department, it had been estimated, under the rigid-vehicle assumption, that the "short-period mode" of longitudinal oscillation had a period of about one second during flight at high dynamic pressure. In another, the fundamental natural frequency of wing bending vibration was calculated to be 1 Hz!

A very systematic scheme for introducing these aeroelastic effects into the equations of motion is to add to the six rigid-body degrees of freedom a set of generalized coordinates q_i associated with the so-called "normal modes" of free vibration in vacuum. These modes are calculated during design under the (usually very accurate) approximations that the structure is linearly elastic and that its deformation vector is always small compared with the vehicle's overall dimensions. Each mode is then characterized by a distinct natural frequency ω_i rad/sec and by a *mode shape vector* $\boldsymbol{\varphi}_i(\mathbf{r})$, which measures the directions and relative magnitudes of displacements of the mass elements dm when in pure simple harmonic motion at this frequency.

In books on structural dynamics (e.g., Bisplinghoff and Ashley, 1962, Chapter 9), it is shown that these modes are governed by

$$M_i\ddot{q}_i + M_i\omega_i^2 q_i = Q_i, \qquad i = 1, 2, \ldots \tag{2.55}$$

There may also be a small "structural damping" term containing $\dot{q}_i$. Here

$$M_i \equiv \int_{\text{Vehicle}} |\boldsymbol{\varphi}_i|^2 \, dm \tag{2.56}$$

is called the *generalized mass* of the ith mode. The generalized force Q_i is an integral over all the distributed and concentrated external loads, each load weighted by scalar multiplication with the mode shape at its place of application. It is important to recognize that (2.55) are wholly consistent with (2.16). Their use is not restricted to small motions in the freedoms described by U, V, W, P, Q, R. For computational efficiency the number of q_i must be truncated to include only enough modes to represent adequately the aeroelastic effects on the maneuver or response being analyzed.

Coupling between the "rigid-body" and aeroelastic coordinates in (2.16) [or (2.39)] and (2.55), respectively, occurs mostly in the aerodynamic terms. Thus it is found that each Q_i receives contributions proportional to angle of attack, roll rate, and/or other quantities defining the overall motion through the atmosphere. Similarly the lift, pitching moment, etc., may be affected by the deformation through q_i and its derivatives $\dot{q}_i$ and $\ddot{q}_i$. In many applications, however, it can occur that the shortest time period associated with the overall motion is long compared with the greatest vibration period $2\pi/\omega_1$. All the $\ddot{q}_i$ terms may then be negligible; for instance, if the ratio of these periods were about 5:1, we can see that the first left-hand term in (2.55) will never exceed 4% of the second. It is then at least theoretically possible to solve algebraically for the q_i in terms of the "rigid-body" coordinates, eliminate all q_i from the aerodynamic terms in (2.16) or (2.39), and arrive at a set of six equations "corrected" for aeroelastic effects.

In practice, normal modes do not constitute the most efficient approach to making such aeroelastic adjustments. Most analysts prefer to calculate directly the deformation due to the instantaneous airloads, using structural "influence coefficients" for the purpose. The forces and moments due to these deformations are then added to the corresponding "rigid-body" terms. For instance, we may think of the stability derivatives, defined and discussed in Chapter 7, as being corrected for aeroelasticity. Once adopted, this scheme also lends itself to the introduction of "inertia elastic" corrections, which are force and moment increments caused by deformation due to inertia loads proportional to various linear and angular accelerations. In any event, neither the apparent number of degrees of freedom nor the total order of the equations of motion is increased. This approach forms the basis of the comprehensive treatment of longitudinal response and dynamic stability published by Roskam *et al.* (1968–69). Among many other papers on this subject, the one by Milne (1964) stands out for its clarity and rigor.

2.8 PROBLEMS

2A Combine what was developed about rotation matrices in Section 2.4 with the definitions (2.20) of the inertia tensor to derive the law of transformation of components, for a Cartesian tensor of second rank, when the axes of observation are given a general rotation about a fixed origin in the body. Prove the existence of principal directions for these axes, about which all components vanish except those for which $i = j$. (Ref.: Halfman 1962, Section 5.3).

2B Sketch-plot versus time the Euler angles for these various maneuvers starting from $\Phi = \Theta = \Psi = 0$ at $t = 0$.

a) Half an inside loop followed by a snap roll at the top
b) Outside loop
c) Coordinated 180° turn
d) Climbing or helical turn

2C The linear velocity of an aircraft's center of mass with respect to the atmosphere (here assumed to be inertial space) is $\mathbf{v}_c$. The angular velocity of an aircraft with respect to the atmosphere is $\boldsymbol{\omega}$. Being vector quantities, these may be resolved in any convenient coordinate system. For example, resolving $\mathbf{v}_c$ in the (x', y', z') inertial reference frame, we get:

$$\mathbf{v}_c = (U_1, V_1, W_1) = \left(\frac{dx'}{dt}, \frac{dy'}{dt}, \frac{dz'}{dt}\right)$$

Resolving this same vector in a reference frame fixed to the aircraft, we get

$$\mathbf{v}_c \equiv (U, V, W)$$

Likewise $\boldsymbol{\omega}$ resolved in the aircraft fixed reference frame is

$$\boldsymbol{\omega} = (P, Q, R)$$

Assume that instruments in the A/C (aircraft) give continuous records of U, V, W, P, Q, R. Knowing these, we are interested in finding the velocity and position components of the A/C resolved in an inertial reference frame, (x', y', z').

a) Write the transformation equations in matrix notation.
b) Assuming that the initial flight conditions of the A/C are known, describe clearly the step-by-step process used with the aid of a high-speed digital computer to calculate

$$x', y', z', \frac{dx'}{dt}, \frac{dy'}{dt}, \frac{dz'}{dt}$$

as functions of time (a block diagram is an effective method).
c) Suppose a ground-based radar site is used to help the above aircraft fire an air-to-air missile. The linear acceleration of a target A/C with respect to the atmosphere is $\mathbf{a}_c$. We are told that the components of $\mathbf{a}_c$ resolved in the inertial (x', y', z') reference frame are known from radar tracking data. However, in order to generate guidance commands for the missile, $\mathbf{a}_c$ must be resolved in a coordinate system fixed to the launch aircraft. Write the transformation relations in matrix notation.

2D The static-performance equations for a flight vehicle can be written quite easily by considering an equilibrium of forces, i.e., neglecting accelerations, and assuming that the only significant forces are gravity; thrust, T; lift, L; and drag, D. The equations are

$$T - D - mg \sin \Theta = 0, \qquad L - mg \cos \Theta = 0$$

where m is the vehicle mass and Θ is the elevation of the velocity vector above the horizon.

For flight vehicles traveling at velocities which are not small compared to low-circular orbital velocity, $\sqrt{gR_e}$, one must add an inertia term to these equations. This term becomes important because the motion is actually not in a straight line, but is circular with a radius r. Correct the static-performance equations by including this term. Use this result for a *brief* study of the following subjects.

a) Entry of a glider ($T = 0$) from a low-circular orbit ($R_e \gg h$).

b) The *corridor of continuous horizontal flight* for cruising vehicles. This corridor, usually plotted on a diagram of altitude h (really density, ρ_∞) versus speed v_c, describes where such cruising flight may be expected to occur. Its upper limit is defined by the fact that maximum achievable values of lift coefficient c_L run around 1. Its lower limit occurs because of excessive loads and aerodynamic heating. This lower limit might be set around

$$\frac{\rho_\infty}{2} v_c^2 = 1500 \text{ lb/ft}^2$$

for lower speeds, and perhaps at a constant value of $\rho_\infty v_c^3$ (the important factor in convective heat transfer) above a Mach number of about 5.

[Semiempirical results from boundary-layer theory indicate that the heat flux to a flat cold surface is proportional to $\rho_\infty^{0.8} v_c^{2.8}$ in the turbulent case and $\rho_\infty^{0.5} v_c^{2.5}$ in the laminar case. Those interested should refer to Schlichting 1960, page 285, or Nielsen 1960, Chapter 6.]

2E Consider once again the weathervane of Problem 1E. To make it more interesting, add a rudder which is free to rotate about a hinge at the outer end of the fin, as shown. (The uncoupled lateral motion of an airplane in steady level flight can be approximated by this fin–rudder combination.)

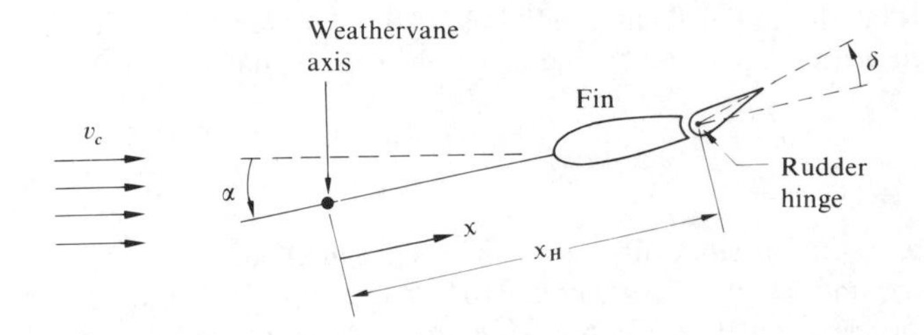

The equations of motion are presented below [cf. (2.54)]. Derive these equations by any convenient method, clearly stating any assumptions.

$$I\ddot{\alpha} + \ddot{\delta}[I_\delta + x_H s_\delta] = M_{\text{Axis}} = q_\infty S\bar{c}[c_{M\alpha}\alpha + c_{M\delta}\delta] - k_1\dot{\alpha}$$

$$I_\delta\ddot{\delta} + \ddot{\alpha}[I_\delta + x_H s_\delta] = H_{\text{Hinge}} = q_\infty S\bar{c}[c_{h\alpha}\alpha + c_{h\delta}\delta] - k_2\dot{\delta}$$

where

$$I = \int_{\text{Fin + rudder}} x^2\, dm = \text{total moment of inertia about axis}$$

$$I_\delta = \int_{\text{Rudder}} [x - x_H]^2\, dm = \text{moment of inertia of rudder about hinge}$$

$$S_\delta = \int_{\text{Rudder}} (x - x_H)\, dm = m_{\text{Rudder}} x_{\text{CG}}$$

Here q_∞ is the dynamic pressure of the oncoming wind, and S and $\bar{c}$ are reference area and length, respectively. Note then that $q_\infty S\bar{c}$ is a reference moment and that terms like $c_{M\alpha}\alpha$, $c_{M\delta}\delta$ represent contributions of α and δ to moments about the axis (similarly for the hinge moments in the second equation). Note also that damping terms are present, representing the effects of bearing friction and aerodynamic contributions. How would you attempt to solve these equations?

2F The equations of motion for a spinning missile or rifle bullet (or, less precisely, a rapidly rolling aircraft) can be approximated as follows:

- The CM moves in a straight line ($\mathbf{v}_c$ = constant).
- There is a constant rolling velocity P_0.
- The quantities Θ, Ψ, Q, R are always small enough to permit linearization.
- The X-force equation and the rolling-moment equation may be assumed automatically satisfied by proper deflections of the control surfaces.
- The pitching moment M_p and yawing moment N consist mainly of moments, proportional to V and W, respectively, which are positive when (as on a finned missile) they tend to restore the x-axis to parallelism with the flight direction.

Under these approximations, show how Q and R can be eliminated from the remaining four equations of motion. With $U \cong$ constant, show how the last two equations can be reduced to

$$\frac{d^2W}{d\tau^2} + \left[\left(\frac{\omega_y}{P_0}\right)^2 - i_1\right]W + (1 + i_1)\frac{dV}{d\tau} = 0$$

$$-(1 + i_2)\frac{dW}{d\tau} + \frac{d^2V}{d\tau^2} + \left[\left(\frac{\omega_z}{P_0}\right)^2 - i_2\right]V = 0$$

Here $\tau \equiv P_0 t$ is a dimensionless time variable, and

$$i_1 \equiv \frac{I_{zz} - I_{xx}}{I_{yy}} > 0, \qquad i_2 \equiv \frac{I_{yy} - I_{xx}}{I_{zz}} > 0, \qquad I_{xz} \cong 0$$

Also:

ω_y = natural frequency of pitching oscillation in the absence of rolling, with the dependence of M_p on W unchanged.

ω_z = natural frequency of yawing oscillation in the absence of rolling, with the dependence of N on V unchanged.

Use the characteristic determinant of this approximate system to study the ranges of the parameters i_1, i_2, (ω_y/P_0), and (ω_z/P_0), where the spinning vehicle is stable or unstable. A bullet has $\omega_y^2, \omega_z^2 < 0$; why is it desirable to have it spin rapidly? (Refs.: Two classics on such questions are Phillips 1948 and Nielsen and Synge 1946. For aeronautical applications, see Seckel 1964, pages 270–274 and reading list pages 486–487.)

2G For an elongated satellite in a spherical gravitational field, show that the variation of gravitational attraction with r can produce a moment about the CM, and that this torque tends to align the longest axis parallel to the radius vector. It is convenient to idealize the satellite into two lumped masses, connected by a massless, rigid bar. (Ref.: Halfman 1962, pages 275–284.)

CHAPTER 3

INTRODUCTION TO VEHICLE AERODYNAMICS

3.1 AERODYNAMIC CONTRIBUTIONS TO X, Z, AND M_p; DIMENSIONLESS COEFFICIENTS DEFINED

By way of introduction to theoretical analysis of the aerodynamic terms in the equations of motion, we take a preliminary look at three airload components and their dependence on some important parameters which characterize the vehicle and its flight régimes. Suppose that the motion occurs steadily through the atmosphere and within a single vertical plane of symmetry. Our interest is then confined to those portions of X, Z, and M_p in (2.40) which are not contributed by the propulsive system. We shall later find it straightforward, however, to generalize our results to other force and moment components.

On the basis of both experimental observation and theory, we can determine the two forces and the pitching moment by the following quantities.

- The vehicle's geometrical shape
- The size, as fixed by such lengths as wingspan b, typical wing chord c, tail length l_t, or $\sqrt{S}$
- Ambient air density ρ_∞
- Airspeed v_c or v_{c0}
- Speed of sound a_∞ in the undisturbed gas
- Dynamic coefficient of viscosity* μ_∞
- Angle of attack. (When questions of trim arise, this concept is broadened to include the angle of incidence between the plane of the wing and the plane of the stabilizer, as well as the elevator setting δ_e.)

We encounter so many definitions of angle of attack that we must be very specific about the one adopted for this chapter. Figure 3.1 illustrates a representative flight condition. Airspeed vector $\mathbf{v}_{c0}$, with subscript zero to remind us that

* When a gas moves in a gliding or laminar fashion, such that nearly plane layers slip past one another with velocity gradient $\partial v/\partial n$ normal to the stream surfaces, the shear stress σ between the layers is given by

$$\sigma = \mu_\infty \, \partial v/\partial n.$$

The viscosity coefficient is a state variable, approximately proportional to $\sqrt{T_\infty}$ over the range of temperatures of the atmosphere.

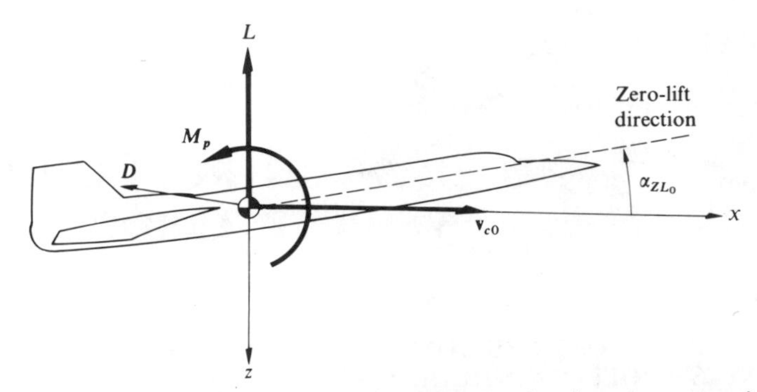

Fig. 3.1. NASA/USAF experimental Lockheed SR-71 showing the approximate zero-lift direction and other quantities needed to describe trimmed flight.

this is rectilinear, equilibrium motion, appears aligned with the x-direction of stability axes. Lift L, drag D, and pitching moment act at and about the CM in the directions discussed in Chapter 1. Also shown dashed is the *zero lift line*, which lies in the plane of symmetry and has a fixed orientation relative to a vehicle of a given shape in a given range of flight Mach number M. This line is defined by the property that, when $\mathbf{v}_{c0}$ is parallel to it, L vanishes. We expect that lift must depend in a simple way on α_{ZL_0}, the instantaneous angle between $\mathbf{v}_{c0}$ and the zero lift line.

We are now in a position to set down three approximate relations, which hold for all streamlined shapes at sufficiently small α_{ZL_0} and which help us to achieve a simple understanding of steady flight.

$$D \cong K_2 + K_3 L^2 \tag{3.1}$$

$$L \cong K_1 \alpha_{ZL_0} \tag{3.2}$$

$$M_p \cong K_4 + K_5 \alpha_{ZL_0} \tag{3.3}$$

Here the five K_i depend, in ways to be examined below, on the first six parameters in the foregoing list; they are therefore constants for a particular flight at a particular speed and atmospheric altitude. The direct proportionality between L and α_{ZL_0} was a fact of vital significance to the achievement of sustained flight. Because of the persistence of such myths as a "sine-squared" law of lift, it did not receive general recognition until surprisingly late in the history of aviation.

Figure 3.1 shows that, in the reference condition, D and L are equal and opposite to the aerodynamic portions of X and Z, respectively. During perturbations of the angle of attack, however, the drag and lift rotate so as to be determined by the instantaneous relative wind (cf. Section 2.3). *Axial force A* and *normal force N* are the symbols employed for airload components along directions attached to the vehicle. These directions might be chosen opposite to the x and z of stability axes, but more commonly they are connected with vehicle geometry, e.g., by locating A in the plane of the wing chord or along the axis of a body of revolution.

According to a universal custom, dimensionless coefficients of lift and drag, C_L and C_D, are defined by dividing out of L and D the product of wing area and dynamic pressure. For nondimensionalizing the pitching moment, an added length factor is required. The convention is to choose the *mean aerodynamic chord*,

$$\bar{c} \equiv \frac{1}{S} \int_{\text{Wingspan}} c^2(y)\, dy \tag{3.4}$$

As with area S, we arbitrarily define $c(y)$ over the central portion of the wing covered by the fuselage by extending the leading and trailing edges to the aircraft centerline. Body length or maximum diameter replaces $\bar{c}$ when we are dealing with vehicles that have no primary lifting surface. In summary,

$$D \equiv \frac{\rho_\infty}{2} v_{c0}^2 S C_D \tag{3.5a}$$

$$L \equiv \frac{\rho_\infty}{2} v_{c0}^2 S C_L \tag{3.5b}$$

$$M_p \equiv \frac{\rho_\infty}{2} v_{c0}^2 S \bar{c} C_m \tag{3.5c}$$

In Problem 3A, we shall appeal to the formalism of dimensional analysis to prove that each coefficient is a function of the following.

- Dimensionless geometrical shape, as characterized by a series of dimensionless groups like body fineness ratio, wing thickness ratio distribution, aspect ratio $Æ \equiv b^2/S$ or $b/\bar{c}$, tail length divided by $\bar{c}$, etc.
- Mach number $\mathrm{M} \equiv v_{c0}/a_\infty$
- Reynolds number $\mathrm{Re} \equiv \rho_\infty v_{c0} \bar{c}/\mu_\infty$
- Angle of attack α_{ZL0}

Reynolds number measures the influence of viscosity on the airloads. For flight in the lower atmosphere, its value ranges from 5×10^5 to 10^9. In this range its influence on drag at all angles and on lift near the stall is much more pronounced than on C_L and C_m at operating α_{ZL0}'s for streamlined vehicles. Indeed we shall see in Chapter 4 that useful estimates of lift and moment can be derived from nonviscous flow theory.

When we combine (3.5) with the approximate relations (3.2)–(3.3), we can write

$$C_L \cong C_{L\alpha} \alpha_{ZL0} \tag{3.6}$$

$$C_m \cong C_{m0} + C_{m\alpha} \alpha_{ZL0} \tag{3.7}$$

Fig. 3.2. Measurements of aerodynamic coefficients at low speed on a typical thin, symmetrical airfoil, the NACA 0006 (code identifies the thickness-chord ratio as 6%). Several Reynolds numbers Re are included, as well as the effect of surface roughness. Note that moment coefficients are given for both the aerodynamic center (A.C., see Section 4.3 for definition and discussion) and the quarter-chord axis; here these axes nearly coincide. [Results courtesy of NASA; see Abbott and von Doenhoff for further details.]

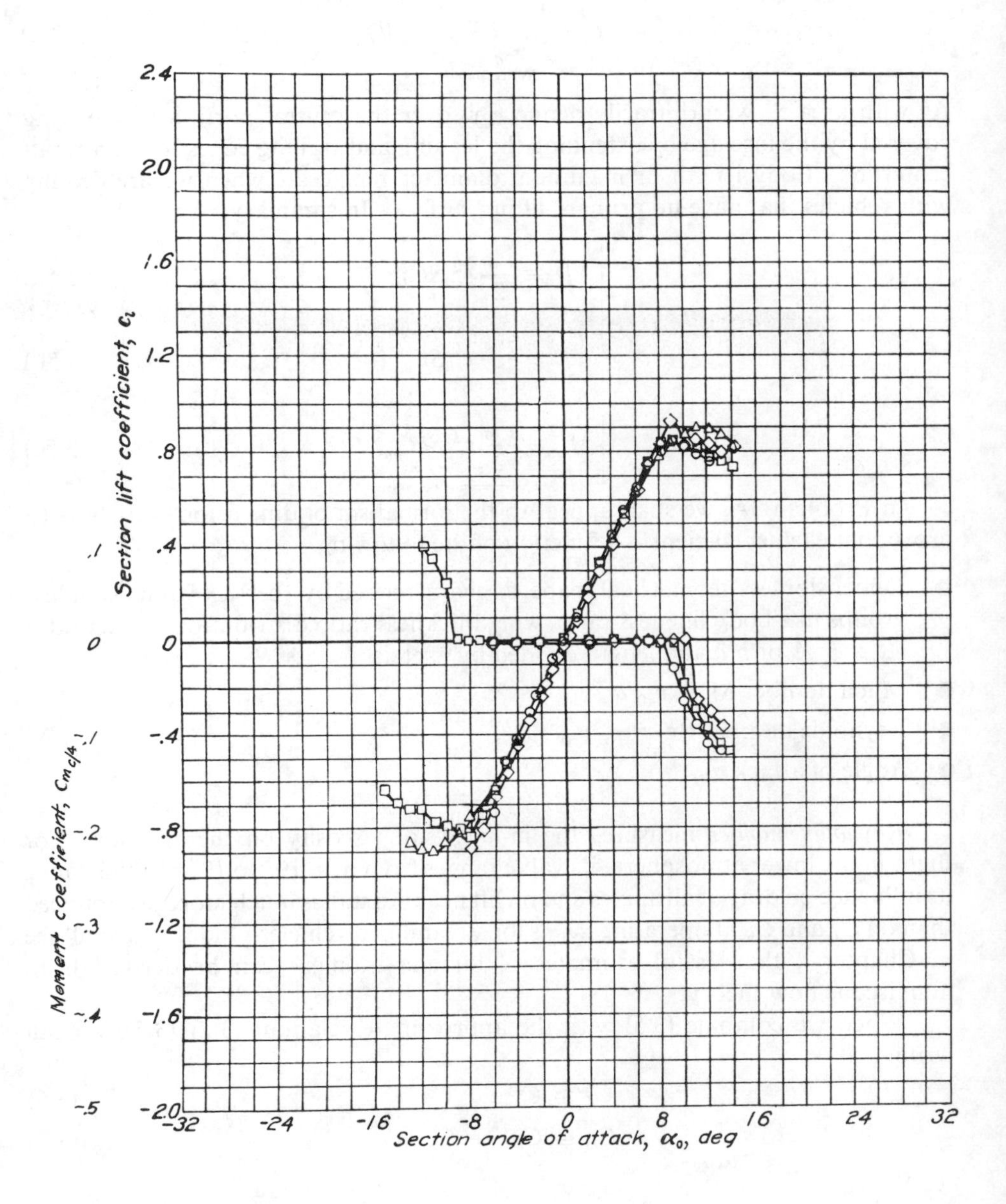

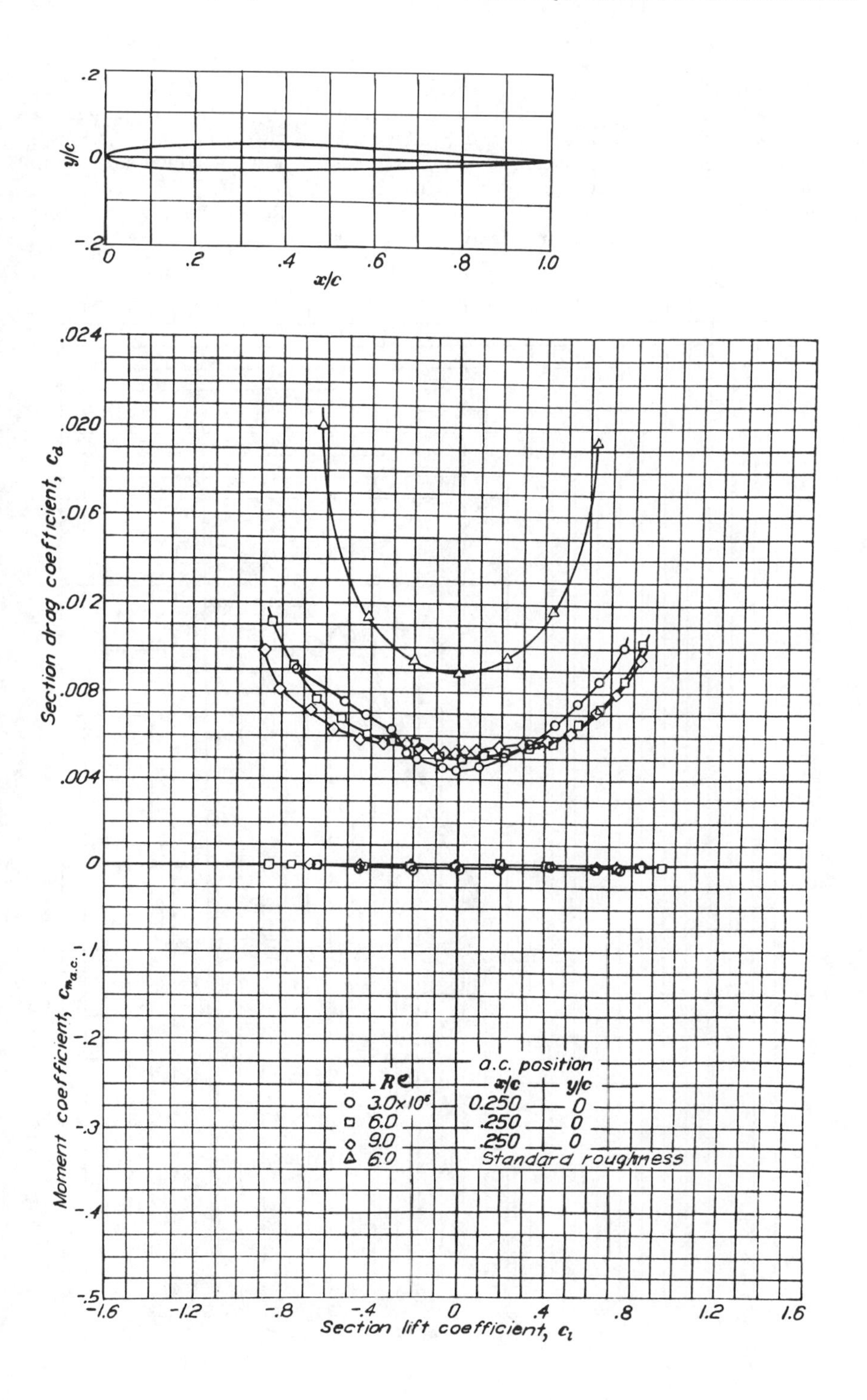

y/c
x/c
Section drag coefficient, c_d
Moment coefficient, $c_{m_{a.c.}}$
Section lift coefficient, c_l
Re
a.c. position
x/c
y/c
○ 3.0×10⁶ 0.250 0
□ 6.0 .250 0
◇ 9.0 .250 0
△ 6.0 Standard roughness

Fig. 3.3. Measurements of aerodynamic coefficients at low speed on a typical thick, cambered airfoil, the NACA 65,3-618 (code identifies the thickness chord ratio as 18%, the design lift coefficient as 0.6, and the minimum pressure due to the thickness distribution at 50% chord). Several Reynolds numbers Re are included, as well as the effect of surface roughness. Note that moment coefficients are given for both the aerodynamic center (AC, see Section 4.3) and the quarter-chord axis. Actual AC positions are specified. [Results courtesy of NASA; see Abbott and von Doenhoff for further details.]

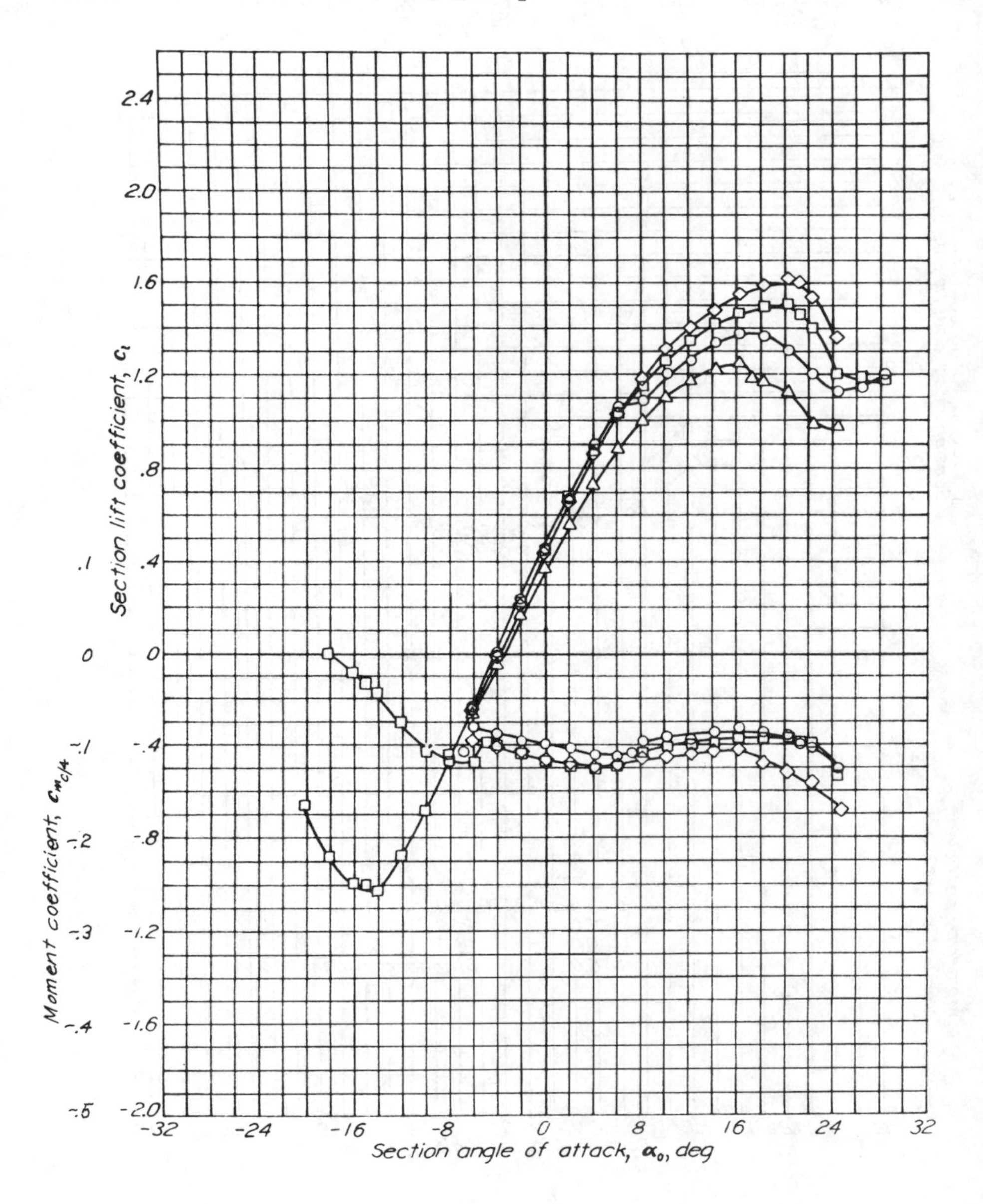

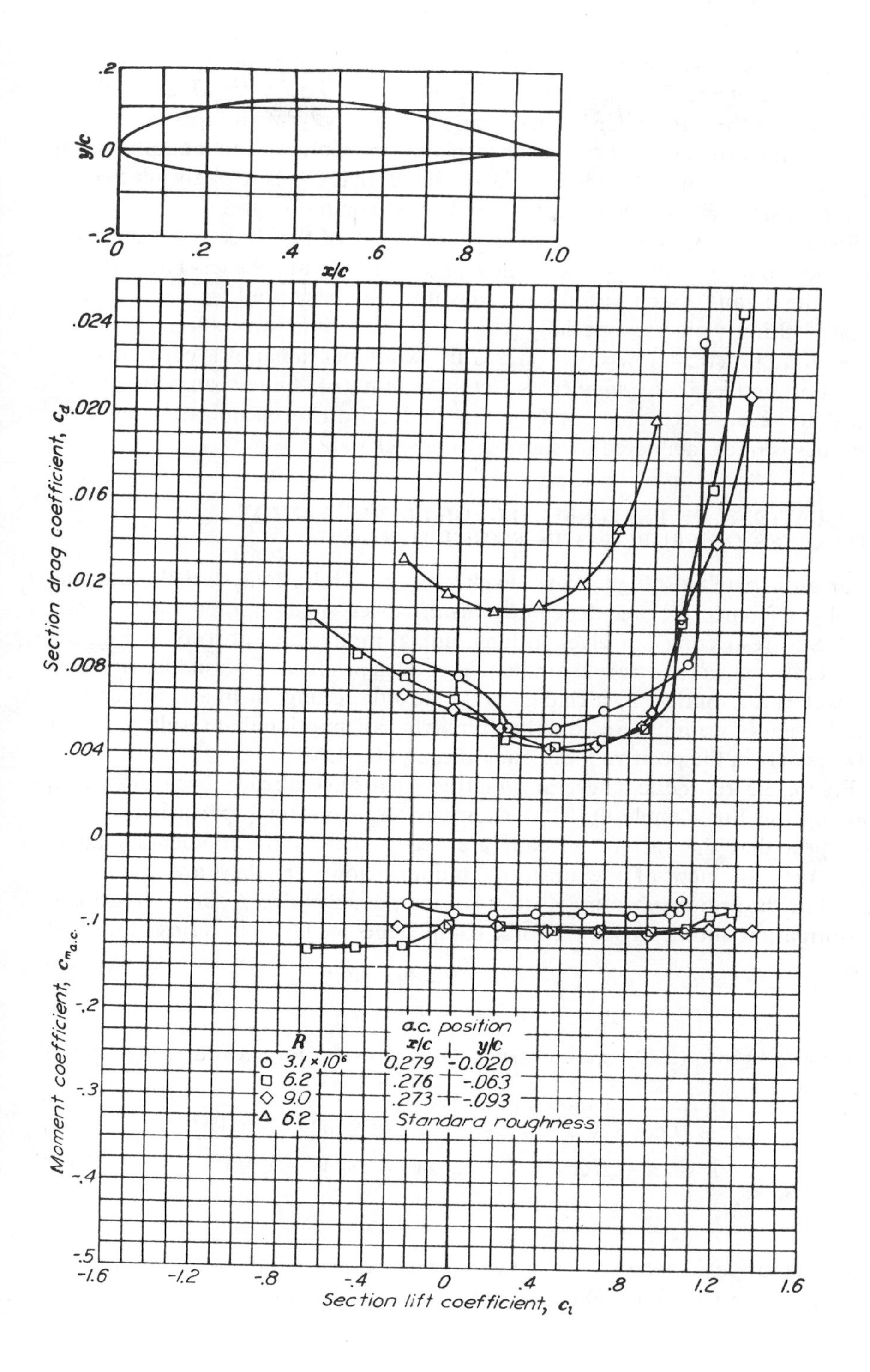

y/c
.2
0
-.2
0
.2
.4
.6
.8
1.0
x/c
Section drag coefficient, c_d
.024
.020
.016
.012
.008
.004
0
Moment coefficient, $c_{m_{a.c.}}$
-.1
-.2
-.3
-.4
-.5
-1.6
-1.2
-.8
-.4
0
.4
.8
1.2
1.6
Section lift coefficient, c_l
R
○ 3.1 × 10⁶
□ 6.2
◇ 9.0
△ 6.2
a.c. position
x/c y/c
0.279 -0.020
.276 -.063
.273 -.093
Standard roughness

Here the *zero-lift moment* C_{m0}, *lift-curve slope* $C_{L\alpha}$, and *moment-curve slope* $C_{m\alpha}$ are constants of the flight condition. That is, they vary with M and vehicle geometry, but they are independent of α_{ZL_0} and are only weak functions of Re. Relative to the problem of trimming an airplane, we mention (1) that another term proportional to δ_e might be added to (3.7), and (2) that $C_{m\alpha}$ is strongly affected by the moment-axis shifts which occur when the CM is moved chordwise.

Figures 3.2, 3.3, and 3.4 illustrate the behavior of the three coefficients for two-dimensional airfoils in subsonic flow and for a typical vehicle. The captions explain the details. Such airfoil data constitute the basic tool for aerodynamic design of lifting surfaces, especially when Æ is large and sweep angle Λ small. Although no practical flow situation is truly two-dimensional, it is convenient to think of properties at any wing cross section in terms of "finite-span corrections" applied to results like those in Figs. 3.2 and 3.3. The optimization of wing properties is often facilitated by starting from such data.

3.2 EQUATIONS OF PERTURBED LONGITUDINAL MOTION; CATEGORIES OF PROBLEMS IN FLIGHT DYNAMICS

We can anticipate most flight-dynamic phenomena, whose analysis will occupy us later, by reference to time-dependent disturbances superimposed on trimmed flight. Still restricting ourselves to longitudinal motion, we portray in Fig. 3.5 one instant along a trajectory that might have been produced by a sudden application of elevator control, by a change in the throttle setting, or by encountering a gust. We adopt stability axes and the simplifying assumption that resultant thrust T always acts in the positive x-direction through the CM.

Figure 3.5 reproduces several quantities that have already been discussed, including the Euler angle Θ. Also shown is angle of attack perturbation $\alpha(t)$. We expect that the various airloads are determined by the momentary angle $[\alpha_{ZL_0} + \alpha]$. In view of the unsteady motion, however, they are sometimes affected by the previous history of $\alpha(t)$, $\mathbf{v}_c(t)$, and possibly $\Theta(t)$. A simple rotational transformation about y gives the force components for (2.40) in terms of lift and drag.

$$X = T - D\cos\alpha + L\sin\alpha \tag{3.8a}$$

$$Z = -D\sin\alpha - L\cos\alpha \tag{3.8b}$$

Inserting (3.8) into (2.40), we construct another useful form of the longitudinal equations.

$$T - D\cos\alpha + L\sin\alpha - mg\sin\Theta = m[\dot{U} + QW] \tag{3.9a}$$

$$-D\sin\alpha - L\cos\alpha + mg\cos\Theta = m[\dot{W} - QU] \tag{3.9b}$$

$$M_p = I_{yy}\dot{Q} \tag{3.9c}$$

Let us suppose that we know how to determine the aerodynamic and propulsive terms from $U(t)$, $W(t)$, $\Theta(t)$, and such "input" quantities as $\delta_e(t)$ and throttle

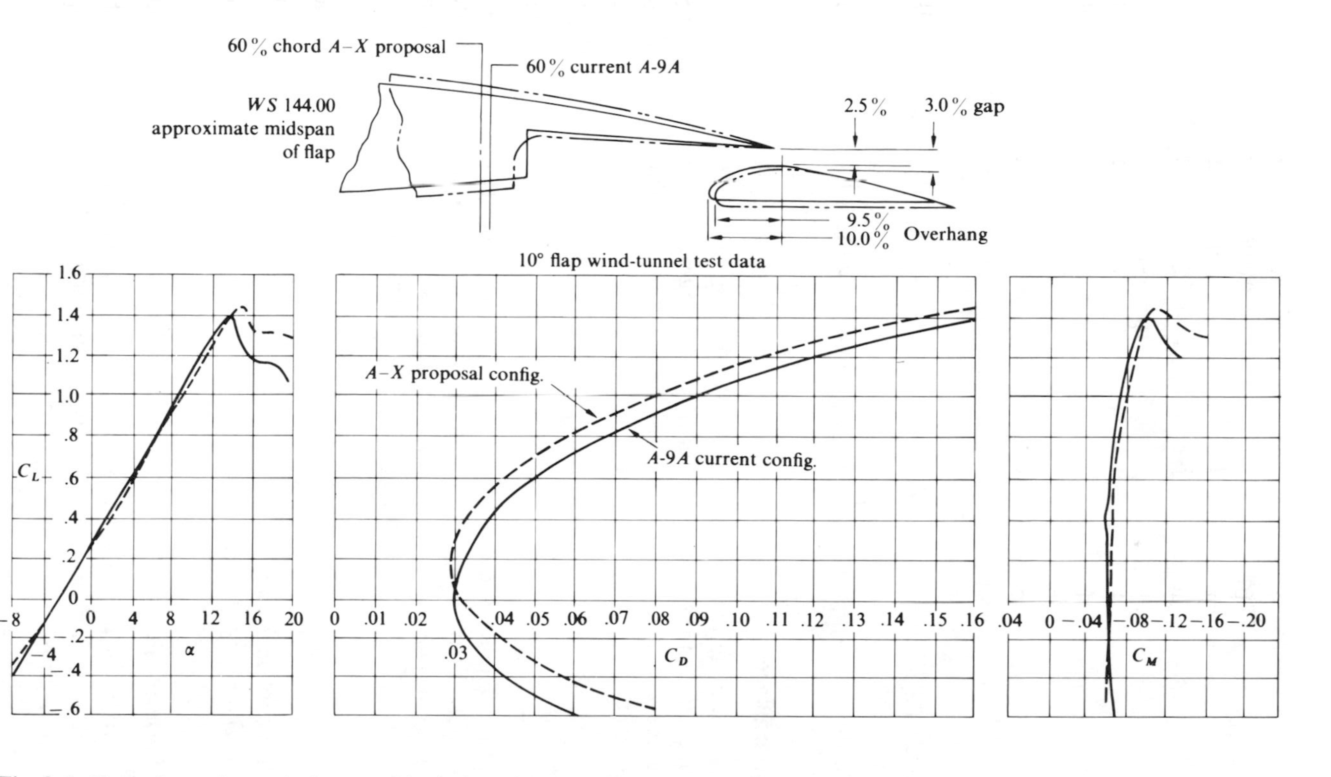

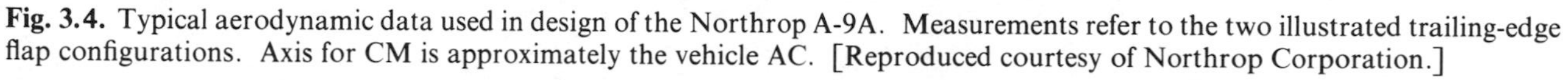
Fig. 3.4. Typical aerodynamic data used in design of the Northrop A-9A. Measurements refer to the two illustrated trailing-edge flap configurations. Axis for CM is approximately the vehicle AC. [Reproduced courtesy of Northrop Corporation.]

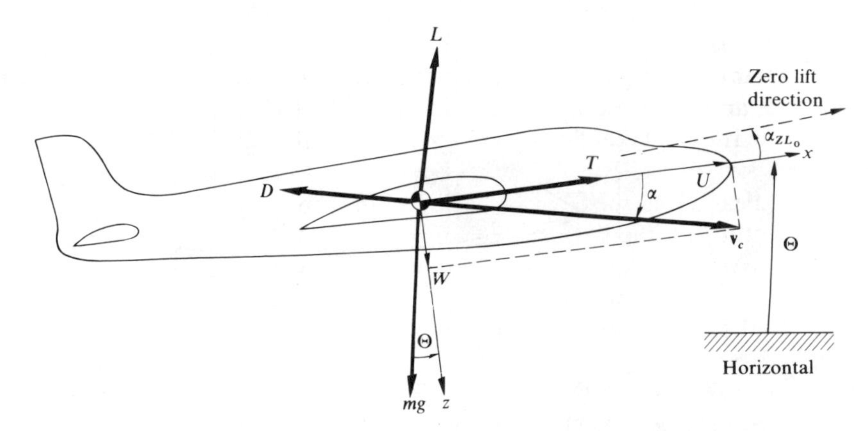

Fig. 3.5. Flight vehicle in a condition of perturbed longitudinal motion.

setting. Here then are some typical design problems that can be analyzed with (3.9).

1. Static Performance

This is the case of unaccelerated flight, in which we can study such questions as maximum horizontal speed, steady rate of climb, gliding performance, range, and endurance. Since $\alpha = 0$ in stability axes, and the pitching moment is trimmed to zero, (3.9) reduce to

$$T = D + mg \sin \Theta \tag{3.10a}$$

$$L = mg \cos \Theta \tag{3.10b}$$

Given suitable engine data, with their dependence on speed and atmospheric state, we can solve (3.10) for v_{c0} as a function of altitude and Θ, which is here the (essentially constant) *climb* or *flight-path angle*. For instance, either a sailplane or an entry glider has $T = 0$, whereupon (3.10a) may be divided by (3.10b) to yield

$$\tan \Theta = -D/L \tag{3.11}$$

(It is assumed, of course, that entry is unaffected by planetary curvature.) Recalling that Θ is positive for flight above the horizontal, we see in (3.11) the vital importance of high L/D for long gliding range. Adjustments are readily made for vertical components of wind speed, known to sailplane pilots as *updrafts* and *downdrafts*.

2. Dynamic Performance

We believe this to be the proper term to identify the study of accelerated motion of the vehicle CM, under circumstances in which the interchange between kinetic and potential energies must be considered. Such phenomena as entry, boost, accelerated climb, "zoom," and the general determination of optimum trajectories fall into this category. Although solution of the full system (3.9)—or their equivalent on a spherical planet—is often required, we can sometimes assume that Θ is instantaneously responsive to elevator control. Equation (3.9c) can then be neglected, and the prescribed $\Theta(t)$ is inserted into (3.9a, b).

The estimation of takeoff and landing distances also falls under the heading of dynamic performance. These distances consist of a segment of airborne flight, required to clear some specified obstacles, plus a horizontal ground roll. Being obviously rectilinear, the latter are calculated from (3.9a) with $Q = 0 = W$ and a term added to account for runway resistance on the wheels.

Until fairly recently one determined the time for an airplane to climb to a given altitude or ceiling statically, by finding the speed for best rate of climb from (3.10), then adding up the times needed for steady climb through a series of atmospheric layers. Many vehicles with T/W in the range between $\frac{1}{4}$ and 1 or greater have been found, however, to perform better when allowance is made for dynamic energy interchange.

3. Static Stability and Control

Traditionally this field has to do with how the primary control surfaces are used for trimming to zero moment about the CM. The elevator, for example, is set to ensure that pitching moment $M_p = 0$ in steady flight. We shall later discuss several detailed questions which arise in connection with stability. Probably the most important concerns how to locate the wings, stabilizer, and vehicle CM so that the moment-curve slope obeys

$$C_{m\alpha} < 0 \tag{3.12}$$

while $|C_{m\alpha}|$ falls within a range associated with desirable flying qualities.

From static considerations one can present a valid rationale for the negative $C_{m\alpha}$. In Chapters 7 and 8, however, we shall describe a more logical justification based on what the pilot desires by way of period and damping from the inherent longitudinal modes of airplane motion.

"Control" has to do generally with angular positioning of the vehicle and with how its movable surfaces are employed, by human or automatic actuation, to cause specified maneuvers.

4. Dynamic Stability and Response

Here we shall analyze disturbed motion from an initial equilibrium flight path. In the past it has usually been assumed that the perturbations to U, V, W, etc., were sufficiently small to permit linearization of the dynamic equations. Great importance is then attached to the homogeneous solutions to these equations, which consist of normal modes of free motion of a vehicle. After some initial input, positive stability is associated with decay over time of all the modal amplitudes. For the treatment of large maneuvers and disturbances, we must abandon the linearizing approximation for one or more of the motion coordinates.

Unless one has recourse to a technique such as "multiple scaling" (Van Dyke 1964, Section 10.4), there is really no rigorous way to draw a line between dynamic stability and dynamic performance. As a practical matter, however, there is no objection to perturbing an assumed rectilinear flight path, so long as the time constants of long-term maneuvers are, say, a factor of ten or more greater than the longest modal period involved in the short-term response. Otherwise the motion should be studied by means of the full nonlinear system (2.39) or (3.9). The latter situation arises, for instance, during the higher portion of the entry trajectory of a lifting glider. The *long-period* or *phugoid* longitudinal mode may then have a period as great as half the total interval required for entry.

5. Guidance and Navigation

This subject concerns manual or automatic procedures for producing a desired CM trajectory. However, we shall not attempt a detailed coverage of it in this book.

3.3 PROBLEMS

3A Consider an aircraft in uniform motion, with speed v_c at angle of attack α in a medium of constant density ρ. A typical length dimension is l. We are interested in finding the drag D on the vehicle. Since the only parameters assumed in the problem are those mentioned above, we expect that

$$D = f(\rho, l, v_c, \alpha)$$

where f is an unknown function. This equation must be dimensionally consistent; i.e., since D has dimensions of a force, the right side of this relation must also have dimensions of a force. For example, suppose we assume that

$$D = k\rho^a l^b v_c^c \alpha^d$$

where k is a constant and a, b, c, d are exponents. Then, in terms of dimensions,

$$[D] = \frac{ML}{T^2} = \left[\frac{M}{L^3}\right]^a [L]^b \left[\frac{L}{T}\right]^c [0]^d$$

Rearrange to get

$$MLT^{-2} = M^a L^{b+c-3a} T^{-c} [0]^d$$

In order for the equality to hold, the exponents of the quantities M, L, and T must match. That is, $a = 1, b + c - 3a = 1, -c = -2$, and d arbitrary.

Solving this system yields $a = 1, b = 2, c = 2$, and hence

$$D = k\rho v_c^2 l^2 \alpha^d$$

or, since d is arbitrary, a more general form is

$$D = k\rho v_c^2 l^2 f(\alpha)$$

Consider the same problem again, but assume that the medium is viscous, with viscosity μ, and compressible, with free-stream density ρ_∞ and speed of sound a_∞. By an argument similar to the above, construct a meaningful functional expression for the drag.

3B Discuss the flight conditions necessary for optimum range of a conventional low-speed glider. How is the analysis complicated when one examines an entry glider, moving so rapidly that the correction terms from Problem 2D must be accounted for?

3C (Example of dynamic performance.) In an effort to set an altitude record, an airplane performs a "zoom" maneuver. Work with longitudinal equations of motion in the form

$$T - D - mg \sin \Theta = m\dot{U} \qquad \text{(what approximation is made here?)}$$

$$-L + mg \cos \Theta = m[\dot{W} - QU]$$

Also assume that:

- The density in the atmosphere is constant (a poor assumption).
- The angle of attack α_{ZL} is held constant throughout (a good assumption; what physical approximations does it involve? What does it say about velocity component W?)
- Thrust T is constant (only fair).
- No Mach-number effects on aerodynamic loads (do you believe it?).

Make the customary aerodynamic substitutions and describe a numerical computation program which could be used to find the maximum value of altitude ($-z'$) when initial values $U(0) = U_0$, $W(0) = 0$, and $\Theta(0) = \Theta_0 > 0$ are specified.

For the case in which $\rho = 0$ (zoom into space), divide the first equation by the second, thus eliminating dt, and solve analytically for $U = f(\Theta)$. Find the speed at maximum altitude for the special case in which $mg = 10{,}000$ lb, $T = 5000$ lb, $U_0 = 5000$ ft/sec, $\Theta_0 = 45°$, and $W(0) = 0$.

3D Reexamine the weathervane of previous problems, which is pictured here at a relatively large angle Ψ from the wind direction.

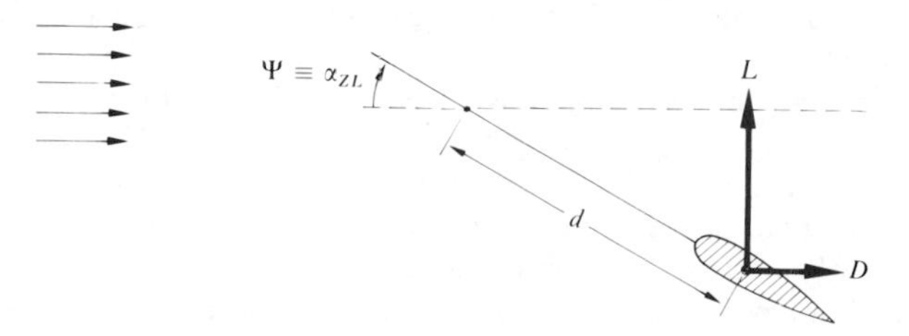

Assume that there is a lift force perpendicular to the wind velocity vector which is proportional to $\sin \alpha_{ZL} \equiv \sin \Psi$. [We shall usually assume $L \sim \alpha$ for small angles of attack, but ideal fluid theory gives us this exact result (e.g., see Karamcheti 1966, Chapter 16, or Ashley and Landahl 1965, Chapter 2).] Thus the restoring moment due to lift is given by

$$N = L(d \cos \Psi) = -(c \sin \Psi)\, d \cos \Psi = -\frac{c}{2} d \sin 2\Psi$$

where a positive indication denotes the clockwise direction of rotation, and c is a constant proportional to $q_\infty S$. Assume that the drag contribution to this moment is relatively small for small values of α and for the case of a large aspect-ratio fin, or include it for better accuracy. In addition, assume a damping moment proportional to $\dot{\Psi}$, so that the equation of motion of the fin becomes

$$I_0 \frac{d^2\Psi}{dt^2} = -a \sin 2\Psi - b\frac{d\Psi}{dt}$$

Discuss the solution of the initial-value problem for $\Psi(t)$, given an initial angular displacement $\Psi(0) = \Psi_0$ at time $t = 0$ and zero initial angular velocity. A digital computer is assumed to be available.

One might first write the problem in state-variable form as a set of first-order differential equations

$$\frac{d\omega}{dt} = -\frac{b\omega}{I_0} - \frac{a}{I_0}\sin 2\Psi, \qquad \frac{d\Psi}{dt} = \omega$$

where ω is obviously the angular rate. Devise a computation scheme to integrate these equations numerically. The essentials of Euler's method, which is a simple example of such a scheme, are as follows.

The problem is to solve a differential equation

$$X' = f(X, t), \qquad 0 \leq t \leq T$$

which is geometrically equivalent to determining a curve $X = X(t)$ passing through a given initial point $(0, X_0)$ and having its slope at each point agree with the slope prescribed by the DE.

$$X_{n+1} \cong X_n + hf(X_n, t_n)$$

where $f(X, t) = X'$ is defined on $0 \leq t \leq T$ and h is the step size for t, chosen for the accuracy desired and the computing time available. Thus $h = t_{n+1} - t_n$ and is taken to be constant for the problem (say $h = 0.1$ sec).

a) Construct a detailed logic flow diagram which would help one to solve this problem on the computer.
b) Discuss the accuracy of this computing scheme (for Euler's method, see Henrici 1962, Chapter 1).
c) Write a program for solving the weathervane problem using the flow diagram from part (a). Design the program to accept arbitrary-time step size, h, and input function. Use language appropriate to whatever computer is available.
d) Choose suitable numerical constants and determine the motion for values of Ψ_0 both within and beyond the linear range of the aerodynamics.

CHAPTER 4

AERODYNAMIC TERMS FOR EQUATIONS OF MOTION; AIRLOADS ON THIN WINGS

4.1 INVISCID FLUID MOTION PAST STREAMLINED SHAPES

It is wholly beyond the scope of Chapters 4 and 10 to give a full systematic review of aerodynamics. We hope rather to present a guide to selected literature, plus an account of how fundamental principles of fluid mechanics are applied to the particular task of estimating airloads like L, D, and pitching moment M_p on streamlined bodies and lifting surfaces at small angle of attack. Most atmospheric flight vehicles are composed of an assemblage of interconnected wing and body elements. It is therefore heartening to realize that the same small-perturbation theoretical methods which have long proved useful for *isolated* elements in subsonic and supersonic flows are now being successfully and efficiently adapted to such "interfering" configurations. Engineers at the forefront of this development would assert that, in the absence of large separated-flow regions and of the transonic interaction which occurs between shocks and boundary layers, aircraft airloads can be predicted with nearly the same accuracy as they can be measured in the wind tunnel.

As for parallel reading, we have been especially impressed by the presentations of fluid-mechanical principles in the books by Batchelor (1967), Liepmann and Roshko (1957), and Shapiro (1953, Volumes I and II). Also we must mention the meticulous exposition of the laws of continuity and momentum in Rauscher (1953). Inviscid, incompressible fluid motion is compendiously treated by Thwaites (1960) and Milne-Thompson (1962), the latter notable for its coverage of two-dimensional flow and complex-variable methods. In order of increasing level of sophistication, the books by Kuethe and Schetzer (1959), Karamcheti (1966), Ashley and Landahl (1965), and Jones and Cohen (1960) are cited relative to the theory of lifting surfaces. A thorough study of the last of these will reward the thoughtful reader, as will an earnest attempt to comprehend the critical classic on hypersonic flow by Hayes and Probstein (1967). Several volumes of the Princeton Series in High Speed Aerodynamics and Jet Propulsion summarize relevant progress up to their dates of publication; Volume VI, edited by Sears (1954), and Vol. VII, edited by Donovan and Lawrence (1957), are especially appropriate to topics in Chapters 4 and 10.

We need to explain the *inviscid approximation* ($\mu_\infty = 0$), which amounts to neglecting any direct influence of shear stresses in the flow field past the wing.

This assumption can be justified formally by resort to the method of matched expansions (cf. Ashley and Landahl, 1965, Chapters 3–4; Van Dyke 1964), but a qualitative discussion will serve our purposes. Since the Reynolds number Re based on chord length runs 10^6 or higher, the rate of shearing $\partial v/\partial n$ must be enormous before there can be appreciable stress between fluid elements. On a streamlined wing at low α_{ZL_0}, for instance, these effects are confined to boundary layers, which are spread like blankets over the upper and lower surfaces and which have thickness δ growing proportionally to some fractional power of streamwise distance measured aft from the leading edge. An idea of scale can be drawn from the observation that, at about 10 ft back along a Boeing-747 wing flying at M $=$ 0.84 and 35,000-ft altitude, this layer is roughly $1\frac{2}{3}$ in. deep. Figure 4.1 sketches an airfoil, showing its upper and lower boundary layers. Both they

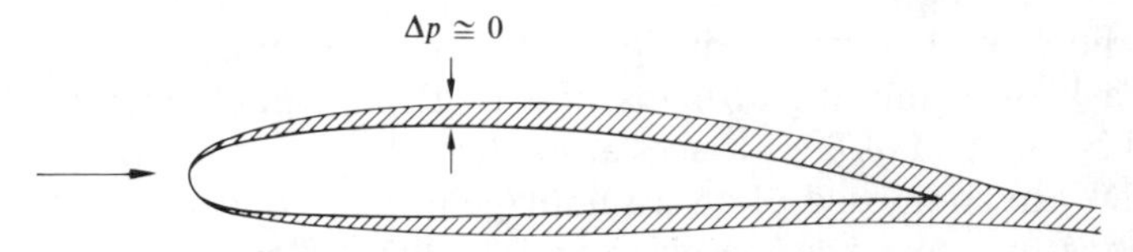

Fig. 4.1. Development of the boundary layer on a subsonic airfoil. The vertical dimensions of the layer and wake (shaded region) are exaggerated.

and the wake formed by their merging at the trailing edge are exaggerated in vertical dimensions.

Significantly the pressure change through the depth of these layers vanishes to the order $(\text{Re})^{-1/2}$ that characterizes theoretical analyses. It follows that airloads like L and M_p, which are almost wholly determined by the distribution of pressure (normal stress) over the wing, can be accurately estimated by letting $\mu_\infty \to 0$ and treating the wing as if it were submerged in inviscid fluid. This idealization is carried out, however, while retaining ingenious artifices to simulate those consequences of viscosity which do remain even at high Re. The first of these is to require that the flow obey the condition of Kutta (1902), who hypothesized that the sharp trailing edge is always the dividing line between stream surfaces flowing smoothly off the upper and lower sides of the wing. A second is to include an estimate of boundary-layer shear stresses when calculating quantities like D, to which they contribute appreciably.

In essence, the boundary layer is collapsed to zero thickness for purposes of aerodynamic theory, but there is presumed to exist a sudden drop in tangential velocity between the innermost streamlines and fluid particles contacting the wing. For drag and heat-transfer prediction, this discontinuity is replaced by a more realistic, finite, but thin viscous layer. Actually, less fluid is able to get by the wing because of the growing boundary layer, within which there is a velocity defect. It is therefore possible to obtain a slightly better estimate of pressure distribution by thickening rearward portions of the wing to account for this "displacement" effect. This step is not often taken in practice, since the displacement is less than a quarter of the already-small thickness δ.

When α_{ZL_0} is too large (typically greater than 15–20°) or when strong shocks are present on the wing, the boundary layer grows and becomes deficient in momentum, to the point that it may "separate" from the surface. There is created a substantial volume of fluid in turbulent motion, and the wing may stall—a situation that has defied rational theoretical analysis. The designer then has no alternative but to turn to the wind tunnel, or to flight testing if a prototype is available. There are several important aerodynamic quantities—angle of stall; D, L, and M_p at high α_{ZL_0}; and drag contributions from the rear portions of certain fuselage shapes—which can be expected to require experimental determination for many decades to come.

4.2 AERODYNAMIC THEORY FOR SMALL DISTURBANCES FROM A UNIFORM FLOW

Figure 4.2 depicts a slender vehicle in flight through the atmosphere. The aerodynamicist's reference frame is adopted, with speed u_0 (equivalent to v_c or v_{c0} in other chapters) of the oncoming stream taken parallel to x. Symbols p, ρ, T denote static pressure, density, and absolute temperature; subscript ∞ identifies their free-stream (ambient) values. The fluid particle velocity vector $\mathbf{q}$ is componentized as follows:

$$\mathbf{q} = (u_0 + u)\mathbf{i} + v\mathbf{j} + w\mathbf{k} \tag{4.1}$$

Henceforth we shall assume that all surfaces of submerged bodies are so gently inclined to the flight direction that the perturbations satisfy

$$|u|, |v|, |w| \ll u_0 \tag{4.2}$$

throughout the flow field. From the governing physics, we can then reason that p, ρ, T, and other state variables also deviate only slightly from p_∞, ρ_∞, T_∞.

Any of the aforementioned textbooks describes the physical principles applicable to the field of $\mathbf{q}$, p, ρ, T, etc. The most important laws are usually quantified in forms closely resembling those we now set down.

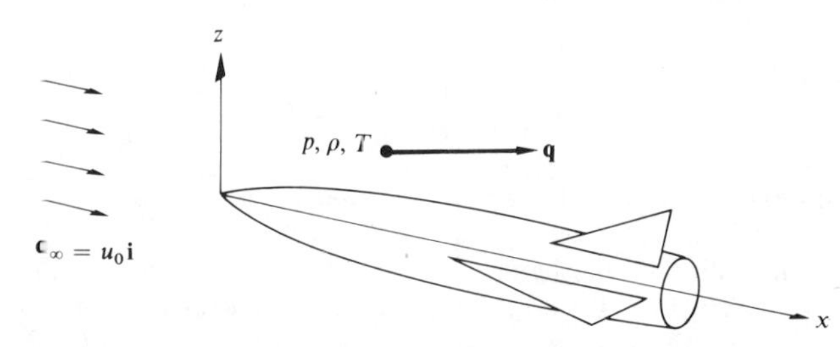

Fig. 4.2. A typical flight vehicle, showing the reference frame used for aerodynamic analysis, the free stream of magnitude u_0 parallel to the x-direction, and flow properties at a general field point.

1. Continuity or Conservation of Mass*

$$\frac{\partial \rho}{\partial t} + \nabla \cdot (\rho \mathbf{q}) = 0 \tag{4.3}$$

2. Newton's Second Law (Euler's Equations)

$$\frac{\partial \mathbf{q}}{\partial t} + (\mathbf{q} \cdot \nabla)\mathbf{q} + \frac{\nabla p}{\rho} = \mathbf{f} \tag{4.4}$$

where $\mathbf{f}$ is the vector of body force per unit mass. Gravity is the only such force of potential significance in aerodynamics. Since even its influence on the airloading of heavier-than-air craft is usually negligible (gravitational buoyancy "holds up the air inside the airplane"), we neglect it here.

3. Thermodynamics

Except for the rise in entropy which takes place because of shocks in steady supersonic flow, the importance of thermodynamic considerations to the fluid motion analyzed here can be summarized by assuming that fluid motion is essentially adiabatic, reversible, and in equilibrium. The thermal equation of state can be written

$$\frac{p}{p_\infty} = \frac{\rho}{\rho_\infty}\frac{T}{T_\infty} \tag{4.5}$$

To first order in small perturbations, the only "thermodynamic quantity" that remains in the final equations is the sound speed,

$$a = \sqrt{\left(\frac{\partial p}{\partial \rho}\right)_s} \tag{4.6}$$

Here s is the specific entropy, which remains constant in reversible state changes without heat transfer. Since it is always true for gas in equilibrium that $p = p(\rho, s)$, for these flows we reason that pressure is a unique function of density, and we replace $(\partial p/\partial \rho)_s$ by $dp/d\rho$ in (4.6). In second-order aerodynamic theory, the specific-heat ratio γ makes its appearance. Since γ is almost invariant in aeronautical situations up to moderate supersonic M, we recall the alternative expressions

$$a = \sqrt{\frac{\gamma p}{\rho}} = \sqrt{\gamma R T} \tag{4.7}$$

where R is the perfect gas constant per unit mass.

* In sections dealing with fluid mechanics, we employ the nabla operator in the conventional way for divergence, gradient, curl, and Laplacian. Thus typical Cartesian components would be

$$\nabla \cdot \mathbf{q} = \frac{\partial(u_0 + u)}{\partial x} + \frac{\partial v}{\partial y} + \frac{\partial w}{\partial z}, \qquad \nabla p = \mathbf{i}\frac{\partial p}{\partial x} + \mathbf{j}\frac{\partial p}{\partial y} + \mathbf{k}\frac{\partial p}{\partial z}, \qquad \nabla \times \mathbf{q} \text{ as in (2.14)}$$

An important consequence of isentropy for an initially uniform flow like that in Fig. 4.2 is that fluid particles have no angular momentum about their CM axes and the motion is irrotational (e.g., see Ashley and Landahl 1965, Section 1.7),

$$\nabla \times \mathbf{q} = 0 \tag{4.8}$$

It follows from (4.8) that the flow kinematics are wholly described by a single scalar function, called the *velocity potential*,

$$\mathbf{q} = \nabla \Phi \tag{4.9}$$

Note that Φ and a *perturbation potential* φ, defined by subtracting the potential $u_0 x$ of the free stream, are valuable tools in the initial formulation of many aerodynamic problems. As derived, for example, in Chapters 1 and 5 of Ashley and Landahl (1965), three equations involving Φ and closely related quantities form the basis of a variety of approximate methods. Subject to no further restrictions than that the gas motion is irrotational and isentropic, they read as follows.

1. Differential Equation for Φ

$$\boxed{\nabla^2 \Phi - \frac{1}{a^2}\left[\frac{\partial^2 \Phi}{\partial t^2} + \frac{\partial}{\partial t}(q^2) + \mathbf{q} \cdot \nabla\left(\frac{q^2}{2}\right)\right] = 0} \tag{4.10}$$

Note that (4.9) enables the formal elimination of $\mathbf{q}$ and $q \equiv |\mathbf{q}|$ here. Because of its nonlinearity, however, (4.10) has not proved useful in the analysis of general flows.

2. Expression for Speed of Sound

$$\boxed{a^2 = a_\infty^2 - (\gamma - 1)\left\{\frac{\partial \Phi}{\partial t} + \tfrac{1}{2}[q^2 - u_0^2]\right\}} \tag{4.11}$$

Equation (4.11) makes it possible for (4.10) to be expressed entirely in terms of the velocity potential, whereupon (4.10) is seen to be quasi-linear, of second order, and of third degree in the derivatives of Φ. The deviations of a^2 from its ambient value a_∞^2 turn out generally to affect the motion only to second order in the small disturbance; there are exceptions in the immediate transonic speed range and in hypersonic flight where u, v, w become of the same order as a_∞.

3. Kelvin's Equation (Bernoulli's Equation for Irrotational Flow)

$$\frac{\partial \Phi}{\partial t} + \tfrac{1}{2}[q^2 - u_0^2] + \int_{p_\infty}^{p} \frac{dp}{\rho} = 0 \tag{4.12}$$

This result is a spatial integral of Euler's equations, whose practical value is for calculating pressure distributions once the flow kinematics are known in terms of Φ or $\mathbf{q}$.

To demonstrate this point, we set down forms applicable to two special cases.

a) When the fluid is incompressible ($\rho = \rho_\infty$), (4.12) becomes the familiar hydraulic equation

$$p - p_\infty + \frac{\rho_\infty}{2}[q^2 - u_0^2] + \rho_\infty \frac{\partial \Phi}{\partial t} = 0 \tag{4.13}$$

(Cf. the steady case, $\partial\Phi/\partial t = 0$, and add the gravity head term $\rho g h$, where h is height above the datum, and where $p = p_\infty$ in the free stream.)

b) The isentropic pressure–density relation

$$\frac{p}{p_\infty} = \left(\frac{\rho}{\rho_\infty}\right)^\gamma \tag{4.14}$$

enables us to integrate the last term of (4.12) and solve explicitly for p in a form applicable to a compressible gas. It is customary to express the result in terms of a dimensionless pressure coefficient,

$$C_p \equiv \frac{p - p_\infty}{(\rho_\infty/2)u_0^2} \tag{4.15}$$

It is found to read

$$C_p = \frac{2}{\gamma \mathrm{M}^2}\left\{\left[1 - \left(\frac{\gamma - 1}{a_\infty^2}\right)\left(\frac{q^2 - u_0^2}{2} + \frac{\partial \Phi}{\partial t}\right)\right]^{\gamma/(\gamma - 1)} - 1\right\} \tag{4.16}$$

where again $\mathrm{M} \equiv u_0/a_\infty$ is the flight Mach number.

Chapter 8 of Liepmann and Roshko (1957) or Chapter 5 of Ashley and Landahl (1965) are two of many sources for detailed derivations of the linearized aerodynamic theory which emerges when the fullest advantage of assumption (4.2) is taken in simplifying (4.10), (4.11), and (4.16). Let us focus attention on three-dimensional flow past a nearly plane wing like that illustrated in Fig. 4.3. When our definition of perturbation potential

$$\Phi \equiv u_0 x + \varphi \tag{4.17}$$

is inserted into (4.10, 11), and when dependence on time is omitted, and terms containing products of derivatives of φ are neglected relative to first-order terms, there results

$$\nabla^2 \varphi - \mathrm{M}^2 \frac{\partial^2 \varphi}{\partial x^2} = 0 \tag{4.18}$$

Here

$$\nabla^2 \equiv \frac{\partial^2}{\partial x^2} + \frac{\partial^2}{\partial y^2} + \frac{\partial^2}{\partial z^2}$$

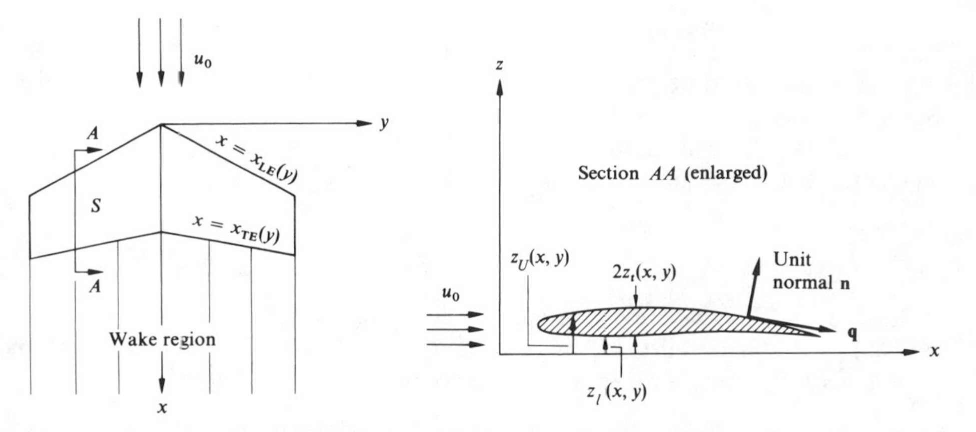

Fig. 4.3. Plan and cross-sectional views of a 3-D, nearly plane lifting surface. Note that S usually refers to the projection of the plan area onto the xy-plane.

is Laplace's operator. (4.18) is the field differential equation behind most aerodynamic theory, although let us again remind ourselves that it fails for many transonic situations and also for high supersonic and hypersonic flow over wings that are not extremely thin and streamlined.

A similar approximate treatment of (4.16) leads to the very simple relation

$$\boxed{C_p = -\frac{2}{u_0}\frac{\partial \varphi}{\partial x}} \tag{4.19}$$

The significance of (4.19) is that deviations of local pressure from p_∞ are due to changes in the magnitude of $\mathbf{q}$ from u_0 [cf. (4.16)]; perturbation $u = \partial\varphi/\partial x$ has a first-order effect on this magnitude, whereas v, w cause only first-order changes in the direction of $\mathbf{q}$. (4.19) is always consistent with (4.18) when one is estimating pressures on isolated lifting surfaces. We shall see, however, that certain refinements are needed in the case of C_p on slender bodies. We also note that C_p, u, v, w are obtained from φ by linear differential operations with constant coefficients. Consequently, all four of these unknowns are governed by the same (4.18) as φ itself. Physically, the content of (4.18) is equivalent to that of linearized acoustics, except that the latter deals with time-dependent wave propagation and is governed by a classical wave equation in x, y, z, and t. (4.18) itself is a wave equation, with x playing the role of t, when $\mathrm{M} > 1$; subsonically ($\mathrm{M} < 1$), it can be transformed to Laplace's equation, and therefore describes an "elliptic" field.

To (4.18) must be appended suitable boundary conditions, which can single out the unique solution appropriate to any given wing shape and angle of attack. One broadly applicable condition is that disturbances produced by the motion must die out in all portions of the field remote from both the wing and its wake, which trails downstream in the x-direction. Normally this requirement is met by making $\varphi \to 0$ when $y \to \pm\infty$, $z \to \pm\infty$, or $x \to -\infty$; the elementary solutions of (4.18) which are superimposed to construct these flows are chosen so as to die out in this way.

In Fig. 4.3 the thin, almost-flat wing is located close and nearly parallel to the xy-plane. The upper and lower surfaces are described by the functions

$$z = \begin{cases} z_U(x, y) \\ z_L(x, y) \end{cases} \tag{4.20}$$

defined for (x, y) anywhere on the area S—here the planform projection onto the xy-plane. The principal remaining boundary condition is that fluid particles in contact with the wing must move parallel to its surface. Mathematically,

$$\mathbf{q} \cdot \mathbf{n} = 0 \qquad \text{at } z = z_U \text{ and } z_L, \qquad \text{for all } (x, y) \text{ on } S \tag{4.21}$$

where $\mathbf{n}$ is a vector locally normal to this surface.

For instance, at $z = z_U$, $\mathbf{n}$ has components proportional to $(-\partial z_U/\partial x)$, $(-\partial z_U/\partial y)$, and 1; with the components $(u + u_0)$, v, w substituted for $\mathbf{q}$, we find that $\mathbf{q} \cdot \mathbf{n} = 0$ is equivalent to the relation

$$w(x, y, z_U) = (u_0 + u)\frac{\partial z_U}{\partial x} + v\frac{\partial z_U}{\partial y}. \tag{4.22}$$

The manner in which "exact" condition (4.22) is simplified so as to render it consistent with (4.18, 19) merits careful discussion. The first step involves recognizing that our requirement for gently inclined surfaces of submerged bodies is tantamount to

$$\left|\frac{\partial z_U}{\partial x}\right|, \qquad \left|\frac{\partial z_U}{\partial y}\right| \ll 1 \tag{4.23}$$

everywhere. In view of (4.2), we should therefore neglect all terms on the right of (4.22) except $u_0\, \partial z_U/\partial x$. The second step consists of expanding $w = \partial\varphi/\partial z$ on the left of (4.22) in Maclaurin series in z about its value at $z = 0+$, which is the "upper side" of the wing's xy-projection,

$$w(x, y, z_U) = w(x, y, 0+) + z_U\frac{\partial w}{\partial z}(x, y, 0+) + \frac{z_U^2}{2!}\frac{\partial^2 w}{\partial z^2}(x, y, 0+) + \cdots \tag{4.24}$$

For consistency, all terms on the right of (4.24) should be dropped except the first—provided that the various rates of change of w with respect to z are not unduly large, which is an unlikely eventuality, in view of the anticipated smooth variation of flow properties throughout the inviscid field.

These two further approximations, together with corresponding steps for the lower surface, yield another version of the boundary condition:

$$\left.\begin{aligned} w(x, y, 0+) &= u_0\frac{\partial z_U}{\partial x} \qquad &(4.25\text{a}) \\ w(x, y, 0-) &= u_0\frac{\partial z_L}{\partial x} \qquad &(4.25\text{b}) \end{aligned}\right\} \quad \text{for all } (x, y) \text{ on } S$$

It is difficult to understand, on physical grounds, why we should wish to apply this condition at $z = 0\pm$ rather than at the actual position of the wing. The reason is that we plan to follow a widespread practice of linearized theory and solve the boundary-value problem (4.18)–(4.25) by distributing, over the projection S, a sheet of elementary singular solutions. In this idealization, there occur discontinuities in flow properties through S, and the surfaces $z = 0+$ and $0-$ are employed to simulate $z = z_U$ and $z = z_L$, respectively. The loading is thereby transferred from the actual wing onto its projected position. For our purposes the success of this approach rests much more on how closely it can predict actual measured load distributions, lifts, and moments rather than intimate, local details of the field.

Another advantage of the form (4.25) is that it enables us to divorce the effects of wing thickness from those of the camber distribution and angle of attack. This separation is accomplished through the substitutions

$$z_U(x, y) = z_1(x, y) + z_t(x, y) \tag{4.26a}$$

$$z_L(x, y) = z_1(x, y) - z_t(x, y) \tag{4.26b}$$

Recognizing the linearity of (4.18), we discover that the complete flow can be built up by summing one solution which satisfies the condition

$$\boxed{w_t(x, y, 0\pm) = \pm u_0 \frac{\partial z_t}{\partial x}, \qquad \text{for all } (x, y) \text{ on } S} \tag{4.27}$$

and another which satisfies

$$\boxed{w(x, y, 0) = u_0 \frac{\partial z_1}{\partial x}, \qquad \text{for all } (x, y) \text{ on } S} \tag{4.28}$$

(Appropriate behavior of φ at infinity is enforced in both cases.) Now $w = \partial\varphi/\partial z$, and differentiation with respect to a coordinate changes an even to an odd function of that coordinate (and conversely). It follows that (4.27) can be met by distributing singularities on S whose potentials φ are even in z and whose effect is to produce a flow with mirror symmetry about the xy-plane. As we would anticipate because of the subscript t, this is the disturbance due to the wing thickness distribution, as if all camber had been removed and $\alpha_{ZL_0} = 0$. From (4.19) and the evenness in z of $\partial\varphi/\partial x$, the flow cannot cause any difference in pressure between the upper and lower surfaces at any (x, y). Hence it contributes nothing to lift or to any other generalized force dependent on such a difference. It may, however, affect drag, and we shall see that it is responsible for a substantial proportion of D when $M > 1$.

Equation (4.28) is the kind of boundary condition we would get by reducing to zero the thickness $2z_t(x, y)$, then calculating the disturbance due to camber and angle of attack. It is this flow field that gives rise to lift, pitching moment, etc., and to which most attention will be paid in what follows.

Our program for the remainder of Chapter 4 is to review a series of practically important solutions, proceeding generally from lower to higher flight speeds and from two- to three-dimensional geometries.

4.3 THIN AIRFOIL IN TWO-DIMENSIONAL, INCOMPRESSIBLE FLOW

We apply the term *airfoil* to a lifting surface of infinite span obtained by projecting indefinitely in the y-direction the cross section, cut by an xz-plane, at any spanwise station on an actual wing. The resulting two-dimensional (2-D) flow disturbance can be represented by a potential $\varphi(x, z)$. A reasonable approximation is achieved by mounting a model between the plane parallel walls of a *two-dimensional wind tunnel*. Although no such flow is ever precisely encountered on a flight vehicle, we shall see several ways in which theoretical or experimental airfoil data are useful in the design process.

We first consider the estimation of lifting pressures. This enables us to neglect thickness, and Fig. 4.4 therefore illustrates only an airfoil camber line, with ordinate $z_1(x)$ positioned relative to the axes in a manner mathematically convenient for subsonic M. Anticipating a simple way of later generalizing the solution for $0 < \mathrm{M} < 1$, we study first the incompressible case, for which (4.18) and (4.28) reduce to

$$\boxed{\frac{\partial^2 \varphi}{\partial x^2} + \frac{\partial^2 \varphi}{\partial z^2} = 0} \tag{4.29}$$

$$w(x, 0) \equiv \frac{\partial \varphi}{\partial z}(x, 0) = u_0 \frac{dz_1}{dx} \qquad \text{for } |x| \leq \frac{c}{2} \tag{4.30}$$

Here c denotes the constant chord, and changes in this projected length due to angle of attack may be neglected because the cosine equals unity to first order in a small angle.

Before directly addressing (4.29)–(4.30), let us examine a few generalities about the 2-D Laplace equation. One interesting observation—which follows from

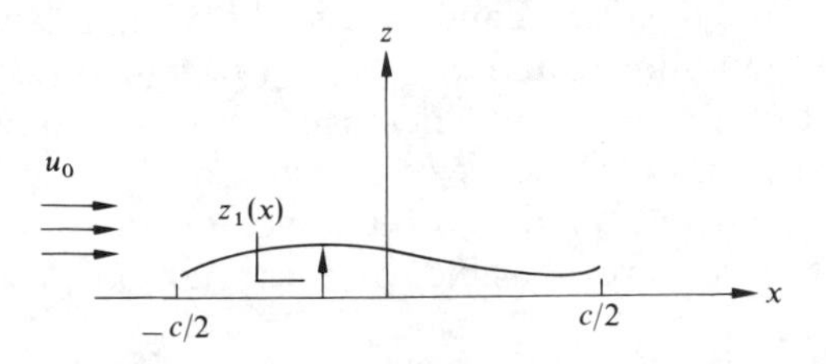

Fig. 4.4. Cross section of a 2-D airfoil mean line, with leading and trailing edges conveniently located for analysis of subsonic flow.

(4.10) when we regard the limit M $\to$ 0 as being applicable to flows in which sound speed a is large compared to all velocities including u_0—is that the linear $\nabla^2\Phi = 0$ holds even when no restriction is placed on the magnitude of flow perturbations. In 2-D, it is useful information that any function of the complex variable

$$\xi = x + iz, \tag{4.31}$$

where $i \equiv \sqrt{-1}$ is the imaginary unit, is a source for solutions of (4.29). Thus, for an appropriately differentiable complex function $F(\xi)$,

$$\frac{\partial^2 F}{\partial x^2} = \frac{d^2 F}{d\xi^2}\left(\frac{\partial \xi}{\partial x}\right)^2 = \frac{d^2 F}{d\xi^2} \tag{4.32a}$$

and

$$\frac{\partial^2 F}{\partial z^2} = \frac{d^2 F}{d\xi^2}\left(\frac{\partial \xi}{\partial z}\right)^2 = -\frac{d^2 F}{d\xi^2} \tag{4.32b}$$

whence we obtain

$$\frac{\partial^2 F}{\partial x^2} + \frac{\partial^2 F}{\partial z^2} = 0 \tag{4.32c}$$

Both real and imaginary parts of F separately satisfy (4.32c). Since the disturbance potential is a real physical quantity, φ can be identified with either part, but it is customary to write

$$F = \varphi + i\psi \tag{4.33}$$

The relations

$$\frac{\partial}{\partial x} = \frac{d}{d\xi} \quad \text{and} \quad \frac{\partial}{\partial z} = i\frac{d}{d\xi}$$

then enable us to develop the following equations for the perturbation components:

$$u = \frac{\partial \varphi}{\partial x} = \frac{\partial \psi}{\partial z} \tag{4.34a}$$

$$w = \frac{\partial \varphi}{\partial z} = -\frac{\partial \psi}{\partial x} \tag{4.34b}$$

$$u - iw = \frac{dF}{d\xi} \tag{4.34c}$$

Although it is not significant to our airfoil theory, we note that the conjugate function $\psi(x, z)$ is called a *stream function*, and that its level lines ψ = constant are the streamline pattern for the same field described by φ [cf. the thorough treatment in Milne-Thompson (1962), who follows the classical British practice of reversing the signs of φ, ψ from those employed here].

Many textbooks contain a systematic presentation of 2-D incompressible motion based on the study of elementary functions $F(\xi)$. Given the resulting family of solutions—notably various idealized flows about an infinite circular cylinder—the technique of conformal transformation of analytic functions then enables one to construct flows about arbitrarily shaped bodies submerged in a free stream. When compared with measurements, such solutions are valuable for defining the bounds of validity of inviscid theory; to us they are also a standard of reference for the linearized approximation. Some problems at the end of this chapter are intended to help you explore further the complex-variable approach. We also call attention to Table A,2 of Jones and Cohen (1960), in which a compendium of selected $F(\xi)$ is given, corresponding to a wide variety of possible fluid motions about thin airfoils with both symmetry and antisymmetry in the z-coordinate.

The airfoil with camber and angle of attack described by a general function $z_1(x)$ is conveniently approached through superposition of fundamental solutions of (4.29) known as 2-D *vortices*. The vortex with its axis or "core" along a spanwise line through (x_0, z_0) is described by

$$\boxed{F_0(\xi) = \frac{i\Gamma_0}{2\pi} \ln [\xi - \xi_0]} \tag{4.35}$$

$\xi_0 \equiv x_0 + iz_0$, and Γ_0 is called the *strength* or *circulation*, being the line integral of $\mathbf{q}$ around any path surrounding the core. Other properties of the vortex are conveniently studied by temporarily centering it at the origin ($\xi_0 = 0$). Such relations as (4.33) and (4.34c) can readily be applied to yield the following.

$$\varphi(x, z) = -\frac{\Gamma_0}{2\pi} \tan^{-1}\left(\frac{z}{x}\right) \equiv -\frac{\Gamma_0 \theta}{2\pi} \tag{4.36a}$$

$$\psi(x, z) = \frac{\Gamma_0}{2\pi} \ln r \tag{4.36b}$$

$$u = \frac{\Gamma_0 z}{2\pi r^2} \tag{4.36c}$$

$$w = -\frac{\Gamma_0 x}{2\pi r^2} \tag{4.36d}$$

$$v_r = 0 \tag{4.36e}$$

$$v_\theta = -\frac{\Gamma_0}{2\pi r} \tag{4.36f}$$

The polar coordinates r, θ play a special role in (4.36), since it is evident from (4.36b, e, f) that this vortex has circular streamlines about the core, with fluid particles circulating clockwise at a speed inversely proportional to radial distance.

A vortex located at any point ($x_0 = x_1, z_0 = 0$) on the x-axis,

$$\varphi = -\frac{\Gamma_0}{2\pi}\tan^{-1}\left(\frac{z}{x - x_1}\right) \tag{4.37}$$

possesses the antisymmetry $\varphi(x, -z) = -\varphi(x, z)$ needed to satisfy (4.30). Our trial solution, using x_1 as a dummy variable of integration, will be to superimpose forms like (4.37) to produce a "sheet" of vortices on the airfoil projection S [that is, on $-(c/2) \leq x_1 \leq (c/2)$]. Γ_0 is replaced by $\gamma(x_1)\,dx_1$, where $\gamma(x_1)$ is circulation per unit chordwise distance. The resulting potential

$$\varphi(x, z) = -\frac{1}{2\pi}\int_{-c/2}^{c/2} \gamma(x_1)\tan^{-1}\left(\frac{z}{x - x_1}\right) dx_1 \tag{4.38}$$

is z-differentiated to yield the vertical velocity

$$w(x, z) = -\frac{1}{2\pi}\int_{-c/2}^{c/2} \gamma(x_1)\left[\frac{x - x_1}{(x - x_1)^2 + z^2}\right] dx_1 \tag{4.39}$$

With $z \to 0$, we use (4.30) to produce the basic integral equation relating the airfoil camber shape and γ,

$$u_0 \frac{dz_1}{dx} = -\frac{1}{2\pi}\oint_{-c/2}^{c/2} \frac{\gamma(x_1)\,dx_1}{x - x_1}, \qquad \text{for all } |x| \leq \frac{c}{2} \tag{4.40}$$

Here we can justify the need for Cauchy-principal-value integration at the singularity where $x = x_1$ either by returning to the complex-variable representation or by the physical observation that there is no reason, when computing w at a given x, to give special preference to contributions from vortices either just behind or just ahead of this station.

Figure 4.5 depicts the vortex sheet with which we are working. We note in passing that, for the streamlined airfoil used here, it does not matter whether the

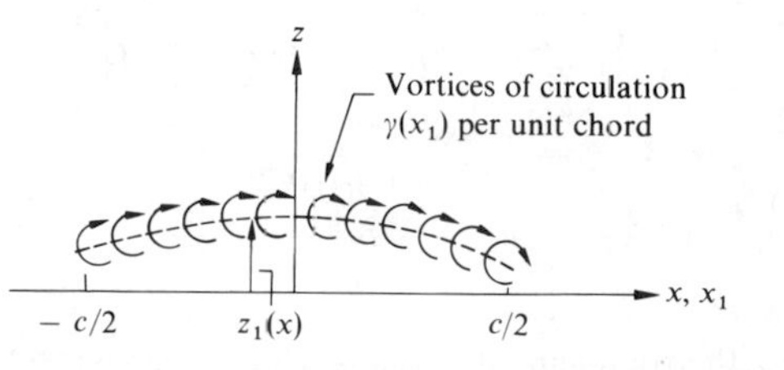

Fig. 4.5. Sheet of distributed 2-D vortices used for solution of the lifting problem in incompressible flow.

sheet is placed right on the x-axis or is slightly bowed so as to coincide with the camber line (cf. the treatment by Kuethe and Schetzer 1959). To first order in the small slopes dz_1/dx, we get the same relation (4.40) if we locate it along $z_1(x)$ and enforce the boundary condition there also.

The significance of γ in this theory is demonstrated by proving its proportionality to the airload per unit area. The vortex sheet gives rise to a discontinuity in velocity perturbation u between its upper and lower sides, as can be seen by calculating the clockwise circulation around the infinitesimal rectangular path shown in Fig. 4.6 and equating this circulation to the value $\gamma(x_1)\,dx_1$ which it must have according to the fundamental properties of vortices. Thus we obtain

$$\begin{aligned}\gamma(x_1)\,dx_1 &= [u_0 + u_U(x_1)]\,dx_1 - \left[w(x_1, 0) + \frac{\partial w}{\partial x}dx_1\right]dz \\ &\quad - [u_0 + u_L(x_1)]\,dx_1 + w(x_1, 0)\,dz \\ &= [u_U(x_1) - u_L(x_1)]\,dx_1 + (\text{H.O.T.}) \end{aligned} \tag{4.41}$$

In view of the linearized Bernoulli relation (4.19), we can cancel dx_1 in (4.41), replace u in terms of the pressures p_U and p_L on the upper and lower surfaces, and arrive at

$$\boxed{p_L - p_U = \rho_\infty u_0 \gamma(x_1)} \tag{4.42}$$

In mathematical terms, we assert that the potential $\tan^{-1}[z/(x - x_1)]$ amounts to a *Green's function* with respect to discontinuities of $u = \partial\varphi/\partial x$ or p through the x-axis. Thus it provides the capability of applying a prescribed loading to the airfoil. We can calculate the camber shape of the airfoil that would sustain such a loading by integrating (4.40). Alternatively, (4.40) becomes an integral equation from which the loading is to be found in the more common situation in which camber and angle of attack are the given quantities. Once γ is known as a function of x_1 (or x), we can find lift, moment, control-surface hinge moment, and the like from weighted chordwise integrations. For instance, the lift per unit

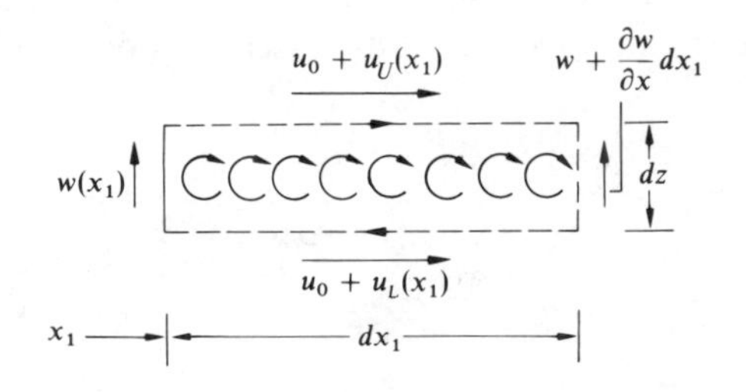

Fig. 4.6. Integration path around an infinitesimal chordwise segment of the 2-D vortex sheet. There is a finite discontinuity in tangential velocity component u between the upper and lower sides, and (4.41) shows this to be proportional to circulation $\gamma(x_1)$ per unit length.

spanwise distance may be written

$$L = \int_{-c/2}^{c/2} [p_L - p_U]\, dx = \rho_\infty u_0 \int_{-c/2}^{c/2} \gamma(x)\, dx \equiv \rho_\infty u_0 \Gamma \tag{4.43}$$

Here Γ stands for the total circulation of all vortices "bound" to the airfoil. (4.43) is the small-perturbation version of the famous relationship between lift and circulation, whose discovery is attributed to Joukowsky and Kutta.

There are several ways of solving (4.40) when γ is the unknown quantity, but we choose a Fourier-series approach due to Glauert (1926) because of its relative simplicity. The airfoil chord is first transformed onto the standard range $0 \le \theta \le \pi$ of an "angle variable" θ through

$$x = -\frac{c}{2}\cos\theta \tag{4.44a}$$

$$x_1 = -\frac{c}{2}\cos\theta_1 \tag{4.44b}$$

$$dx_1 = \frac{c}{2}\sin\theta_1\, d\theta_1 \tag{4.44c}$$

It is an easy matter to show that leading edge $x = -(c/2)$, midchord $x = 0$, and trailing edge $x = c/2$ are plotted at $\theta = 0$, $\pi/2$, and π, respectively. The integral equation becomes

$$w_1(\theta) = -\frac{1}{2\pi}\oint_0^\pi \frac{\gamma(\theta_1)\sin\theta_1\, d\theta_1}{\cos\theta_1 - \cos\theta} \tag{4.45}$$

where we are using the abbreviation w_1 in place of $u_0(dz_1/dx)$.

Solution of (4.45) is facilitated by reference to an integral introduced into aerodynamics by Glauert (1926),

$$\oint_0^\pi \frac{\cos(n\theta_1)\, d\theta_1}{\cos\theta_1 - \cos\theta} = \frac{\pi \sin n\theta}{\sin\theta}, \qquad n = 0, 1, 2, \ldots \tag{4.46}$$

For general n, (4.46) is efficiently proved by the theory of residues, but we also cite Rauscher (1953, Chapter IX) for a straightforward, painstaking derivation.

A glance at (4.45, 6) shows that choosing γ proportional to the ratio $\cos\theta_1/\sin\theta_1$ makes $n = 1$ and is therefore a possible way of representing the loading on an airfoil with constant slope. To be specific,

$$z_1 = -\alpha x \tag{4.47}$$

is the equation of a flat plate or the mean ordinate of an uncambered symmetrical profile. Either is inclined at a positive angle of attack α to its flight direction. Since $w_1 = -u_0\alpha$ in this case, we specialize (4.45) to

$$u_0\alpha = \frac{1}{2\pi}\oint_0^\pi \frac{\gamma(\theta_1)\sin\theta_1\, d\theta_1}{\cos\theta_1 - \cos\theta} \tag{4.48}$$

and examine the trial solution

$$\gamma(\theta_1) \stackrel{?}{=} 2u_0\alpha \frac{\cos\theta_1}{\sin\theta_1} \tag{4.49a}$$

Re-transformation by (4.44b) recovers

$$\gamma(x) \stackrel{?}{=} -2u_0\alpha \frac{x}{\sqrt{(c/2)^2 - x^2}} \tag{4.49b}$$

where the subscript 1 on x has now been dropped as unnecessary.

When (4.49b) is plotted, as by the dashed line in Fig. 4.7, we encounter a curious chordwise load distribution, which not only seems to produce no lift but

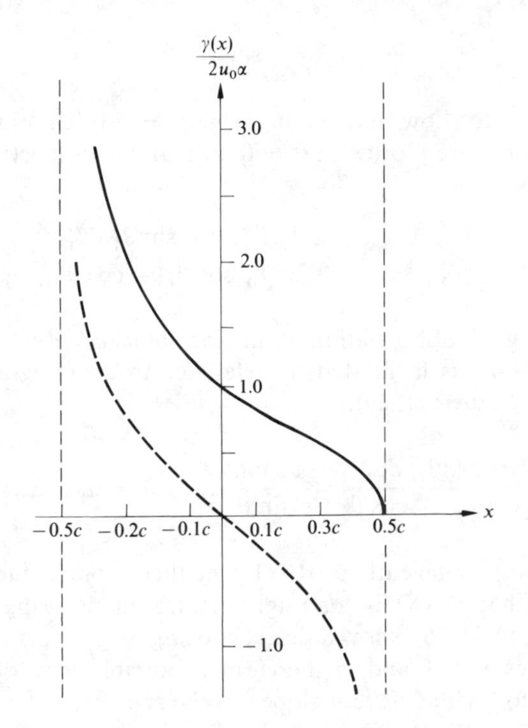

Fig. 4.7. Chordwise plots of two dimensionless circulation distributions related to the 2-D flat plate in incompressible flow. The lower curve [Eq. (4.49b)] is dashed because it does not satisfy Kutta's condition, as does the solid curve of Eq. (4.50b).

approaches infinite values at both leading and trailing edges. The resolution of this difficulty appeals to Kutta's condition—briefly mentioned in Section 4.1—that the effects of viscosity can be properly simulated only if the motion is required to be smooth where the flow passes off the trailing edge. We find that this can be accomplished by insisting that $\gamma(c/2)$ [that is, $\gamma(\theta_1 = \pi)$] be nonsingular. We observe that a term proportional to $1/\sin\theta_1$, corresponding to $n = 0$ in (4.46), may be added to γ in arbitrary amounts without affecting the form of w_1 which results from (4.45). It can, moreover, easily be seen that such a term places a net circulation Γ on the airfoil, thus contributing the sought-after lift. For the plate or symmetrical airfoil, l'Hôpital's rule as $\theta_1 \to \pi$ shows that the desired behavior is achieved by revising (4.49a) as follows:

$$\gamma(\theta_1) = 2u_0\alpha\left(\frac{1}{\sin\theta_1} + \frac{\cos\theta_1}{\sin\theta_1}\right) \equiv 2u_0\alpha \cot\left(\frac{\theta_1}{2}\right) \tag{4.50a}$$

Figure 4.7 graphs versus x the circulation

$$\gamma(x) = 2u_0\alpha\sqrt{\frac{(c/2) - x}{(c/2) + x}} \tag{4.50b}$$

obtained by returning to the physical plane. Except for the singularity at $x = -c/2$, the result (4.50b) is much more satisfactory. When compared with the distribution of $(p_L - p_U)$, measured on such an airfoil at the same lift coefficient C_L, $\rho_\infty u_0 \gamma(x)$ from (4.50b) agrees fairly accurately up to within a few percent of c behind the leading edge. In contrast to the zero-thickness idealization (4.47), any real airfoil has a blunted leading edge. If Laplace's equation is solved exactly, under Kutta's condition, for the blunted case, not only does the pressure singularity disappear but also $\gamma(x)$ is found to come back again in precisely the form (4.50b) as the leading-edge radius is reduced to zero and α is made small. The infinity here is the price that must be paid for the series of approximations underlying the linearized model. Through the exact approach based on conformal transformation, this imperfection can be removed for 2-D incompressible flow. On practical wings at realistic Mach numbers, however, it is not so readily avoidable. Aerodynamicists have learned to be comfortable with such singularities when they are integrable and their effects are confined to the vicinity of certain lines or surfaces in the field.

The solution to (4.45) for general camber-line shape is conveniently expressed, while preserving the Kutta condition, by adding to (4.50a) a Fourier sine series,

$$\gamma(\theta_1) = 2u_0\left(\alpha \cot\frac{\theta_1}{2} + \sum_{n=1} A_n \sin n\theta_1\right) \tag{4.51}$$

When (4.51) is substituted, the nth resulting integral can be associated through trigonometric identities with (4.46) for the two integers $(n - 1)$ and $(n + 1)$. After some further manipulation, we derive the corresponding series for w_1,

$$w_1(\theta) = u_0\left(-\alpha + \sum_{n=1} A_n \cos n\theta\right) \tag{4.52}$$

The pair of equations (4.51, 2) is really our desired result. For any piecewise-differentiable function $w_1 \equiv u_0\, dz_1/dx$, expressed in terms of θ through (4.44a), each of the constants $(-\alpha)$, A_n can be individually calculated by the standard Fourier-series procedure of multiplying the two sides of (4.52) by $\cos m\theta$ (for $m = 0, 1, 2, \ldots$) and integrating between 0 and π. Thus we obtain

$$\alpha = -\frac{1}{\pi}\int_0^\pi \frac{w_1(\theta)}{u_0}\, d\theta \tag{4.53a}$$

$$A_m = \frac{2}{\pi}\int_0^\pi \frac{w_1(\theta)}{u_0} \cos m\theta\, d\theta, \qquad m \geq 1 \tag{4.53b}$$

When the results of (4.53) are inserted into (4.51), all properties of the corresponding load distribution are revealed. For instance, the coefficient of lift is found, through (4.43), by the following steps:

$$C_L \equiv \frac{L}{(\rho_\infty/2)u_0^2 c} = \frac{2}{c}\int_{-c/2}^{c/2} \frac{\gamma(x)}{u_0}\, dx$$

$$= \int_0^\pi \frac{\gamma(\theta)}{u_0} \sin\theta\, d\theta \tag{4.54a}$$

and, with (4.51),

$$\boxed{C_L = 2\pi\left(\alpha + \frac{A_1}{2}\right)} \tag{4.54b}$$

We observe that C_L depends on just two of the Fourier coefficients of the function $w_1(\theta)$. Moreover, as may be seen by adding any constant to $w_1(\theta)$ in (4.53b), A_1 (like the other A_m) is determined exclusively by the camber-line shape and is unaffected by rotating a particular airfoil relative to its direction of flight. We may therefore differentiate (4.54b) with respect to α, the resulting constant being our first example of the lift-curve slope introduced in (3.6),

$$\boxed{C_{L\alpha} = 2\pi} \tag{4.55}$$

Comparing (3.6) with (4.54, 5), we note that the angle of attack measured from the zero-lift attitude of the airfoil is

$$\alpha_{ZL} = \alpha + \frac{A_1}{2} \tag{4.56}$$

Thus the angle α is not necessarily the same as α_{ZL}. Indeed, α is a more or less arbitrary angle determined from the way in which $z_1(x)$ is specified; changing α is the way to vary incidence on a given airfoil. The zero-lift attitude is $\alpha = -A_1/2$, usually a negative angle because most subsonic camber lines are concave downward and therefore have $A_1 > 0$.

The data plotted in Chapter 3 show how important are the limitations of linear equations like (4.54b). They do prove reasonably accurate for normal flight near the design point—even better when minor empirical adjustments are made to $C_{L\alpha}$ and other constants. The designer must always be alert, however, to problems with stall and other nonlinearities. On low-speed airfoils, the range between positive-α and negative-α stalling is smaller the lower the thickness ratio and the sharper the leading edge.

Let us close this section by discussing a few more results of linearized theory for 2-D airfoils at low M.

1. Ideal Incidence

From (4.51), we note that when $\alpha = 0$ there is no loading singularity at the leading edge. This condition is referred to as *ideal incidence*, because the stream surface separating the fluid which passes over the top and bottom meets the camber surface along its nose, permitting both the actual flow near the airfoil and the boundary-layer development to be especially smooth. Except on a symmetrical profile, there is finite lift at the ideal angle,

$$C_{Li} = \pi A_1 \tag{4.57}$$

Since the curve of measured 2-D drag versus α tends to have its minimum near this condition, keeping $C_{Li} > 0$ helps to improve performance in cruising flight. Accordingly, 3-D wings are cambered to produce positive lift at an approximately ideal incidence, which often corresponds to the "design point" mentioned in Chapter 1.

2. Pitching Moment

The theoretical pitching-moment coefficient per unit span about an axis at $x = x_0$ is given by

$$C_m \equiv \frac{M_p}{(\rho_\infty/2)u_0^2 c^2} = -\frac{2}{\rho_\infty u_0^2 c^2}\int_{-c/2}^{c/2} [\rho_\infty u_0 \gamma(x)][x - x_0]\, dx$$

$$= \frac{\pi}{2}\left\{\alpha\left(\frac{4x_0}{c} + 1\right) + A_1 \frac{2x_0}{c} + \frac{A_2}{2}\right\} \tag{4.58}$$

Within the range of unseparated flow, C_m is predicted even more successfully than C_L.

3. Aerodynamic Center

From the standpoint of trim and stability, a useful property of all streamlined lifting surfaces is the existence of an *aerodynamic center* (AC), or axis location about which the pitching moment at a given flight condition is independent of incidence. In the present case, (4.58) locates the center at the quarter-chordline $x_0 = -c/4$, this being the only x_0 which eliminates the term proportional to α. The corresponding coefficient,

$$C_{m\text{AC}} = -\frac{\pi}{4}(A_1 - A_2) \tag{4.59}$$

is negative on normally cambered airfoils. $C_{m\text{AC}}$ is evidently equal to C_{m0}, the pitching moment at zero lift. Moreover, we can rearrange (4.54b), (4.58), and (4.59) to obtain the following useful relation between lift and moment about any axis:

$$\boxed{C_m = C_{m0} + C_L\left(\frac{x_0}{c} + \frac{1}{4}\right)} \tag{4.60}$$

4. Drag

The most disappointing feature of conventional subsonic airfoil theory is that it yields zero drag, $C_D = 0$. We shall not elaborate this result, but remark only that it can be derived by momentum considerations, relating the streamwise force component exerted by the body on the fluid to net x-momentum efflux from a control volume surrounding it. The small resistance experienced by a real airfoil can be ascribed to a combination of skin friction and pressure-distribution distortions due to the displacement effects of its boundary layers. In fact, fairly successful attempts at estimating C_D have been based on such models (cf. Abbott and von Doenhoff, 1959, Section 5.13). It is speculated that a truly inviscid fluid would exert no drag, and attempts have even been made at verification by moving small airfoils through liquid Helium II in the superfluid state.

4.4 COMPRESSIBILITY CORRECTION AT SUBSONIC AND SUPERSONIC SPEEDS

Before proceeding to other regimes of Mach number and to three dimensions, let us describe some useful generalities which inhere in the full linearized problem statement (4.18), (4.25), and the C_p relation (4.19). Because of the simplicity of the governing partial differential equation, there exist reference values of M such that, if a lifting surface theory has been developed for these values, wing flows can be calculated at any other M in the same regime. For $M < 1$, the subsonic reference is conveniently taken as incompressible, $M \cong 0$. Supersonically, a reference at $M = \sqrt{2}$ has certain advantages. In the former regime, the process of transforming from the reference solution to properties of the given wing is called *compressibility correction*.

The physical discussion of this transformation process on pages 30–31 of Jones and Cohen (1960) is very illuminating. Liepmann and Roshko (1957) develop it in terms of the broader concept of "similarity laws." We suggest both these texts as background for our rather succinct presentation.

Equations (4.25) are rewritten so as to emphasize their role as boundary conditions on the dependent variable φ.

$$\frac{\partial \varphi}{\partial z}(x, y, 0\pm) = \begin{cases} u_0 \dfrac{\partial z_U}{\partial x} \\ u_0 \dfrac{\partial z_L}{\partial x} \end{cases} \quad \text{for all } (x, y) \text{ on } S \tag{4.61}$$

The upper- and lower-surface slopes on the right are to be regarded as prescribed functions of (x, y). Looking first at $M < 1$, we introduce into (4.18) and (4.61) the affine transformation

$$x = \sqrt{1 - M^2}\, x_0, \qquad y = y_0, \qquad z = z_0 \tag{4.62}$$

By $\varphi_0(x_0, y_0, z_0)$ we denote that function which, at each point of $x_0 y_0 z_0$-space, equals $\varphi(x, y, z)$ at the corresponding point located by means of (4.62). Since

$$\frac{\partial}{\partial x} = \frac{1}{\sqrt{1 - M^2}} \frac{\partial}{\partial x_0} \tag{4.63}$$

φ_0 must satisfy Laplace's equation

$$\frac{\partial^2 \varphi_0}{\partial x_0^2} + \frac{\partial^2 \varphi_0}{\partial y_0^2} + \frac{\partial^2 \varphi_0}{\partial z_0^2} = 0 \tag{4.64}$$

We therefore conclude that there must be some 3-D incompressible flow which, through (4.62), is equivalent to the subsonic flow.

More detail on the equivalence is found by (1) noting that φ_0, like φ, must die out to zero far from the wing and its wake, and by (2) transforming boundary conditions (4.61). When the derivative

$$\frac{\partial \varphi}{\partial z} = \frac{\partial \varphi_0}{\partial z_0} \tag{4.65}$$

is substituted into (4.61), we discover that surface slopes must be equal at corresponding points on the upper and lower sides of the original wing and its equivalent at $M = 0$. It follows that, at each chordwise cross section, the new wing has the same profile shape, angle of attack, percentage camber, thickness ratio, etc. The other clue as to the shape of the equivalent wing comes from the effect of (4.62) on its projected area S: All spanwise dimensions are unaltered, but each chord is stretched in the proportion

$$c_0 = \frac{c}{\sqrt{1 - \mathrm{M}^2}} \tag{4.66}$$

and the aspect ratio is reduced by the factor $\sqrt{1 - \mathrm{M}^2}$.

To summarize, if theoretical or experimental means are available for calculating the low-M properties of 3-D lifting surfaces generally, properties can be found at any subsonic M at which the field equation (4.18) is valid. It is simply a matter of determining $\varphi_0(x_0, y_0, z_0)$ for the reduced-Æ surface with the same cross-sectional shapes, then finding $\varphi = \varphi_0$ through (4.62). Pressure and airloads result from the transformation of (4.19),

$$C_p = -\frac{2}{u_0}\frac{\partial \varphi}{\partial x} = \frac{1}{\sqrt{1 - \mathrm{M}^2}}\left(-\frac{2}{u_0}\frac{\partial \varphi_0}{\partial x_0}\right) = \frac{C_{p0}}{\sqrt{1 - \mathrm{M}^2}} \tag{4.67}$$

The appearance of airspeed u_0 in (4.67) suggests one not unique, but convenient, way of physically envisioning the two flows: Both can be considered to have the same free-stream velocity, but in the second fluid (or liquid) the sound speed is so high as to make compressibility effects negligible.

When $\mathrm{M} > 1$, (4.62) may be replaced by

$$x = \sqrt{\mathrm{M}^2 - 1}\, x_2, \qquad y = y_2, \qquad z = z_2 \tag{4.68}$$

Equation (4.18) is thereby transformed into

$$\frac{\partial^2 \varphi_2}{\partial x_2^2} - \frac{\partial^2 \varphi_2}{\partial y_2^2} - \frac{\partial^2 \varphi_2}{\partial z_2^2} = 0 \tag{4.69}$$

which describes supersonic flow at the reference $\mathrm{M} = \sqrt{2}$. Again (4.61) shows that sectional parameters of the lifting surface are unchanged. When one goes to $\mathrm{M} = \sqrt{2}$, the chordwise dimensions of S are decreased if the actual flight is at $\mathrm{M} > \sqrt{2}$ and stretched for $\mathrm{M} < \sqrt{2}$. The equivalence between pressure coefficients at corresponding points is

$$C_p = \frac{(C_p)_{\mathrm{M}=\sqrt{2}}}{\sqrt{\mathrm{M}^2 - 1}} \tag{4.70}$$

It is an easy matter to work out the relationships among other flow properties in either speed regime. For instance, C_L and similar dimensionless loading coefficients are all weighted integrals of $[C_{pL} - C_{pU}]$ over equivalent nondimensionalized planform areas. Hence they are transformed in the same way as C_p itself; e.g., for $\mathrm{M} < 1$,

$$C_L = \frac{(C_L)_{\mathrm{M}=0}}{\sqrt{1 - \mathrm{M}^2}} \tag{4.71}$$

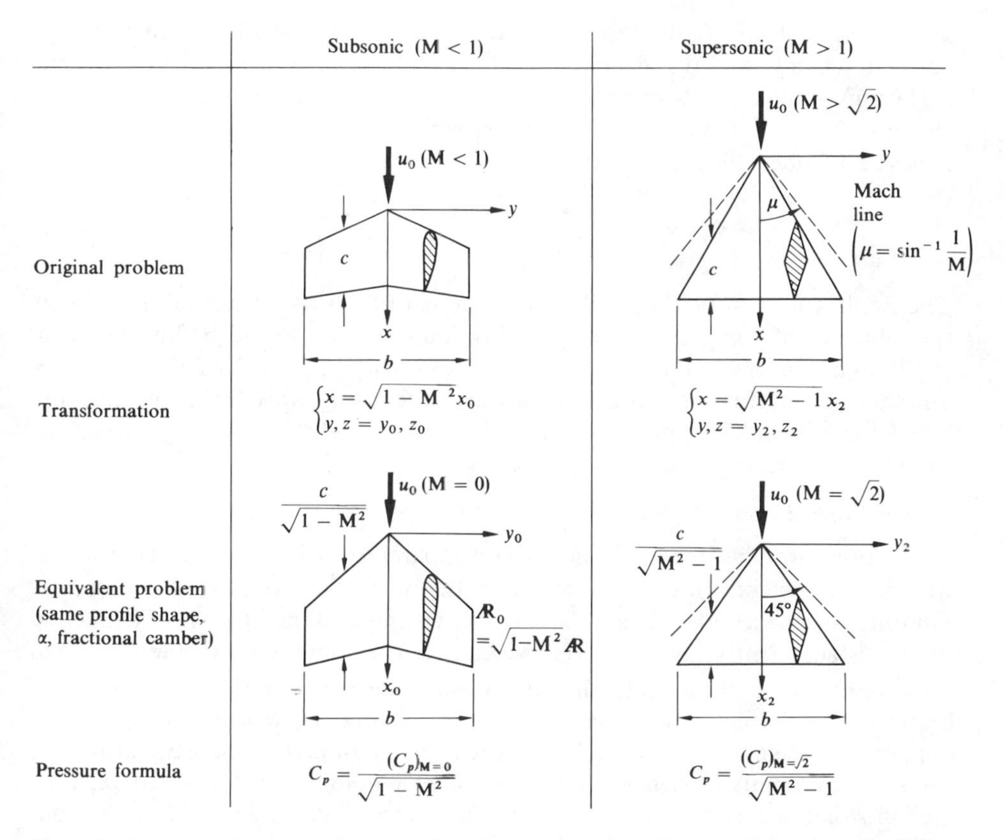

Fig. 4.8. Interpretation of the relationships between actual subsonic and supersonic lifting surfaces and the surfaces equivalent to them through the 3-D linearized compressibility correction.

Figure 4.8 illustrates typically equivalent wings and the manner in which their dimensions are distorted in the two regimes of M.

Special cases of the foregoing results deserve mention.

1. Two-Dimensional Airfoils

In 3-D, as we change the flight M of the actual wing, the planform shape and Ꮭ of its reference wing are progressively altered. To acquire such information as the M-dependence of $C_{L\alpha}$, therefore, we would have to conduct tests or calculations on a series of models. In 2-D, all Ꮭ's are infinite, however, and there is no effect of size on the dimensionless airloads of a given profile at a fixed M. Accordingly, the subsonic (4.67), the supersonic (4.70), or any formula like (4.71) connects properties of the same 2-D airfoil. A single set of tests or analyses at the reference M would therefore cover either the entire subsonic or supersonic range of validity of linearized theory.

For 2-D flows the literature contains more exact rules for compressibility correction, and several refined similarity laws have proved useful for 2-D and 3-D wings in the essentially nonlinear transonic and hypersonic regimes. Perhaps the most successful example of the former is due to Karman and Tsien (see, e.g., Jones and Cohen 1960, page 32):

$$C_p = \frac{C_{p0}}{\sqrt{1 - \mathrm{M}^2} + \frac{1}{2}(1 - \sqrt{1 - \mathrm{M}^2})C_{p0}} \tag{4.72}$$

The implications of (4.72) are that, in regions of "positive pressure" ($C_{p0} > 0$), the influence of compressibility is less pronounced than would be inferred from (4.67), and conversely for "suction" regions in which $C_{p0} < 0$. This rule serves quite accurately for most practical airfoils, when well away from the stall and below the critical Mach number—the subsonic M at which, at some point on the surface, the local speed $|\mathbf{q}|$ first exceeds a.

2. The Sonic Limit, M → 1

Correction formulas like (4.67), (4.70), etc. apparently yield meaningless results as $\mathrm{M} \to 1$. In general, this is a reflection of the inadequacy of (4.18) for transonic fluid motion over both 2-D and larger-AR 3-D wings. When the AR of the actual surface is sufficiently low, however, we can restore some sense to the linearized sonic limit by observing that the equivalent wing aspect ratio $\sqrt{|1 - \mathrm{M}^2|}\ AR$ becomes indefinitely small. As we shall see during subsequent study of such slender wings, aerodynamic coefficients tend to be directly proportional to AR. Thus for the family of slender deltas and other planforms with pointed vertices, the following linearized result has been well verified, independent of M (see Jones and Cohen 1960, pages 96–97):

$$C_{L\alpha} = \frac{\pi}{2} AR \tag{4.73}$$

Now angle of attack is preserved by the transformations, so we can reason from (4.71), (4.73), and the relation

$$(AR)_{\mathrm{M}=0} = \sqrt{1 - \mathrm{M}^2}\, AR \tag{4.74}$$

that the lift-curve slope of any such wing near $\mathrm{M} = 1$ is predicted to be

$$(C_{L\alpha})_{\mathrm{M} \cong 1} = \frac{\pi}{2} AR \tag{4.75}$$

The same formula is obtained by approaching from the supersonic side. The physical implications of this limiting process are that the disturbances produced by any streamlined configuration in transonic flight are very "three-dimensional," and that, for wings of low AR (and slender bodies also) the principal flow perturbation occurs in yz-planes normal to the free stream. As we shall see, the idea of 2-D "crossflow" has proved to be a useful approximation when analyzing such cases.

Naturally, we must turn to measurements or to more exact theory to establish conditions of validity for results like (4.75). An example of the latter is the requirement (cf. Ashley and Landahl 1965, page 236, and their cited references),

$$\mathcal{R}\, \tau^{1/3}[\mathrm{M}^2(\gamma + 1)]^{1/3} \ll 1 \tag{4.76}$$

Here γ is the specific-heat ratio and τ is a measure of the order of streamwise surface slopes on the wing, such as the thickness ratio in a nonlifting problem.

4.5 THIN AIRFOIL IN TWO-DIMENSIONAL SUPERSONIC FLOW

Figure 4.9 shows the type of airfoil that will be analyzed at $\mathrm{M} > 1$. For mathematical convenience we place the leading edge along the y-axis. Since there is no advantage here to separating the thickness and lifting cases, we return to boundary conditions of the form (4.25). The problem to be solved reads

$$(\mathrm{M}^2 - 1)\frac{\partial^2 \varphi}{\partial x^2} - \frac{\partial^2 \varphi}{\partial z^2} = 0 \tag{4.77}$$

$$w(x, 0+) = u_0 \frac{dz_U}{dx} \equiv w_U(x) \tag{4.78a}$$

$$\text{for } 0 \le x \le c$$

$$w(x, 0-) = u_0 \frac{dz_L}{dx} \equiv w_L(x) \tag{4.78b}$$

Despite apparent similarities in the governing equations, supersonic flows are markedly different from subsonic ones. The former are characterized, for instance, by what have been colorfully called "zones of forbidden signals." In the linearized approximation, any impulsive disturbance propagates away from its origin at constant speed a_∞ relative to the fluid at rest. At the same time it is

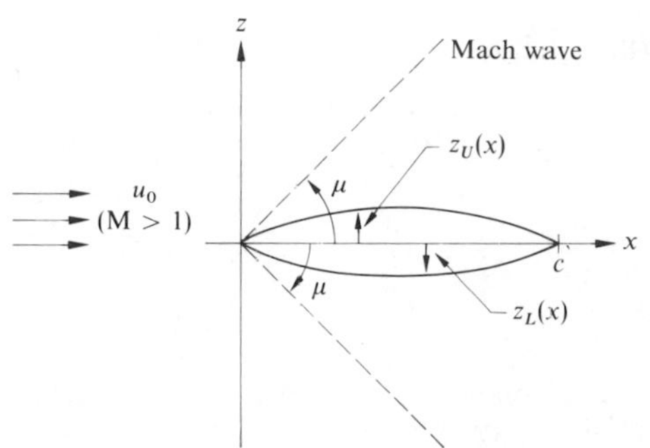

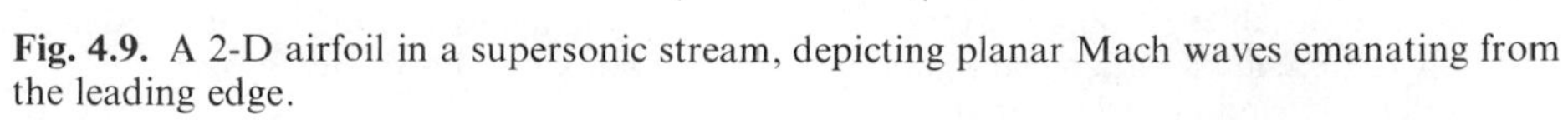
Fig. 4.9. A 2-D airfoil in a supersonic stream, depicting planar Mach waves emanating from the leading edge.

"blown aft" in our reference frame at speed $u_0 > a_\infty$. Consequently, the effects of a disturbance source at a fixed (x, y, z) are confined to a right-circular cone, with its vertex at the source and its axis parallel to positive x. By a familiar construction of the volume swept by a spherical wave whose radius increases at a_∞ and whose center moves downstream at u_0 (cf. Jones and Cohen 1960, pages 120–124), the semi-apex angle of this "Mach cone" is found to be

$$\mu \equiv \sin^{-1}\frac{1}{\mathrm{M}} \equiv \tan^{-1}\frac{1}{\sqrt{\mathrm{M}^2 - 1}} \tag{4.79}$$

The dashed lines in Fig. 4.9 are 2-D Mach waves from the leading edge, marking the forward boundary of the perturbation due to the airfoil. They are planes with equations

$$x = \pm\sqrt{\mathrm{M}^2 - 1}\, z \tag{4.80}$$

which form the envelope of Mach cones originating from all points along the y-axis. Because of forbidden signals, φ is identically zero ahead of this boundary.

Since nothing can proceed upstream and the airfoil itself is a solid obstacle, we conclude that no disturbance originating at its upper surface can be felt at its lower surface, and conversely. It follows that the two are "independent," in the sense that a change in $z_L(x)$ cannot affect the pressure distribution on $z_U(x)$. Moreover, we cannot expect to enforce Kutta's condition here, since the shape and location of the trailing edge exert no influence on the flow ahead. Indeed, it is useful to imagine that the actual airfoil is the front end of a 2-D barrier extending all the way to $x = +\infty$, the functions z_U and z_L then being specified for all positive x.

The observation that conditions at a given chordwise station are felt only downstream suggests that x is a "timelike" variable and leads us to an efficient way of solving (4.77)–(4.78): Laplace transformation* on x. In view of the definition

$$\mathscr{L}\{\varphi(x, z)\} \equiv \bar{\varphi}(s, z) = \int_0^\infty e^{-sx}\varphi(x, z)\, dx \tag{4.81}$$

we can write transforms such as

$$\mathscr{L}\left\{\frac{\partial^2\varphi}{\partial z^2}\right\} = \frac{\partial^2\bar{\varphi}}{\partial z^2} \tag{4.82}$$

$$\mathscr{L}\left\{\frac{\partial^2\varphi}{\partial x^2}\right\} = s^2\bar{\varphi} \tag{4.83}$$

* Although the Laplace transform will have its greatest utility when we encounter systems of linear, ordinary differential equations in time, we introduce it here for a problem in two independent variables. You are expected to have some familiarity with the technique. In addition to recent books on automatic control, accounts can be found in such texts as Churchill (1944) and Doetsch (1944).

No initial-condition terms at $x = 0$ appear in (4.83), because we have established that $\varphi = 0$ along the vertical line through the leading edge; strictly, the behavior of $\partial\varphi/\partial x$ as $x \to 0+$ for $z = 0$ must be examined *a posteriori*.

Let us study the flow in the upper half-space $z \geq 0+$. The lower side of the airfoil can then be treated by inspection. The transformed problem statement reads

$$\frac{\partial^2 \bar{\varphi}}{\partial z^2} - (\mathrm{M}^2 - 1)s^2 \bar{\varphi} = 0 \tag{4.84}$$

$$\frac{\partial \bar{\varphi}}{\partial z}(s, 0+) = \mathscr{L}\{w_U(x)\} \equiv \bar{w}_U(s) \tag{4.85}$$

It is customary in routine applications of this method to regard s as a positive real number or, at most, to require its real part to be ≥ 0. Of the two possible solutions $\sim \exp[\pm\sqrt{\mathrm{M}^2 - 1}\, sz]$ of (4.84), this specification—plus the condition that disturbances die out at great distances from the airfoil—forces us to choose the negative sign when $z \geq 0+$. The constant A in the transformed result must depend on s. It is an easy matter to substitute

$$\bar{\varphi} = A(s) \exp[-\sqrt{\mathrm{M}^2 - 1}\, sz] \tag{4.86}$$

into (4.85) and obtain

$$\bar{\varphi} = -\frac{\bar{w}_U(s)}{\sqrt{\mathrm{M}^2 - 1}\, s} \exp[-\sqrt{\mathrm{M}^2 - 1}\, sz] \tag{4.87}$$

Transformation of the Bernoulli relation (4.19) shows that the pressure $\bar{C}_p$ is related to $\bar{\varphi}$ by a simple factor $(2s/u_0)$.

$$\bar{C}_p = \frac{2\bar{w}_U(s)}{u_0\sqrt{\mathrm{M}^2 - 1}} \exp[-\sqrt{\mathrm{M}^2 - 1}\, sz] \tag{4.88}$$

For arbitrary z, the inversion of (4.87) depends on the following basic formulas:

$$\mathscr{L}^{-1}\left\{\frac{\bar{f}(s)}{s}\right\} = \int_0^x f(x_1)\, dx_1 \tag{4.89}$$

$$\mathscr{L}^{-1}\{e^{-bs}\bar{f}(s)\} = \begin{cases} 0, & x < b \\ f(x - b), & x \geq b \end{cases} \tag{4.90}$$

$\bar{f}$ is the transform of $f(x)$, the variable z going through the process as a parameter; b is a real constant, here equal to $\sqrt{\mathrm{M}^2 - 1}\, z$. Thus we get, for the perturbation potential,

$$\boxed{\varphi(x, z) = -\frac{1}{\sqrt{\mathrm{M}^2 - 1}} \int_0^{(x - \sqrt{\mathrm{M}^2 - 1}\, z)} w_U(x_1)\, dx_1, \qquad \text{for } x \geq \sqrt{\mathrm{M}^2 - 1}\, z,\ z \geq 0+} \tag{4.91}$$

and for the pressure coefficient

$$C_p(x, z) = \frac{2}{\sqrt{M^2 - 1}} \frac{w_U(x - \sqrt{M^2 - 1}\,z)}{u_0} \tag{4.92}$$

At the approximate location $z = 0+$ of the upper surface, (4.91) and (4.92) reduce to

$$\varphi(x, 0+) = -\frac{1}{\sqrt{M^2 - 1}} \int_0^x w_U(x_1)\, dx_1 \tag{4.93}$$

$$C_p(x, 0+) = \frac{2}{\sqrt{M^2 - 1}} \frac{w_U(x)}{u_0} \equiv \frac{2}{\sqrt{M^2 - 1}} \frac{dz_U}{dx} \tag{4.94}$$

It is evident from (4.94) that there is direct proportionality between the pressure disturbance and the small angle through which any streamline adjacent to the airfoil surface has been turned from its free-stream direction. Positive dz_U/dx produces "compressive turning," because it constricts the stream tubes, thus causing the fluid to slow down and the pressure and density to rise in a steady supersonic flow. Equation (4.94) is an elementary instance of the unique relationships between variables that occur in "simple wave" situations; the shock-expansion theory of Section 4.6 is one generalization.

Analysis of the field below the airfoil is analogous to the foregoing, except that conditions as $z \to -\infty$ lead to choosing positive signs for the exponentials of $\sqrt{M^2 - 1}\,sz$. Important results are

$$\varphi(x, z) = \frac{1}{\sqrt{M^2 - 1}} \int_0^{(x + \sqrt{M^2 - 1}\,z)} w_L(x_1)\, dx_1 \tag{4.95}$$

$$C_p(x, z) = -\frac{2}{\sqrt{M^2 - 1}} \frac{w_L(x + \sqrt{M^2 - 1}\,z)}{u_0} \tag{4.96}$$

both for $z \leq 0$ and $x \geq \sqrt{M^2 - 1}\,|z|$. The lower-surface pressure is

$$C_p(x, 0-) = -\frac{2}{\sqrt{M^2 - 1}} \frac{dz_L}{dx} \tag{4.97}$$

An interesting interpretation of (4.91) and (4.95) is that all properties are predicted to remain constant on the Mach waves $x \mp \sqrt{M^2 - 1}\,z = \text{constant}$ emanating from the two surfaces. What happens along any given wave is fixed by the local slope at which it leaves the surface. The ensuing pattern would seem to be propagating forever without decay or dispersion, thus violating our sense of what should actually happen at remote distances. For more realistic 3-D wings there is, of course, a dying out due to dispersion; also, other effects of atmospheric viscosity, winds, and vertical property gradients can no longer be ignored. Nevertheless, supersonic disturbances are remarkably persistent, the sonic boom being one undesirable manifestation of their penetration to the ground.

We remark in passing that supersonic flows can equally well be analyzed by starting from the wave-equation nature of (4.77). By comparison with (4.31) for the 2-D Laplace's equation, any suitable function of $(x - \sqrt{M^2 - 1}\,z)$ only—or of $(x + \sqrt{M^2 - 1}\,z)$ only—must automatically satisfy (4.77). From our knowledge of the waves, we choose

$$\varphi = F(x - \sqrt{M^2 - 1}\,z) \tag{4.98}$$

for $z \geq 0+$ and use the condition (4.78a) to reproduce (4.91) quite straightforwardly. That pressures and streamline slopes are in direct proportion follows from

$$C_p = -\frac{2}{u_0}\frac{\partial\varphi}{\partial x} = -\frac{2}{u_0}\frac{d\varphi}{d(x - \sqrt{M^2 - 1}\,z)} = \frac{2}{\sqrt{M^2 - 1}\,u_0}\left(\frac{\partial\varphi}{\partial z}\right) \tag{4.99}$$

The various aerodynamic loads experienced by the airfoil can be estimated from

$$C_{pU} = \frac{2}{\sqrt{M^2 - 1}}\frac{dz_U}{dx}, \qquad C_{pL} = -\frac{2}{\sqrt{M^2 - 1}}\frac{dz_L}{dx} \tag{4.100a, b}$$

These are valid over essentially the entire area, supersonic leading edges being sharp or only slightly blunted because of the severe drag penalty that would otherwise accrue from the formation of strong shocks ahead of them.

As a first example, let us take the flat plate at incidence, for which both z_U and z_L are given by (4.47). (4.100a, b) predict a constant overpressure on the lower side and an equal suction on the top. The resultant force is therefore $N = [p_L - p_U]c$, per unit span, in a direction normal to the airfoil plane and acting through its midchord line. Its coefficient,

$$C_N \equiv \frac{N}{(\rho_\infty/2)u_0^2 c} = (C_{pL} - C_{pU}) = \frac{4\alpha}{\sqrt{M^2 - 1}} \tag{4.101}$$

is projected perpendicular to the flight direction by $\cos\alpha$ to produce C_L and opposite by $\sin\alpha$ to produce the *wave drag* C_{DW}. With the small-angle approximation,

$$C_L = \frac{4\alpha}{\sqrt{M^2 - 1}} \tag{4.102}$$

$$C_{DW} = \frac{4\alpha^2}{\sqrt{M^2 - 1}} \tag{4.103}$$

(It would be delightful if the theoretical lift–drag ratio of $1/\alpha$ could be approached in practice, but skin friction and the need for profile thickness to accommodate the structure preclude any such possibility!) The pitching moment follows directly from (4.101) and the location of the center of loading at $x = c/2$.

Certain linearized characteristics of all supersonic airfoils may be inferred from this solution because, as explained in Section 4.3, the effects of changing angle of attack are simulated by adding the proper amount of flat-plate loading to that on the basic profile at some standard incidence angle. Thus we can just differentiate (4.102) to get the lift-curve slope

$$\boxed{C_{L\alpha} = \frac{4}{\sqrt{M^2 - 1}}, \qquad M > 1} \tag{4.104}$$

The supersonic aerodynamic center must be at the midchord line, where any added lift due to changed α acts.

It is instructive to compare (4.104) with its subsonic counterpart, derived from (4.55) and (4.71),

$$\boxed{C_{L\alpha} = \frac{2\pi}{\sqrt{1 - M^2}}, \qquad M < 1} \tag{4.105}$$

Turning to loads on an arbitrary airfoil, we must take components of the pressure force at each chordwise station along the upper and lower sides. Cosines of small angles are replaced by unity to arrive at the following simple expressions for lift and moment about $x = x_0$:

$$C_L = \frac{1}{c}\int_0^c (C_{pL} - C_{pU})\,dx = -\frac{2}{c\sqrt{M^2 - 1}}\int_0^c \left(\frac{dz_L}{dx} + \frac{dz_U}{dx}\right) dx \tag{4.106}$$

$$C_m = \frac{2}{c^2\sqrt{M^2 - 1}}\int_0^c \left(\frac{dz_L}{dx} + \frac{dz_U}{dx}\right)(x - x_0)\,dx \tag{4.107}$$

When computing drag, we must be more careful to recognize that the constant pressure p_∞, which we subtracted when we defined C_p, can exert no resultant force on a closed body. Nevertheless, some supersonic airfoils do have blunt bases, and the following formula assumes that any such base is exposed only to the uniform p_∞:

$$C_{DW} = \frac{1}{c}\int_0^c \left(-C_{pL}\frac{dz_L}{dx} + C_{pU}\frac{dz_U}{dx}\right) dx = \frac{2}{c\sqrt{M^2 - 1}}\int_0^c \left[\left(\frac{dz_L}{dx}\right)^2 + \left(\frac{dz_U}{dx}\right)^2\right] dx \tag{4.108}$$

With these general results in hand, it is instructive to separate, by means of (4.26), the contributions of thickness from those of camber and angle of attack.

When (4.26) are substituted into (4.106) and (4.108), we evaluate integrals and do some rearranging to find that

$$C_L = \frac{4\overline{\alpha(x)}}{\sqrt{M^2 - 1}} = \frac{4}{\sqrt{M^2 - 1}}\left[\frac{z_1(0)}{c} - \frac{z_1(c)}{c}\right] \tag{4.109}$$

$$\begin{aligned} C_{DW} &= \frac{4}{c\sqrt{M^2 - 1}}\int_0^c\left[\left(\frac{dz_1}{dx}\right)^2 + \left(\frac{dz_t}{dx}\right)^2\right]dx \\ &\equiv \frac{4}{\sqrt{M^2 - 1}}\left[\overline{\alpha^2(x)} + \overline{\left(\frac{dz_t}{dx}\right)^2}\right] \end{aligned} \tag{4.110}$$

Here $\alpha(x) \equiv dz_1/dx$, and the bar above a quantity means that its average value over the chord is taken. Unlike the subsonic case, lift is seen to be generated only by having the leading edge $z_1(0)$ of the mean line above the trailing edge. Relationship (4.110) shows how the influence on drag of the two basic characteristics of airfoil geometry can be separated, even though C_D is nonlinearly dependent on z_U and z_L.

The importance of wave drag is exemplified by applying (4.110) to the double-wedge profile, which was used on some early supersonic wings. It has a doubly symmetrical diamond profile, with the maximum thickness t at midchord. Hence its mean line is the flat plate, and

$$z_t(x) = \begin{cases} \dfrac{t}{c}x, & 0 \le x \le \dfrac{c}{2} \\ \dfrac{t}{c}(c - x), & \dfrac{c}{2} \le x \le c \end{cases} \tag{4.111}$$

Its drag formula

$$C_{DW} = \frac{4}{\sqrt{M^2 - 1}}\left[\alpha^2 + \left(\frac{t}{c}\right)^2\right] \tag{4.112}$$

illustrates the compelling need to keep thickness ratios as small as possible, since the penalty is generally proportional to the square of this parameter. Some amusing and practically useful exercises in the calculus of variations can be posed by seeking profile shapes of minimum supersonic wave drag with such constraints as fixed thickness ratio or fixed cross-sectional area.

Because of forbidden signals, there are certain finite wings to which the C_p formulas (4.94) and (4.97) are wholly or partially applicable, whereupon elementary integrations yield all their aerodynamic coefficients. Figure 4.10 depicts four examples in plan view, with the traces of Mach cones from the tips shown as dashed lines. It is to be understood that each streamwise cross section of any

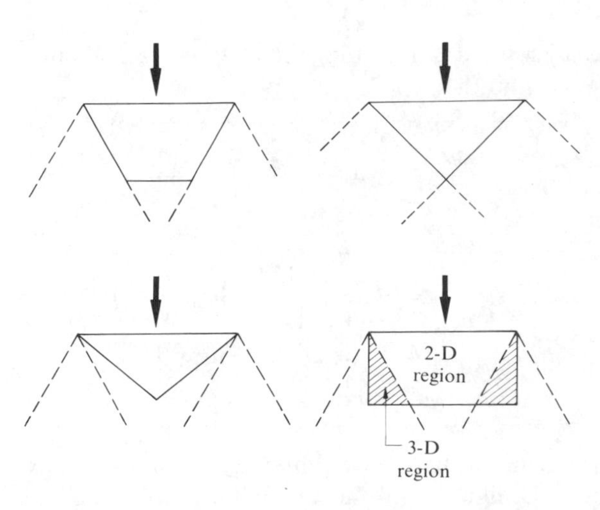

Fig. 4.10. Examples of finite wings which can be treated by 2-D theory in supersonic flight. (See text for discussion.)

particular wing is identical up to where it is cut off by the raked trailing edge—otherwise the apparent two-dimensionality of the flow ahead of that edge would be spoiled by spanwise variations of z_U and z_L. The flat plate $z_1 = -\alpha x$ used to calculate $C_{L\alpha}$ meets this test.

Three of the planforms shown are effectively 2-D over their entire areas, but only at and above the M at which the Mach lines going inboard from the tips coincide with the trailing edges. At lower M, we see that signals can pass around these edges ("subsonic trailing edges"), so that communication occurs between the upper and lower surfaces. The fourth case is a rectangular planform. On it there are 3-D regions, between the Mach lines and tips, in which our simple results are not valid; with increasing M, however, the region of invalidity becomes a smaller fraction of the total area, to a point at which larger-Æ aerodynamic coefficients can be estimated fairly accurately by (4.107), (4.109), (4.110), etc.

Finally, there exist relationships between coefficients of a particular planform in forward and reversed flow at the same M (see, e.g., Ursell and Ward 1950). These would, for instance, allow us immediately to determine $C_{L\alpha}$ for the delta with supersonic leading edges, since it is equal to that of the triangular shape in Fig. 4.10.

4.6 SHOCK-EXPANSION METHODS FOR THICKER SUPERSONIC AIRFOILS

The condition $M\tau \ll 1$, which governs all linearized supersonic theory, is more restrictive in practice than corresponding conditions for subsonic wings. Fortunately, various higher approximations are available. A comprehensive review will be found in Lighthill (1960). We confine our account here to the use of oblique shocks and Prandtl–Meyer expansions for computing surface pressures on airfoils. This straightforward and useful method is discussed, with sample computations, in Lighthill's Section E,3. There are many 2-D situations in which shock expansion is "exact" with respect to its airload predictions.

Two well-known flows provide the tools of the method. A valuable source of derivations, plus tables and charts for both, based on the perfect-gas state equation, is Liepmann and Roshko (1957).

1. Oblique Shock Waves

One of the first supersonic phenomena to be observed and analyzed was the nearly discontinuous wave called the *shock*. It may be thought of as the ultimate aggregate of a steepening train of compressive sound waves originating, e.g., from an explosion or piston displaced rapidly into the fluid. Relative to the undisturbed medium, it always propagates faster than a_∞ and causes positive jumps in all the state properties, as well as in the particle velocity component parallel to the shock-front motion. Like sound, a shock always moves toward the resting fluid in directions locally normal to its surface; an example is the planar wave generated in the one-dimensional device called a *shock tube*.

Shocks appear in steady flows, such as that past an airfoil, only within supersonic regions and only when the geometry calls for a compression of the gas. They may be *normal*—perpendicular to the oncoming stream tube, as in front of a blunt nose or at the surface of a transonic wing—or *oblique*. In the inviscid idealization, the normal shock is assumed to have zero thickness. The change in properties p, ρ, T, $\mathbf{q}$ through its surface can be predicted with high precision by means of four physical laws: conservation of mass, momentum and thermodynamic energy, and the thermal equation of state. The second law of thermodynamics renders the result unique by ruling out an otherwise-possible discontinuity through which entropy would decrease.

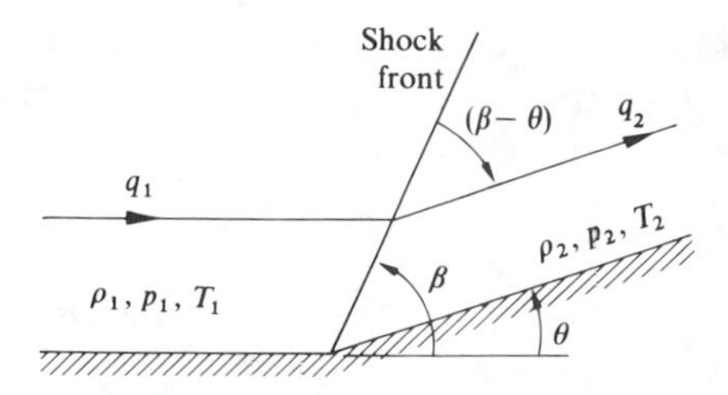

Fig. 4.11. Oblique shock wave, originating from a finite corner in a supersonic flow.

The oblique case, Fig. 4.11, is easiest analyzed by the "simple-sweep" transformation, whereby a normal shock is observed in a second reference frame with constant velocity tangential to its plane discontinuity surface. As derived in any gas-dynamics text, the relations between conditions behind the wave (subscript 2) and those ahead of it (subscript 1) are conveniently expressed in terms of shock angle β and upstream $M_1 \equiv q_1/a_1$. Typical of such relations are those for pressure and density,*

$$\frac{p_2 - p_1}{p_1} = \frac{2\gamma}{\gamma + 1}(M_1^2 \sin^2 \beta - 1) \tag{4.113a}$$

$$\frac{\rho_2}{\rho_1} = \frac{(\gamma + 1)M_1^2 \sin^2 \beta}{(\gamma - 1)M_1^2 \sin^2 \beta + 2} \tag{4.113b}$$

The usefulness of this wave in airfoil theory arises at places like wedge-shaped leading edges at which a sudden, finite compressive turning is required of the stream tube. This turn, through an angle θ such as that shown in the figure, can be—and in fact usually is—accomplished by an "attached" oblique shock. For a given profile, it is θ rather than β that is the known quantity. Tables of the implicit relationship

$$\tan \theta = 2 \cot \beta \frac{[M_1^2 \sin^2 \beta - 1]}{[(\gamma + \cos 2\beta)M_1^2 + 2]} \tag{4.113c}$$

are therefore needed. With both angles available, the law of mass conservation yields an expression connecting the resultant flow speeds,

$$\frac{q_2}{q_1} = \frac{\rho_1 \sin \beta}{\rho_2 \sin(\beta - \theta)} \tag{4.113d}$$

which may be combined with (4.113b). Of interest is the order of magnitude of the rise in specific entropy,

$$\frac{s_2 - s_1}{R} = O\left[\left(\frac{p_2 - p_1}{p_1}\right)^3\right] \tag{4.113e}$$

which shows that a "weak" shock is a nearly isentropic phenomenon.

As expected, (4.113c) indicates a vanishing of the turning angle either when the shock degenerates into an infinitesimal Mach wave

$$\beta \equiv \mu = \sin^{-1}\frac{1}{M_1} \tag{4.114}$$

or when it becomes normal ($\beta = \pi/2$). The strength, as measured by pressure or density ratio at fixed M_1, increases steadily as we progress from the former limit to the latter. In between there is a certain wave which gives the greatest turning, θ_{max}. Since β for a prescribed θ is double-valued, we must turn to experiment to

* All shock formulas reproduced here have been specialized for a perfect gas with constant specific heat ratio γ. More refined equations of state are often used; they would add nothing but complexity to our development.

establish that it is always the weaker, lower-β shock that is found attached to a wing. When the surface turning calls for θ in excess of the θ_{max} allowable at that M—e.g., at a blunt leading edge or below a highly deflected control surface—the shock "detaches," moving forward and becoming locally normal. Since there can be a serious drag penalty associated with the entropy rise through a detached shock, designers of high-performance vehicles try to prevent their appearance.

2. Prandt–Meyer Expansions

The expansion is the "inverse" of the shock, in the sense that it occurs when the bounding surface is so shaped as to allow supersonic stream tubes to increase in area and the fluid to expand. Figure 4.12 comprises two examples, illustrating

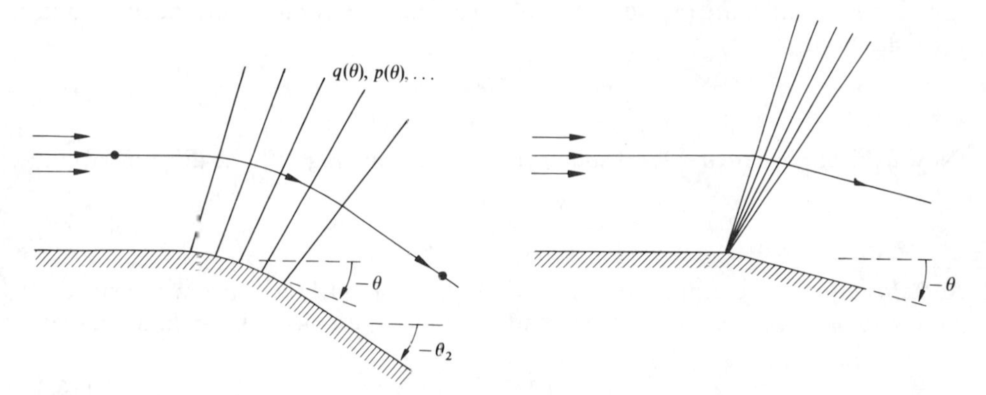

Fig. 4.12. Two situations in which a Prandtl–Meyer expansion would be observed in a uniform supersonic flow. Note that the sharp turn on the right may be thought of as the limit of the continuous turning on the left when the radius of curvature on the wall approaches zero.

patterns of expansion waves which arise when the flow ahead is uniform and when there are no obstacles nearby to cause reflections.

An excellent discussion appears in Liepmann and Roshko (1957, Sections 4.7–4.10). By assuming that each wave in the "fan" is a planar Mach wave in a stream at the local M just ahead, and by observing that the whole process is isentropic, we can develop a simple theory as an extension of the results of the previous section. For consistency with the shock analysis, define angle θ to be positive for compressive turning. The continuously curving surface at the left side of the figure may be thought of as the limit of a series of infinitesimal facets, each inclined at a negative $d\theta$ from the one ahead of it. The change in pressure along the stream tube passing around this corner can be adapted from (4.94), which we write

$$C_p \equiv \frac{\Delta p}{(\rho/2)q^2} = \frac{2}{\sqrt{M^2 - 1}} \frac{dz_U}{dx} \tag{4.115}$$

We identify dz_U/dx with $d\theta$, Δp with the infinitesimal pressure change dp through any wave, and dynamic pressure $\rho q^2/2$ with $\gamma p\mathrm{M}^2/2$ based on the Mach number just ahead of the wave. Thus we obtain a basic differential relation for $d\theta < 0$,

$$\frac{dp}{p} = \frac{\gamma \mathrm{M}^2}{\sqrt{\mathrm{M}^2 - 1}} d\theta \tag{4.116a}$$

A similar differential equivalent of (4.19) (also an exact consequence of isentropic conservation of momentum along the tube),

$$\frac{dp}{p} = -\gamma \mathrm{M}^2 \frac{dq}{q} \tag{4.116b}$$

enables us to eliminate pressure from (4.116a) and thus connect turning angle and particle speed.

$$d\theta = -\sqrt{\mathrm{M}^2 - 1}\, \frac{dq}{q} \tag{4.116c}$$

Now $q \equiv \mathrm{M}a$, where a is local sound speed. Through logarithmic differentiation,

$$\frac{dq}{q} = \frac{d\mathrm{M}}{\mathrm{M}} + \frac{da}{a} \tag{4.116d}$$

and this equation permits the elimination of q from (4.116c). From the steady-flow version of (4.11), differentiated and combined with (4.116d), we find that

$$\frac{dq}{q} = \frac{d\mathrm{M}}{\mathrm{M}} \left(1 + \frac{\gamma - 1}{2} \mathrm{M}^2\right)^{-1} \tag{4.116e}$$

(4.116e) and (4.116c) yield finally

$$-d\theta = \frac{\sqrt{\mathrm{M}^2 - 1}}{1 + \dfrac{\gamma - 1}{2} \mathrm{M}^2} \frac{d\mathrm{M}}{\mathrm{M}} \tag{4.116f}$$

It is an easy matter to integrate (4.116f) between any two stations 1 and 2 along the stream tube which follows the turning surface,

$$-[\theta_2 - \theta_1] = \left[\sqrt{\frac{\gamma + 1}{\gamma - 1}} \tan^{-1} \sqrt{\left(\frac{\gamma - 1}{\gamma + 1}\right)(\mathrm{M}^2 - 1)} - \tan^{-1} \sqrt{\mathrm{M}^2 - 1}\right]\Bigg|_{\mathrm{M}_1}^{\mathrm{M}_2} \tag{4.117}$$

The form of (4.117) suggested to Prandtl and Meyer that the following function be tabulated:

$$\nu(\mathrm{M}) \equiv \sqrt{\frac{\gamma + 1}{\gamma - 1}} \tan^{-1} \sqrt{\left(\frac{\gamma - 1}{\gamma + 1}\right)(\mathrm{M}^2 - 1)} - \tan^{-1} \sqrt{\mathrm{M}^2 - 1} \tag{4.118}$$

Given the initial state (say, 1) and Mach number in the uniform flow ahead of the expansive corner, the dependence of local M on position can then be determined from

$$\theta_1 - \theta = \nu(\mathrm{M}) - \nu(\mathrm{M}_1) \tag{4.119}$$

[Note that θ_1 will sometimes be zero, but the difference $(\theta_1 - \theta)$ must be positive.]

Once M is calculated, the state of the gas follows from the isentropy of the flow. Thus the pressure at θ, in terms of known p_1, can be calculated from the familiar perfect-gas relation,

$$\frac{p}{p_1} = \left[\frac{1 + \dfrac{\gamma - 1}{2}\mathrm{M}_1^2}{1 + \dfrac{\gamma - 1}{2}\mathrm{M}^2}\right]^{\gamma/(\gamma - 1)} \tag{4.120}$$

For air with specific heat ratio $\gamma = 1.4$, extensive tables of isentropic formulas like (4.120), oblique shock formulas, and the function $\nu(\mathrm{M})$ are found in NACA Report 1135 (Ames Research Staff 1953).

Figure 4.13 depicts several applications of the shock-expansion method to

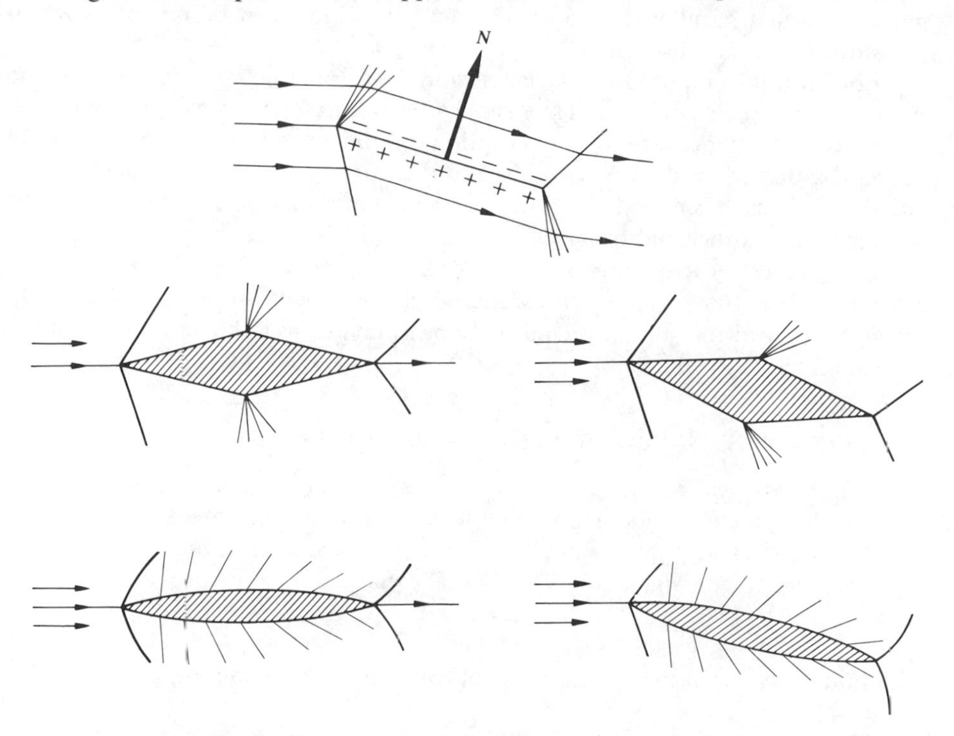

Fig. 4.13. Examples illustrating qualitatively the use of shock-expansion methods to determine near flow fields and airloads on 2-D supersonic airfoils.

airfoils. It is instructive to choose values for flight Mach number, angle of attack, and other parameters, then use the tables or a suitable computer program to calculate numerically the aerodynamic properties of some of these. Of particular interest might be the influence of $M(t/c)$ in modifying the linearized prediction that the lift-curve slope is $4/\sqrt{M^2 - 1}$ and the AC is at midchord for each case.

In these examples, any sudden compressive turning is assumed to be accomplished by an attached shock. A sudden expansion, as at the upper leading edge of the flat plate, is achieved by centering the fan of Prandtl–Meyer waves along one spanwise line. The presence of wave drag is obvious, in every instance, from the net downstream resultant of the pressure forces. The origins of lift and pitching moment can also be discerned on those airfoils with $\alpha \neq 0$.

The flat plate deserves special mention, since its loading is again seen to be the consequence of a uniform pressure difference which exerts a force normal to its plane, acting through midchord. Turning angle θ is $+\alpha$ on the lower surface and $-\alpha$ on the upper. The principal difference from linearized theory is that the pressure disturbances after the shock and the expansion no longer vary linearly with α as they do when $M\alpha \ll 1$. Conditions at and behind the trailing edge are determined by the combined requirements that the upper and lower streamlines must be turned parallel to one another and that there can be no difference in pressure across the wake.

For airfoils composed of flat faces, flow fields may be constructed to some distance from their surfaces. Thus shocks propagate outward as planar fronts, whereas—consistent with our conception of the expansion as composed of outward-going Mach waves—we observe that all quantities are constant along radial lines in each centered fan. Ultimately, however, the shocks and expansions run into one another and progressively cancel, with an accompanying pattern of reflected waves propagating away from the interaction region. A fortunate circumstance is that, in most cases, these reflections tend to be very weak. Up through $O(\theta^2)$, airfoil surface pressure can be estimated as if the reflections did not exist at all (cf. Lighthill 1960).

4.7 THREE-DIMENSIONAL WINGS IN SUBSONIC FLIGHT

Since the subcritical regime for $M < 1$ can be covered by compressibility correction, we here focus on incompressible flow. We study only the steady loading due to camber and angle of attack. Thickness effects will be the subject of a suggested problem assignment. Differential equation (4.18) is specialized to

$$\nabla^2 \varphi = 0 \tag{4.121}$$

with boundary condition (4.28) applicable over the sort of planform area S shown in Fig. 4.3.

From among the host of analysis schemes that have been proposed and mechanized, we choose the integral equation based on superposition of the antisymmetric singular solutions known as *pressure doublets*. This approach generalizes directly our vortex-sheet representation of the 2-D airfoil, although unfortunately the 3-D mathematical statement leads to nothing as simple as the relations (4.51, 52).

A logical starting point is the elementary solution of Laplace's equation known as a point source,

$$f_s(x, y, z) = -\frac{H_0}{4\pi\sqrt{x^2 + y^2 + z^2}} \equiv -\frac{H_0}{4\pi r} \tag{4.122}$$

If f_s is identified with φ, for instance, the equipotential surfaces are seen to be spheres centered at the origin. The perturbation streamlines radiate spherically, the velocity along them being

$$v_r = \frac{\partial f_s}{\partial r} = \frac{H_0}{4\pi r^2} \tag{4.123}$$

The constant H_0 is called the *source strength*. By integrating (4.123) over any spherical surface $r = a$, we see that H_0 equals the volume of fluid per unit time emanating outward from the origin. To obtain the doublet, we operate on the source, whose field is symmetrical in z, to get the desired z-antisymmetrical one by differentiating f_s with respect to z. We also transfer the center to an arbitrary point $(x_1, y_1, 0)$ on the plane of the wing.

$$\begin{aligned} f_D(x, y, z; x_1, y_1) &= -\frac{H_0}{4\pi}\frac{\partial}{\partial z}\frac{1}{\sqrt{(x - x_1)^2 + (y - y_1)^2 + z^2}} \\ &= \frac{H_0 z}{4\pi[(x - x_1)^2 + (y - y_1)^2 + z^2]^{3/2}} \end{aligned} \tag{4.124}$$

We shall prove that (4.124) is the Green's function for a discontinuity through the xy-plane of whatever physical quantity is represented by f_D. Accordingly, our objectives are met most directly by using doublets to represent C_p, or rather u, which is directly proportional to C_p through the linearized Bernoulli relation (4.19). This step is permissible because, as noted below (4.19), u and C_p are also solutions of (4.121). A sheet of such pressure doublets, distributed over S and having strength $h(x_1, y_1)$ per unit of area, is characterized by

$$u(x, y, z) = -\frac{1}{4\pi}\iint_S h(x_1, y_1)\frac{\partial}{\partial z}\frac{1}{\sqrt{(x - x_1)^2 + (y - y_1)^2 + z^2}}\,dx_1\,dy_1 \tag{4.125}$$

The point-function nature of this expression can be demonstrated by calculating its limiting value as z is brought down to $0+$, or up to $0-$, on the area S. In the expression

$$u(x, y, 0+) = \frac{1}{4\pi}\operatorname*{Lim}_{z\to 0+}\left\{z\iint_S h(x_1, y_1)\frac{dx_1\,dy_1}{[(x - x_1)^2 + (y - y_1)^2 + z^2]^{3/2}}\right\} \tag{4.126}$$

the factor z causes all contributions of the integral to vanish, except for those in the immediate vicinity of $(x_1 = x,\ y_1 = y)$, where the integrand is strongly singular. Let ϵS denote a little area around this point, and assume that h is a smoothly varying function. Then, for small enough ϵS, the mean-value theorem permits us to write

$$u(x, y, 0+) = \frac{h(x, y)}{4\pi} \lim_{z \to 0+} \left\{ z \iint_{\epsilon S} \frac{dx_1\, dy_1}{[(x - x_1)^2 + (y - y_1)^2 + z^2]^{3/2}} \right\} \tag{4.127}$$

It can be shown that the correct result is obtained from (4.127) independently of the precise shape chosen for ϵS, so long as we require that $\epsilon \gg z$ while going to the limit. For convenience, we pick a circle of radius ϵ and transform to the polar coordinate

$$r_1 = \sqrt{(x_1 - x)^2 + (y_1 - y)^2} \tag{4.128}$$

with origin at the singularity. Since the area element $dx_1\, dy_1$ should then be replaced by $2\pi r_1\, dr_1$, we get

$$\begin{aligned} u(x, y, 0+) &= \frac{h(x, y)}{4\pi} \lim_{z \to 0+} \left\{ z \int_0^{\epsilon} \frac{2\pi r_1\, dr_1}{(r_1^2 + z^2)^{3/2}} \right\} \\ &= \frac{h(x, y)}{2} \lim_{z \to 0+} \left\{ \frac{z}{|z|} - \frac{z}{\sqrt{\epsilon^2 + z^2}} \right\} = \frac{h(x, y)}{2} \end{aligned} \tag{4.129}$$

The same process for $z \to 0-$ furnishes

$$u(x, y, 0-) = -\frac{h(x, y)}{2} \tag{4.130}$$

so we have demonstrated that the sheet gives rise to a discontinuity in u (or C_p) through S. Accordingly, we introduce

$$\begin{aligned} h(x, y) &= u(x, y, 0+) - u(x, y, 0-) \\ &\equiv \gamma(x, y) \end{aligned} \tag{4.131}$$

and rewrite (4.125)

$$u(x, y, z) = -\frac{1}{4\pi} \iint_S \gamma(x_1, y_1) \frac{\partial}{\partial z} \left(\frac{1}{\sqrt{(x - x_1)^2 + (y - y_1)^2 + z^2}} \right) dx_1\, dy_1 \tag{4.132}$$

As in the 2-D case, it can be reasoned that γ is the component of bound circulation, about a spanwise axis, per unit chordwise distance. (4.19) and (4.131) yield, as before,

$$p_L - p_U = \rho_\infty u_0 \gamma \tag{4.133}$$

These discoveries verify our judgment in distributing the pressure doublets only on S and not over the downstream wake extension of S, where certain perturbation quantities like φ itself are known to be discontinuous. Were we to have included area behind the trailing edge in (4.125), we would have caused a physically impossible pressure jump across the wake.

The interested reader will be rewarded by working out the streamline pattern for our basic doublet, starting from (4.124) as a representation of $u = \partial\varphi/\partial x$. He will find a flow resembling that of a closely separated pair of equal and opposite vortices, extending downstream from an origin $(x_1, y_1, 0)$ to $x = +\infty$. At this origin they are connected by an infinitesimal spanwise vortex segment, whose orientation normal to the free stream generates lift. In fact, this is a limiting case, with infinitesimal spanwise dimensions, of the *horseshoe vortex*, used by earlier investigators such as Falkner (1948) to approximate the flow over a loaded surface by a finite-element method.

The camber shape for that wing which carries a specified distribution of $(p_L - p_U)$ can be computed by (somewhat tedious) integrations and differentiations of (4.132, 133). Treatment of the inverse case of known $z_1(x, y)$ also requires us to manipulate (4.132) first into an integral expression for $w = \partial\varphi/\partial z$ on $z = 0$. The y-component of the irrotationality condition (4.8),

$$\frac{\partial w}{\partial x} = \frac{\partial u}{\partial z} \tag{4.134}$$

provides a tool for this purpose, leading to the formula

$$\begin{aligned} w(x, y, z) &\equiv \int_{-\infty}^{x} \frac{\partial w}{\partial x_0}(x_0, y, z)\, dx_0 \\ &= \int_{-\infty}^{x} \frac{\partial u}{\partial z}(x_0, y, z)\, dx_0 \end{aligned} \tag{4.135}$$

Here the vanishing of disturbances far upstream has been accounted for. When (4.132) is inserted into (4.135), we get after inverting order

$$w(x, y, z) = -\frac{1}{4\pi} \iint_S \gamma(x_1, y_1) \int_{-\infty}^{x} \frac{\partial^2}{\partial z^2} \frac{1}{\sqrt{(x_0 - x_1)^2 + (y - y_1)^2 + z^2}}\, dx_0\, dx_1\, dy_1 \tag{4.136}$$

Regardless of which half space we start from, the same limiting w is found from (4.136) as $z \to 0$. In the light of boundary condition (4.28), the result can be written formally as follows:

$$\boxed{\begin{aligned} w(x, y, 0) &= u_0 \frac{\partial z_1}{\partial x} \\ &= \frac{1}{4\pi} \iint_S \!\!\!\!\!\!\times\!\!\times\, \gamma(x_1, y_1) K\, dx_1\, dy_1, \qquad \text{for all } (x, y) \text{ on } S \end{aligned}} \tag{4.137}$$

In (4.137) the *kernel function*

$$\boxed{K(x - x_1, y - y_1) = \frac{1}{(y - y_1)^2}\left[1 + \frac{x - x_1}{\sqrt{(x - x_1)^2 + (y - y_1)^2}}\right]} \tag{4.138}$$

is what we compute formally, after $z \to 0$, from the inner integration and differentiations in (4.136).

According to classical mathematics, K contains a nonintegrable singularity along the line $y_1 = y$. The author believes that Mangler (1951) was the first to clarify the origin of such singularities in linearized aerodynamic theory and to give the rules by which they should be numerically or analytically integrated. In the present case, the y_1 integration is carried out according to the following generalization of the Cauchy principal value:

$$\bigstar\!\!\int_A^B \frac{F(y_1)}{(y - y_1)^2} dy_1$$

$$= \lim_{\epsilon \to 0} \left\{ \int_A^{y-\epsilon} \frac{F(y_1)}{(y - y_1)^2} dy_1 + \int_{y+\epsilon}^B \frac{F(y_1)}{(y - y_1)^2} dy_1 - \frac{2F(y)}{\epsilon} \right\} \tag{4.139}$$

Here $A < y < B$, and F is any continuous, differentiable function. When Simpson's formula or the like is used on (4.139), one finds the term $2F(y)/\epsilon$ more and more nearly canceling the contributions from the limits near $y_1 = y$ as ϵ's magnitude is reduced. Moreover, when an indefinite integral is available for the integrand, Mangler (1951) has shown that the correct procedure is simply to evaluate it at limits A and B, taking the absolute value of the argument for any logarithmic term.

Before addressing the general solution of (4.137), let us mention that several of the most popular approximate schemes for predicting loadings on wings can be derived from (4.137) by suitable approximations to K and/or to the manner in which γ varies over the planform. A few of these are discussed below or in the problems. Since (4.137, 138) are "exact," within the framework of linearized theory, we prefer to concentrate on them. It might be mentioned that separated solutions of Laplace's equation can be employed as an alternative route when S is a circle or an ellipse (cf. Jones and Cohen 1960, Section A,7).

Multhopp's (1950) approach to (4.137) is the natural extension of our 2-D airfoil analysis. In effect, he transforms S into a square of side π by defining two angle variables η and θ as follows:

$$y = -\frac{b}{2}\cos\eta \tag{4.140a}$$

$$x = \tfrac{1}{2}[x_{TE}(y) + x_{LE}(y)] - \tfrac{1}{2}c(y)\cos\theta \tag{4.140b}$$

where

$$c \equiv x_{TE} - x_{LE} \tag{4.141}$$

is the chord length at spanwise station y and b is wingspan. Note that the relationship (4.140) will be single-valued for nearly every isolated lifting surface of practical interest. Although θ is dependent on both x and y, the Jacobian of this transformation shows the area element in (4.137) to be simply

$$dx_1\, dy_1 = \frac{b}{4} c(\eta_1) \sin \theta_1 \sin \eta_1 \, d\theta_1 \, d\eta_1 \tag{4.142}$$

Clearly the wingtips are plotted into $\eta = 0, \pi$ and the leading and trailing edges into $\theta = 0, \pi$, respectively.

The manner of representing the unknown loading $\gamma(\theta_1, \eta_1)$ as a double series is suggested by the form of (4.51), by Kutta's condition along $\theta_1 = \pi$, and by the knowledge that γ has singular behavior along the leading edge similar to that encountered in 2-D flow. Prandtl's lifting-line idealization also indicates that γ's spanwise variation can be efficiently described by a Fourier sine series in η_1. We therefore try*

$$\gamma = \frac{4\pi b u_0}{c(\eta_1)} \left\{ \cot(\theta_1/2) \sum_{m=1} a_{0m} \sin m\eta_1 + 4 \sum_{m=1} \sum_{n=1} \frac{a_{nm}}{2^{2n}} \sin n\theta_1 \sin m\eta_1 \right\} \tag{4.143}$$

There are other considerations behind the choice of (4.143): the 2^{2n} in the denominator of the double series ensures that the various constants a_{nm} have more nearly the same orders; the factor in front carries the correct dimensions, cancels the 4π in (4.137), and removes the inconvenient $c(\eta_1)$ in the area element (4.142).

A digital computer is required for solving (4.137) in practice. Several programs now exist (cf. Lamar 1968) which do this job with substitutions similar to (4.143) and (4.142). The rationale consists of forcing (4.137) to be satisfied at a finite network of stations (say, $M \times N$) on the interior of S (often selected at equal intervals of the variables θ and η). The necessary double integrations are performed numerically, with special attention both to the leading edge at which $\cot(\theta_1/2)$ has an integrable singularity and to $\eta_1 = \eta$, at which (4.139) applies. In the most straightforward version, known as *collocation*, the series in (4.143) would be truncated at $m = M$ and $n = (N - 1)$, and the retained constants a_{nm} would be calculated by inverting the simultaneous linear algebraic equations relating them to values of $\partial z_1/\partial x$ at the chosen stations.

Another scheme, which often proves more accurate, is to overdetermine this system of equations by keeping an excess of a_{nm}. Then the method of least squares is used to find a solution optimum in the sense that the a_{nm} so found cause the difference-squared between the two sides of each equation to be a minimum when averaged over S.

Since nearly every lifting surface has an xz-plane of symmetry, it is

* Note that (4.134) differs in certain respects from Multhopp's series; it is taken from Watkins *et al.* (1959).

advantageous to divide γ and $\partial z_1/\partial x$ into portions that are even and odd in y (and η). Equation (4.137) may then be solved for each separately. The even part, which is determined by retaining only odd m's in (4.143) and collocating on the right half of S, contains such information as total lift due to angle of attack and camber, pitching moment, and location of AC. The odd part—with the even m's—gives rolling moment due to rolling velocity, the linear effects of sideslip, etc.

Let us now assume the availability of a means for computing the a_{nm} coefficients of a given surface at a given subsonic flight condition. In view of (4.133), all aerodynamic loads except drag can then be predicted very directly. We begin by studying the running lift $l(y)$, which is a quantity of special importance to someone designing the structure of a large-$\mathcal{R}$ wing. Symbolically,

$$l(y) = \int_{\text{Chord}} [p_L - p_U]\, dx$$

$$= \rho_\infty u_0 \int_{x_{\text{LE}}(y)}^{x_{\text{TE}}(y)} \gamma(x, y)\, dx \equiv \rho_\infty u_0 \Gamma(y) \tag{4.144}$$

where Γ denotes total circulation bound to the wing at station y. If we transform to the angle variables, we discover, as in the 2-D case, that only $n = 0$ and $n = 1$ contribute to lift,

$$\Gamma(\eta) = \int_0^\pi \gamma(\theta, \eta) \frac{c(\eta)}{2} \sin\theta\, d\theta$$

$$= 2\pi b u_0 \int_0^\pi \left[\cot\frac{\theta}{2} \sum_{m=1} a_{0m} \sin m\eta + 4 \sum_{m=1} \sum_{n=1} \frac{a_{nm}}{2^{2n}} \sin n\theta \sin m\eta \right] \sin\theta\, d\theta$$

$$= 2\pi b u_0 \left\{ \pi \sum_{m=1} a_{0m} \sin m\eta + \frac{\pi}{2} \sum_{m=1} a_{1m} \sin m\eta \right\} \tag{4.145}$$

Introducing the definition

$$A_m \equiv 2\pi^2 \left(a_{0m} + \frac{a_{1m}}{2} \right) \tag{4.146}$$

we recover a series familiar in the history of aerodynamics,

$$\boxed{\Gamma(\eta) = b u_0 \sum_{m=1} A_m \sin m\eta} \tag{4.147}$$

The *running lift coefficient* is

$$c_l \equiv \frac{l(\eta)}{(\rho_\infty/2) u_0^2 c(\eta)} = \frac{2b}{c(\eta)} \sum_{m=1} A_m \sin m\eta \tag{4.148}$$

and the total lift

$$L = \int_{-b/2}^{b/2} \rho_\infty u_0 \Gamma(y)\, dy$$

$$= \rho_\infty u_0^2 b \int_0^\pi \left[\sum_{m=1} A_m \sin m\eta \right] \frac{b}{2} \sin \eta \, d\eta$$

$$= \rho_\infty u_0^2 b^2 \frac{\pi}{4} A_1 \tag{4.149a}$$

Thus L and its coefficient

$$\boxed{C_L = \frac{\pi}{2}\frac{b^2}{S} A_1} \tag{4.149b}$$

are entirely determined by the lead term in (4.147). It is worth remarking that (4.149), as well as many 2-D results derived in foregoing sections, confirm our assertions in Chapter 3 about the dependence of airloads on dynamic pressure and wing dimensions.

Another quantity directly related to $l(y)$ is the rolling moment

$$L_r = -\int_{-b/2}^{b/2} l(y) y \, dy \tag{4.150}$$

When (4.140a), (4.144), and (4.147) are introduced, we find L_r and its moment coefficient to be given by A_2 alone,

$$\boxed{C_l \equiv \frac{L_r}{(\rho_\infty/2) u_0^2 S b} = \frac{\pi}{8}\frac{b^2}{S} A_2} \tag{4.151}$$

The pitching moment, positive nose-up about a spanwise axis at $x = x_0$, is calculated from

$$M_p = -\iint_S (p_L - p_U)(x - x_0)\, dx\, dy \tag{4.152}$$

into which (4.133) and the complete series (4.143) must be substituted. The AC is located from that value of x_0 which causes M_p due only to a change in angle of attack α to vanish. There is no convenient analog to (4.149) or (4.151), which expresses this moment for a wing of general planform. For a rectangular surface of constant chord c_0, however, it is easy to work out the following:

$$C_m \equiv \frac{M_p}{(\rho_\infty/2) u_0^2 S c_0}$$

$$= \pi^2 \frac{b}{c_0} \left[\frac{a_{01}}{2} + \frac{a_{21}}{16} + \frac{x_0}{c} (a_{01} + a_{11}) \right] \tag{4.153}$$

The 3-D wing must pay a penalty for the generation of its lift in the form of the streamwise force called *induced drag* D_i. This drag is normally of the same order as, and incremental to, the resistance due to skin friction and pressure distortions by the boundary layer. It is predicted quite successfully by linearized theory. Our approach* is the traditional one of observing that the wing does on the air a quantity of work $D_i u_0$, per unit time, when observed in a frame attached to the atmosphere at rest. Since the inviscid fluid is both incompressible and nondissipative, this work must produce an equal amount of kinetic energy of particle motion in the vicinity of the wake. We study the structure of this wake by deriving its disturbance potential, which describes a flow antisymmetrical in z and having known discontinuities of u and v through portions of the xy-plane. Connected with these is a discontinuity in φ itself that can be calculated from

$$\begin{aligned}\Delta\varphi(x, y) &\equiv \varphi(x, y, 0+) - \varphi(x, y, 0-)\\ &= \int_{x_{LE}(y)}^{x} [u_U(x_1, y) - u_L(x_1, y)]\, dx_1\\ &= \int_{x_{LE}(y)}^{x} \gamma(x_1, y)\, dx_1\end{aligned} \tag{4.154}$$

The integral here starts from the leading edge because the pressure doublet sheet causes no jumps in any flow property ahead of that line. We have seen that $\gamma = 0$ behind the trailing edge, $x > x_{TE}(y)$; $\Delta\varphi$ across the wake is therefore given by [cf. (4.144)]

$$\Delta\varphi_W = \int_{x_{LE}(y)}^{x_{TE}(y)} \gamma(x_1, y)\, dx_1 = \Gamma(y), \qquad \text{for } x \geq x_{TE}(y), |y| \leq b/2 \tag{4.155}$$

The physical significance of the independence of $\Delta\varphi_W$ from x is that, at great distances downstream, the wake is a 2-D vortex sheet causing motion only in yz-planes. It resembles the vortex sheet used in the earlier incompressible airfoil theory, the circulation per unit distance (spanwise, in this case) being

$$\Delta v = \frac{\partial}{\partial y}\Delta\varphi = \frac{d\Gamma}{dy} \tag{4.156}$$

From this analogy we could determine the whole flow field. In the atmospheric

---

* We remark that Sears, in private communication, has quite rightfully criticized this approach on the grounds that the flow at $x = \infty$ is actually unsteady, with the wake moving downward under the influence of its own induced velocities, and that (4.157) and (4.161) involve second-order quantities. He does not, however, challenge the results, and his refined analysis verifies the elliptic-loading result (4.167).

reference frame the only particle velocities would be $v(y, z)$ and $w(y, z)$, and D_i would simply be the kinetic energy per unit x-distance,

$$D_i = \int_{\text{Unit length of wake}} (\tfrac{1}{2}\rho_\infty q^2)\, dV = \frac{\rho_\infty}{2}\int_{-\infty}^{\infty}\int_{-\infty}^{\infty} (v^2 + w^2)\, dy\, dz \qquad (4.157)$$

A cross section of the remote wake is shown in Fig. 4.14, in which the conventional approximation is made that these vortices remain in the plane $z = 0$. We

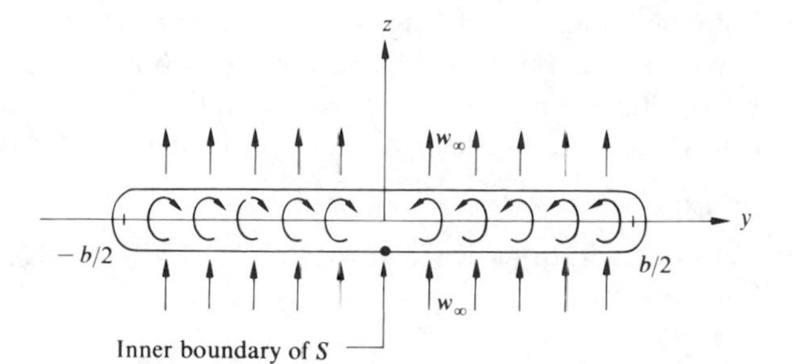

Fig. 4.14. Cross section of the wake far downstream of a planar wing. This wake is approximated by a vortex sheet between the wing tips in the xy-plane. The "upwash" w_∞ is continuous when we pass through the wake surface.

repeat Sears's criticism that the wake is free and should actually be permitted to move downward and deform under the action of its self-induced "downwash" velocities. For the purpose of working out the energy due to D_i, Sears has proved the fortunate result (fortunate at least for the case of downwash independent of y) that it does not matter whether this motion is accounted for.

An easy way to evaluate the integral in (4.157) is by means of Gauss' theorem (e.g., Karamcheti, page 131), which applies to any differentiable vector **A** defined over a volume V bounded by a closed surface S.

$$\int_V (\nabla \cdot \mathbf{A})\, dV = \oint_S \mathbf{A} \cdot \mathbf{n}\, dS \qquad (4.158)$$

Here **n** is the unit vector normal to S and outward from V. The term on the left becomes the desired kinetic energy if we replace **A** by the quantity $(\rho_\infty \varphi\, \nabla\varphi)/2$, whereupon

$$\nabla \cdot \mathbf{A} = \frac{\rho_\infty}{2}\nabla \cdot (\varphi\, \nabla\varphi) = \frac{\rho_\infty}{2}[\nabla\varphi \cdot \nabla\varphi + \varphi\, \nabla^2\varphi] = \frac{\rho_\infty}{2} q^2 \qquad (4.159)$$

The second term in brackets equals zero, for φ is known to satisfy Laplace's equation. The right side of (4.158) contains

$$\mathbf{A} \cdot \mathbf{n} = \frac{\rho_\infty}{2}\varphi(\nabla\varphi \cdot \mathbf{n}) = \frac{\rho_\infty}{2}\varphi\frac{\partial\varphi}{\partial n} \qquad (4.160)$$

where $\partial\varphi/\partial n$ is just the outward velocity component. Substituting (4.158)–(4.160) into (4.157), we find it convenient to separate S into three parts.

- Two parallel planes, a unit distance apart and each perpendicular to the x-axis. Since the flow is 2-D, $\partial\varphi/\partial n = 0$ over each of these.
- An outer boundary consisting of a large cylindrical surface, surrounding the wake and with generators parallel to the wake vortices. It is readily shown that, as this boundary recedes toward infinite distance from the wake, the disturbance dies out such that its contribution to the S-integral vanishes.
- The inner boundary pictured in Fig. 4.14. This surface is wrapped tightly around the vortex sheet on $-(b/2) \le y \le (b/2)$ but isolates the vortex singularities from the interior of V. $\varphi = \varphi(x, y, 0+)$ on its upper side and $\varphi(x, y, 0-)$ on its lower side. $\partial\varphi/\partial n = \mp w(x, y, 0)$ on these two sides, the minus sign applying to the upper side.

The foregoing considerations lead us to write (4.157)–(4.158), replacing dS by $1 \cdot dy$, as follows:

$$D_i = \frac{\rho_\infty}{2}\int_{-b/2}^{b/2} \varphi(x, y, 0+)[-w(x, y, 0)]\,dy + \frac{\rho_\infty}{2}\int_{-b/2}^{b/2} \varphi(x, y, 0-)[w(x, y, 0)]\,dy$$

$$= -\frac{\rho_\infty}{2}\int_{-b/2}^{b/2} \Delta\varphi w(x, y, 0)\,dy = -\frac{\rho_\infty}{2}\int_{-b/2}^{b/2} \Gamma(y)w_\infty(y)\,dy \tag{4.161}$$

In the last member of (4.161), we account for the two-dimensionality of the remote wake flow by eliminating the dependence on x of the vertical velocity ("upwash" w_∞) there. The relation between $w_\infty(y)$ and the wake vortex strength $d\Gamma/dy$ is precisely that between w and γ expressed by (4.40). The appropriate change of variables yields

$$w_\infty(y) = -\frac{1}{2\pi}\int_{-b/2}^{b/2} \frac{d\Gamma}{dy_1}\frac{dy_1}{y - y_1} \tag{4.162}$$

An especially meaningful form is obtained for D_i when we introduce the variable η and the series (4.147). From (4.162) and an integral similar to (4.46), we first calculate

$$w_\infty(\eta) = -u_0 \sum_{m=1} mA_m \frac{\sin m\eta}{\sin \eta} \tag{4.163}$$

With (4.147) and (4.163), (4.161) finally yields

$$D_i = -\frac{\rho_\infty b}{4}\int_0^\pi \Gamma(\eta)w_\infty(\eta)\sin\eta\,d\eta = \frac{\pi\rho_\infty}{8}u_0^2 b^2 \sum_{m=1} mA_m^2 \tag{4.164}$$

The coefficient of induced drag is

$$C_{Di} = \frac{\pi}{4}\frac{b^2}{S}\sum_{m=1} mA_m^2 \tag{4.165}$$

Although not discussed above in connection with compressibility correction, we mention that the work-energy reasoning still holds true for small-perturbation subsonic flow. Hence the relationship between C_{Di} and the spanwise distribution of bound circulation implicit in (4.165) may be used for any $M < 1$, as long as we stay in the definitely subcritical range. No mention has been made here of how the induced-drag force is distributed across the wingspan—important information in the area of structural loads. Lamar (1968) and Wagner (1969) contain excellent treatments of this question.

As might be expected, (4.165) shows that D_i is always positive. It was first observed from the lifting-line approximation that (4.149) and (4.165) together point to a strategy for minimizing the drag due to a given quantity of lift. Evidently the lifting surface should be so configured as to make $A_m = 0$ for $m > 1$, while A_1 is adjusted to furnish the C_L desired at the design point. The associated spanwise load distribution is known as *elliptic*, because

$$\Gamma = bu_0A_1 \sin\eta = bu_0A_1\sqrt{1 - \left(\frac{2y}{b}\right)^2} \tag{4.166}$$

plots versus coordinate y as half an ellipse, with semi-axes $(b/2)$ and $\Gamma(0)$. It is fortunate that representative wings at cruising angles of attack tend naturally to assume a loading close to (4.166)—provided that the twist and sweep angles are not excessive and that there is no control-surface deflection. An untwisted wing of elliptic planform shape and $Æ$ greater than about 5 sustains almost perfect elliptic loading. Although certain World War II aircraft adopted this feature, it is not commonly used because such a configuration is inconvenient to manufacture and the 2:1 straight-tapered planform achieves nearly the same end.

By substituting (4.149b), we can rewrite the minimum drag from (4.165) as follows:*

$$(C_{Di})_{\min} = \frac{\pi}{4}\frac{b^2}{S}A_1^2 = \frac{C_L^2}{\pi Æ} \tag{4.167}$$

where $Æ \equiv (b^2/S)$. The form of (4.167) suggested an empirical formula that is useful for fitting the measured curves of total drag versus lift over the normal operating range of α_{ZL_0} for many aircraft,

$$C_D \cong C_{D\infty} + \frac{C_L^2}{\pi e Æ} \tag{4.168}$$

* By setting $L = W$, we can use (4.167) to confirm (1.1), which we introduced for the purpose of discussing drag after takeoff.

The *zero-lift drag* $C_{D\infty}$ is a function of M, flap setting, etc., that is best determined from the wind tunnel. The *efficiency factor* e(M) is a number, $0 < e \leq 1$, proposed by Oswald (1932) as an index of how closely any design approaches the ideal $(C_{Di})_{\min}$ behavior. Equation (4.168) is often the basis for preliminary estimates of vehicle performance.

Let us close this section by describing one special case of the lifting-surface integral equation (4.137) which has more than routine interest. This approximation pertains to a low-Æ, pointed planform like that in Fig. 4.15. If (4.137) is

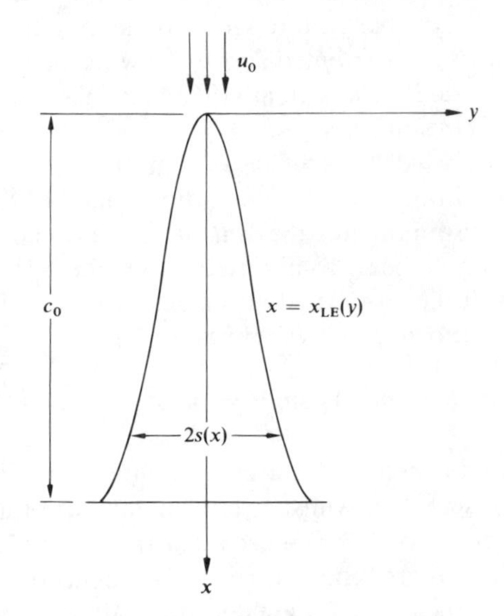

Fig. 4.15. Plan view of a very low-Æ lifting surface in a subsonic airstream.

integrated by parts in the spanwise variable y_1 and the area S is imagined to extend an infinitesimal distance beyond the boundaries of the doublet sheet to where $\gamma = 0$, we obtain

$$u_0 \frac{\partial z_1}{\partial x}$$

$$= \frac{1}{4\pi} \iint_S \frac{\gamma(x_1, y_1)}{(y - y_1)^2} \left[1 + \frac{x - x_1}{\sqrt{(x - x_1)^2 + (y - y_1)^2}} \right] dy_1 \, dx_1$$

$$= -\frac{1}{4\pi} \oint_0^{c_0} \oint_{-s(x_1)}^{s(x_1)} \frac{\partial \gamma}{\partial y_1} \left[\frac{1}{y - y_1} + \frac{\sqrt{(x - x_1)^2 + (y - y_1)^2}}{(x - x_1)(y - y_1)} \right] dy_1 \, dx_1 \quad (4.169)$$

The limits of the final integral here contain the planform dimensions from the figure; Cauchy principal-value integrations are required at the poles $x_1 = x$ and $y_1 = y$.

The assumption that we introduce into (4.169) comes from observing that, over most of the low-$\mathcal{R}$ area, $|x - x_1| \gg |y - y_1|$. Hence it may be acceptable to write

$$\sqrt{(x - x_1)^2 + (y - y_1)^2} \cong |x - x_1| \tag{4.170a}$$

which simplifies the kernel as follows:

$$\left[\frac{1}{y - y_1} + \frac{\sqrt{(x - x_1)^2 + (y - y_1)^2}}{(x - x_1)(y - y_1)}\right] \cong \begin{cases} \dfrac{2}{y - y_1}, & \text{for } x > x_1 \\ 0 \quad , & \text{for } x < x_1 \end{cases} \tag{4.170b}$$

The physical interpretation of (4.170b) is that the field at a given chordwise station x can be influenced only by properties of the doublet sheet (and therefore by wing geometry) ahead of that station. Although we do not try to justify this result here as a "rational" approximation, it can also be arrived at by systematic expansion in the small parameter $\mathcal{R}$.

In the light of (4.154) for $x \leq c_0$, we can insert (4.170b) into (4.169) and manipulate as follows:

$$\begin{aligned} u_0 \frac{\partial z_1}{\partial x} &\cong -\frac{1}{2\pi} \oint_0^x \oint_{-s(x_1)}^{s(x_1)} \frac{\partial \gamma}{\partial y_1} \frac{dy_1}{y - y_1} dx_1 \\ &= -\frac{1}{2\pi} \oint_{-s(x)}^{s(x)} \frac{1}{y - y_1} \left[\frac{\partial}{\partial y_1} \int_{x_{LE}(y_1)}^{x} \gamma \, dx_1\right] dy_1 \\ &= -\frac{1}{2\pi} \oint_{-s(x)}^{s(x)} \frac{\partial \, \Delta\varphi(x, y_1)}{\partial y_1} \frac{dy_1}{y - y_1} \end{aligned} \tag{4.171}$$

The coordinate x enters (4.171) merely as a parameter. Indeed, we see that the spanwise derivative $(\partial \, \Delta\varphi/\partial y)$ across any x is governed by the identical integral equation of the 2-D vortex sheet that was earlier encountered in (4.40) and (4.162). Fourier-series relations between the two dependent quantities in (4.171) can be developed, as in Section 4.3, by adopting the transformation

$$y = -s(x) \cos \eta \tag{4.172}$$

which leads to

$$u_0 \frac{\partial z_1}{\partial x}(x, \eta) = -\frac{1}{2\pi} \oint_0^\pi \left[\frac{\partial \, \Delta\varphi}{\partial y_1}(x, \eta_1)\right] \frac{\sin \eta_1 \, d\eta_1}{\cos \eta_1 - \cos \eta} \tag{4.173}$$

Recognizing that the integral of $(\partial \, \Delta\varphi/\partial y_1)$ across the span must vanish, one finds that an appropriate series substitution in (4.173) would consist of one singular term proportional to $\cos \eta_1 / \sin \eta_1$ plus a sum of $A_m \sin m\eta_1$ with $m > 1$.

The lead term here is much the most important, since it corresponds to a slender wing which is cambered only in the chordwise direction, so that

$$u_0 \frac{\partial z_1}{\partial x} = -u_0 \alpha(x) \tag{4.174}$$

For this we can tentatively substitute into (4.173)

$$\frac{\partial \, \Delta\varphi}{\partial y_1} = A(x) \frac{\cos \eta_1}{\sin \eta_1} \tag{4.175}$$

and use (4.46), with $n = 1$, to get

$$A(x) = 2u_0 \alpha(x) \tag{4.176}$$

Now the derivative can be integrated with respect to y_1 or η_1, with account taken that $\Delta\varphi$ vanishes at the edges $y_1 = \pm s(x)$,

$$\Delta\varphi = 2u_0 \alpha(x) s(x) \sin \eta = 2u_0 \alpha(x) \sqrt{s^2(x) - y^2} \tag{4.177}$$

From (4.155) we determine the bound circulation

$$\Gamma(y) = \Delta\varphi(c_0, y) = 2u_0 \alpha_{\mathrm{TE}} \sqrt{\left(\frac{b}{2}\right)^2 - y^2} \tag{4.178}$$

and thence the running lift

$$l(y) = 2\rho_\infty u_0^2 \alpha_{\mathrm{TE}} \sqrt{\left(\frac{b}{2}\right)^2 - y^2} \tag{4.179}$$

and the total lift

$$L = \int_{-b/2}^{b/2} l(y)\, dy = \left(\frac{\rho_\infty}{2} u_0^2 b^2 \alpha_{\mathrm{TE}}\right) \frac{\pi}{2} \tag{4.180}$$

It is remarkable that this narrow lifting surface, with its camber slope α constant across any spanwise section, has an elliptic distribution of lift, (4.179), whose magnitude is fixed entirely by α just ahead of the trailing edge. From (4.180) we work out the coefficients

$$\boxed{C_L = \frac{\pi}{2} \frac{b^2}{S} \alpha_{\mathrm{TE}}} \tag{4.181}$$

$$\boxed{C_{L\alpha} = \frac{\pi}{2} \mathcal{R}} \tag{4.182}$$

Equation (4.182) verifies what we anticipated in our discussion of linearized compressibility corrections in the transonic range. For $\mathcal{R} < 1$ it agrees well with lift-curve slopes measured on pointed models at subsonic, transonic, and supersonic Mach numbers.

Further understanding of these results may be achieved by examining the running lift $l'(x)$ per unit chordwise distance. The total force ahead of any station x would be

$$L'(x) = \int_0^x \int_{-s(x_1)}^{s(x_1)} \rho_\infty u_0 \gamma(x_1, y_1)\, dy_1\, dx_1 \tag{4.183}$$

For the case without spanwise camber, we calculate, with (4.177),

$$\begin{aligned} l'(x) = \frac{dL'}{dx} &= \rho_\infty u_0 \int_{-s(x)}^{s(x)} \gamma(x, y_1)\, dy_1 \\ &= \rho_\infty u_0 \int_{-s(x)}^{s(x)} \frac{\partial\, \Delta\varphi}{\partial x}(x, y_1)\, dy_1 \\ &= 2\rho_\infty u_0^2 \frac{d}{dx} \int_{-s(x)}^{s(x)} \alpha(x)\sqrt{s^2(x) - y^2}\, dx \\ &= \rho_\infty u_0^2 \frac{d}{dx}[\pi s^2(x)\alpha(x)] \end{aligned} \tag{4.184}$$

This last relation is exactly what we would obtain by momentum considerations, assuming that the fluid motion was 2-D and incompressible in planes normal to the direction of flight. This "slender-body approximation," due to Munk (1924) and Jones (1946), is the subject of a problem assignment at the end of this chapter.

4.8 THREE-DIMENSIONAL WINGS IN SUPERSONIC FLIGHT

Jones and Cohen (1960) and Heaslet and Lomax (1954) are definitive references describing the linearized treatment of supersonic wings. It is our intention merely to indicate certain tools that have proved useful. The discussion in Section 4.4 shows that we can pick any particular $M > 1$ when studying solutions of the differential equation (4.18). As before, a suitable condition like (4.28) fixes the vertical velocity over S.

Recall that, in the case of thin, pointed airfoils above the speed of sound, the flows adjacent to the upper and lower surfaces proceed independently of one another. This behavior carries over to certain 3-D wings called *simple planforms*, of which the delta with supersonic leading edges is an example. A remarkable consequence is that solutions to the thickness problem for a simple planform can be adapted immediately to determine loadings due to camber and angle of attack.

For this reason, and because of its simplicity, we look first at the flow due to thickness.

The boundary condition (4.27) can be combined with an obvious result of the z-symmetry of the continuous fluid motion for portions of $z = 0$ off the planform projection S to obtain

$$\frac{\partial \varphi}{\partial z}(x, y, 0\pm) = \begin{cases} \pm u_0 \dfrac{\partial z_t}{\partial x}, & \text{for } (x, y) \text{ on } S \\ 0 \quad , & \text{for } (x, y) \text{ off } S \end{cases} \tag{4.185}$$

The advantage of (4.185) is that we are no longer faced with a "mixed" boundary-value problem; that is, a single property of φ is now prescribed on the whole xy-plane.

A sheet of potential sources is the method of choice for solving (4.18) under (4.185). An isolated source, of unit strength and centered at the origin, has its disturbance described by [cf. (4.122)]

$$\varphi_S(x, y, z) = -\frac{1}{2\pi\sqrt{x^2 - (\mathrm{M}^2 - 1)(y^2 + z^2)}} \tag{4.186}$$

To be rigorous, we should be guided by the law of forbidden signals to set $\varphi_S = 0$ except within the downstream Mach cone, whose equation follows from the vanishing of argument of the denominator in (4.186). This end can also be achieved by specifying $x \geq 0$ and taking the real part of (4.186). The source sheet, with strength $h(x_1, y_1)$ per unit area of the $x_1 y_1$-plane, evidently has the potential

$$\varphi(x, y, z) = -\frac{1}{2\pi} \iint_{S'} \frac{h(x_1, y_1)\, dx_1\, dy_1}{\sqrt{(x - x_1)^2 - (\mathrm{M}^2 - 1)[(y - y_1)^2 + z^2]}} \tag{4.187}$$

The area S' consists of those portions of the sheet whose signals can be felt at (x, y, z). In general, S' is bounded at the rear by the hyperbolic intersection between the xy-plane and the forward Mach cone from (x, y, z). When $z \to 0$, however, this boundary degenerates to a pair of Mach lines. For instance, a typical S' is shown in Fig. 4.16 for a swept-back wing whose S has been covered with sources.

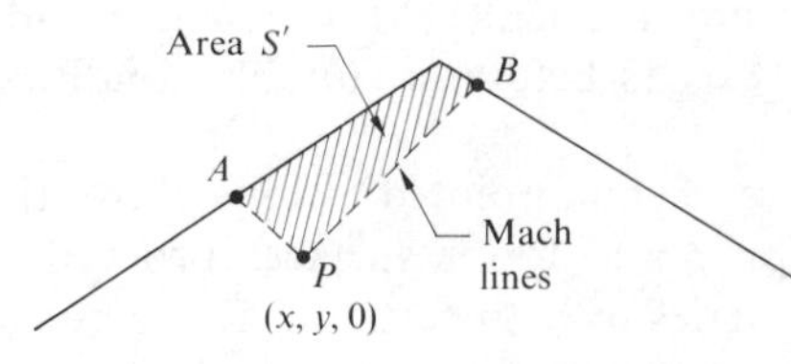

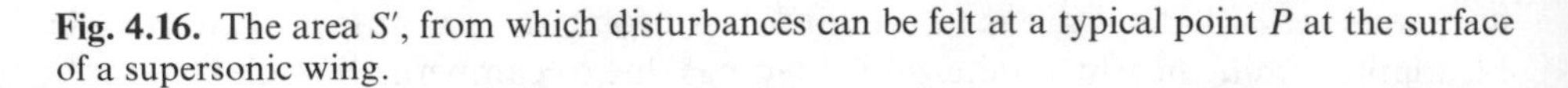
Fig. 4.16. The area S', from which disturbances can be felt at a typical point P at the surface of a supersonic wing.

As in subsonic flow, the sources turn out to be Green's functions for the velocity component perpendicular to any planar area over which they are distributed. In the present case, it can be shown (see Heaslet and Lomax 1954 for a rigorous proof) that

$$\pm h(x, y) = 2\frac{\partial \varphi}{\partial z}(x, y, 0\pm) \tag{4.188}$$

Hence (4.185) is satisfied by setting

$$h(x_1, y_1) = 2u_0 \frac{\partial z_t}{\partial x_1} \tag{4.189}$$

on the planform and $h = 0$ elsewhere. We are thus led to the desired result,

$$\varphi(x, y, z) = -\frac{u_0}{\pi}\iint_{S'} \frac{\dfrac{\partial z_t}{\partial x_1}(x_1, y_1)}{\sqrt{(x - x_1)^2 - (M^2 - 1)[(y - y_1)^2 + z^2]}}\,dx_1\,dy_1 \tag{4.190}$$

Given z_t, we can calculate all properties of the near flow field by integrations and differentiations of (4.190). For instance, the pressure, which is known to be the same at the upper and lower surfaces, can be determined from (4.19) and an integration by parts on x_1 as follows:

$$\begin{aligned} C_p(x, y, 0) &= \frac{2}{\pi}\frac{\partial}{\partial x}\iint_{S'} \frac{\dfrac{\partial z_t}{\partial x_1}(x_1, y_1)}{\sqrt{(x - x_1)^2 - (M^2 - 1)(y - y_1)^2}}\,dx_1\,dy_1 \\ &= \frac{2}{\pi}\left\{\int_A^B \frac{\dfrac{\partial z_t}{\partial x}(x_{LE}(y_1), y_1)}{\sqrt{[x - x_{LE}(y_1)]^2 - (M^2 - 1)(y - y_1)^2}}\,dy_1 \right. \\ &\qquad \left. + \iint_{S'} \frac{\dfrac{\partial^2 z_t}{\partial x_1^2}\,dx_1\,dy_1}{\sqrt{(x - x_1)^2 - (M^2 - 1)(y - y_1)^2}}\right\} \end{aligned} \tag{4.191}$$

Here A and B designate, respectively, the points at which the leftward- and rightward-going Mach lines through $(x, y, 0)$ intersect the leading edge (see Fig. 4.16). Some supersonic wings are composed of slab surfaces with $(\partial^2 z_t/\partial x^2) = 0$; for them we note that the loading is established entirely by the leading-edge geometry. Another special case of (4.190, 191) consists of locating a straight leading edge along $x = 0$ and dropping the dependence of thickness slope on y_1; one can then recover, without difficulty, 2-D formulas equivalent to (4.91) and (4.94).

The solution (4.190) is valid for essentially any planar wing with $M\tau \ll 1$. When we undertake to adapt it to the lifting problem, we first restrict ourselves to simple planforms like those in Fig. 4.17. The lack of communication between

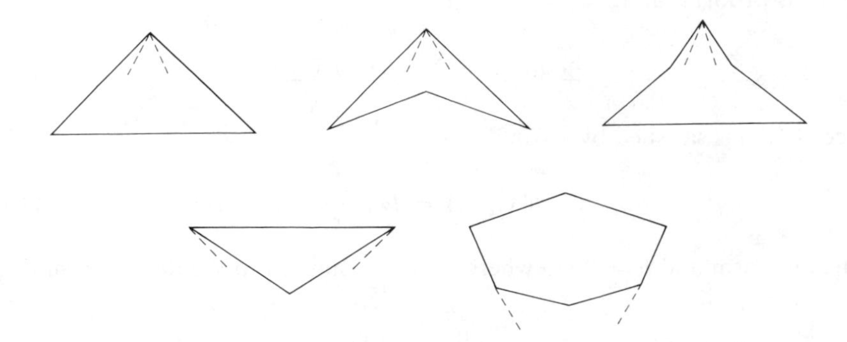

Fig. 4.17. Representative simple planforms in supersonic flow. The dashed lines are traces of Mach cones on the xy-plane.

their upper and lower sides leads us to the following procedure for determining the flows for arbitrary $z_U(x, y)$ and $z_L(x, y)$: the perturbation potential at and above the upper surface is given directly by (4.190) with $\partial z_t/\partial x_1$ replaced by $\partial z_U/\partial x_1$, at least for all points (x, y, z) which are not in communication with disturbed regions of the xy-plane exterior to planform S (e.g., the wake). Similarly $C_{pU} \equiv C_p(x, y, 0+)$ for (x, y) on S is expressed by (4.191) with the same substitution for z_t. Near the lower surface, at which $z \leq 0-$, $\partial z_t/\partial x_1$ must be replaced by $(-\partial z_L/\partial x_1)$ on the grounds that positive $\partial z_L/\partial x_1$ amounts to expansive turning of the streamlines there, whereas positive $\partial z_t/\partial x$ produces compression. $C_{pL} \equiv C_p(x, y, 0-)$ requires substituting $z_t = -z_L$ in the various derivatives of (4.191).

The load distribution on a simple planform, taken as usual in the direction to produce positive lift, would therefore be

$$
\begin{aligned}
\Delta C_p &\equiv C_{pL} - C_{pU} \\
&= -\frac{2}{\pi}\left\{\frac{\partial}{\partial x}\iint_{S'} \frac{\dfrac{\partial z_L}{\partial x_1}(x_1, y_1)}{\sqrt{(x - x_1)^2 - (M^2 - 1)(y - y_1)^2}}\, dx_1\, dy_1 \right. \\
&\qquad \left. + \frac{\partial}{\partial x}\iint_{S'} \frac{\dfrac{\partial z_U}{\partial x_1}(x_1, y_1)}{\sqrt{(x - x_1)^2 - (M^2 - 1)(y - y_1)^2}}\, dx_1\, dy_1 \right\}
\end{aligned}
\tag{4.192}
$$

These integrals can in some cases be simplified by steps such as those leading to the last member of (4.191). It should be clear that only the camber and angle of attack, as embodied in $z_1(x, y)$, can contribute to the net loading. Equation (4.192) accordingly reduces to

$$\Delta C_p(x, y) = -\frac{4}{\pi}\frac{\partial}{\partial x}\iint_{S'} \frac{\dfrac{\partial z_1}{\partial x_1}(x_1, y_1)}{\sqrt{(x - x_1)^2 - (\mathrm{M}^2 - 1)(y - y_1)^2}}\, dx_1\, dy_1 \tag{4.193}$$

(for simple planforms)

An elementary example is that of Puckett and Stewart (1947; see also pages 198–202 of Jones and Cohen 1960), who first treated the flat delta wing with supersonic leading-edge sweep angle Λ. The Mach lines through the vertex,

$$y = \frac{\pm x}{\sqrt{\mathrm{M}^2 - 1}} \tag{4.194}$$

are found to divide the planform into distinct regions of loading. With α the angle of attack,

$$\Delta C_p = \begin{cases} \dfrac{4\alpha}{\pi\sqrt{\mathrm{M}^2 - \sec^2\Lambda}}\left\{\cos^{-1}\left[\dfrac{\tan\Lambda - \dfrac{y}{x}(\mathrm{M}^2 - 1)}{\sqrt{\mathrm{M}^2 - 1}\left(1 - \dfrac{y}{x}\tan\Lambda\right)}\right] + \cos^{-1}\left[\dfrac{\tan\Lambda + \dfrac{y}{x}(\mathrm{M}^2 - 1)}{\sqrt{\mathrm{M}^2 - 1}\left(1 + \dfrac{y}{x}\tan\Lambda\right)}\right]\right\}, & \text{for } |y| \le \dfrac{x}{\sqrt{\mathrm{M}^2 - 1}} \\ \dfrac{4\alpha}{\sqrt{\mathrm{M}^2 - \sec^2\Lambda}}, & \text{for } \dfrac{x}{\tan\Lambda} \ge |y| \ge \dfrac{x}{\sqrt{\mathrm{M}^2 - 1}} \end{cases} \tag{4.195}$$

As might be expected from the physics of the situation, this flow is "conical," ΔC_p being constant along radiating lines from the vertex with $(y/x) = \text{constant}$. In the regions between the leading edge and the dividing Mach lines, the loading is uniform and is easily seen to agree with what one would find from "simple-sweep" theory by assuming this to be a portion of an infinite wing, swept back at the same Λ. Integration of ΔC_p over the total area yields a familiar but surprising [cf. (4.102)] result,

$$C_L = \frac{4\alpha}{\sqrt{\mathrm{M}^2 - 1}} \tag{4.196}$$

Thus the theoretical lift-curve slope of a delta with supersonic edges equals that of a 2-D airfoil—something which is not achieved by any 3-D wing in subsonic flow. Further information on related planforms and on subsonic-leading-edge deltas is summarized in Section A,14 of Jones and Cohen (1960).

Evvard (1950) and Krasilshchikova (1951; see also Landahl and Ashley 1965, Section 8-4) extended the foregoing analysis to the lifting problem for surfaces of more general planform. In essence, they demonstrated an exact cancellation between (1) the signals to a point $(x, y, 0+)$ on S which come from the lower side around the wingtip and subsonic leading edge, and (2) those to $(x, y, 0+)$ from a certain portion of the upper side itself. Their remarkable theorem states that S' in equations like (4.193) can be replaced, for a "non-simple" wing, by a reduced area S''. Examples of S'' are presented in Fig. 4.18; for wings with noninteracting tips,

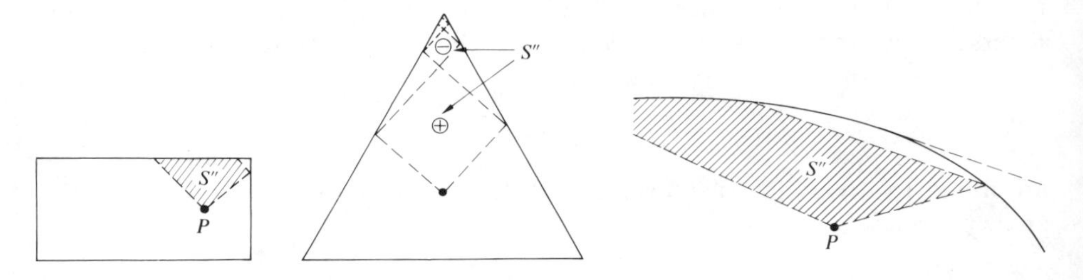

Fig. 4.18. Typical cases of the area S'' used for applying (4.193) to supersonic wings with subsonic leading edges. On the delta planform, the sign in (4.193) is to be changed when integrating over the area marked ⊖. All dashed lines are Mach lines.

S'' is just S' with a portion removed which is bounded by a "reflection" from the tip of the outward-going Mach line from $(x, y, 0+)$.

Since (4.190, 191) are valid for the thickness flow due to any supersonic planform, they can be taken together with (4.193) and the Evvard-Krasilshchikova generalization to cover most near-field problems of practical concern. There are, in fact, numerous other mechanizations of linearized theory which have been put forward and which improve the efficiency of solution for particular types of information and particular wing shapes; the references and the literature citations contained therein cover the majority of these.

Supersonic drag can also be predicted—and minimized—with considerable success by the small-perturbation idealization. If we again adopt the work–energy approach which proved so useful at $M < 1$, it can be shown (cf. Ashley and Landahl 1965, Chapter 9) that, exclusive of viscous effects,

$$\begin{aligned} D_{\text{Nonviscous}} &\equiv D_w + D_v \\ &= -\rho_\infty \iint\limits_{S_2} \frac{\partial \varphi}{\partial x}\frac{\partial \varphi}{\partial r}\, dS_2 + \frac{\rho_\infty}{2} \iint\limits_{S_3} \left[\left(\frac{\partial \varphi}{\partial y}\right)^2 + \left(\frac{\partial \varphi}{\partial z}\right)^2\right] dS_3 \end{aligned} \tag{4.197}$$

The control volume for drag is illustrated in Fig. 4.19, with S_3 a plane normal to x

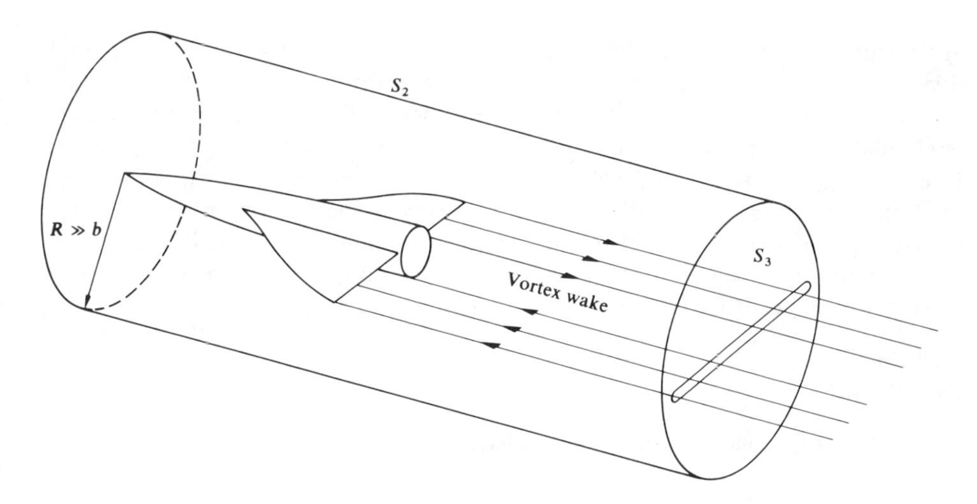

Fig. 4.19. Control volume used for predicting drag of a supersonic vehicle. (See text for description.)

far downstream in the wake and S_2 a cylinder with generators parallel to x and the flight direction. When the radius of S_2 is made very large, D_v (*vortex drag*) becomes identical with the subsonic induced drag D_i associated with the wake kinetic energy, as can be seen from (4.157). Indeed, once the spanwise distribution of lift or circulation is known, D_v can be calculated by (4.165).

Note that D_w is the *wave drag* due to acoustic energy radiated laterally in the disturbance fields generated by thickness and the lifting airloads. The magnitude of D_w often exceeds D_v; it generally contains portions proportional to the squares of thickness ratio, α_{ZL_0}, and fractional camber. The total resistance of a wing or, for that matter, a complete flight vehicle can be written

$$D = D_{\text{Viscous}} + D_w + D_v \tag{4.198}$$

Here the viscous component can be estimated fairly satisfactorily, as in subsonic flow, from a mean coefficient of skin friction integrated over the entire wetted area.

Designers, starting with the F-106, the B-58, and other aircraft of the 1950's, have tried to take advantage of increasingly sophisticated schemes for maximizing the design-point L/D of supersonic lifting surfaces. Today this formalism has been absorbed into the huge computer programs which predict the linearized flow over assemblages of wings, bodies, and empennages at both $M > 1$ and $M < 1$. An attempt to summarize the state of this art as of late 1971 can be found in the article by Ashley and Rodden (1972). Although it will be some years before the complete flow-tangency boundary conditions associated with an arbitrary streamlined vehicle can be satisfied in an unexceptionable fashion, we discern two approaches which are progressing toward this goal. The first—typified by Woodward's theory (1968) and recently applied to the B-58 and other realistic cases by Bradley and Miller (1971)—consists basically of "plating" the configuration with small quadrilateral areas containing uniform source and doublet sheets. The second is the "vortex lattice" method of Albano and Rodden (1969), wherein the disturbance is simulated by a pattern of singularities resembling discrete horseshoe vortices. In both instances, the boundary conditions are enforced at a finite set of stations over the vehicle surface and/or along the axes of bodies of revolution. The distribution of loading is approximated by summing force contributions from each singularity element. Lift, pitching moment, drag, and other quantities of interest can often be found to an accuracy of a very few percent.

4.9 PROBLEMS

4A Prove that Γ_0 is indeed equal to the line integral

$$\oint \mathbf{q} \cdot d\mathbf{s}$$

for any clockwise path surrounding the core of the vortex described by (4.35).

4B a) Review the relationships for two-dimensional, incompressible flow between the following:

- the complex potential function, $F(\xi)$
- the velocity potential
- the stream function
- the velocity components of the flow
- the streamlines of the flow

For the flow described by each of the following, sketch the equipotential lines and the streamlines with arrows showing flow direction.

1) $F(\xi) = i\dfrac{\Gamma_0}{2\pi} \ln \xi$

2) $F(\xi) = K\xi^2$

3) $F(\xi) = \dfrac{c}{2\pi\xi}$

4) $F(\xi) = \mu_0\xi + \dfrac{Q}{2\pi}[\ln(\xi + a) - \ln(\xi - a)]$

b) Sketch the streamlines and calculate the velocities for the flows given by the velocity potentials below. Indicate stagnation points, if they exist, and also show clearly the direction of the velocity along the streamlines. (1) $\Phi = k(x - y)$, (2) $\Phi = kxy$, (3) $\Phi = k(x^2 - y^2)$.

(Note that the constants k are applied for reasons of correct dimensions; they do not affect the pattern of streamlines.)

4C a) The parabolic-arc camber line is described by

$$z_1(x) = 4\frac{h}{c^2}\left[\left(\frac{c}{2}\right)^2 - x^2\right]$$

Use subsonic thin-airfoil theory to determine how the fractional camber h/c affects the following: (1) angle of zero lift; (2) ideal angle of attack and C_{Li}; (3) pitching moment at zero lift (C_{m0}) and $C_{m\alpha}$.

b) Find the airfoil with finite C_{Li} and $C_{m0} = 0$ and sketch it.

c) Since separation and transition to turbulent flow are brought on by adverse pressure gradients, it would seem desirable to design an airfoil with loading that is constant along the chord in order to prevent these unfavorable effects. In order to investigate the practicality of such an airfoil, perform the following computations:

1) Use 2-D thin-airfoil theory (incompressible flow) to design an airfoil that will have constant pressure on its upper and lower surfaces and that will produce nonzero lift.

2) Determine the velocities and camber-line shape for constant loading, and sketch as functions of x.

[*Hint:* Don't use the trigonometric-series formulation. Discuss the validity of the result with respect to thin-airfoil theory. Why is this camber line not used in actual wing design?]

4D Develop details of the solution to the thickness problem for a 2-D airfoil in incompressible flow, by analogy with the lifting solution worked out in Section 4.3.

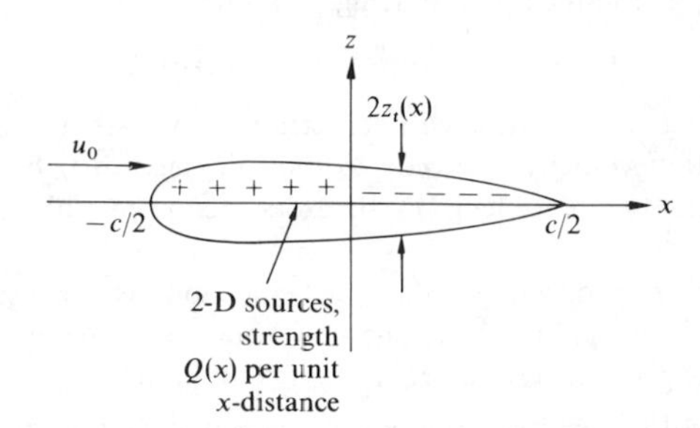

The basic tool is the illustrated sheet of sources, placed along the x-axis between $-c/2$ and $+c/2$, with "strength" $Q(x)$ per unit distance. The flow due to a concentrated source at the origin is

$$\varphi + i\psi = \frac{Q_0}{2\pi}\ln\xi$$

where $\xi = x + iz$ and Q_0 is its "strength." This source has a velocity pattern

$$v_r = \frac{Q_0}{2\pi r}, \qquad v_\theta = 0$$

It discharges a volume of fluid Q_0 cubic ft per unit time per unit span in the y-direction.

a) Show by consideration of continuity (conservation of mass), remembering that the profile surface $z = \pm z_t(x)$ must be a streamline of the flow, that

$$Q(x) = 2u_0 \frac{dz_t}{dx}$$

(The small-perturbation approximation is used here.)

b) Show that the perturbation velocity near the airfoil surface, from which pressure C_p is found, is given by

$$u_t(x) = \frac{1}{2\pi} \oint_{-c/2}^{c/2} \frac{Q(x_1)\, dx_1}{x - x_1} = \frac{u_0}{\pi} \oint_{-c/2}^{c/2} \frac{\frac{dz_t(x_1)}{dx_1} dx_1}{x - x_1}$$

c) Suppose $x = -(c/2) \cos \theta$, and dz_t/dx is represented by the Fourier series

$$\frac{dz_t(\theta)}{dx} = A_0 \frac{1 + \cos \theta}{\sin \theta} + \sum_{n=1} A_n \sin n\theta$$

Find the Fourier series for u_t and C_p. Why do you suppose no $(1 - \cos \theta)/\sin \theta$ term is used in dz_t/dx?

d) Find the pressure distribution on an airfoil with no camber and with $\alpha_{ZL} = 0$, but with

$$z_t(x) = \frac{8z_0}{c^3}\left[\left(\frac{c}{2}\right)^2 - x^2\right]^{3/2}$$

4E Allen (1945) showed that a more accurate approximation to the load distribution on a subsonic airfoil might be obtained by replacing (4.42) with

$$p_L - p_U = \rho_\infty[u_0 + u_t(x)]\gamma(x)$$

Here $\gamma(x)$ is to be calculated by some such linearized theory as that of Section 4.3. Note that $u_t(x)$ is the x-perturbation velocity due only to the thickness distribution $z_t(x)$; it might be determined from the results of Problem 4D, by exact incompressible-flow theory, or even by experiment.

Discuss the rationale underlying Allen's scheme and its consistency with the small-disturbance approximation. How would one go about mechanizing it, from results of this chapter, for an airfoil of given profile shape and angle of attack?

4F a) Choose any two-dimensional airfoil for which suitable data, taken in a low-speed wind tunnel, can be found. Compare linearized-theory and measured values of lifting pressure distribution (at one small angle of attack), C_L, AC location, and moment at zero lift, C_{m0}. Discuss sources of discrepancy. (Good sources of data are, for instance, Abbott, von Doenhoff, Stivers 1945, and Kuethe and Schetzer 1959, Appendix C.)

Consideration might be given to comparing the results of fully linearized incompressible theory and Allen's refinement discussed in Problem 4E.

b) Do the same for a supersonic airfoil, adding the wave drag, C_{Dw}, at one value of C_L. [*Note:* NASA publications are a good source of supersonic airfoil data.]

4G Derive the basic differential equation of acoustics by considering the potential φ of small disturbances in a fluid originally at rest ($u_0 = 0$). This is found by showing that the continuity equation

$$\frac{\partial \rho}{\partial t_0} + \nabla_0 \cdot (\rho \mathbf{q}) = 0$$

may be approximated by

$$\frac{1}{\rho_\infty}\frac{\partial\rho}{\partial t_0} + \nabla_0^2\varphi = 0 \tag{1}$$

and that Bernoulli's equation

$$\frac{\partial\varphi}{\partial t_0} + \int_{p_\infty}^{p}\frac{dp}{\rho} + \frac{q^2}{2} = 0$$

may be approximated by

$$\frac{\partial\varphi}{\partial t_0} + \frac{a_\infty^2}{\rho_\infty}(\rho - \rho_\infty) = 0 \tag{2}$$

A differentiation of (2) with respect to t_0 makes it possible to eliminate ρ from (1), leaving the classical wave equation

$$\frac{\partial^2\varphi}{\partial t_0^2} - a_\infty^2\nabla_0^2\varphi = 0$$

Then consider the transformation of independent variables from x_0, y_0, z_0, t_0 to a new system fixed to a wing moving with speed u_0 in the negative x_0-direction. For instance, show that

$$\frac{\partial(\ )}{\partial t_0} = \frac{\partial(\ )}{\partial t}\frac{\partial t}{\partial t_0} + \frac{\partial(\ )}{\partial x}\frac{\partial x}{\partial t_0} = \frac{\partial}{\partial t} - u_0\frac{\partial}{\partial x}$$

By assuming the flow to be steady in the x, y, z reference frame, show that one recovers the same linearized partial differential equation (4.18).

4H It is generally accepted that subsonic compressibility corrections become inaccurate at or slightly above the *critical Mach number* M_{crit}. This quantity is defined as the flight M which, at a given angle of attack, causes the local maximum speed somewhere in the flow just to equal the speed of sound. It can be proved that this maximum q will always be found on the wing surface. Considering what you know about shock waves in steady flow, explain physically why small-perturbation theory might fail above M_{crit}.

Given $C_{p_{min}}$, the lowest pressure occurring in the flow at a given M, derive linearized and exact formulas for estimating M_{crit}. Why is the first of these two methods often unreliable?

4I a) Using values predicted by small-perturbation methods, compare zero-lift wave drags for the two symmetrical airfoils shown.

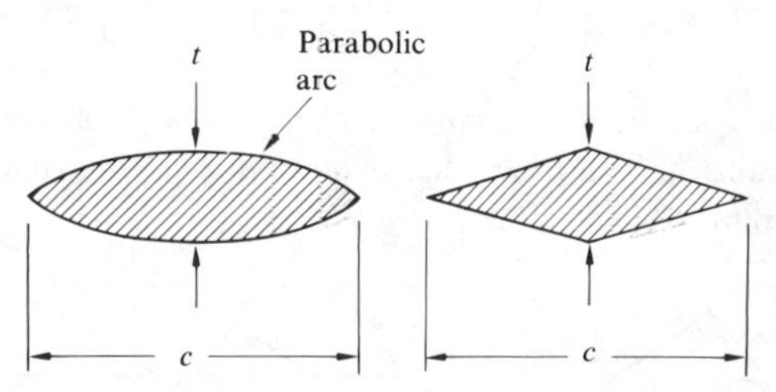

b) Take the symmetrical double-wedge airfoil at $M = 3$ and $t/c = \tan 5°$ (right side of figure). Compare numerically your drag estimate in (a) with what you would obtain by the shock-expansion technique.

c) Some supersonic airfoils have the trailing edge cut off by a plane normal to the mean surface, so as to form a flat "base" of area S_B per unit span. The average pressure p_B

acting over this base must be estimated by semi-empirical or experimental means. Discuss quantitatively the contributions of such a base to the airloads determined in Section 4.5. In particular, why is

$$C_{DB} \equiv \frac{(p_\infty - p_B)S_B}{(\rho_\infty/2)u_0^2 S}$$

a proper definition for the "coefficient of base drag"?

4J Explain how the properties of flow past a swept-back wing of large span can be obtained from those of a 2-D airfoil with the same cross-sectional profile, by starting from the airfoil in a stream $u_0 \cos \Lambda$ and adding a uniform velocity component $u_0 \sin \Lambda$ parallel to the span.

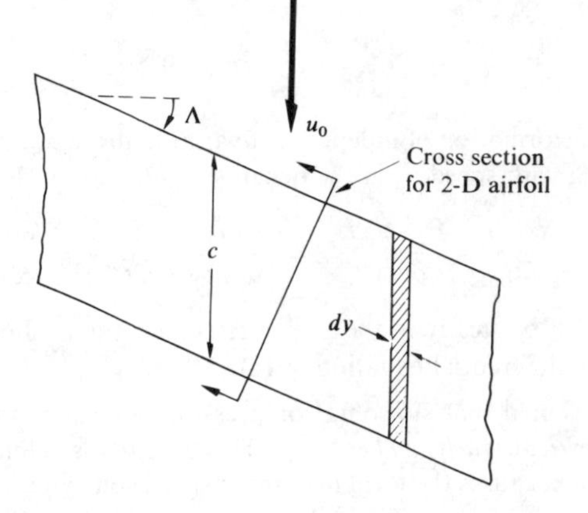

In particular, demonstrate the following:

a) The Mach number which governs "compressibility effects" on the wing is M cos Λ, where $M = u_0/a_\infty$.

b) Neglecting thickness effect in incompressible flow, the lift and pitching moment (about midchord) on the illustrated strip of width dy spanwise are

$$dL = (\pi\rho_\infty u_0^2 c \cos \Lambda)\alpha\, dy, \qquad dM_p = \left(\pi\rho_\infty u_0^2 \frac{c^2}{4} \cos \Lambda\right)\alpha\, dy$$

c) Use the result of (b) to estimate roughly the effects on $C_{L\alpha}$ of changing Λ. Note that reference area S remains fixed during each operation, but what happens to $Æ = b^2/S$, and how might this affect $C_{L\alpha}$?

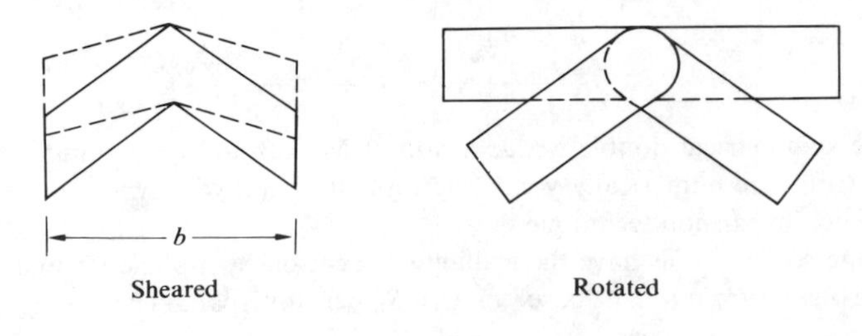

4K The following rule has been proposed for estimating the lift-curve slope of swept wings at low speed:

$$\frac{dC_L}{d\alpha} \cong \frac{\text{Æ}}{\text{Æ} + \left(\dfrac{a_0}{\pi}\right)\cos\Lambda}\, a_0 \cos\Lambda$$

where a_0 is the 2-D slope for the corresponding airfoil section. Usually $a_0 \cong 2\pi$ or slightly less. Investigate the limits of this rule for very low and very high aspect ratios. Generalize it to arbitrary subsonic (subcritical) Mach number.

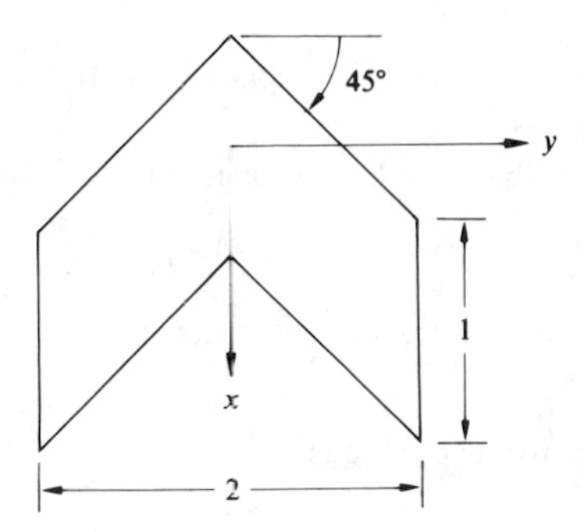

4L For the constant-chord swept-back wing shown in the figure above, Hsu solved the lifting-surface integral equation at M = 0 to find

$$\gamma = \frac{4\pi u_0 b}{c}\left\{\cot\frac{\theta}{2}\sum_{m=1} a_{0m}\sin m\eta + \sum_{n=1}\sum_{m=1}\frac{4}{2^{2n}}a_{nm}\sin n\theta\sin m\eta\right\}$$

When there is no camber and the angle of attack equals 1 radian, the coefficients a_{nm} turn out approximately as follows.

n \ m	1	2	3
0	0.0487	0.00906	0.00189
1	−0.0254	−0.0156	−0.00396
2	−0.0372	−0.0170	−0.0156

Find and plot the spanwise lift distribution. Determine $C_{L\alpha}$, the AC location, and C_{m0}.

4M One very useful scheme for finding the spanwise loading across subsonic wings of relatively large Æ and small sweep is Prandtl's *lifting-line theory*. Although it can be made more rigorous by matched-expansion methods, a simple development is based on the next figure.

The angle of attack from zero lift α_{ZL}, which this section would have in purely 2-D flow, is reduced by the "induced angle" $\alpha_i \cong w_\infty/2u_0$ due to the presence of the trailing vortex wake. $w_\infty(y)$ is the (here downward) vertical velocity on the wake far downstream. By comparison

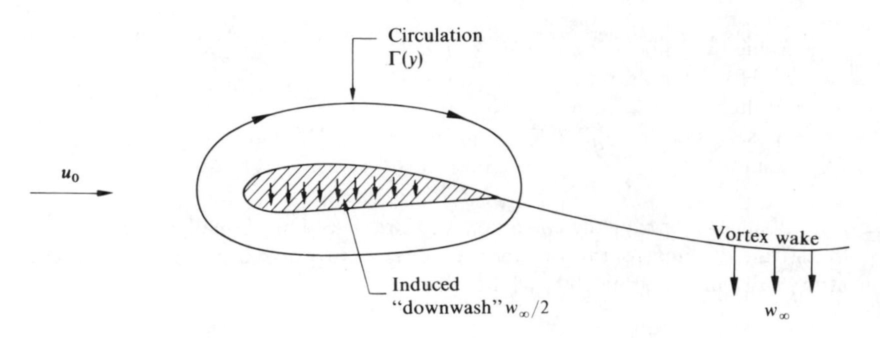

Wing profile section at spanwise station y

with the 2-D thin-airfoil vortex sheet, we have shown in (4.162) that

$$w_\infty(y) = \frac{1}{2\pi} \oint_{-b/2}^{b/2} \frac{\frac{d\Gamma}{dy_1} dy_1}{y - y_1}$$

a) By equating the two alternative expressions

$$\rho_\infty u_0 \Gamma \quad \text{and} \quad \frac{\rho_\infty}{2} u_0^2 c \cdot (C_{L\alpha})_{2\text{-D}}[\alpha_{ZL} - \alpha_i]$$

both of which describe the lift per unit span in the assumed model, derive the lifting-line equation

$$\Gamma(y) = \tfrac{1}{2} u_0 c (C_{L\alpha})_{2\text{-D}} \left[\alpha_{ZL} - \frac{1}{4\pi u_0} \oint_{-b/2}^{b/2} \frac{\frac{d\Gamma}{dy_1} dy_1}{y - y_1} \right]$$

b) Transform to the angle variable $y = -(b/2)\cos\eta$, substitute an appropriate Fourier series for $\Gamma(\eta)$, and use the previous result

$$\oint_0^\pi \frac{\cos m\eta_1}{\cos\eta_1 - \cos\eta} d\eta_1 = \frac{\pi \sin m\eta}{\sin\eta}$$

to show how solution of the lifting-line formulation can be reduced to solving linear algebraic equations. (Remember that c and α_{ZL} may vary from station to station.)

c) Use your result to describe wing shapes and twist distributions that might be expected to sustain the very desirable "elliptic load distribution"

$$l(\eta) \sim \sin\eta \sim \sqrt{(b/2)^2 - y^2}$$

Find how the 3-D $C_{L\alpha}$ varies with $A\!\!R$ for a flat wing with elliptic planform

$$c = \frac{c_0}{b/2}\sqrt{\left(\frac{b}{2}\right)^2 - y^2}$$

d) Discuss the utility of the lifting-line approach for predicting stability derivatives like rolling moment due to rolling and *dihedral effect* (rolling moment due to sideslip on a wing with dihedral). How about determining the onset of stall, finding whether the wingtip will stall before or after the root, and correcting undesirable stalling characteristics?

e) For purposes of estimating 3-D pitching moment, lifting-line theory is often supplemented by the assumptions:

- $l(y)$ acts at the quarter-chord point of each profile section.
- The section pitching moment at zero lift, described by its coefficient c_{m0} based on dynamic pressure and local chord-squared $c^2(y)$, is unaffected by 3-D.

Quantify these ideas, and show how they lead logically to the definition (3.4) of $\bar{c}$ as the "natural" length for use in making C_m dimensionless.

4N The boundary condition for the thickness problem on a 3-D wing in incompressible flow reads [cf. (4.27)]

$$\left.\begin{aligned} \frac{\partial \varphi}{\partial z}(x, y, 0+) &= u_0 \frac{\partial z_t}{\partial x} \\ \frac{\partial \varphi}{\partial z}(x, y, 0-) &= -u_0 \frac{\partial z_t}{\partial x} \end{aligned}\right\} \quad \text{for } (x, y) \text{ on } S$$

Both the symmetry condition here and the requirement that $\varphi \to 0$ at infinity are met by a velocity-potential source

$$\varphi = -\frac{H_0}{4\pi R}$$

centered on the xy-plane. In particular, a source of "strength" $h(x_1, y_1)$ per unit area, centered on a little area element at $x = x_1$, $y = y_1$, has the potential

$$d\varphi(x, y) = \frac{-h(x_1, y_1)\, dx_1\, dy_1}{4\pi\sqrt{(x - x_1)^2 + (y - y_1)^2 + z^2}}$$

Place a sheet of these sources over area S. Prove that the local normal velocity is proportional to the local strength (+ above and − below), and show that the relationship is $h = 2u_0\, \partial z_t/\partial x$. Thus we arrive at expressions for potential, local surface pressure $p_L = p_U$, and other quantities of interest in terms of integrals over $\partial z_t/\partial x$. Why are no sources needed except right on S? Why no wake? Why no lift and drag? What would these results be useful for?

4O On *any* thin wing in hypersonic flight, the second-order relation between the local over-pressure and the local normal fluid velocity w_0 (outward away from the surface) can be simplified to

$$p - p_\infty = \rho_\infty a_\infty w_0 + \frac{\gamma + 1}{4} \rho_\infty w_0^2$$

(perfect gas; ambient speed of sound a_∞). For rectangular planforms with symmetrical profile sections but general thickness distribution, investigate lift, wave drag, AC, and pitching moment. Wherever convenient, separate the effects of thickness and angle of attack.

4P A useful physical approach to the estimation of airloads on pointed wings of low *Æ*, slender bodies of revolution, wing–body combinations, and the like is as follows: The flow disturbance they create is largely confined to yz-planes normal to the flight direction, so that u

is a higher-order small quantity than v and w. As viewed in any thin slab of fluid, parallel to the yz-plane and fixed in the atmosphere at rest, the resulting flow appears to be 2-D and unsteady. The situation in such a slab has been compared with what happens when a sharp arrowhead passes perpendicularly through a thin sheet of rubber. The chordwise lift distribution and total lift are then reactions to the rate of change of downward z-momentum generated in the fluid slabs passing by the body. Per unit distance in the x-direction, it is given that the "apparent momentum," generated by the motions and bodies shown in the figure, in an unbounded mass of fluid at rest at infinity, has the values shown.

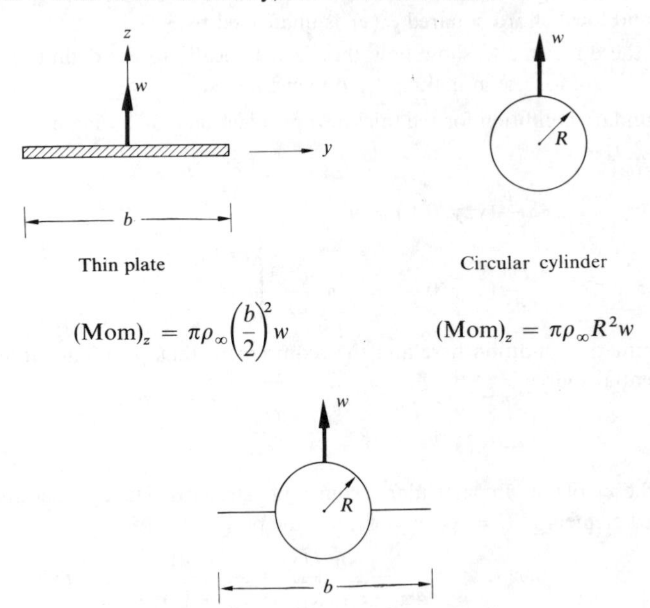

Thin plate

$$(\text{Mom})_z = \pi\rho_\infty\left(\frac{b}{2}\right)^2 w$$

Circular cylinder

$$(\text{Mom})_z = \pi\rho_\infty R^2 w$$

Cylinder with thin wing

$$(\text{Mom})_z = \pi\rho_\infty\left[\left(\frac{b}{2}\right)^2 - R^2 + \frac{4R^4}{b^2}\right]w$$

a) Show that $C_{L\alpha}$ for the narrow delta wing is $(\pi/2)$ Æ, and that $C_{L\alpha} = 2$ for a slender body of revolution when the reference area S chosen for defining lift coefficient is just the area of the body's cut-off base. Explain physically why these results appear to be independent of flight Mach number, and speculate about their limitations.

b) For some specified delta-wing body combination, estimate the force distribution along the length in a longitudinal maneuver.

c) Discuss qualitatively how this approach might be useful for studying wing wakes and effects of the vortices which peel off the leading edges of slender deltas at angle of attack.

CHAPTER 5

PROPULSIVE TERMS—ONE-DIMENSIONAL ANALYSIS OF JET AND ROCKET PROPULSION

5.1 TURBOJET OPERATION; STEADY-FLOW ENERGY EQUATION; PARTITION OF THRUST AND DRAG

Every flight vehicle except the glider relies for both performance and economy of operation on the effectiveness of its "reaction propulsion." It is equipped with one or more devices for imparting momentum to a fluid stream, which combines, in some proportions, atmospheric air and the combustion products of fuel or propellants. The detailed working of these devices makes a complicated and fascinating story. Their historical progress toward lighter weight, higher thrust, and improved efficiency has always given impetus to, but placed constraints upon, what the airframe designer could achieve. In view of their importance to our understanding and exploitation of the vehicle equations of motion, we are lucky in having a simplified way of analyzing them (at least under normal conditions near the design point), which serves quite accurately to estimate thrust and fuel consumption. We are indebted to Kerrebrock (1973) for this approach, as well as for cutting through the thicket of notation that seems to surround even elementary treatments of the field.

The turbojet will be our point of departure. Not only has it been the most popular engine of its era, but its cycle contains, as special cases, those of the ramjet and the liquid-propellant rocket. We can also study the turbofan by direct extension of our one-dimensional turbojet model. We must, however, relegate propellers to the assigned problems. Whether powered by a gas turbine or a reciprocating piston engine, the propeller should properly be viewed as a 3-D lifting surface moving on a helical path. Along with the "cascade" representation of internal rotating machines, it belongs to an advanced domain of aerodynamic theory.

From the substantial literature on jet propulsion, we single out Kerrebrock (1973), Hill and Peterson (1965), and Zucrow (1958) as textbook sources for the serious reader who wishes to pursue the subject. Shapiro (1953) is a gold mine of relevant gas-dynamic fundamentals, notably the general momentum and energy theorems which we shall review in this section.

Our first tool for idealizing turbojet performance is the *steady-flow energy equation*, which embodies the conservation laws of thermodynamic energy and momentum. It describes the state change experienced by a mass $\dot{m}$ per unit time, passing steadily through an apparatus with clearly defined "inlets" and "outlets" for the working fluid. No specification need be made about the details of processes between these stations, except to assert that a constant amount of heat $\dot{Q}$ is added per unit time and power is withdrawn at a rate $\mathbb{P}$. [$\mathbb{P}$ is called *shear work* in Hill and Peterson (1965, Section 2–2), since it is associated with shear stresses such as those that give rise to torque in a rotating shaft.] It is necessary that conditions be 1-D at each inlet (I) and outlet (II), in the sense that all entering or leaving fluid particles are moving in the same direction, with average speeds q_I and q_{II}, respectively. We remark that, in most realizations of 1-D flow, the static pressure p is truly constant across any cross section of the channel, whereas other properties are assigned their averaged values. The averaged absolute temperature and specific enthalpy h are of particular concern to us here. All such quantities will be given subscripts to identify the station at which they are observed.

For the case of a single inlet and outlet, the energy equation reads

$$\dot{m}\left[\left(h_{II} + \frac{q_{II}^2}{2}\right) - \left(h_I + \frac{q_I^2}{2}\right)\right] = \dot{Q} - \mathbb{P} \tag{5.1}$$

Every term here is, of course, expressed in mechanical energy units. We shall usually find it convenient to deal with *stagnation properties*, such as the stagnation or total enthalpy

$$h_t = h + \frac{q^2}{2} \tag{5.2}$$

To be precise, h_t is the enthalpy to which the fluid at the given station would rise, if brought to rest by an adiabatic, 1-D steady-flow process. In terms of it, (5.1) becomes

$$\dot{m}(h_{tII} - h_{tI}) = \dot{Q} - \mathbb{P} \tag{5.3}$$

These forms are derived in the cited references, as well as in any text on engineering thermodynamics (e.g., Reynolds 1965, Chapter 5). They do imply that the working fluid is a pure, homogeneous substance, a fact which causes some difficulty when we analyze devices wherein combustion is taking place. We must make the sometimes-questionable assumption that all mixing and chemical reactions come to equilibrium between each inlet and outlet. Indeed, our simplest approximation to the jet engine combustor treats it as adding only heat to the air passing through, without changing the mass flow or the composition significantly.

Figure 5.1 presents a schematic of the turbojet, pictured as if it were built up of five or six distinct, connected, 1-D components. The fluid state at each point of connection is numbered as indicated, subscript 0 ($\equiv \infty$ on some symbols in previous chapters) referring to ambient atmospheric properties. We observe that the inlet/diffuser and the exhaust nozzle are shown as parts of the engine, although the former and often portions of the latter are regarded in practice as the vehicle designer's responsibility. The *afterburner* (also called *augmenter* or *reheat device*) is shown dashed, since it is used, mainly in military engines, for temporary

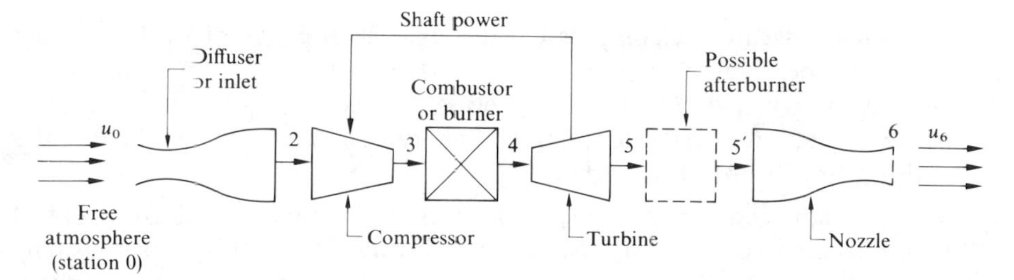

Fig. 5.1. Schematic of the elements of a turbojet engine, showing the subscript numbers used for identifying various locations during cycle analysis.

augmentation of thrust, at the penalty of greatly increased fuel consumption and noise.

The turbojet operates on the *Brayton cycle*, which, under ideal circumstances, consists of the following steps.

- 0–2, adiabatic-reversible (isentropic) compression through an inlet or diffuser, which slows down the air for delivery to the front face of the compressor. (See Fig. 5.2.)

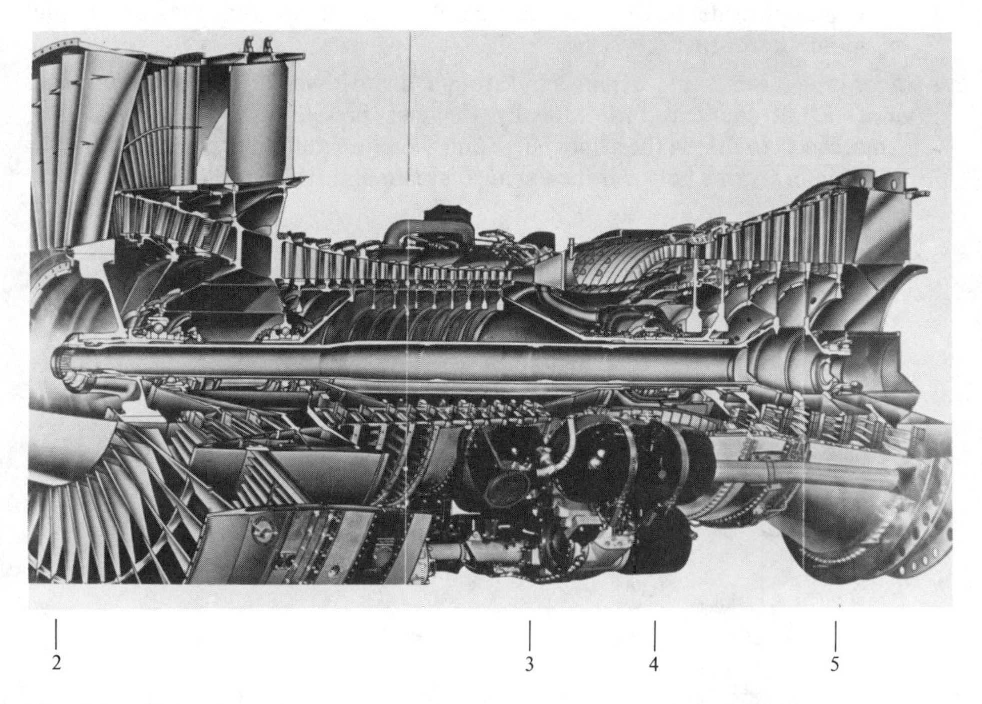

Fig. 5.2. JT9-D axial flow turbofan, cut away to reveal the details of its components. These should be examined in the light of the idealizations adopted in the text. (Reproduced courtesy of Pratt & Whitney Aircraft).

- 2–3, further isentropic compression through the addition of work $\mathbb{P}_c/\dot{m}$ per unit mass per unit time (note that $\mathbb{P}_c < 0$ because of the thermodynamic concept of positive work). This process occurs in a compressor, usually composed of many axial-flow stages of alternating fixed "stator" blade rows and rotating "rotor" blade rows.
- 3–4, heat addition at rate $\dot{Q}_B/\dot{m}$ per unit mass, by complete combustion of fuel in a burner or combustor. No work is involved, and this step is usually assumed to occur with zero pressure drop—a thermodynamic impossibility, but an acceptable approximation when the Mach number in the burner is low enough. The fuel is commonly a mixture of saturated hydrocarbons, injected at a rate $\dot{m}_f \ll \dot{m}$.
- 4–5, isentropic expansion through a turbine, which ideally withdraws work at rate $\mathbb{P}_T = -\mathbb{P}_c$ in order to drive the compressor. Power for such auxiliaries as hydraulic pumps, generators, and fuel pumps must frequently be subtracted from $\mathbb{P}_T$. The mechanical arrangement involves one or more axial-flow stages mounted on one or two shafts that connect directly and coaxially with the compressor.
- 5–5′, if applicable, further complete combustion, at constant pressure, of fuel $\dot{m}_a$ in an afterburner.
- 5 or 5′–6, isentropic expansion through an exhaust nozzle to flow speed $u_6 > u_0$ at the exit face. Ideally the exit pressure should be $p_6 = p_0$, "matched" to that in the atmosphere and therefore minimizing the irreversibility of interactions between the exhaust stream and its surroundings.

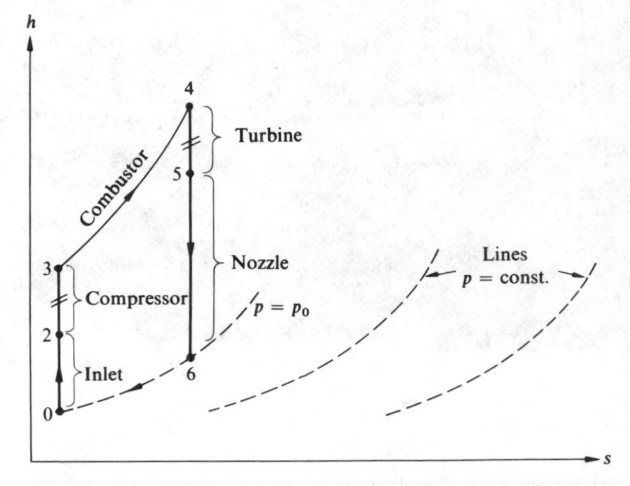

Fig. 5.3. Brayton cycle for ideal turbojet engine (without afterburner), superimposed on Mollier diagram of specific enthalpy h versus specific entropy s. For more detail, including numerical values and addition of an afterburner, see Hill and Peterson (1965, Fig. 6-12).

Figure 5.3 shows the ideal Brayton cycle overlaid on a state diagram of specific enthalpy h versus specific entropy s. Shown dashed are lines of constant pressure, which for perfect gas are simply exponential curves displaced parallel to one another and with the lower pressures on the right. Adiabatic-reversible processes occur along vertical lines. You will find it useful to write (5.1) and (5.3) for each step. In the absence of power for auxiliaries, the lines 2–3 and 4–5 turn out to have equal lengths. Moreover, if the states at 2, 3, 4, 5 represent stagnation, then the excess in length of 5–6 over 0–2 will be seen to account for u_6 being greater than u_0; this difference is the principal source of turbojet thrust. In Section 5.2, we shall analyze this situation in more detail and develop relations between performance and the important parameters that characterize this "loss-free" engine.

We need first to say something about the thrust and efficiencies of reaction propulsion systems. Let us consider any ducted machine, which somehow accelerates a 1-D mass flow $\dot{m}$ from speed u_0 in the undisturbed stream to u_6 at the nozzle exit. We study its momentum balance, in a frame attached to the machine, by reference to the control volume illustrated in Fig. 5.4. The lateral boundary of

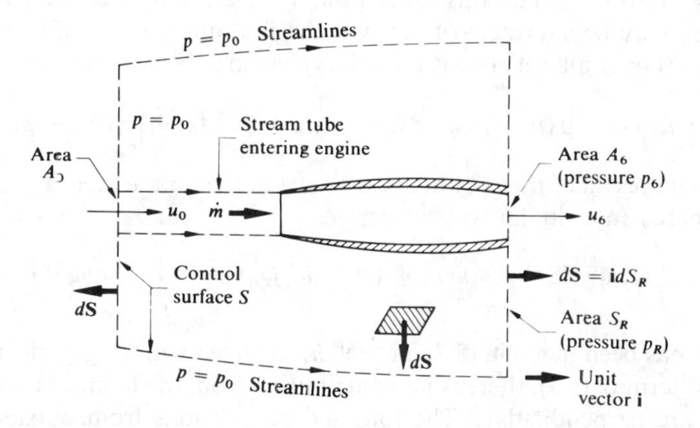

Fig. 5.4. Control volume used to study thrust and drag of a reaction propulsion system. (See text for discussion.)

this volume is a stream surface so remote that the pressure is p_0 ($\equiv p_\infty$) everywhere. It is bounded in front by a plane normal to u_0 and so far ahead that $p = p_0$ also there. Given that the area cut out of this plane by the stream tube entering the inlet is A_0, then clearly

$$\dot{m} = \rho_0 u_0 A_0 \tag{5.4}$$

The aft boundary is another normal plane, so placed that it contains area A_6 of the exit face. Excluded from the volume are all mechanical parts of the propulsion system and duct. We also neglect any small angle of attack between the axis and direction of flight.

We tentatively define two components of the streamwise force applied by the system to the fluid as follows:

D (*nacelle drag*) is the upstream resultant of all pressure and frictional forces exerted across the outside of the duct or nacelle.

T (*net thrust*) is the downstream resultant of all forces exerted by everything inside on the gas which passes through.

Conservation of streamwise momentum (cf. Shapiro 1953, Chapter 2) requires that

$$T - D - \int_S p\mathbf{i} \cdot d\mathbf{S} = \int_S u(\rho\mathbf{q} \cdot d\mathbf{S}) \tag{5.5}$$

Here $\mathbf{i}$ is a unit vector in a direction opposite to the flight direction and u the corresponding velocity component. Over the boundary S of our control volume, each area element dS has an outward vector $d\mathbf{S}$ associated with it.

It is convenient to assume that the aft lip of the duct is sharp, like a wing trailing edge, although we could account for blunting or truncation by introducing "base drag" into D. Under this restriction, S is essentially a closed surface, and a constant p_0 may be subtracted from p in the first integral of (5.5). This integral then vanishes over all but the aft boundary, leaving

$$-\int_S p\mathbf{i} \cdot d\mathbf{S} = -\oint_S (p - p_0)\mathbf{i} \cdot d\mathbf{S} = -(p_6 - p_0)A_6 - \int_{S_R} (p_R - p_0)\, dS_R \tag{5.6}$$

where S_R denotes all of that boundary external to the nozzle area A_6, and p_R is the pressure there. In a similar way we derive

$$\int_S u(\rho\mathbf{q} \cdot d\mathbf{S}) = -\dot{m}u_0 + (\dot{m} + \dot{m}_f)u_6 + \int_{(S - A_0 - A_6)} \rho u\mathbf{q} \cdot d\mathbf{S} \tag{5.7}$$

Allowance has been made in (5.7) for fuel $\dot{m}_f$, injected with negligible momentum. In the final term of (5.7), there is no contribution from the lateral boundary where $\mathbf{q}$ and $d\mathbf{S}$ are perpendicular. The fore and aft portions from outside the "slip-stream" do not generally cancel, however, since we expect some momentum defect to be produced by the drag.

We substitute (5.6, 7) into (5.5) and transfer all quantities except $(T - D)$ to the right. At this point, an arbitrary decision is made about what constitutes thrust and what drag. Customarily, the various terms are assigned in the following way.

$$T = (p_6 - p_0)A_6 + (\dot{m} + \dot{m}_f)u_6 - \dot{m}u_0 \tag{5.8}$$

$$D = \int_{S_R} (p_0 - p_R)\, dS_R - \int_{(S - A_0 - A_6)} \rho u\mathbf{q} \cdot d\mathbf{S} \tag{5.9}$$

Equation (5.8) would be specialized for ideal operation by neglecting the added fuel mass and setting $p_6 = p_0$,

$$\boxed{\text{Ideal } T \cong \dot{m}(u_6 - u_0)} \tag{5.10}$$

It is mentioned that the name *net thrust* is applied to T as envisioned in (5.8) and (5.10). *Gross thrust* F does not include the ram term $\dot{m}u_0$; when examining the literature, take care as to which is meant.

A few comments should be appended here about the correctness and applicability of (5.8) and (5.9). Evidently their derivation comprehends a device, such as a ducted fan, turbojet, or ramjet, which acts on a 1-D streamtube and is housed in some isolated body resembling a nacelle on a jet transport. It takes only a modest stretch of the imagination, however, to generalize Fig. 5.4 to cover one or more engines buried within the fuselage or wing of a complete airplane; D would then become the *vehicle drag*. Something like a turbofan, with a pair of coaxial slipstreams, might be handled with four rather than two momentum-flux terms on the right of (5.8). The rocket motor represents the limit at which $\dot{m} = 0$ and $\dot{m}_f$ is interpreted as the entire rate of propellant utilization.

The unshrouded propeller falls outside our present framework, but its performance has been analyzed with some success by an analogous idealization based on the "actuator disc" concept (cf. Glauert 1926 or Rauscher 1953). In fact, we could apply (5.10) to a propeller by defining u_6 as the average speed back at a station in the slipstream at which $p_6 = p_0$.

The (5.8, 9) partition between T and D would agree closely with our initial definitions if there were no "spillage," that is, if the upstream mass flux $\dot{m}$ occupied the full inlet area and was divided from the outer flow by a stagnation ring at the inlet leading edge. But there are off-design conditions in which this is far from the case; e.g., at low forward speeds a jet pumps much more $\dot{m}$ than is contained within a tube extended forward from the inlet. Higher "losses" are experienced, because regions of separated flow can be encountered, either inside the diffuser or around the outer nacelle surface. The definitions of thrust and drag often become controversial. Arguments between the engine and airframe manufacturers tend to be exacerbated by the great expense and uncertainty of all methods that have been tried to date for measuring T or D separately in flight.

The idea of *work efficiency* of a reaction propulsion system must be approached with some caution. When operating a rocket in space, for example, the apparent power output can be made almost anything one wishes, since there exists no uniquely appropriate frame of observation. Within the sensible atmosphere, however, it is natural to calculate useful work or *thrust horsepower* relative to the air at rest, so that it is proportional to Tu_0. Similarly, the chemical energy available from perfect combustion of the fuel constitutes a suitable "input" or denominator for an overall efficiency expression. Thus we arrive at

$$\eta_{\text{Overall}} = \frac{Tu_0}{\dot{m}_f h_f} \tag{5.11}$$

where h_f, in mechanical-energy units, is the *heating value* or specific enthalpy increase for the completed fuel–air reaction.

η_{Overall} is often divided into a mechanical or *propulsive* factor and a *thermal* factor. These would be defined, respectively, by

$$\eta_p \equiv \frac{(\text{Power delivered to vehicle})}{\begin{bmatrix}(\text{Power delivered to vehicle}) + \\ (\text{Kinetic energy of slipstream air})\end{bmatrix}}$$

$$= \frac{Tu_0}{Tu_0 + \frac{1}{2}(\dot{m} + \dot{m}_f)(u_6 - u_0)^2} \tag{5.12}$$

and

$$\eta_{\text{th}} \equiv \frac{\begin{bmatrix}(\text{Power delivered to vehicle}) + \\ (\text{Kinetic energy of slipstream air})\end{bmatrix}}{(\text{Heating value of fuel used per unit time})}$$

$$= \frac{Tu_0 + \frac{1}{2}(\dot{m} + \dot{m}_f)(u_6 - u_0)^2}{\dot{m}_f h_f} \tag{5.13}$$

We gain physical insight by adopting in (5.12, 13) the approximations (5.10) and $\dot{m}_f \ll \dot{m}$. A little manipulation yields

$$\eta_p \cong \frac{2u_0}{u_6 + u_0} \tag{5.14}$$

$$\eta_{\text{th}} \cong \frac{u_0(u_6 - u_0)}{fh_f} \tag{5.15}$$

where $f \equiv (\dot{m}_f/m)$ is the *fuel–air ratio* of the engine. For atmospheric airbreathing devices, η_{th} tends to be less than 30%, and there appears to be little chance of appreciable improvements. On the other hand, (5.14) shows that mechanical efficiency can be brought close to 100% by measures that minimize u_6. We can reason that the best way to do this is to act on the greatest possible mass flow. The most mechanically efficient choice would always seem to be a large-diameter propeller, but unfortunately its merits are far outweighed above 300–350 mph by the aerodynamic noise and the power losses due to its tips moving through the air at transonic or supersonic speeds. On recent designs a trend can be seen back toward the propeller's advantages with the appearance of turbofans having increasing bypass ratios.

5.2 ONE-DIMENSIONAL ANALYSIS OF IDEAL RAMJET AND TURBOJET

Three quantities are in common usage to characterize the performance of a given engine. The first of these is the thrust itself, for which we adopt here the formula (5.10). All that is required for T are the flight speed and estimates of mass flow and exhaust speed.

The second and third performance indices are, except for a difference in units, the inverses of one another. For general comparisons between systems, we favor the *specific impulse*, defined by

$$\boxed{I_{sp} \equiv \frac{T}{\dot{m}_f}} \tag{5.16}$$

Unlike power and the efficiencies defined in Section 5.1, I_{sp} is independent of the frame of observation. It tells how much impulse—thrust-force multiplied by time of application—can be obtained per unit mass of propellant. The logical dimensions are force divided by mass per unit time, which combination is equivalent to a velocity (for example, ft-sec^{-1}). Indeed, on a rocket engine we shall see that I_{sp} equals the average downstream velocity of the exhaust gases. In practice, however, this quantity is quoted in pounds of force divided by pounds of mass per second, units which are often described as "seconds." Actually, it should be multiplied by the standard value of gravitational acceleration in order to recover I_{sp} as a velocity.

The *specific fuel consumption*, often associated with jet engines, is the inverse

$$\boxed{\text{SFC} \equiv \frac{1}{I_{sp}} = \frac{\dot{m}_f}{T}} \tag{5.17}$$

Evidently, it shows how much propellant or fuel must be burned per unit time for each unit of thrust delivered. It is generally quoted in pounds of fuel per hour per pound of thrust, a choice which has the advantage of yielding numbers near unity for typical turbojets. High I_{sp}, and correspondingly low SFC, are primary goals of the engine designer.

Let us begin by estimating these three properties for the ideal ramjet. As is apparent from the schematic in Fig. 5.5, this device resembles a turbojet with the

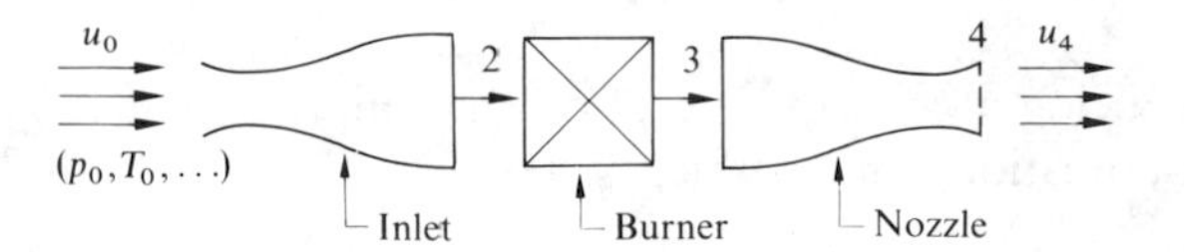

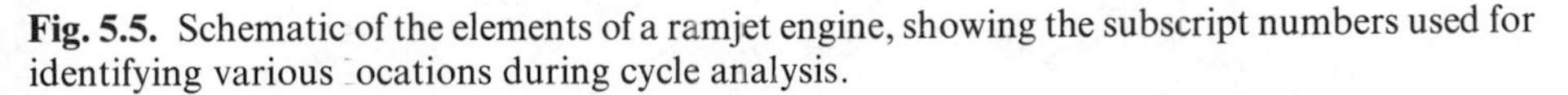

Fig. 5.5. Schematic of the elements of a ramjet engine, showing the subscript numbers used for identifying various locations during cycle analysis.

turbine and compressor removed. It relies on rapid motion through the air to produce "ram compression" and thus make possible combustion at the high pressures needed for effective operation. Following Kerrebrock (1973), we examine the ratios of pressure and absolute temperature across each component of the engine. For simplicity's sake, we also acquiesce in his rather extreme approximation that the working fluid behaves throughout as a perfect gas with constant γ

and c_p. This step allows us to use two formulas governing 1-D steady motion of such a gas:

- For isentropic expansion or compression (cf. 4.120),

$$\frac{p_t}{p} = \left(1 + \frac{\gamma - 1}{2} M^2\right)^{\gamma/(\gamma - 1)} \tag{5.18}$$

where M is Mach number at a station with static pressure p.

- For adiabatic or isentropic expansion or compression,

$$\frac{T_t}{T} = 1 + \frac{\gamma - 1}{2} M^2 \tag{5.19}$$

(5.19) is just a specialization of (5.3), with $\dot{Q} = 0 = \mathbb{P}$, $\Delta h = c_p \Delta T$, and $a^2 = \gamma RT$.

Symbols π and τ, respectively, with appropriate subscripts, denote ratios of stagnation pressure and stagnation temperature. Their values would both be unity, for instance, across an ideal inlet, since it is supposed to produce isentropic compression without work or heat addition.

Starting from the free stream at M_0, p_0, and T_0, we adapt the foregoing equations and definitions to the three processes depicted in Fig. 5.5.

- *The Inlet*

$$\frac{p_{2t}}{p_0} = \left(1 + \frac{\gamma - 1}{2} M_0^2\right)^{\gamma/(\gamma - 1)} \tag{5.20}$$

$$\frac{T_{2t}}{T_0} = 1 + \frac{\gamma - 1}{2} M_0^2 \tag{5.21}$$

- *The Burner*

$$\pi_b \equiv \frac{p_{3t}}{p_{2t}} = 1 \tag{5.22}$$

this value being a consequence of our assumption that the drop in pressure during combustion is negligible.

$$\tau_b \equiv \frac{T_{3t}}{T_{2t}} > 1 \tag{5.23}$$

τ_b is a basic parameter of ramjet operation; it is determined by the fuel–air ratio and limited by the capability of available materials to withstand loading at high temperatures.

- *The Exhaust Nozzle*

$$\frac{p_{4t}}{p_4} = \frac{p_{3t}}{p_4} = \left(1 + \frac{\gamma - 1}{2} M_4^2\right)^{\gamma/(\gamma - 1)} \tag{5.24}$$

$$\frac{T_{4t}}{T_4} = \frac{T_{3t}}{T_4} = 1 + \frac{\gamma - 1}{2} M_4^2 \tag{5.25}$$

We have previously required that this nozzle be designed for matched operation, so that $p_4 = p_0$. In view of our finding that $p_{4t} = p_{3t} = p_{2t}$, it follows from (5.20) and (5.24) that

$$\left(1 + \frac{\gamma - 1}{2} M_0^2\right)^{\gamma/(\gamma-1)} = \left(1 + \frac{\gamma - 1}{2} M_4^2\right)^{\gamma/(\gamma-1)} \tag{5.26}$$

and therefore that

$$M_4 = M_0 \tag{5.27}$$

This remarkable result says that the exit Mach number equals (or, on an actual ramjet, is less than) the M_0 at which the vehicle flies. We conclude that the only way u_4 can exceed u_0 is through an increase in sound speed due to heating. The rise in slipstream temperature is given by

$$\frac{T_4}{T_0} = \frac{T_{4t}\left(1 + \frac{\gamma - 1}{2} M_0^2\right)}{T_{2t}\left(1 + \frac{\gamma - 1}{2} M_4^2\right)} = \frac{T_{4t}}{T_{2t}} = \frac{T_{3t}}{T_{2t}} = \tau_b \tag{5.28}$$

Because a^2 is proportional to temperature in the perfect gas,

$$\frac{u_4}{u_0} = \frac{M_4 a_4}{M_0 a_0} = \sqrt{\frac{T_4}{T_0}} = \sqrt{\tau_b} \tag{5.29}$$

From (5.10), with u_6 replaced by u_4, the ideal thrust is

$$\boxed{T = \dot{m} u_0(\sqrt{\tau_b} - 1)} \tag{5.30}$$

Near the design point, which is always supersonic for ramjets, the inlet is often configured so as to capture nearly all the air in a tube which is the forward projection of its entry area A_I. One such arrangement consists of a circular duct with a coaxial conical centerbody; for "internal compression," this cone would be pushed forward to the point at which its conical shock passed slightly ahead of the diffuser leading edge. In such a case the mass flow would be

$$\dot{m} \cong \rho_0 u_0 A_I \tag{5.31}$$

(5.30) and (5.31) then express thrust in terms of the single engine parameter τ_b.

To determine specific impulse, we first apply (5.3) to the hypothetically complete combustion process. The ramjet has no moving parts capable of exchanging power with the fluid, and

$$\dot{Q} = \dot{m}_f h_f = f \dot{m} h_f \tag{5.32}$$

in terms of previously defined quantities. Since enthalpy is assumed proportional to temperature of the gas, we generalize (5.3) to include the mass of the fuel and obtain

$$\dot{m}c_pT_{2t} + \dot{m}_fh_f = (\dot{m} + \dot{m}_f)c_pT_{3t} \tag{5.33}$$

After substitution of (5.21) and division by $\dot{m}c_pT_{2t}$, we get

$$1 + \frac{fh_f}{c_pT_0\left(1 + \frac{\gamma - 1}{2}M_0^2\right)} = (1 + f)\tau_b \tag{5.34}$$

The fuel–air ratio is always just a few percent ($f \ll 1$), so a close approximation to f would be

$$f \cong \frac{\tau_b - 1}{h_f\Big/\left[c_pT_0\left(1 + \frac{\gamma - 1}{2}M_0^2\right)\right]} \tag{5.35}$$

(5.16), (5.30), and (5.35) lead finally to

$$\boxed{I_{sp} = \frac{T}{f\dot{m}} = \frac{u_0h_f}{c_pT_0\left(1 + \frac{\gamma - 1}{2}M_0^2\right)}\left(\frac{\sqrt{\tau_b} - 1}{\tau_b - 1}\right)} \tag{5.36}$$

SFC can be found from the inverse, (5.17). [We remark that (5.33) and the succeeding formulas turn out the same whether we regard the heating as coming from some "external source" or as an enthalpy rise due to the mixture of air and fuel vapor, in proportions $1:f$, proceeding to reactive equilibrium.]

A meaningful way to study "full-throttle" ramjet operation is to think of the fuel consumption as being governed by metallurgical considerations, which set an upper limit on the temperature T_{3t}. For flight in the stratosphere region of the standard atmosphere, where $T_0 = 390°\text{R}$, fixed T_{3t} implies that the parameter

$$\tau_\lambda \equiv \frac{T_{3t}}{T_0} = \tau_b\left(1 + \frac{\gamma - 1}{2}M_0^2\right) \tag{5.37}$$

is also constant. Accordingly τ_b must drop as M_0 increases, up to a theoretical maximum flight Mach number at which $\tau_b \to 1$ and $T \to 0$ as a consequence of the fuel being shut off. At the other end of the speed scale, T vanishes when $u_0 \to 0$ because of the inlet's inability to store any mechanical energy of ram compression. The specific impulse also goes to zero there, irrespective of how much fuel the engine might try to burn.

For these reasons the ramjet has seen service mainly as a "sustainer" or second-stage booster on vehicles which had other propulsive devices for acceleration up to the point at which it is efficient. The Bomarc winged interceptor missile was a very successful early application. The ramjet tends to be competitive in an intermediate range of speeds and altitudes, above those of the turbojet, whose metallurgical constraints are more severe because of the heated moving parts, but below where the rocket, with its independence of atmospheric air, can go. During the late 1960's, the technology received a stimulus from demonstrations that ramjet combustion is possible in a stream of fluid moving supersonically past the burner (see Ferri 1968). This development holds promise of reducing high-temperature materials problems, thus making possible larger T_{3t} and operation at M_0 in the range 10–20. The point is not that the stagnation temperature is reduced, but that rates of heat transfer to the burner walls are much less in the lower static pressures associated with supersonic combustion.

Figure 5.6 plots ideal thrust per unit mass flow and dimensionless I_{sp} versus M_0 in the stratosphere for a representative subsonic-burning ramjet. The parameter from (5.37) is assigned constant value $\tau_\lambda = 9$, corresponding to a $T_{3t} = 3510°R$, which is quite conservative by today's standards. The peak M_0 at which fuel flow must cease is about 6.32 in this case. Interestingly, we can show from (5.36) that I_{sp} has a maximum value at $M_0 = 3.59$, but remains finite through $\tau_b = 1$; the portion of the curve above this limit is, of course, a mathematical curiosity. It is also easily proved that $T/\dot{m}$ has its maximum at

$$(M_0)_{\max} = \sqrt{\frac{2}{\gamma - 1}(\tau_\lambda^{1/3} - 1)} = 2.32 \tag{5.38}$$

Since the mass flow itself would be a function of M_0, however, we expect peak thrust to occur at some higher speed.

We can now carry out a parallel examination of the ideal turbojet, without afterburning, by just adding a compressor and turbine to what has already been done. Turning to the schematic in Fig. 5.1, we observe that (5.20, 21) describe the inlet directly. The remaining components require some new definitions and analysis, as follows.

- *The Compressor*

$$\tau_c \equiv \frac{T_{3t}}{T_{2t}} > 1, \qquad \pi_c = \frac{p_{3t}}{p_{2t}} \tag{5.39a, b}$$

Because the process is isentropic,

$$\tau_c = \pi_c^{(\gamma-1)/\gamma} \tag{5.40}$$

- *The Burner*

$$\tau_b \equiv \frac{T_{4t}}{T_{3t}}, \qquad \pi_b \equiv \frac{p_{4t}}{p_{3t}} = 1 \tag{5.41a, b}$$

- *The Turbine*

$$\tau_T \equiv \frac{T_{5t}}{T_{4t}} < 1, \qquad \pi_T \equiv \frac{p_{5t}}{p_{4t}} \tag{5.42a, b}$$

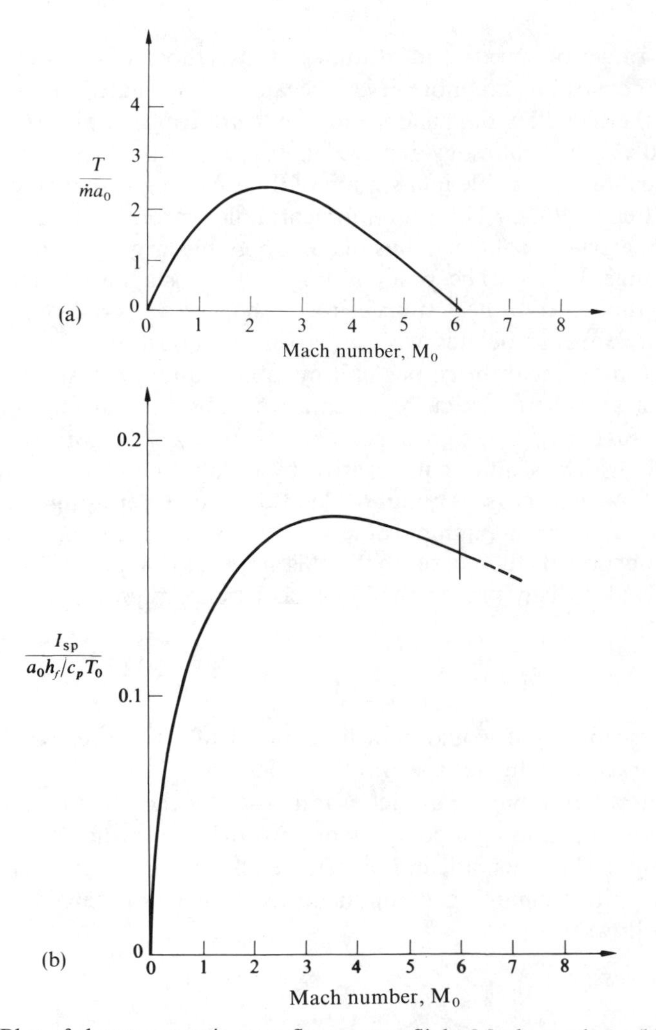

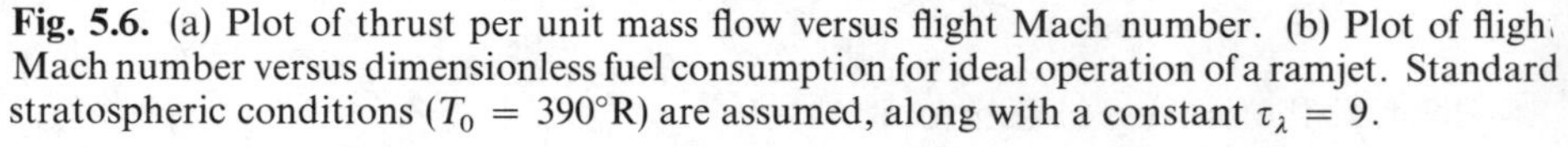

Fig. 5.6. (a) Plot of thrust per unit mass flow versus flight Mach number. (b) Plot of fligh. Mach number versus dimensionless fuel consumption for ideal operation of a ramjet. Standard stratospheric conditions ($T_0 = 390°\text{R}$) are assumed, along with a constant $\tau_\lambda = 9$.

Again, isentropy at constant γ gives

$$\tau_T = \pi_T^{(\gamma-1)/\gamma} \tag{5.43}$$

- *The Nozzle*

$$\frac{p_{6t}}{p_6} = \frac{p_{5t}}{p_6} = \left(1 + \frac{\gamma - 1}{2}\,\mathrm{M}_6^2\right)^{\gamma/(\gamma-1)} \tag{5.44}$$

$$\frac{T_{6t}}{T_6} = \frac{T_{5t}}{T_6} = 1 + \frac{\gamma - 1}{2}\,\mathrm{M}_6^2 \tag{5.45}$$

By cascading (5.20, 21) and the foregoing relations, we can relate the inlet and exit temperatures and pressures,

$$T_6\left(1 + \frac{\gamma - 1}{2}M_6^2\right) = T_{5t} = T_{2t}\tau_c\tau_b\tau_T = T_0\left(1 + \frac{\gamma - 1}{2}M_0^2\right)\tau_c\tau_b\tau_T \quad (5.46)$$

$$p_6\left(1 + \frac{\gamma - 1}{2}M_6^2\right)^{\gamma/(\gamma-1)} = p_{2t}\pi_c\pi_b\pi_T = p_0\left(1 + \frac{\gamma - 1}{2}M_0^2\right)^{\gamma/(\gamma-1)}\pi_c\pi_T \quad (5.47)$$

The condition of matched pressures permits $p_6 = p_0$ to be canceled from (5.47). We raise the resulting equation to the power $(\gamma - 1)/\gamma$, substitute (5.40) and (5.43), and obtain

$$1 + \frac{\gamma - 1}{2}M_6^2 = \left(1 + \frac{\gamma - 1}{2}M_0^2\right)\tau_c\tau_T \quad (5.48)$$

Dividing (5.46) by (5.48) then produces the simple result

$$\frac{T_6}{T_0} = \tau_b \quad (5.49)$$

which [cf. (5.28)] relates combustion with the slipstream temperature rise exactly as for the ramjet.

To calculate thrust we also need the ratio of slipstream speeds, given by

$$\frac{u_6}{u_0} = \frac{M_6}{M_0}\sqrt{\frac{T_6}{T_0}} \quad (5.50)$$

From (5.48) we derive the ratio of Mach numbers

$$\frac{M_6}{M_0} = \sqrt{\frac{\left(1 + \frac{\gamma - 1}{2}M_0^2\right)\tau_c\tau_T - 1}{\left(\frac{\gamma - 1}{2}\right)M_0^2}} \quad (5.51)$$

whereupon (5.49)–(5.51) yield

$$\frac{u_6}{u_0} = \sqrt{\frac{2\tau_b}{(\gamma - 1)M_0^2}\left[\left(1 + \frac{\gamma - 1}{2}M_0^2\right)\tau_c\tau_T - 1\right]} \quad (5.52)$$

The ideal expression (5.10) therefore becomes

$$T = \dot{m}u_0\left(\frac{u_6}{u_0} - 1\right)$$

$$= \dot{m}u_0\left\{\sqrt{\frac{2\tau_b}{(\gamma - 1)M_0^2}\left[\left(1 + \frac{\gamma - 1}{2}M_0^2\right)\tau_c\tau_T - 1\right]} - 1\right\} \quad (5.53)$$

The key piece of information that we have not yet incorporated is the relationship between turbine and compressor powers. Steady-flow energy, (5.3), tells us that $|\mathbb{P}_c|$ is proportional to $(T_{3t} - T_{2t})$ and $\mathbb{P}_T$ is proportional to $(T_{4t} - T_{5t})$ when we neglect c_p-variations through these devices. In the absence of any drain for auxiliaries, it follows from the equality of these powers, by division with T_{2t}, etc., that

$$\frac{T_{3t}}{T_{2t}} - 1 = \frac{T_{4t} - T_{5t}}{T_{2t}} \tag{5.54a}$$

In terms of the total-temperature ratios, we can manipulate (5.54a) to produce

$$\tau_T = 1 - \frac{\tau_c - 1}{\tau_c \tau_b} \tag{5.54b}$$

If we regard τ_c and τ_b as the primary independent parameters, we can eliminate τ_T from (5.53) through (5.54b).

$$\boxed{T = \dot{m}u_0\left\{\sqrt{\left(1 + \frac{2}{(\gamma - 1)M_0^2}\right)(\tau_b - 1)(\tau_c - 1) + \tau_b} - 1\right\}} \tag{5.55}$$

(5.55) is checked by its reduction to the ramjet, (5.30), when $\tau_c = 1$.

We must be able to demonstrate the turbojet's capability for generating "static thrust" at $u_0 = 0$, and we do this by inserting $u_0 = M_0 a_0$ in (5.55) and taking the limit as $M_0 \to 0$.

$$\boxed{T_{\text{Static}} = \dot{m}a_0\sqrt{\left(\frac{2}{\gamma - 1}\right)(\tau_b - 1)(\tau_c - 1)}} \tag{5.56}$$

The static mass flow does not vanish in (5.56). It must be estimated, however, by analyzing the 3-D pumping characteristics of the compressor in the absence of inlet ram compression.

Further insight can be gathered, as with the ramjet, by defining

$$\tau_\lambda \equiv \frac{T_{4t}}{T_0} = \tau_c \tau_b \left(1 + \frac{\gamma - 1}{2} M_0^2\right) \tag{5.57}$$

For a given level of design refinement of a given class of engines, operating at the stratospheric temperatures at which most turbojets perform best, a nearly constant τ_λ serves as a sort of figure of merit. Thus the turbojets on early generations of subsonic transports have turbine inlet temperatures ranging from about 1800°F to 2400°F. More recent designs, including the "gas-generator" cores of turbofans, have reached nearly 2800°F without elaborate provisions for cooling the turbine blades, whereas advanced materials and cooling schemes have been demonstrated to above 4000°F. With $T_0 \cong 390$°R, these numbers suggest a range $\tau_\lambda = 6$ to 11 or so for design-point operation. On the other hand, flight near sea level in the tropical summer can reduce τ_λ below 4 in some cases.

The substitution of τ_λ from (5.57) in place of τ_b reflects the fact that τ_c tends to be an inherent characteristic of a particular design, whereas full-throttle τ_b is determined by fuel-flow controls which will not permit T_{4t} to exceed a prescribed limit.

Evidently (5.55) and (5.56) become

$$T = \dot{m}u_0\left\{\left[\left(1 + \frac{2}{(\gamma - 1)M_0^2}\right)(\tau_c - 1)\left(\left[\frac{\tau_\lambda}{\tau_c\left(1 + \frac{\gamma - 1}{2}M_0^2\right)}\right] - 1\right) + \frac{\tau_\lambda}{\tau_c\left(1 + \frac{\gamma - 1}{2}M_0^2\right)}\right]^{1/2} - 1\right\} \tag{5.58}$$

$$T_{\text{Static}} = \dot{m}a_0\sqrt{\left(\frac{2}{\gamma - 1}\right)(\tau_c - 1)\left(\frac{\tau_\lambda}{\tau_c} - 1\right)} \tag{5.59}$$

To predict fuel utilization, we adopt the same reasoning underlying (5.33) and write

$$\dot{m}c_pT_{3t} + \dot{m}_fh_f = (\dot{m} + \dot{m}_f)c_pT_{4t} \tag{5.60}$$

admitting here the special weakness of assuming that c_p has everywhere the same value as in ambient air. From formulas for the evolution of engine temperatures, we take

$$T_{3t} = \tau_c\left(1 + \frac{\gamma - 1}{2}M_0^2\right)T_0 \tag{5.61a}$$

and

$$T_{4t} = \tau_c\tau_b\left(1 + \frac{\gamma - 1}{2}M_0^2\right)T_0 \tag{5.61b}$$

Together with the approximation that the fuel–air ratio $f \ll 1$, we can use (5.61) in (5.60) to solve for

$$\begin{aligned} f &\cong \frac{c_pT_0}{h_f}\left(1 + \frac{\gamma - 1}{2}M_0^2\right)\tau_c(\tau_b - 1) \\ &= \frac{c_pT_0}{h_f}\left(1 + \frac{\gamma - 1}{2}M_0^2\right)\tau_c\left[\frac{\tau_\lambda}{\tau_c\left(1 + \frac{\gamma - 1}{2}M_0^2\right)} - 1\right] \end{aligned} \tag{5.62}$$

In terms of τ_λ, the specific impulse from (5.16) then turns out to be

$$\begin{aligned} I_{sp} &= \frac{T}{f\dot{m}} \\ &= \frac{u_0h_f}{c_pT_0}\left\{\frac{\sqrt{\left(1 + \frac{2}{(\gamma - 1)M_0^2}\right)(\tau_c - 1)\left[\frac{\tau_\lambda}{\tau_c\left(1 + \frac{\gamma - 1}{2}M_0^2\right)} - 1\right] + \frac{\tau_\lambda}{\tau_c\left(1 + \frac{\gamma - 1}{2}M_0^2\right)}} - 1}{\tau_\lambda - \tau_c\left(1 + \frac{\gamma - 1}{2}M_0^2\right)}\right\} \end{aligned} \tag{5.63}$$

By a limiting process similar to that behind (5.56), we calculate the value for zero forward speed,

$$(I_{sp})_{\text{Static}} = \frac{a_0 h_f}{c_p T_0}\left[\frac{\sqrt{\dfrac{2}{\gamma - 1}(\tau_c - 1)\left(\dfrac{\tau_\lambda}{\tau_c} - 1\right)}}{\tau_\lambda - \tau_c}\right] \tag{5.64}$$

As before, specific fuel consumption SFC is found by inverting the right sides of (5.63) or (5.64).

Provided that reasonable estimates of $\dot{m}$, τ_c, τ_λ, and other quantities are available, (5.58) and (5.63) are suitable for studying the effects of changing altitude and M_0, partial-throttle operation, different types of fuel, and the like. Since we have hypothesized here that all turbojet components are functioning as well as they possibly can, rough upper bounds are thus determined for T and I_{sp} associated with a given set of parameters.

Now addressing the subject of the *design point*, we can think of τ_λ, M_0, $\dot{m}u_0$ and $(u_0 h_f/c_p T_0)$ as fixed quantities. It is logical to ask whether there is a choice of compressor—that is, of τ_c and π_c—which is in some sense the optimum. In the assigned problems, typical plots of (5.58) and (5.63) will be constructed. Here we remark only that increasing τ_c, with other quantities held fixed, causes the ideal I_{sp} to go up monotonically but leads to a maximum of T (or $T/\dot{m}$). We can differentiate the radical in (5.58) partially with respect to τ_c, equate to zero, and find a peak in thrust versus τ_c at

$$(\tau_c)_{\max} = \frac{\sqrt{\tau_\lambda}}{1 + \dfrac{\gamma - 1}{2} M_0^2} \tag{5.65a}$$

In view of (5.57), this can also be expressed as

$$(\tau_c)_{\max} = \frac{\tau_b}{1 + \dfrac{\gamma - 1}{2} M_0^2} \tag{5.65b}$$

The corresponding pressure ratio follows by raising (5.65) to the power $\gamma/(\gamma - 1)$. Normally $(\tau_c)_{\max}$ from (5.65) is greater than unity; if not, the analysis suggests that there is no need for a compressor and that ramjet propulsion should be investigated for this flight condition.

When (5.65a) is inserted into (5.58), simplification yields

$$T_{\max} = \dot{m}u_0\left\{\sqrt{1 + \frac{2}{(\gamma - 1)M_0^2}[\sqrt{\tau_\lambda} - 1]^2} - 1\right\} \tag{5.66}$$

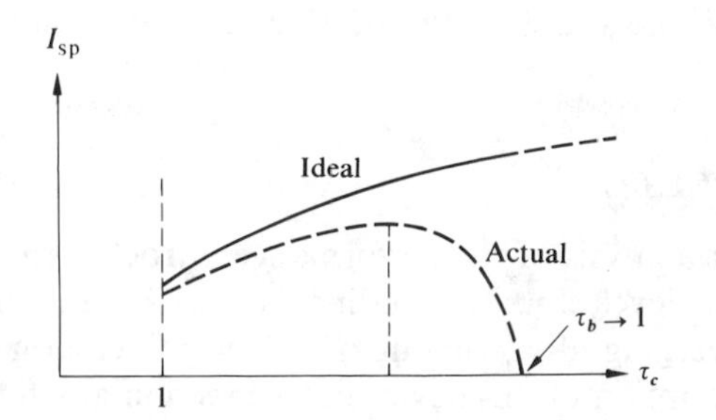

Fig. 5.7. Qualitative variation of specific impulse with ratio of compressor temperature τ_c for idealized and actual turbojet cycles.

The corresponding specific impulse *at this design point only* is

$$(I_{sp})_{T_{max}} = \frac{u_0 h_f}{c_p T_0} \left\{ \frac{\sqrt{1 + \frac{2}{(\gamma - 1)M_0^2}(\sqrt{\tau_\lambda} - 1)^2}}{\sqrt{\tau_\lambda}(\sqrt{\tau_\lambda} - 1)} - 1 \right\} \tag{5.67}$$

Figure 5.7 sketches qualitatively the variations of ideal T and I_{sp} versus τ_c. The dashed line indicates the way in which I_{sp} might be expected to depend on compressor size for an actual engine, with losses accounted for. Both ideal T and actual I_{sp} fall to zero near a value of τ_c given by

$$\tau_c\left(1 + \frac{\gamma - 1}{2} M_0^2\right) = 1$$

which for the specified M_0 indicates that the limit on T_{4t} is forcing the fuel flow to be shut off. (The data on ideal I_{sp} at τ_c's greater than this are, of course, meaningless.) What is significant is that the peak in the actual I_{sp} curve usually falls at considerably higher τ_c than the peak in either the ideal or the actual curve of thrust. The designer therefore has the option of selecting a relatively small number of compressor stages, and thus creating a relatively lightweight turbojet optimized for high thrust performance, or of going to a larger, heavier compressor which will improve fuel consumption. Heavy engines with "good specifics" are favored for long-range cruising aircraft; they are typified particularly by certain early British and Russian models.

Small, high-thrust designs are attractive for such applications as lift engines for VTOL aircraft, in which the maximum attainable ratio of static thrust to weight is the vital consideration. Sacrifices in terms of greater fuel consumption are acceptable, because such an engine is operated for only a brief period until the vehicle accelerates forward and the wing takes over the lifting function. The development of successful turbofans has rendered this tradeoff an oversimplification, however, for the designer now has bypass ratio as another parameter which he can manipulate in the search for a "best" combination among thrust, weight, and I_{sp}.

5.3 ROCKET PROPULSION

As we approach our analysis of the performance of rocket engines, we can come close to achieving the level of understanding we seek by thinking in terms of 1-D flow through a converging–diverging nozzle. The principal information that we must add to the description of this device can be read in any fluid-mechanics text, and concerns how the state of high pressure and temperature p_c, T_c is produced in the "reservoir." Given these data, plus something about the chemical constitution of the working fluid, we can treat the rocket just like an exhaust nozzle configured to attain a very high exit speed.

The typical liquid-propellant rocket is bell-shaped and has, as its reservoir, a combustion chamber into which are injected either a single *monopropellant* or (more commonly) two reagents in nearly stoichiometric proportions. Stored in tanks built into the vehicle, the propellants are brought to the desired p_c by the use of separately driven turbopumps, by maintaining the entire tankage at high pressure, or possibly by a combination of these methods. In *bipropellant* rockets, the injectors are so arranged as to cause swift, intimate mixing between the fuel and the oxidizer. After ignition or spontaneous reaction—the latter in the case of "hypergolic" fuel—this mixture is burned to chemical equilibrium; the pressure remains close to p_c, but temperature rises to a value T_c which is fixed by the energy released from the exothermic reaction together with the specific heat of the products of combustion.

In a solid-propellant rocket, the entire case forms the combustion chamber. Initially this case is almost filled with a solid mixture of fuel, oxidizer, and other chemicals, cast into a form known as a *grain*. After ignition, T_c and (to some degree) p_c are determined by propellant chemistry, whereas the mass flow of combustion products depends also on the size and shape of the grain as it is consumed.

When comparing the merits of these two principal classes of rockets, we note that the solids tend to require a somewhat higher proportion of the gross weight for structure, because the whole "fuel tank" must withstand pressure p_c. This disadvantage should be balanced, however, against the greater complexity and the weight of liquid pumps and pump motors. These pumps provide the opportunity to shut off and restart the liquid rocket, and make possible the control of the direction and magnitude of its thrust. Recently, solids have been tested which also possess similar capabilities. Many of the best liquid propellants, such as O_2 and H_2, are *cryogenic*, requiring storage and handling at very low temperatures. Most solids, on the other hand, have the property of storability, and can be allowed to stand for long periods without special environmental provisions or loss due to evaporation.

Several other schemes have been demonstrated or proposed for creating suitable reservoir conditions. Nuclear rockets, such as those tested under the NERVA Program (Rice and Arnold 1969), employ a gas such as H_2, which is compressed and then brought to T_c by passage through a reactor operating close to critical mass. Electromagnetically heated plasmas and certain lasers constitute further potential sources of working fluid. Several other ways are also known for converting electrical to mechanical energy, ranging all the way to the acceleration of heavy ions directly through an electrostatic field. Many of these devices hold promise of very high I_{sp}, a fact that makes them candidates for long-term interplanetary missions in which fuel is at a premium. In most cases studied to date, however, the bulk of such necessary equipment as electrical power generators has resulted in rather low values of the ratio between thrust and total propulsion-system weight.

A fairly accurate estimate of T and I_{sp} for most rockets can be obtained by assuming that a 1-D isentropic expansion occurs between the reservoir conditions and the nozzle exit. We adopt subscripts c and e, respectively, for these stations. Recognizing that no mass is picked up and that propellant mass flux $\dot{m}_f = \dot{m}$ here, we modify the thrust formula (5.8) to obtain

$$T = \dot{m}u_e + A_e(p_e - p_0) \tag{5.68}$$

where A_e is the exit area. The steady-flow energy equation (5.1) furnishes, in principle, an expression for exhaust speed,

$$u_e = \sqrt{2(h_c - h_e)} \tag{5.69}$$

Before (5.68, 69) can be adapted to relate performance to quantities which can be readily measured or estimated, more needs to be said about both the chemical and thermodynamic behavior of gases passing through the nozzle.

One reasonable procedure is to assume that the propellants arrive at a state close to chemical equilibrium—not necessarily a state of complete combustion—while they are still nearly stagnated. They then accelerate down the channel and are thereby cooled very rapidly, so that a "frozen" condition is reached, in which the reaction rates are too slow to permit any further significant composition change before the gas is discharged. Such a mixture will behave approximately as a perfect gas with constant R_c per unit mass; we note that R_c is proportional to the "universal gas constant" divided by the mean molecular weight of the combustion products. If we make the final assumption that an overall average specific heat c_{pc} and an average γ_c can be assigned to the mixture, we can write

$$h_c - h_e = c_{pc}T_c\left(1 - \frac{T_e}{T_c}\right) = \frac{\gamma_c}{\gamma_c - 1} R_c T\left[1 - \left(\frac{p_e}{p_c}\right)^{(\gamma_c - 1)/\gamma_c}\right] \tag{5.70}$$

In (5.70) we have again employed familiar relations governing perfect gases and isentropic state changes.

Substitution of (5.70) into (5.69) brings us closer to an estimate for thrust.

$$u_e = \sqrt{\frac{2\gamma_c R_c T_c}{\gamma_c - 1}\left(1 - \frac{T_e}{T_c}\right)}$$

$$= \sqrt{\frac{2\gamma_c R_c T_c}{\gamma_c - 1}\left[1 - \left(\frac{p_e}{p_c}\right)^{(\gamma_c - 1)/\gamma_c}\right]} \tag{5.71}$$

As an interesting aside, we remark that $(T_c - T_e)/T_c$ in (5.71) is a sort of "Carnot efficiency" for the expansion process. The desirability of making this ratio as close as possible to unity is clear. One approach is evidently to achieve very high chamber pressure p_c, and rockets with p_c near 3000 psi are under development for the Space Shuttle and other applications. It is also advantageous to have p_e both very low and matched to the surrounding ambient p_0. Accordingly, rockets are constructed with the largest ratio between exit area A_e and throat area A^* that is consistent with the available volume and limitations on weight and structure. Potentially higher performance is associated with higher altitude, the vacuum of space, in which $p_0 \cong 0$, being the ideal environment.

To complete this simplified picture, we must introduce the controlling effects of conditions at the nozzle throat (section of minimum area, identified by superscript *) and the role of the distribution of cross-sectional area in isentropic flow. Thus, under present assumptions, mass flow is given by the familiar 1-D formulas (e.g., Sutton 1956)

$$\dot{m} = \rho^* A^* u^*$$

$$= p_c A^* \sqrt{\frac{\gamma_c}{R_c T_c}\left(\frac{2}{\gamma_c + 1}\right)^{(\gamma_c + 1)/(\gamma_c - 1)}} \tag{5.72}$$

For instance, (5.72) might be employed to size A^*, once the reservoir state and the desired mass flow were known. Among the many available isentropic relations,

$$\frac{A^*}{A_e} = \left(\frac{\gamma_c + 1}{2}\right)^{1/(\gamma_c - 1)}\left(\frac{p_e}{p_c}\right)^{1/\gamma_c}\sqrt{\left(\frac{\gamma_c + 1}{\gamma_c - 1}\right)\left[1 - \left(\frac{p_e}{p_c}\right)^{(\gamma_c - 1)/\gamma_c}\right]} \tag{5.73}$$

would serve to determine the exit area in terms of a design-point value of p_e.

We now combine (5.68), (5.71), and (5.73) to derive a working expression for rocket thrust.

$$\boxed{\frac{T}{p_c A^*} = \sqrt{\frac{2\gamma_c^2}{\gamma_c - 1}\left(\frac{2}{\gamma_c + 1}\right)^{(\gamma_c + 1)/(\gamma_c - 1)}\left[1 - \left(\frac{p_e}{p_c}\right)^{(\gamma_c - 1)/\gamma_c}\right]} + \frac{A_e}{A^*}\left(\frac{p_e}{p_c} - \frac{p_0}{p_c}\right)} \tag{5.74}$$

Specific impulse for this device is merely T divided by mass flow. Equations

(5.68), (5.71), and (5.72) therefore yield

$$I_{sp} \equiv \frac{T}{\dot{m}} = \sqrt{\frac{2\gamma_c R_c T_c}{\gamma_c - 1}\left[1 - \left(\frac{p_e}{p_c}\right)^{(\gamma_c - 1)/\gamma_c}\right]} + \frac{A_e}{A^*}\left(\frac{p_e}{p_c} - \frac{p_0}{p_c}\right)\sqrt{\frac{R_c T_c}{\gamma_c}\left(\frac{\gamma_c + 1}{2}\right)^{(\gamma_c + 1)/(\gamma_c - 1)}} \tag{5.75}$$

The advantages of high reservoir pressure and temperature can be inferred from examination of (5.74) and (5.75). Not quite so obvious is the desirability of low molecular weight in the exhaust gases, but it does enhance I_{sp} in proportion to its inverse square root, as can be seen from the R_c in both terms of (5.75).

A theoretical maximum value of I_{sp}, for firing in space, is obtained by setting $p_e = 0 = p_0$.

$$(I_{sp})_{max} = \sqrt{\frac{2\gamma_c R_c T_c}{\gamma_c - 1}} \tag{5.76}$$

The quantities in (5.76) are mainly determined by the combustion reaction of the particular propellants. The Pratt & Whitney Aeronautical Handbook (1968) furnishes an extensive table of specific impulses, based on equilibrium stoichiometric burning, for liquid rocket propellants. Table 5.1 shows some representative data for p_c = 100 psi, A_e/A^* = 40, and firing into vacuum.

Table 5.1

Combination	T_c, °R	$(I_{sp})_{max}$, "seconds"
$H_2 + O_2$	5610	454
$H_2 + F_2$	6820	475
$N_2H_4 + H_2O_2$	5700	338
RP-1* + O_2	6030	351

* "Rocket Propellant Type 1," consisting mainly of kerosene

It is worth mentioning that an increase in temperature tends to drive an exothermic reaction in the direction of dissociating the products of combustion, whereas higher p_c drives it the other way. Since complete combustion is favorable for extracting all possible energy from the fuel, we discover yet another reason for high chamber pressure.

From Table 5.1 we see that liquid oxygen (LOX) and kerosene, which have been used in booster stages and certain missiles, have a high heating value, but tend to be penalized in terms of I_{sp} by the molecular weight of the CO_2 produced and by incomplete burning. For this mixture in many first stages, such as Saturn V. matched for p_c = 1000 psi, the sea-level, optimum I_{sp} = 299 "seconds."

LOX is relatively easy to manufacture, pump, and insulate; it therefore became operational more than a decade earlier than liquid hydrogen. The latter must not only be kept at temperatures close to absolute zero, but its specific gravity of 0.07 calls for a huge volume of tankage compared to other fuels. Nevertheless, LOX and LH_2 are today's propellants of choice for peacetime operation in which quick response and long-term storability are not required. This is especially so for upper booster stages, in which considerations of orbital mechanics demand the highest possible I_{sp}. Theoretical maximum values can be closely approached in the low-p_0 environment; for instance, 427 "seconds" in vacuum is quoted for the J2 engine used in upper Saturn V stages, whereas high-p_c Space Shuttle designs anticipate nearly 450 "seconds."

Evidently fluorine is even more attractive than LOX as an oxidizer, but the highly corrosive nature of this element and the generation of hydrofluoric acid as the combustion product constitute strong deterrents to its use.

5.4 ONE-DIMENSIONAL ANALYSIS OF TURBOJET WITH REALISTIC COMPONENT EFFICIENCIES

In this section we retain the 1-D methodology but undertake to refine the description of turbojet and ramjet performance by accounting for the most significant deviations from ideal operation of each component. We shall first go over the subject of the turbojet in some detail, and then cover the ramjet as a special case. No discussion of the drag, as expressed by (5.9), will be attempted.

The principal losses occurring within typical engine installations arise from the following.

- Increase in entropy during diffusion from the free stream to the compressor intake.
- Imperfect compression and expansion in the compressor and turbine, resulting in power losses and an increase in entropy through each of these elements. Friction with the walls, effects of 3-D flow and mixing, heat transfer, and the formation of separated wakes during off-design conditions are important causes. The compressor is the more critical component, because the generally rising pressure as the fluid passes through its stages is unfavorable to boundary layer growth, promoting thickening and separation from blades and other parts.
- Incomplete combustion and a reduction in pressure across the burner.
- Underexpansion or overexpansion in the nozzle, so that the pressure p_6 does not match the ambient p_0 and there is greater irreversibility in the slipstream mixing process.

Omitting the afterburner, let us go through the schematic of Fig. 5.1 and define all new quantities needed to characterize the nonideal turbojet.

1. The Inlet

Diffusion always involves wall friction and 3-D flow effects that cause the average stagnation pressure p_{2t} to be less than the free-stream value

$$p_{0t} = p_0\left(1 + \frac{\gamma - 1}{2} M_0^2\right)^{\gamma/(\gamma - 1)} \tag{5.77}$$

In supersonic flight, an even greater rise in entropy occurs because of the pattern of oblique and normal shocks which is intentionally produced to abet the deceleration process. Away from the design point and during throttling of the engine, these shocks may move into positions in which they cause separated flow regions and spillage from the inlet lip. Both drag and greater internal losses are the result.

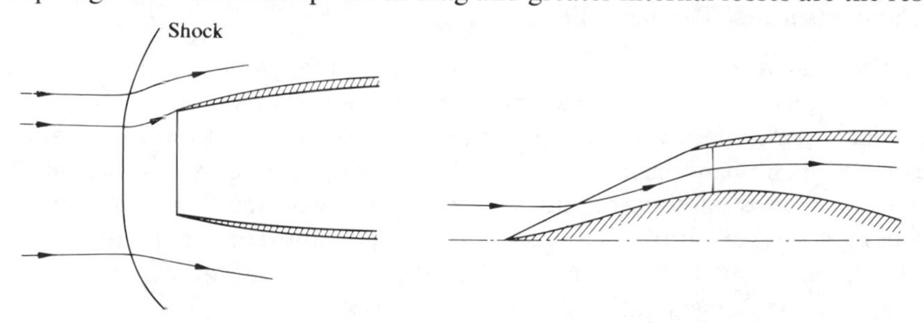

Fig. 5.8. Two extreme cases of inlets operating in supersonic flight. (See text for discussion.)

Figure 5.8 sketches two extreme cases of supersonic inlets. On the left a normal shock has been forced out ahead of the cowl. The stream tube behind it enters subsonically, but not until after a drop in p_t has taken place corresponding to that through a normal shock at flight M_0. This is about the highest steady-flow rise in entropy that can occur for given M_0. At the right of the figure is the cross section of a typical axisymmetrical design, which employs a single conical shock followed by some isentropic compression and a much-weakened normal shock inside the duct. Similar 2-D arrangements may use two or three successive oblique shocks ahead of the cowl. In either case the overall rise in entropy is substantially less than that for the single normal shock.

Figure 5.9 presents typical inlet performance, expressed in terms of the pressure ratio

$$\pi_d \equiv \frac{p_{2t}}{p_{0t}} = \frac{p_{2t}}{p_0 \tau_r^{\gamma/(\gamma - 1)}} \tag{5.78}$$

Here

$$\tau_r \equiv 1 + \frac{\gamma - 1}{2} M_0^2 \tag{5.79}$$

We note that constant $\gamma = 1.4$ is an acceptable approximation for 1-D inlet analysis up to $M_0 = 4$ or so. The inlet temperature ratio $\tau_d \cong 1$, since the state change up to the compressor face is very nearly adiabatic.

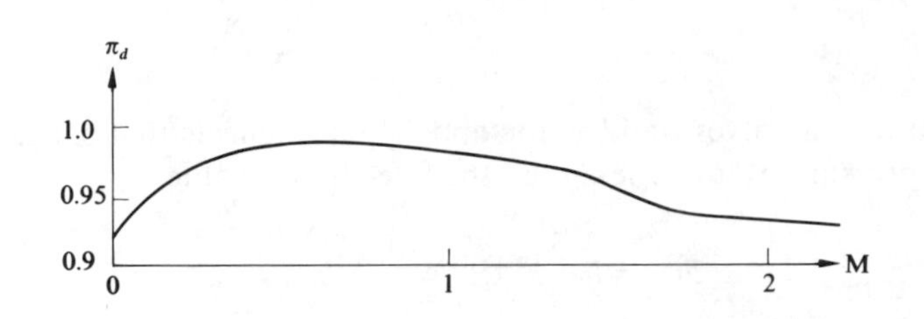

Fig. 5.9. Typical variation of stagnation pressure ratio with Mach number for an actual 2-D inlet, employing multiple oblique shocks and considerable "external compression" at supersonic speeds. Note that altitude increases with Mach number from values near sea level below 0.55 to 50,000 ft at 2.2.

2. Compressor and Turbine Efficiencies

So-called *adiabatic efficiencies* η_c and η_T are defined for each of these components by adopting as a standard of reference the ideal power that would be involved in isentropic operation at the same pressure ratio. Deviations from the ideal occur because of such phenomena as drag of blades, bearing friction, and various heat-transfer processes between the machinery and the working fluid. We retain the approximation that enthalpy changes can be proportioned to temperature changes, but the values assigned to c_p and γ are averages for the thermodynamic states in the device under consideration and are appropriately subscripted.

For the compressor,

$$\eta_c \equiv \frac{(\text{Ideal compressor power absorbed for the given pressure ratio } \pi_c)}{(\text{Actual compressor power for the same } \pi_c)}$$

$$= \frac{\dot{m}\,\Delta h_{\text{Ideal}}}{\dot{m}\,\Delta h_{\text{Actual}}} \cong \frac{\dot{m}[(T_{3t})_{\text{Ideal}} - T_{2t}]}{\dot{m}(T_{3t} - T_{2t})}$$

$$= \frac{(\tau_c)_{\text{Ideal}} - 1}{\tau_c - 1} = \frac{\pi_c^{(\gamma_c - 1)/\gamma_c} - 1}{\tau_c - 1} \tag{5.80}$$

In the last member of (5.80), the pressure and temperature ratios would be those actually observed at a given operating condition. π_c has, of course, been related to $(\tau_c)_{\text{Ideal}}$ isentropically for the mean specific-heat ratio. Adiabatic efficiencies approaching 90% are being achieved under optimum circumstances by modern compressors, although a time average of 85% might be more realistic as the aircraft flies typical mission patterns.

For the turbine,

$$\eta_T \equiv \frac{(\text{Actual turbine power output for the given } \pi_T)}{(\text{Ideal turbine power output for the same } \pi_T)}$$

$$= \frac{-(\dot{m} + \dot{m}_f)\,\Delta h_{\text{Ideal}}}{-(\dot{m} + \dot{m}_f)\,\Delta h_{\text{Actual}}}$$

$$\cong \frac{(\dot{m} + \dot{m}_f)(T_{4t} - T_{5t})}{(\dot{m} + \dot{m}_f)[T_{4t} - (T_{5t})_{\text{Ideal}}]} = \frac{1 - \tau_T}{1 - (\tau_T)_{\text{Ideal}}}$$

$$= \frac{1 - \tau_T}{1 - \pi_T^{(\gamma_T - 1)/\gamma_T}} \tag{5.81}$$

Again the symbols agree with those defined in Section 5.2. Values of η_T approaching 95% have been demonstrated.

We remark that other definitions of compressor and turbine efficiencies are employed in different fields, such as with stationary steam turbines, but that they are not so suitable for our purposes here. We shall later have to relate the actual turbine power output to that absorbed in the compressor; in so doing, we shall find that the single independent parameter π_c characterizes the whole system once the various efficiencies, auxiliary power consumption, etc., are prescribed.

3. The Burner

Whenever heat is added to gas in motion, the stagnation pressure drops. This we prove by reproducing one of the familiar 1-D flow formulas from the table given by Shapiro (1953, page 231):

$$\frac{dp_t}{p_t} = -\frac{\gamma M^2}{2}\frac{dT_t}{T_t} \tag{5.82}$$

Since the whole purpose of combustion is to raise total temperature, $dT_t > 0$, and it follows that the ratio

$$\pi_b \equiv \frac{p_{4t}}{p_{3t}} \tag{5.83}$$

is less than one. Its nearness to unity is a measure of how well the designer has coped with (5.82) and such other effects as wall friction and injector drag. There is an obvious advantage to keeping the Mach number—hence the speed—of flow as low as possible, but this objective must be traded off against the need for driving high mass flow through an engine of minimum cross-sectional area.

Account is taken of incompleteness in the fuel–air reaction and of heat lost to the walls through conduction and radiation by introducing a "burner efficiency"

$$\eta_b = \frac{(\dot{m} + \dot{m}_f)h_{4t} - \dot{m}h_{3t}}{\dot{m}_f h_f} \cong \frac{c_{pT}[T_{4t} - T_{3t}]}{f h_f} \tag{5.84}$$

(Typical values are $\eta_b = 95$–98%.) An averaged specific heat has been used in the last member of (5.84), along with $f \ll 1$. The latter approximation is sometimes acceptable, since the stoichiometric ratio is 0.067 for jet propellants in air and turbojets usually run very "lean," that is, with an excess of available oxygen (see Hill and Peterson 1965, Fig. 7–25, page 218).

4. The Exhaust Nozzle

The process of nozzle expansion is nearly isentropic, and we adopt relations like (5.44,45) unmodified. As for the impact on thrust of the mismatch between p_6 and p_0, this has already been accounted for in the derivation of (5.8), which is rewritten as

$$T = \dot{m}[(1 + f)u_6 - u_0] + A_6(p_6 - p_0) \tag{5.85}$$

It is in (5.85) that the highly irreversible slipstream behavior makes itself indirectly felt. Because of the inverse relationship between u_6 and p_6, whereby high exhaust speed is associated with low pressure, the two terms in (5.85) work against one another. Detailed calculation shows a maximum of T, other things being the same, near the matched condition. For practical reasons such as limiting the engine diameter, however, typical nozzles tend to be "choked" at their exits most of the time, so that $M_6 \cong 1$.

We have discussed ten interrelated ratios—π_d, π_c, τ_c, η_c, π_b, τ_b, η_b, π_T, τ_T, and η_T—which, together with p_6, u_6, and quantities like τ_r describing the flight condition, characterize one-dimensionally the installed engine cycle. They are adequate for the construction of formulas for T and I_{sp}, whose derivation we now present.

In generalization of (5.46) and (5.47), exhaust temperature and pressure may be expressed as follows:

$$T_6\left(1 + \frac{\gamma_T - 1}{2} M_6^2\right) = T_{6t} = T_0\tau_r\tau_c\tau_b\tau_T \tag{5.86}$$

$$p_6\left(1 + \frac{\gamma_T - 1}{2} M_6^2\right)^{\gamma_T/(\gamma_T - 1)} = p_{6t} = p_0\tau_r^{\gamma/(\gamma - 1)}\pi_d\pi_c\pi_b\pi_T \tag{5.87}$$

Here we suppose that the average γ_T through the turbine also applies to the nozzle expansion. τ_r is given by (5.79), and it is convenient to define a pressure ratio

$$\pi_r \equiv \tau_r^{\gamma/(\gamma - 1)} \tag{5.88}$$

We shall again use (5.50), so M_6 is eliminated between (5.86) and (5.87) to get

$$\frac{T_6}{T_0} = \frac{\tau_r\tau_c\tau_b\tau_T}{\left(\dfrac{p_0}{p_6}\pi_r\pi_d\pi_c\pi_b\pi_T\right)^{(\gamma_T - 1)/\gamma_T}} \tag{5.89}$$

Either of these formulas can also be solved for M_6, and we choose (5.87),

$$\left(\frac{M_6}{M_0}\right)^2 = \left(\frac{\gamma - 1}{\gamma_T - 1}\right)\left\{\frac{\left(\dfrac{p_0}{p_6}\pi_r\pi_d\pi_c\pi_b\pi_T\right)^{(\gamma_T - 1)/\gamma_T} - 1}{\tau_r - 1}\right\} \tag{5.90}$$

Three other compressor and turbine ratios can be eliminated in terms of τ_c, which we have seen to be one of the fundamental parameters of the turbojet. We do this by equating actual output of power by the turbine to the power absorbed in the compressor, noting that it would be easy here to account for auxiliaries, but that little is gained from the added complication:

$$[\dot{m} + \dot{m}_f][h_{4t} - h_{5t}] = \dot{m}[h_{3t} - h_{2t}] \tag{5.91}$$

With the averaged specific heats substituted, (5.91) becomes

$$[1 + f]c_{pT}[T_{4t} - T_{5t}] = c_{pc}[T_{3t} - T_{2t}] \tag{5.92}$$

and can be manipulated to yield

$$\tau_T \equiv \frac{T_{5t}}{T_{4t}} = 1 - \frac{1}{1+f}\frac{c_{pc}}{c_{pT}}\left(\frac{\tau_c - 1}{\tau_c \tau_b}\right) \tag{5.93}$$

The last member of (5.81), together with (5.93), produces

$$\begin{aligned} \pi_T^{(\gamma_T - 1)/\gamma_T} &= 1 - \frac{1 - \tau_T}{\eta_T} \\ &= 1 - \frac{c_{pc}/c_{pT}}{(1+f)\eta_T}\left(\frac{\tau_c - 1}{\tau_c \tau_b}\right) \end{aligned} \tag{5.94}$$

For the compressor, (5.80) gives directly

$$\pi_c^{(\gamma_c - 1)/\gamma_c} = 1 + \eta_c(\tau_c - 1) \tag{5.95}$$

Finally, we adopt the ratio $\tau_\lambda = \tau_r \tau_c \tau_b$ from (5.57), and also substitute (5.89), (5.90), (5.93), (5.94), and (5.95) into (5.50) to obtain the speed ratio:

$$\frac{u_6}{u_0} = \frac{M_6}{M_0}\sqrt{\frac{T_6}{T_0}} = \left\{ \frac{\tau_r}{\tau_r - 1}\left(\frac{\gamma - 1}{\gamma_T - 1}\right)\left[\frac{\tau_\lambda}{\tau_r} - \frac{c_{pc}/c_{pT}}{1+f}(\tau_c - 1)\right] \right.$$
$$\left. \times \left[1 - \frac{1}{\left(\dfrac{p_0}{p_6}\pi_r \pi_d \pi_b\right)^{(\gamma_T - 1)/\gamma_T}\left[1 - \dfrac{c_{pc}/c_{pT}}{(1+f)\eta_T}\left(\dfrac{\tau_r(\tau_c - 1)}{\tau_\lambda}\right)\right] \times [1 + \eta_c(\tau_c - 1)]^{[\gamma_c/(\gamma_c - 1)][(\gamma_T - 1)/\gamma_T]}}\right]\right\}^{1/2} \tag{5.96}$$

When put together with a suitably modified (5.85),

$$\boxed{\frac{T}{\dot{m}} = u_0\left[(1+f)\frac{u_6}{u_0} - 1\right] + \frac{A_6}{\dot{m}}(p_6 - p_0)} \tag{5.97}$$

(5.96) furnishes the desired prediction of thrust. An alternative version of the last term in (5.97) can be derived by means of continuity and the perfect gas law. Thus,

$$\dot{m} = \frac{\dot{m}_6}{1+f} = \frac{\rho_6 u_6 A_6}{1+f} \tag{5.98}$$

and

$$\rho_6 = \frac{\rho_6}{\rho_0}\rho_0 = \frac{p_0}{RT_0}\left(\frac{p_6}{p_0}\frac{T_0}{T_6}\right) \tag{5.99}$$

We combine (5.98), (5.99), and the relation $a_0^2 = \gamma RT_0$ to obtain

$$\begin{aligned}\frac{A_6}{\dot{m}} = \frac{1+f}{\rho_6 u_6} &= (1+f)\frac{u_0^2}{u_6}\left(\frac{RT_0}{u_0^2}\right)\frac{1}{p_0}\left(\frac{p_0}{p_6}\frac{T_6}{T_0}\right)\\ &= \left(\frac{1+f}{\gamma \mathrm{M}_0^2}\right)\frac{u_0^2}{u_6}\frac{T_6}{T_0}\frac{1}{p_6}\end{aligned} \tag{5.100}$$

Substitution into (5.97) gives rise to

$$\boxed{\frac{T}{\dot{m}} = u_0\left\{(1+f)\frac{u_6}{u_0} - 1 + \left(\frac{1+f}{\gamma \mathrm{M}_0^2}\right)\frac{u_0}{u_6}\frac{T_6}{T_0}\left(1 - \frac{p_0}{p_6}\right)\right\}} \tag{5.101}$$

which, with (5.96) and an adjusted (5.89), predicts thrust without explicit reference to the exit area. f in (5.97) or (5.101) can be computed from (5.102) below.

We remark that, as in the ideal turbojet analysis at its design point, a compressor parameter τ_c can be calculated from (5.97) or (5.101) which maximizes T for fixed values of τ_λ and of the other ratios and efficiencies which remain in these formulas.

Specific impulse, as expressed by the first two members of (5.63), requires us to relate f to the same set of independent parameters. Clearly (5.84) is the needed equation, and it also brings in h_f and combustion efficiency for the first time. When (5.84) is solved explicitly for f and appropriate substitutions are made, we find that

$$\boxed{f = \frac{\dfrac{c_{pT}}{c_p}(\tau_\lambda - \tau_r\tau_c)}{\left(\dfrac{\eta_b h_f}{c_p T_0}\right) - \dfrac{c_{pT}}{c_p\tau_\lambda}}} \tag{5.102}$$

(Note that the assumption $f \ll 1$ would be tantamount to neglecting the second term in the denominator, but here it is retained for consistency.)

The final result reads

$$\boxed{I_{\mathrm{sp}} = \frac{T}{f\dot{m}} = \frac{\dfrac{T}{\dot{m}}\left(\dfrac{\eta_b h_f}{c_p T_0} - \dfrac{c_{pT}}{c_p}\tau_\lambda\right)}{\dfrac{c_{pT}}{c_p}(\tau_\lambda - \tau_r\tau_c)}} \tag{5.103}$$

the thrust again being taken from (5.97) or (5.101). Since the first term always dominates the numerator brackets in (5.103), we observe the approximate proportionality of I_{sp} to both combustor efficiency and heating value of the fuel.

It is now straightforward to specialize the foregoing to describe a ramjet which has losses in the diffusion and combustion processes. All ratios relating to the compressor and turbine are replaced by unity, and a single pair of averaged properties γ_T, c_{pT} is assumed to characterize the gas during combustion and the nozzle expansion. These steps reduce (5.89) and (5.96) to

$$\frac{T_6}{T_0} = \frac{\tau_r \tau_b}{\left(\dfrac{p_0}{p_6}\pi_r \pi_d \pi_b\right)^{(\gamma_T - 1)/\gamma_T}} \equiv \frac{\tau_\lambda}{\left(\dfrac{p_0}{p_6}\pi_r \pi_d \pi_b\right)^{(\gamma_T - 1)/\gamma_T}} \tag{5.104}$$

and

$$\frac{u_6}{u_0} = \left\{\frac{\tau_\lambda}{\tau_r - 1}\left(\frac{\gamma - 1}{\gamma_T - 1}\right)\left[1 - \frac{1}{\left(\dfrac{p_0}{p_6}\pi_r \pi_d \pi_b\right)^{(\gamma_T - 1)/\gamma_T}}\right]\right\}^{1/2} \tag{5.105}$$

The thrust can be found by inserting (5.104,105) into (5.101), as can I_{sp} from (5.103) with $\tau_c = 1$ in the denominator.

5.5 ACTUAL TURBOJET PERFORMANCE

It is instructive to think of the turbojet as a "single-control" engine, in the sense that only throttle position is under the pilot's immediate command. By contrast, a variable-pitch propeller powered by a reciprocating engine makes possible independent adjustments, within limits, of the carburetor valve settings and propeller speed. The turbojet throttle activates a fuel pump, which in turn regulates the values of $\dot{m}_f$ and T_{4t}. Although small changes in $\dot{m}$, τ_c, τ_T, and many other quantities occur when the throttle is moved, τ_λ is the parameter in our formulas for thrust and specific impulse which most directly reflects such an action. In particular, peak thrust is tied to the maximum permissible τ_λ at the given speed and altitude. Since even a few seconds of exposure to excessive turbine-inlet temperature can cause extensive damage, modern turbojets are provided with automatic overrides which, near full power, limit the fuel flow in response to the measured value of T_{4t}.

Recalling the cycle analyses of Section 5.4, let us now consider some numerical estimates of typical performance. The engine for which these data are computed is the General Electric YJ93-GE-3, of which six powered the B-70 airplane. We are indebted to Mr. B. Y. Rolf for most of the information used in this example; Messrs. J. M. Summa and K. V. Krishna Rao of Stanford University carried out the calculations.

Table 5.2 presents approximate values of the quantities needed for application of (5.101) and (5.103). Parameters assumed constant correspond to M = 3 and 60,000 feet altitude. We remark that the YJ93 engine bypasses 7% of the inlet air for the purpose of cooling its outer surfaces. The bypassed air is not accounted for in the present calculations. Rather, we assume that each YJ93 acts on the full working fluid delivered to it by the B-70 inlet system.

Table 5.2. YJ93 engine parameters used in calculations

$\tau_c = 1.462$	$\gamma_c = 1.400$
$\tau_b = 1.545$	$\gamma_T = 1.316$
$\pi_c = 3.284$	$c_{pc} = 0.2511$ Btu/lb°R
$\pi_b = 0.926$	$c_{pT} = 0.2855$ Btu/lb°R
$\pi_T = 0.366$	$h_f = 18{,}400$ Btu/lb
$\eta_c = 0.804$	$A_0 = 8.816$ ft^2
$\eta_b = 0.980$	$A_{\text{Throat}} = 5.951$ ft^2
$\eta_T = 0.897$	$A_6 = 14.743$ ft^2
$T_{4t} = 2446$°R	

M_0	1.0	2.0	3.0
τ_r	1.200	1.800	2.800
π_r	1.893	7.825	36.738
π_d	1.000	0.900	0.722
p_0/p_6	6.417	1.723	0.458
f	0.0279	0.0225	0.0134

Figures 5.10 and 5.11 present the calculated performance of a single YJ93 (afterburner off), plotted versus Mach number, in terms of thrust and specific inpulse. Over the M ranges here employed, no allowance is made for variations in any of the internal efficiencies. Inlet pressure recovery is estimated in accordance with the Aircraft Industries Association (AIA) standard curve. Further, the exhaust nozzle is assumed to remain in the wide-open position, with isentropic flow conditions throughout all the calculations. To assess the accuracy of our 1-D analysis, thrust was calculated with afterburner in operation at M = 3 and 60,000 feet altitude. Agreement between the 1-D value and that supplied by General Electric (15,400 lb) was within 5%, and the resulting augmentation ratio (thrust with afterburner/thrust without afterburner) was found to be approximately 2.0.

One interesting property of turbojet operation that can be deduced from our formulas and sample calculations goes by the name *kinematic similarity*. Consider a given engine, at full throttle and propelling a given aircraft in the standard stratosphere, where T_0 = const. We remark, incidentally, that this is the altitude range at which $\tau_\lambda \equiv (T_{4t}/T_0)$ is largest and that the favorable influence of high τ_λ on thrust

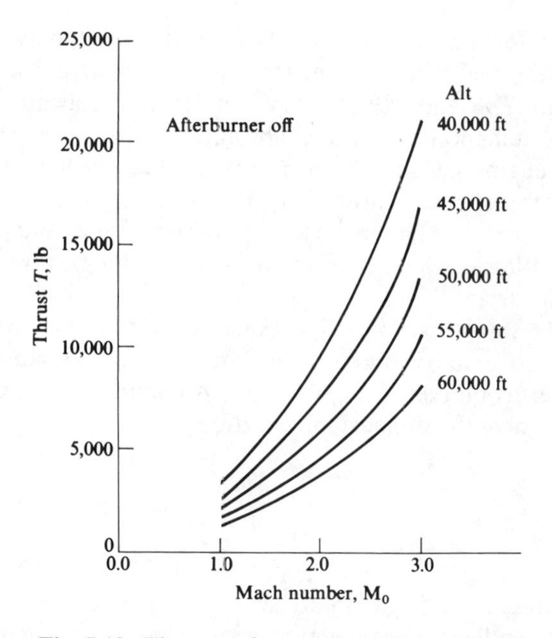

Fig. 5.10. Thrust performance of YJ93 engine.

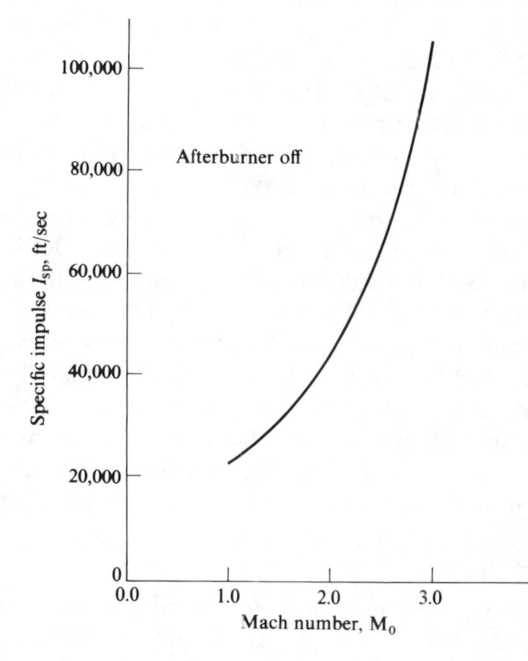

Fig. 5.11. Specific impulse of YJ93 engine.

performance therefore makes it the best place for jet aircraft to fly. With the throttle setting, τ_λ, T_0, M_0, and other parameters fixed, we see from (5.96), (5.101), and (5.103) that both $T/\dot{m}$ and I_{sp} are also constants. As altitude is changed, all pressure ratios, temperature ratios, absolute temperatures, and flow speeds throughout the engine keep the same values, while mass flow, fuel flow, density, and static pressure at each station vary in direct proportion with the ambient density ρ_0 or pressure p_0. The entire pattern of streamlines and particle velocity is independent of altitude, which means that the kinematic behavior of the flow may be regarded as invariant.

Evidently the thrust itself is proportional to ρ_0, since all curves of $T/\dot{m}$ versus M_0 (study Fig. 5.10) collapse onto a single line. It is equally satisfactory to present thrust data in terms of $T/\dot{m}$, T/ρ_0, or T/p_0. A common practice, however, is to work with T/δ, where the dimensionless ratio

$$\delta \equiv \frac{p_0}{p_{sl}} \tag{5.106}$$

refers ambient pressure to its standard sea-level value $p_{sl} = 14.7$ psi.

As we shall see later in connection with vehicle performance, constant-M_0 flight in the stratosphere is a way of getting the optimum performance from both the aircraft and its propulsion system. This approach involves setting the speed and throttle so that $T = D$ and $L = W$ at the lift coefficient C_L which corresponds to a maximum lift–drag ratio L/D. As weight W goes down because of the fuel consumed, an upward drift in altitude is (ideally) permitted, in such a way that C_L remains fixed while all four forces go down in proportion to ρ_0. Evidently L/D stays at its peak value. Moreover, in view of kinematic similarity, the jets can continue to operate at the same efficient design point.

We complete our discussion of turbojets and related devices with an introduction to the 3-D characteristics of rotating components. Rather oversimplified considerations of dimensional analysis suggest that there are actually two Mach numbers—more precisely, two distributions of Mach number—which must be important in determining the dimensionless details of flow through a compressor, fan, or gas turbine. The first of these describes the "through-flow," and a typical value would be proportional to the mean axial speed at some station divided by the square root of absolute temperature there. The second involves the effects of angular velocity of rotation and might, for instance, be Mach number associated with the component of circumferential velocity at a blade tip; any such quantity will be dependent on $N/\sqrt{T}$, where N is the angular velocity, usually in units of revolutions per minute (RPM).

The foregoing concepts can be used to explain the origin of two parameters commonly employed in turbojet design. To be specific, consider the case of a given compressor. By a series of proportionalities, the axial and tip Mach numbers can be transformed into these parameters as follows.

1. Corrected Weight Flow

A typical station might be the entry to the first stage, which we have identified with subscript 2. M_2 there depends on the mass flow; thus, since the area A_2 is fixed,

$$M_2 \sim \frac{u_2}{\sqrt{T_2}} \sim \frac{\rho_2 u_2 A_2}{\rho_2 \sqrt{T_2}} \sim \frac{\dot{m}}{(p_2/T_2)\sqrt{T_2}} \sim \frac{\dot{m}\sqrt{T_2/T_{sl}}}{\delta_2} = \frac{\dot{m}\sqrt{\theta_2}}{\delta_2} \tag{5.107}$$

Here δ is defined in (5.106), and $\theta \equiv T/T_{sl}$ is a corresponding temperature ratio. The parameter $\dot{m}$ quoted in lbm $(\sec)^{-1}$ is called the *weight flow*, and (5.107) shows how this weight flow must be "corrected" to become a measure of axial Mach number.

2. Corrected RPM

We have seen that N is adjusted to be proportional to M_{Tip} by division with $a = \sqrt{\gamma R T}$ at the corresponding station. It is therefore not surprising that N is "corrected" according to the rule.

$$M_{Tip} \sim \frac{N}{\sqrt{\theta}} \tag{5.108}$$

For instance, θ might be replaced by θ_2 at the entry.

The efficiency, pressure ratio π_c, etc., of any compressor are found to depend primarily on the corrected $\dot{m}$ and N_c. A *compressor map* consists of a plot of π_c versus corrected weight flow, with curves of constant $N_c/\sqrt{\theta_2}$, η_c, and other information superimposed. Figure 5.12 presents a typical such map, including the

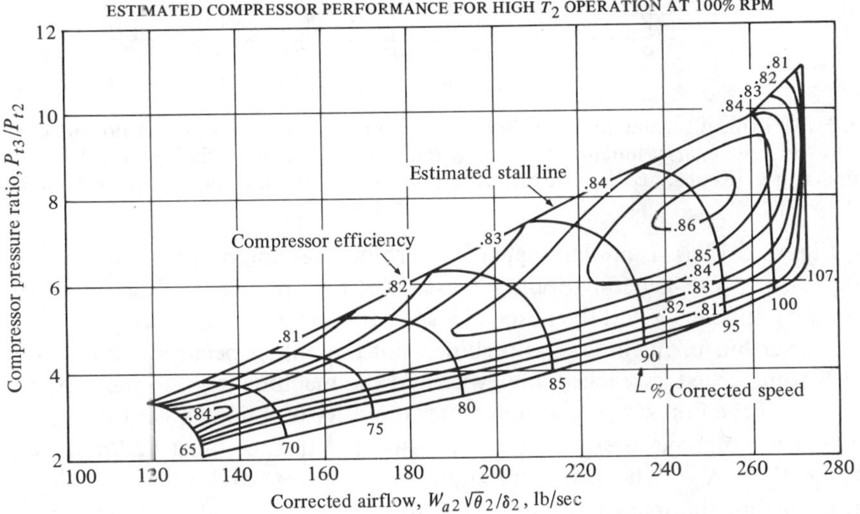

Fig. 5.12. Typical compressor map for the J93 jet engine, of which six were used to propel the XB-70. The "operating line" parallels the "estimated stall line" and runs approximately through the peaks of the closed curves of constant compressor efficiency. (Reproduced courtesy of Aircraft Engine Group, General Electric Company.)

line along which the machine is normally expected to operate and another along which a significant amount of "surging" or of stalling in the most critical stage begins to appear. As expected, these conditions arise when π_c is driven to an inconsistently high level for the mass of gas passing through. The separation between the stall–surge and operating lines provides one measure of how sensitive the engine is to off-design operation.

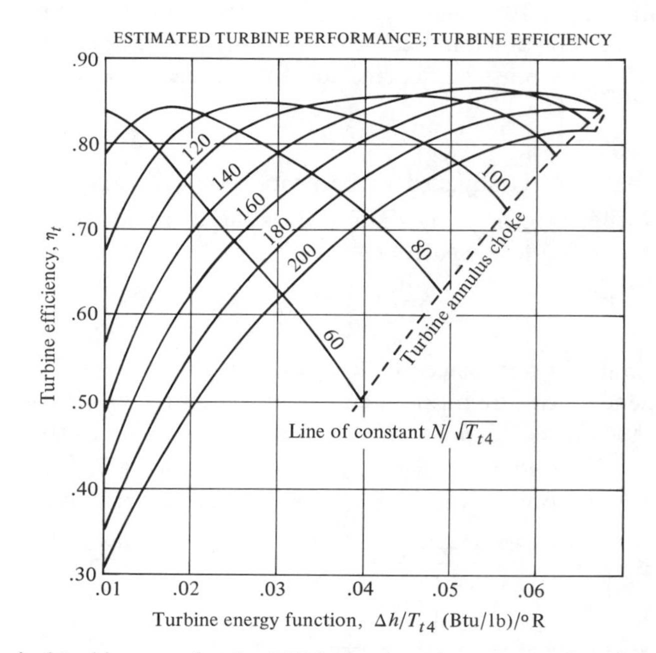

Fig. 5.13. Typical turbine map for the J-93 jet engine, shown on a plot of turbine efficiency versus "turbine energy function," which is the ratio of specific enthalpy drop to inlet total temperature. (Reproduced courtesy of Aircraft Engine Group, General Electric Company.)

Figure 5.13 is a similar map of a turbine that might be connected to this compressor. The general topology resembles that in Fig. 5.12, except that the ordinate is efficiency and the abscissa a special "turbine energy function." Stalling is not a serious problem for gas turbines, but there does occur a "choke limit" at which sonic speed is reached somewhere in the machine (e.g., in the stator blade passages of the first stage) and no greater quantity of fluid can get through.

During turbojet design, the compressor and turbine must be "matched" by setting $N_c = N_T$. The geometry, number of stages, and other details are so arranged that the mass flows are compatible and that, under all circumstances, enough power can be furnished to the compressor and auxiliaries. Kerrebrock (1973) uses these considerations and several other simplifying assumptions, such as $M_6 = 1$, to develop fairly simple formulas for I_{sp} and T/p_0A_2 as functions

M_0, T_0, and T_{4t}. These are worthy of study by readers with a particular interest in the subject of propulsion.

5.6 PROBLEMS

5A In cross-section view, an element of a lightly loaded propeller turning at angular velocity ω on an airplane flying at speed u_0 looks like the figure shown here.

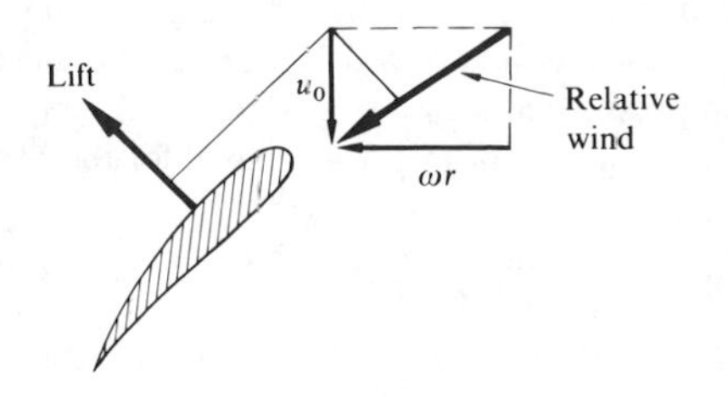

Remember that each blade is just like a twisted wing following a spiral flight path. Given that the flow at each section is 2-D and $L/D \to \infty$, show why the torque power supplied by engine to propeller equals thrust power delivered by propeller to airplane (i.e., 100% mechanical efficiency). Discuss some sources of power loss that must reduce this efficiency.

As an example, assume that at each element along the propeller blade (i) L/D has the same fixed value and (ii) the "inflow" speed of slipstream fluid passing through the propeller disk is a fixed multiple $r > 1$ of the flight speed u_0. Try to find an expression for the mechanical efficiency of the propeller, which is the ratio of thrust power to torque power.

5B Remembering the temperature limit of the turbine and the relationship of thrust to atmospheric density, explain why jet aircraft have such serious trouble with length of takeoff run on very hot days.

5C Extend the method of ideal, one-dimensional analysis to cover a turbofan engine operating at its design condition. A fraction r of the mass flow entering the inlet is bypassed around the central "hot-gas generator"; $r/(1 - r)$ is called the *bypass ratio*. The bypassed air is compressed in a ratio π_{cB} by an axial compressor connected to the main shaft driven by the turbine; it is then reexpanded (ideally) down to pressure p_0 at the exit of the bypass nozzle. Discuss why this device might be expected to have (1) higher static thrust, and (2) specific impulse dropping off more rapidly with speed than a comparable turbojet.

As a check, the formula for ideal thrust, when fuel–air ratio $f \ll 1$, should read

$$\frac{T}{\dot{m}u_0} = r\left[\sqrt{\frac{\tau_r\tau_{cB} - 1}{\tau_r - 1}} - 1\right]$$

$$+ [1 - r]\left[\sqrt{\frac{\tau_r(\tau_b - 1)(\tau_c - 1) + \tau_b(\tau_r - 1) - \left(\dfrac{r}{1 - r}\right)\tau_r(\tau_{cB} - 1)}{\tau_r - 1}} - 1\right]$$

where τ_r is given by (5.79). What happens when $r \to 0$ and $r \to 1$?

5D Using the results of 5C, try to determine whether there is an "optimum" value of the bypass parameter r, intermediate between 0 and 1, such that $T/\dot{m}u_0$ reaches a maximum while all other dimensionless parameters are held constant.

5E There are several methods besides the afterburner of providing thrust augmentation for turbojets. One such method is that of water injection at the compressor inlet through a series of spray nozzles. This ideal process of water evaporation through the compressor can be thought of as a heat-extractor operation.

a) Show by means of a Mollier diagram (h–S) and a compressor performance chart how water injection improves thrust. (Why doesn't the h–S diagram show the thrust increase due to increased mass flow?) List the factors which produce the increase in thrust. [*Hint:* An expression for the compressor work in terms of the total temperature at the inlet and the pressure ratio across the compressor might be helpful for the Mollier diagram. Recall that $(\Delta h)_{\text{Compressor}} = (\Delta h)_{\text{Turbine}}$.]

b) In most applications of water injection, an alcohol–water mixture is used. Can you give reasons why?

c) Where does water injection give the highest augmented thrust ratio? Why? Is the magnitude of the thrust greater here?

5F Preliminary experiments have been conducted using pure hydrogen H_2 as the fuel for a turbojet engine. Bearing in mind the fact that the heating value h_f is approximately 52,000 Btu/lbm of this fuel, as compared with 18,500 Btu/lbm for more conventional jet propellants, study quantitatively what can be achieved. Review some advantages and disadvantages of this proposal, assuming that the H_2 would be stored in the vehicle in liquid form.

5G A nuclear rocket with a chamber pressure of 1000 psia uses gaseous hydrogen as a fluid. Compare the thrust and specific impulse performance for the following cases.

a) The rocket uses a solid reactor ($T_c = 2500$°K).

b) The rocket uses a gas core reactor ($T_c = 5000$°K).

(Assume ideal operation in space.)

5H Design a rocket engine to propel a single-stage sounding rocket (initial weight 10,000 lb, of which 90% is propellant) vertically to great height. Initial thrust-to-weight ratio should be about 1.2 and chamber pressure 2000 psi. Assume ideal operation of $H_2 + O_2$ propellant, *except* for the effect of differences between p_e and p_0 as functions of altitude. Plot a few points on the acceleration time history; estimate altitude at burnout and maximum altitude.

5I There are three definitions of inlet or diffuser efficiency in common usage:

a) $\pi_d = \dfrac{p_{2t}}{p_{0t}}$

b) η_N defined by

$$\frac{p_{2t}}{p_0} = \left(1 + \eta_N \frac{\gamma - 1}{2} M_0^2\right)^{\gamma/(\gamma-1)}$$

c) *Isentropic efficiency* η_d, defined by

$$\eta_d = \frac{(h_{2t})_{\text{Isen}} - h_0}{h_{2t} - h_0} = \frac{(T_{2t})_{\text{Isen}} - T_0}{T_{2t} - T_0}$$

where $(T_{2t})_{\text{Isen}}$ represents the state that would be reached by isentropic compression to the *actual* pressure p_{2t}.

Derive the following alternative formulas for η_d:

$$\eta_d = \frac{\left(\dfrac{p_{2t}}{p_0}\right)^{(\gamma-1)/\gamma} - 1}{\dfrac{\gamma - 1}{2} M_0^2} = \frac{\left[(\pi_d)^{(\gamma-1)/\gamma}\left(1 + \dfrac{\gamma - 1}{2} M_0^2\right) - 1\right]}{\dfrac{\gamma - 1}{2} M_0^2}$$

Show that it is also logical to call η_d the *kinetic-energy efficiency*, in the sense that it connects the original kinetic energy per unit mass of the oncoming airstream to the KE that would be obtained by isentropically re-expanding the gas from state 2 to atmospheric pressure.

Given the experimentally determined variation of π_d with flight Mach number for an inlet, explain how corresponding plots of the other two efficiencies are obtained. Make a sketch plot in a particular case, and decide which definition you would prefer to use if you were in the inlet business.

5J The general formula derived for jet thrust of an air-breathing engine is

$$T = A_6(p_6 - p_0) + (\dot{m} + \dot{m}_f)u_6 - \dot{m}u_0$$

Explain physically why the ideal operating condition occurs when pressures p_6 and p_0 are matched to one another. Then attempt, perhaps with the assistance of a cited reference, to arrive at a quantitative demonstration.

5K It is obviously important to determine what contributions large fleets of aircraft make to atmospheric pollution. As examples, consider the following supersonic types.

	YF-12	US SST (hypothetical)	European SST (hypothetical)
Estimated number of vehicles	50	250	300
Utilization (hours/day)	2	6	6
Cruise L/D	6	9	8
No. of engines	2	4	4
Cruise altitude, ~ ft	70,000	70,000	60,000
Average cruise weight, ~ lb	100,000	600,000	400,000
Fuel used		Kerosene ($C_{12}H_{24}$ ~ approx.)	

Estimate the emission of H_2O and CO_2, considering stoichiometric combustion. Assume that the average residence time of H_2O and CO_2 in the stratosphere (above 36,000 ft) is one year. How much additional H_2O and CO_2 would be placed on a steady basis in comparison to the total H_2O and CO_2 present in the stratosphere? (Ref.: United States Air Force Handbook of Geophysics 1960)

5L a) A high-altitude, high-speed vehicle is propelled by a ramjet. Find a formula for this engine's thrust under the following assumptions:

$$p_0 \cong p_6; \qquad \gamma_T = \gamma = \text{constant}; \qquad f \ll 1$$

Efficiencies $\pi_d \cong \pi_b \cong 0.95$; controls so arranged that $\tau_\lambda \equiv T_{3t}/T_0$ is held constant, independent of M; mass flow proportional to $\rho_0 u_0$. Check your result from the fact that it should have the *form*

$$T = \rho_0\left[\sqrt{\frac{AM_0^2 + BM_0^4}{C + DM_0^2}} - FM_0^2\right]$$

where A, B, C, D are constants. What is I_{sp}?

b) The ramjet in (a) propels the vehicle with *fixed* C_D up a rectilinear stratospheric flight path inclined 30° above the horizon. T_0 remains constant in this region, but density varies with altitude, according to

$$\rho_0 = \rho_{Ref} e^{-(h - h_{Ref})/K}$$

Write an equation of motion, starting from Mach number $M = 2$ at $h = h_{Ref}$. For the special case in which the weight term can be neglected relative to thrust, drag, and acceleration, see if you can separate your equation in terms of M_0 and h. Then discuss solving for the history of M_0 versus altitude (and therefore versus time). Remember:

$$\frac{du_0}{dt} = \frac{du_0}{ds}\frac{ds}{dt}$$

CHAPTER 6

SMALL-PERTURBATION RESPONSE AND DYNAMIC STABILITY OF FLIGHT VEHICLES

6.1 EQUATIONS OF MOTION; AERODYNAMIC APPROXIMATIONS; STABILITY DERIVATIVES

Our purpose in this chapter is to review the dramatic simplifications of the vehicle equations of motion that can be accomplished by means of the same assumption regarding limited disturbances from an equilibrium state which proved so fruitful for the aerodynamic theory of Chapter 4. We shall try to suggest in Chapter 7 some ways whereby the results contribute toward aiding the designer's task. One notable instance is the connection between the characteristic properties of the linearized equations and the flying or "handling qualities" perceived by pilots—a discovery which has gone a long way toward quantifying the assurance of pilot acceptance early in the conceptual stage of new designs.

For a rigid aircraft such as that portrayed in Fig. 2.1, we start with system (2.39). Sometimes we must associate with (2.39) a set of relations like (2.29) and (2.37) connecting the linear and angular velocity to an earth or atmospheric reference frame, but for small-perturbation analyses these can usually be set aside once the equilibrium condition is chosen. Section 3.2 listed categories of problems whose treatment rests on such equations of motion. Among them, dynamic stability and control are the most amenable to the present approach.

We have found it an amusing exercise to investigate, by the "method of multiple scaling" (cf. Kevorkian, 1966, or Van Dyke, 1964, Section 10.4), the *formal* circumstances in which dynamic stability can be analyzed separately from what we call dynamic performance, which deals with longer-term properties of time-dependent trajectories. Although the subject is beyond our scope here, we remark that one test permitting such a separation in the longitudinal equations is the case in which $(L/D) \gg 1$. Thus the same feature which enhances efficient cruising and gliding also validates certain of the approximations which we shall adopt. There are undoubtedly other order-of-magnitude limitations that can be placed on specific results of the theory, and a productive topic for research might be to seek them out systematically.

We follow Etkin (1959, 1972), Seckel (1964), and other classical presentations by first perturbing about a rectilinear flight path, with the wings level ($\Phi = 0$) and the constant velocity vector

$$\mathbf{v}_{c0} = u_0\mathbf{i} \tag{6.1}$$

inclined above the horizontal at an angle θ_0. We observe that the initial equilibrium could just as well involve a steady coordinated turn, constant rate of roll, curving entry trajectory, or some other maneuver. A few of these cases will be taken up in later discussion and assigned problems. In the assumed reference condition, all aerodynamic moments about the CM must vanish, and the steady forces (X_0, Y_0, Z_0) are governed by a suitable reduction of (2.39) or (2.40),

$$X_0 - mg \sin \theta_0 = 0; \qquad Y_0 = 0; \qquad Z_0 + mg \cos \theta_0 = 0 \qquad (6.2a,b,c)$$

When the vehicle encounters turbulence or when its path is intentionally altered by limited applications of the aerodynamic controls, deviations can occur in the motion coordinates which are functions of time. Those needed here we introduce as follows:

$$U = u_0 + u(t); \qquad V = v(t); \qquad W = w(t) \qquad (6.3a,b,c)$$

$$P = p(t); \qquad Q = q(t); \qquad R = r(t) \qquad (6.3d,e,f)$$

$$\Phi = \varphi(t); \qquad \Theta = \theta_0 + \theta(t) \qquad (6.3g,h)$$

For the translational velocity components, our fundamental approximation [cf. (4.2)] is

$$|u|, |v|, |w| \ll u_0 \qquad (6.4a)$$

The angles of pitch and bank being dimensionless, we also require

$$|\varphi|, |\theta| \ll 1 \text{ rad} \qquad (6.4b)$$

whereas restrictions like

$$\left| \frac{pb}{2u_0} \right| \ll 1 \qquad (6.4c)$$

are imposed on the angular velocities. Since $b/2$ is the wing semispan, (6.4c) can be interpreted as meaning that the instantaneous helix traced by the tip during rolling is always only slightly inclined to the flight direction.

In a similar fashion, the forces and moments experience time-dependent perturbations defined by

$$X = X_0 + \Delta X(t); \qquad Y = \Delta Y(t); \qquad Z = Z_0 + \Delta Z(t) \qquad (6.5a,b,c)$$

$$L_r = \Delta L_r(t); \qquad M_p = \Delta M_p(t); \qquad N = \Delta N(t) \qquad (6.5d,e,f)$$

The incremental forces are assumed small compared with $|Z_0|$, which nearly equals the equilibrium lift. A quantity such as $|Z_0|\, b$ or $|Z_0|\, \bar{c}$, where $\bar{c}$ is mean aerodynamic chord, serves as a suitable bound for ΔL_r, ΔM_p, and ΔN.

Perturbation equations of motion are constructed by first substituting (6.3) and (6.5) into (2.39). The large equilibrated terms are then removed from the system by subtracting (6.2a) and (6.2c) from the corresponding force relations in (2.39). In the process of eliminating the steady gravitational forces, we employ

(6.4b) to justify

$$\sin \Theta \equiv \sin (\theta_0 + \theta) \cong \sin \theta_0 + \theta \cos \theta_0 \tag{6.6a}$$

$$\cos \Theta \equiv \cos (\theta_0 + \theta) \cong \cos \theta_0 - \theta \sin \theta_0 \tag{6.6b}$$

Finally all nonlinear inertia terms are dropped from the right sides, and we recover

$$\Delta X - \theta[mg \cos \theta_0] = m\dot{u} \tag{6.7a}$$

$$\Delta Y + \varphi[mg \cos \theta_0] = m[\dot{v} + u_0 r] \tag{6.7b}$$

$$\Delta Z - \theta[mg \sin \theta_0] = m[\dot{w} - u_0 q] \tag{6.7c}$$

$$\Delta L_r = I_{xx}\dot{p} - I_{xz}\dot{r} \tag{6.7d}$$

$$\Delta M_p = I_{yy}\dot{q} \tag{6.7e}$$

$$\Delta N = -I_{xz}\dot{p} + I_{zz}\dot{r} \tag{6.7f}$$

Equations (6.7) can readily be reorganized into the state-vector form (2.41), with all first derivatives of $\{u, v, w, p, q, r\}$ explicitly displayed. For this purpose we recommend the matrix representation used by Etkin [1972, see, e.g., his (5.10,23)]. We warn, however, that at this stage such a step would imply that ΔX, etc., were independent of the motion-coordinate derivatives.

Incidentally, we suggest that you look into extending (6.7) and subsequent developments to the asymmetric vehicle, for which I_{xy} and I_{yz} are not zero. In view of recent design proposals like Jones's (1972), such generalizations may soon acquire practical importance.

Similarly, the foregoing treatment can also be applied to the flight-path and Euler-angle formulas (2.29) and (2.37). With $\psi(t)$ the perturbed azimuth angle, we thus obtain

$$\dot{\varphi} = p + r \tan \theta_0; \qquad \dot{\theta} = q; \qquad \dot{\psi} = r \sec \theta_0 \tag{6.8a,b,c}$$

$$\dot{x}' = [u_0 + u] \cos \theta_0 - \theta u_0 \sin \theta_0 + w \sin \theta_0 \tag{6.8d}$$

$$\dot{y}' = v + \psi u_0 \cos \theta_0 \tag{6.8e}$$

$$\dot{z}' = -[u_0 + u] \sin \theta_0 - \theta u_0 \cos \theta_0 + w \cos \theta_0 \tag{6.8f}$$

Next we face the crucial and sometimes controversial question of how to linearize the aerodynamic and propulsive terms. Specifically, are steady-flow relations like those derived for wings in Chapter 4 suitable? In a rigorous sense, the answer is no. Since the motion is unsteady, so must be the flow; unsteady aerodynamic theory tells us therefore (cf. Bisplinghoff and Ashley, 1962, Chapter 4) that a typical airload $A = A_0 + \Delta A$ depends on the past history as well as the present value of each coordinate. In symbolic form, a linear expression for the contribution of, say, $w(t)$ to $\Delta A(t)$ would read

$$\Delta A(t) = \int^{t} w(\tau) h_{Aw}(t - \tau)\, d\tau \tag{6.9}$$

Here τ is a dummy integration variable, and h_{Aw} is the *unit impulse response*, giving the "output" ΔA due to an artificially defined impulsive change in the "input" w which takes place at the time when the argument of h_{Aw} vanishes. The lower limit in (6.9) might be $\tau = 0$ or whatever other instant is chosen for the start of the disturbed motion. If it is set at zero, the convolution theorem of Laplace transformation shows us that (6.9) has a very simple transform.*

$$\Delta\bar{A}(s) = \bar{w}(s)\bar{h}_{Aw}(s) \tag{6.10}$$

A useful property (cf. Doetsch 1944, page 187) is that small values of s are associated with the behavior of the system at large t—in the present case, with the airloads due to very slow variations in $w(t)$. Accordingly, if the motion is appropriately "slow" and if $\bar{h}_{Aw}$ is a regular function near $s = 0$ (which is not always true), one might think of replacing $\bar{h}_{Aw}(s)$ in (6.10) with its Maclaurin series expansion about $s = 0$. Since each multiplication by a factor s corresponds to $(d/dt)(\ \ )$ in the time domain, the inverse of the resulting expansion amounts to a relationship of the following form:

$$\Delta A(t) = A_w w(t) + A_{\dot{w}}\dot{w}(t) + A_{\ddot{w}}\ddot{w}(t) + \ldots \tag{6.11}$$

The constant coefficients in (6.11) have interpretations such as

$$A_w = \bar{h}_{Aw}(0) \tag{6.12a}$$

$$A_{\dot{w}} = \left.\frac{d\bar{h}_{Aw}}{ds}\right|_{s=0} \tag{6.12b}$$

In the light of the aforementioned connection between $s = 0$ and $t \to \infty$, moreover, A_w is the factor connecting w and ΔA when the disturbance varies so gradually that unsteady effects are negligible. Similarly, $A_{\dot{w}}$ is the "quasi-steady" coefficient which gives the correction to $A_w w$ due to whatever rate of change of w does occur, and so forth. As an illustration, we might think of A as the lift force and replace w with the angle-of-attack perturbation $\alpha \equiv (w/u_0)$. A_w would have as its principal part $(C_{L\alpha}\rho_\infty u_0^2 S/2)$ and is thus proportional to the steady lift-curve slope of the vehicle.

Considerations like those leading to (6.11) provide a modern rationale for the concept of "stability derivative," which was introduced into aeronautics by Bryan (1911). Our particular approach may become unwieldy if carried to absurd extremes (cf. the discussion in Rodden and Giesing, 1970), but Bryan's approximation has generally served well for dynamic analysis of typical cruising aircraft.

---

* We recall the introduction of these transformations and the literature citations in Chapter 4. $\Delta\bar{A}(s)$, for instance, is found from $\Delta A(t)$ through an equation like (4.81), with x replaced by t. In systems analysis $\bar{h}_{Aw}(s)$ would be called a "transfer function." Expressions like (6.10) permit unsteady effects to be included when dynamic stability and response are being investigated by Laplace or Fourier transformation, but at the price of much more complicated input–output relations.

In the light of (5.11), therefore, we treat ΔA as depending on the instantaneous values of the motion coordinates and some of the more important first derivatives, writing

$$\Delta A = F(t; u, \dot{u}, v, \dot{v}, w, \dot{w}, p, \dot{p}, q, \dot{q}, r, \dot{r}, \delta_a, \delta_e, \delta_r, \ldots)$$
$$\cong A_u u + A_{\dot{u}} \dot{u} + A_v v + \cdots + A_p p + \cdots + A_{\delta_r} \delta_r + \cdots \quad (6.13)$$

In (6.13), time t is included explicitly to allow for such possibilities as the gradual change in air density along a climbing flight path. The angular displacements δ_a, δ_e, δ_r of aileron, elevator, and rudder, respectively, remind us that loads arise from applying the controls. The coefficients in the third member of (6.13) constitute a linearization, justified both by (6.11) and by the smallness of u, v, etc. The Maclaurin expansion gives a representative definition,

$$A_u \equiv \left(\frac{\partial A}{\partial u}\right)_{\text{Reference flight condition}} \quad (6.14)$$

Figure 6.1 depicts A_u as the slope, at $u = 0$, on a plot of A versus U taken from steady-flow data, with the proviso that all other motion coordinates are held at their equilibrium values while u is being varied.

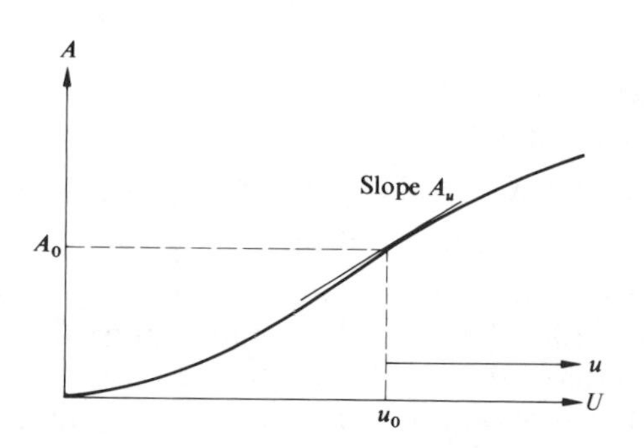

Fig. 6.1. Plot of typical airload A versus x-component U of the relative wind, shown for case in which all other motion coordinates are held constant at their equilibrium values. The stability derivative A_u is the slope at flight speed u_0.

A_u, A_v, etc., are dimensional forms of the derivatives, each being constant for a fixed configuration, speed, and altitude. Although representative formulas will be derived in Sections 6.3 and 6.4, we are already able to make certain generalizations relative to a symmetrical vehicle. Some are obvious consequences of symmetry itself, while others result from linearity or the appeal to physical observations. Four such are the following:

1) Lateral forces and moments (i.e., those which act out of the plane of symmetry) are identically zero during purely longitudinal motion. Therefore, derivatives of Y, L_r, and N with respect to u, w, q, δ_e, and their time rates must vanish.

So long as $p = r = v = 0$, this statement does not rely on small perturbations, but holds for arbitrary motion in the symmetry plane. It is invalidated, however, by such power-plant effects as that of a swirling propeller slipstream flowing past the empennage.

2) To first order in the coordinates, derivatives of X, Z, and M_p with respect to v, p, r, δ_a, δ_r and their time rates are zero. This follows from the fact that each perturbation in longitudinal force is an even function of each lateral coordinate; the zero slope at the reference condition is illustrated for vertical force due to sideslip in Fig. 6.2.

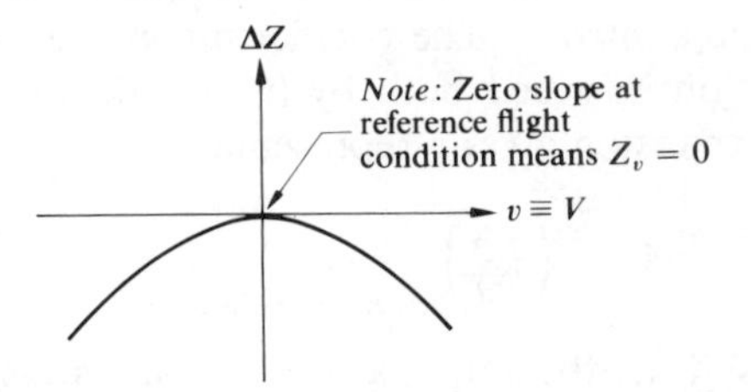

Fig. 6.2. Plot versus sideslip velocity component v of the incremental vertical force ΔZ, used to illustrate the vanishing of derivatives of longitudinal airloads with respect to lateral coordinates.

3) Of all derivatives involving linear or angular accelerations, only $M_{\dot{w}}$ and $Z_{\dot{w}}$ normally deserve retention.

4) For most aircraft, X_q, X_{δ_e}, $X_{\dot{\delta}_e}$, $Z_{\dot{\delta}_e}$, Y_{δ_a}, $Y_{\dot{\delta}_a}$, $Y_{\dot{\delta}_r}$, $L_{\dot{\delta}_r}$, and $N_{\dot{\delta}_a}$ are negligible.

We rely on Section 4-14 of Etkin (1959) for approximations (3) and (4). Their apparent arbitrariness again prompts us to recall that only by resort to experience, measurement, and rigorous theory applicable to each particular situation can the analyst decide which derivatives must be retained—or, for that matter, whether the whole quasi-steady apparatus is accurate enough for his purposes.

With the foregoing omissions, we can now set down aerodynamic and propulsive linearizations suitable for (6.7).

$$\Delta X = X_u u + X_w w \tag{6.15a}$$

$$\Delta Y = Y_v v + Y_p p + Y_r r + Y_{\delta_r}\delta_r \tag{6.15b}$$

$$\Delta Z = Z_u u + Z_w w + Z_{\dot{w}}\dot{w} + Z_q q + Z_{\delta_e}\delta_e \tag{6.15c}$$

$$\Delta L_r = L_v v + L_p p + L_r r + L_{\delta_a}\delta_a + L_{\dot{\delta}_a}\dot{\delta}_a + L_{\delta_r}\delta_r \tag{6.15d}$$

$$\Delta M_p = M_u u + M_w w + M_{\dot{w}}\dot{w} + M_q q + M_{\delta_e}\delta_e + M_{\dot{\delta}_e}\dot{\delta}_e \tag{6.15e}$$

$$\Delta N = N_v v + N_p p + N_r r + N_{\delta_a}\delta_a + N_{\delta_r}\delta_r + N_{\dot{\delta}_r}\dot{\delta}_r \tag{6.15f}$$

When (6.15) are substituted into (6.7), it is no surprise to find (6.7) separating into two uncoupled sets: three for the longitudinal or symmetrical unknowns u, w, $q \equiv \dot{\theta}$ and three in the lateral v, p, r. Anticipating the study of maneuvers, we write these with the control-surface terms on the right, in the role of specified inputs. When the controls are "reversible" and unrestrained, however, it may be necessary to treat δ_a, δ_e, δ_r as additional unknowns and supply three hinge-moment equations for their motion (cf. Etkin 1959, Section 4.14).

With necessary additions from (6.8) to ensure determinateness, we assemble the following systems.

1. Longitudinal Dynamic Stability and Response

$$\left[m\frac{d}{dt} - X_u\right]u - X_w w + [mg\cos\theta_0]\theta = 0 \tag{6.16a}$$

$$-Z_u u + \left[(m - Z_{\dot{w}})\frac{d}{dt} - Z_w\right]w - \left[(mu_0 + Z_q)\frac{d}{dt} - mg\sin\theta_0\right]\theta = Z_{\delta_e}\delta_e \tag{6.16b}$$

$$-M_u u - \left[M_{\dot{w}}\frac{d}{dt} + M_w\right]w + \left[I_{yy}\frac{d^2}{dt^2} - M_q\frac{d}{dt}\right]\theta = \left[M_{\dot{\delta}_e}\frac{d}{dt} + M_{\delta_e}\right]\delta_e \tag{6.16c}$$

$$\dot{\theta} = q \tag{6.16d}$$

2. Lateral Dynamic Stability and Response

$$\left[m\frac{d}{dt} - Y_v\right]v - Y_p p + [mu_0 - Y_r]r - [mg\cos\theta_0]\varphi = Y_{\delta_r}\delta_r \tag{6.17a}$$

$$-L_v v + \left[I_{xx}\frac{d}{dt} - L_p\right]p - \left[I_{xz}\frac{d}{dt} + L_r\right]r = \left[L_{\dot{\delta}_a}\frac{d}{dt} + L_{\delta_a}\right]\delta_a + L_{\delta_r}\delta_r \tag{6.17b}$$

$$-N_v v - \left[I_{xz}\frac{d}{dt} + N_p\right]p + \left[I_{zz}\frac{d}{dt} - N_r\right]r = N_{\delta_a}\delta_a + \left[N_{\dot{\delta}_r}\frac{d}{dt} + N_{\delta_r}\right]\delta_r \tag{6.17c}$$

$$\dot{\varphi} = p + r\tan\theta_0 \tag{6.17d}$$

$$\dot{\psi} = r\sec\theta_0 \tag{6.17e}$$

It is straightforward to construct state-vector versions of (6.16) or (6.17). By reference to (2.41), (2.43), (2.45), etc., we observe that the longitudinal system has state four, with

$$\{\mathscr{X}\} = \begin{Bmatrix} u \\ w \\ q \\ \theta \end{Bmatrix} \tag{6.18}$$

Formally, the control $\{\mathscr{U}\}$ contains the two elements $\dot{\delta}_e$, δ_e. The lateral system has state five, except that ψ may be decoupled and calculated as an afterthought; the controls are $\dot{\delta}_a$, δ_a, $\dot{\delta}_r$, and δ_r.

6.2 DIMENSIONLESS EQUATIONS OF MOTION

Several authors on dynamic stability (Kolk 1961, Seckel 1964, page 220) argue persuasively against complete nondimensionalization. They point out, for instance, that physical time must be kept as the independent variable when an analog mechanization is to be coupled with human pilots in a simulator. We respect this view, and we refer you to their books for the treatment of numerous examples. In the present context, however, the dimensionless equations are justified by the succinctness with which their solutions display large amounts of information.

Our cue for normalizing aerodynamic terms in (6.16)–(6.17) is taken, to the extent possible, from the natural dependence of lift, moment, etc., on density, airspeed, and geometry. This relationship was introduced at (3.5) and later verified theoretically. During perturbed motion, the instantaneous speed

$$\begin{aligned} v_c &= \sqrt{U^2 + V^2 + W^2} \\ &= \sqrt{(u_0 + u)^2 + v^2 + w^2} \end{aligned} \tag{6.19}$$

is what determines airloads. Although $v_c \cong u_0$, we shall find that their difference sometimes affects even linearized phenomena. A potential source of confusion is the custom of using $\bar{c}/2$ as the longitudinal reference length and semispan $b/2$ for the lateral case, whereupon a few symbols (notably density ratio μ and dimensionless time) acquire different meanings, depending on which case is under study. To avoid the difficulty here, we propose subscripts s (symmetrical) and a (antisymmetrical) to distinguish them.

Table 6.1 lists the quantities and groups that we shall divide out of constants or variables having the dimensions indicated in the first column. The application of this approach to various terms in (6.16) and (6.17) then leads to the many new coefficients and other symbols defined in the two parts of Table 6.2.

For completeness, we have added the aerodynamic hinge moments $H_{(\ldots)}$ experienced by the aileron, elevator, and rudder. The corresponding area and

Table 6.1. Typical nondimensionalizing factors

Physical quantity	Reference quantity or group		Physical quantity	Reference quantity or group	
	Longitudinal motion	Lateral motion		Longitudinal motion	Lateral motion
Time	$\bar{c}/2u_0$	$b/2u_0$	Speed	$v_c \cong u_0$	$v_c \cong u_0$
Mass	$\rho_\infty S\bar{c}/2$	$\rho_\infty Sb/2$	Pressure	$\rho_\infty v_c^2/2$	$\rho_\infty v_c^2/2$
Length	$\bar{c}/2$	$b/2$	Force	$\rho_\infty v_c^2 S/2$	$\rho_\infty v_c^2 S/2$
Area	S	S	Moment	$\rho_\infty v_c^2 S\bar{c}/2$	$\rho_\infty v_c^2 Sb/2$

Table 6.2. Summary of dimensionless notation used for aircraft dynamic stability and response

I. *Longitudinal case*

Dimensional quantity	Dimensionless quantity
X, Z	$C_x = \dfrac{X}{\dfrac{\rho_\infty}{2} v_c^2 S}, \qquad C_z = \dfrac{Z}{\dfrac{\rho_\infty}{2} v_c^2 S}$
M_p	$C_m = \dfrac{M_p}{\dfrac{\rho_\infty}{2} v_c^2 S\bar{c}}$
H_e	$C_{he} = \dfrac{H_e}{\dfrac{\rho_\infty}{2} v_c^2 S_e c_e}$
u, w	$\hat{u} = u/u_0, \qquad \alpha = \tan^{-1}\dfrac{W}{U} \cong \dfrac{w}{u_0}$
q	$\hat{q} = \dfrac{q\bar{c}}{2u_0}$
m	$\mu_s = \dfrac{m}{\rho_\infty S \dfrac{\bar{c}}{2}}$
I_{yy}	$i_{yy} = \dfrac{I_{yy}}{\rho_\infty S\left(\dfrac{\bar{c}}{2}\right)^3}$
t	$\hat{t}_s = u_0 t/(\bar{c}/2)$
$(\dot{\ldots}) = \dfrac{d}{dt}(\ldots)$	$D_s(\ldots) = \dfrac{d}{d\hat{t}_s}(\ldots) = \dfrac{\bar{c}}{2u_0}(\dot{\ldots})$

(*continued*)

Table 6.2 (*continued*)

	II. *Lateral case*
Dimensional quantity	Dimensionless quantity
Y	$C_y = \dfrac{Y}{\frac{\rho_\infty}{2} v_c^2 S}$
L_r, N	$C_l = \dfrac{L_r}{\frac{\rho_\infty}{2} v_c^2 Sb}, \quad C_n = \cdots$
H_a, H_r	$C_{ha} = \dfrac{H_a}{\frac{\rho_\infty}{2} v_c^2 S_a c_a}, \quad C_{hr} = \cdots$
v	$\beta = \sin^{-1} \dfrac{v}{v_c} \cong \dfrac{v}{u_0}$
p, r	$\hat{p} = \dfrac{pb}{2u_0}, \quad \hat{r} = \dfrac{rb}{2u_0}$
m	$\mu_a = \dfrac{m}{\rho_\infty S \frac{b}{2}}$ (Notational ambiguity here)
I_{xx}, I_{xz}, I_{zz}	$i_{xx} = \dfrac{I_{xx}}{\rho_\infty S\left(\frac{b}{2}\right)^3}, \quad i_{xz} = \ldots, \quad i_{zz} = \ldots$
t	$\hat{t}_a = u_0 t/(b/2)$ (Notational ambiguity here)
$(\dot{\ldots}) = \dfrac{d}{dt}(\ldots)$	$D_a(\ldots) = \dfrac{d}{d\hat{t}_a}(\ldots) = \dfrac{b}{2u_0}(\dot{\ldots})$ (Notational ambiguity here)

chord behind the control-surface hingeline are customarily used to make these dimensionless, for example, S_a and c_a, respectively, for the aileron. These would become total area and chord on an all-movable surface, whereas other appropriate definitions are needed for spoilers, extendable-chord flaps, and the like.

We also call attention to the conventional definitions of angle of attack α and angle of sideslip β. Thus

$$\alpha \equiv \tan^{-1}\left(\frac{W}{U}\right) \cong \frac{w}{u_0} \tag{6.20}$$

is the perturbation angle between the x-axis and relative wind that would be seen by an observer looking normal to the plane of symmetry. The middle member of (6.20) is meaningful even for large disturbances. In any event, α always differs by an additive term from the "angle from zero lift" α_{ZL} employed in Chapters 3 and 4. Similarly,

$$\beta = \sin^{-1}\left(\frac{V}{v_c}\right) \cong \frac{v}{u_0} \tag{6.21}$$

is observed from above, but in a plane containing v_c and V.

It seems unprofitable to discuss every entry in Table 6.2, since their significances will emerge as we develop the dimensionless equations and stability-derivative formulas. Instead, we begin by going through four representative longitudinal derivatives to exemplify how terms in (6.16) are handled and how the variable airspeed is accounted for. Subscript zero here identifies the reference flight condition.

1. Normal Force Due to Angle of Attack

$$\begin{aligned} Z_w \equiv \left(\frac{\partial Z}{\partial w}\right)_0 &= \left.\frac{\partial[C_z \rho_\infty v_c^2 S/2]}{\partial w}\right|_0 \\ &= \left(\frac{\rho_\infty v_c^2 S}{2}\right)_0 \left(\frac{\partial C_z}{\partial w}\right)_0 + \left(C_z \frac{\rho_\infty}{2} S\right)_0 \left(\frac{\partial v_c^2}{\partial w}\right)_0 \\ &= \left(\frac{\rho_\infty v_c^2 S}{2}\right)_0 \left(\frac{1}{u_0}\frac{\partial C_z}{\partial \alpha}\right)_0 + \left(C_z \frac{\rho_\infty}{2} S\right)_0 (2w)_0 \\ &= \frac{\rho_\infty u_0 S}{2} C_{z\alpha} + \text{(zero)} \end{aligned} \tag{6.22}$$

These steps are presented *in extenso* to clarify such questions as how (6.20) is used to divide out u_0 and obtain the wholly dimensionless $C_{z\alpha}$. Note also that, after we differentiate each factor dependent on w, we can finally replace v_c by u_0 and w by 0. Equation (6.19), with $W \equiv w$, leads to the latter replacement, illustrating the fact that variations in v_c due to v and w do not affect the linearized derivatives, whereas we shall see that u does do so.

2. Pitching Moment Due to Pitch Velocity

$$M_q \equiv \left(\frac{\partial M_p}{\partial q}\right)_0 = \left.\frac{\partial[C_m \rho_\infty v_c^2 S\bar{c}/2]}{\partial\left[\dfrac{2u_0}{\bar{c}} \cdot \dfrac{q\bar{c}}{2u_0}\right]}\right|_0$$

$$= \frac{\left[\dfrac{\rho_\infty}{2} v_c^2 S\bar{c}\right]_0}{\dfrac{2u_0}{\bar{c}}} \left(\frac{\partial C_m}{\partial \hat{q}}\right)_0 = \frac{\rho_\infty u_0 S\bar{c}}{4} C_{mq} \tag{6.23}$$

Note here that v_c is independent of q. Furthermore, the caret is omitted from subscript $\hat{q}$ when writing the wholly dimensionless C_{mq}; this general usage avoids difficult typesetting and other inconveniences.

3. Normal Force Due to Plunging Acceleration

$$Z_{\dot{w}} \equiv \left(\frac{\partial Z}{\partial \left(\dfrac{dw}{dt} \right)} \right)_0 = \frac{\partial [C_z \rho_\infty v_c^2 S/2]}{\partial \left[\dfrac{2u_0^2}{\bar{c}} \dfrac{d(w/u_0)}{d(2u_0 t/\bar{c})} \right]} \Bigg|_0$$

$$= \frac{\left[\dfrac{\rho_\infty}{2} v_c^2 S \right]_0}{\dfrac{2u_0^2}{\bar{c}}} \left(\frac{\partial C_z}{\partial \left[\dfrac{d\alpha}{d\hat{t}_s} \right]} \right)_0$$

$$= \frac{\rho_\infty S \bar{c}}{4} C_{z\dot{\alpha}} \tag{6.24}$$

4. Normal Force Due to Forward Velocity

$$Z_u \equiv \left(\frac{\partial Z}{\partial u} \right)_0 = \frac{\partial [C_z \rho_\infty v_c^2 S/2]}{\partial \left[u_0 \cdot \dfrac{u}{u_0} \right]} \Bigg|_0$$

$$= \left(\frac{\rho_\infty v_c^2 S}{2u_0} \right)_0 \left(\frac{\partial C_z}{\partial \hat{u}} \right) + \left(\frac{C_z \rho_\infty S}{2u_0} \right)_0 u_0 \left(\frac{\partial v_c^2}{\partial u} \right)_0$$

$$= \frac{\rho_\infty u_0 S}{2} C_{zu} + \frac{C_{z0} \rho_\infty S}{2} \cdot 2[u_0 + u] \Big|_0$$

$$= \frac{\rho_\infty u_0 S}{2} C_{zu} - \rho_\infty u_0 S C_{L0} \tag{6.25}$$

It is useful here to introduce the trimmed coefficient of lift C_{L0} in place of C_{z0}, which is its negative. The first-order influence of u on airspeed changes [i.e., the last term in (6.25)] may be visualized because this component lengthens the vector $\mathbf{v}_c$ in direct proportion to itself, while v and w, when small, rotate $\mathbf{v}_c$ without significantly altering its magnitude.

Our results (6.22), (6.24), and (6.25) can be applied toward the nondimensionalization of the entire equation (6.16b), which we go through as an example. Each term has dimensions of force, and must therefore be divided by the group $(\rho_\infty v_c^2 S/2)$, where $v_c \rightarrow u_0$ in the reference condition. Thus the product $Z_w w$, in the light of (6.22), is treated as follows:

$$\frac{1}{\dfrac{\rho_\infty}{2} u_0^2 S} Z_w w = \frac{1}{\dfrac{\rho_\infty}{2} u_0^2 S} \left[\frac{\rho_\infty u_0 S}{2} C_{z\alpha} \right] u_0 \frac{w}{u_0} = C_{z\alpha} \alpha \tag{6.26}$$

after evident cancellations. Other terms whose transformation is not obvious are the following:

$$\frac{1}{\frac{\rho_\infty}{2}u_0^2S}\,m\frac{dw}{dt} = \frac{2m}{\rho_\infty S\bar{c}}\frac{2d(w/u_0)}{d(2u_0t/\bar{c})} = 2\mu_s D_s\alpha \tag{6.27}$$

($mu_0\,d\theta/dt$ is dealt with in an identical fashion.)

$$\frac{1}{\frac{\rho_\infty}{2}u_0^2S}[mg\sin\theta_0]\theta = \frac{\left[-Z_0\dfrac{\sin\theta_0}{\cos\theta_0}\right]\theta}{\frac{\rho_\infty}{2}u_0^2S}$$

$$= \left(\frac{L_0}{\frac{\rho_\infty}{2}u_0^2S}\right)[\tan\theta_0]\theta = [C_{L0}\tan\theta_0]\theta \tag{6.28}$$

[Here we have employed (6.2c) and, in turn, substituted trimmed lift for the force $-Z_0$.]

$$\frac{1}{\frac{\rho_\infty}{2}u_0^2S}Z_{\delta_e}\,\delta_e = \frac{\partial\left(\dfrac{Z}{\frac{\rho_\infty}{2}u_0^2S}\right)}{\partial\delta_e}\delta_e = C_{z\delta_e}\,\delta_e(\hat{t}_s) \tag{6.29}$$

When all members are collected in the same order as the original (6.16b), we finally obtain the equation numbered (6.30b) in the system below.

The remaining dimensionless equations of motion can be constructed by analogous manipulations, which we leave as an exercise. We arrange the results in the same format as (6.16, 17).

1. Longitudinal Dynamic Stability and Response

$$[2\mu_sD_s - 2C_{L0}\tan\theta_0 - C_{xu}]\hat{u} - C_{x\alpha}\alpha + C_{L0}\theta = 0 \tag{6.30a}$$

$$[2C_{L0} - C_{zu}]\hat{u} + [(2\mu_s - C_{z\dot{\alpha}})D_s - C_{z\alpha}]\alpha - [(2\mu_s + C_{zq})D_s - C_{L0}\tan\theta_0]\theta = C_{z\delta_e}\,\delta_e(\hat{t}_s) \tag{6.30b}$$

$$-C_{mu}\hat{u} - [C_{m\dot{\alpha}}D_s + C_{m\alpha}]\alpha + [i_{yy}D_s^2 - C_{mq}D_s]\theta = [C_{m\dot{\delta}_e}D_s + C_{m\delta_e}]\,\delta_e(\hat{t}_s) \tag{6.30c}$$

$$D_s\theta = \hat{q} \tag{6.30d}$$

2. Lateral Dynamic Stability and Response

$$[2\mu_a D_a - C_{y\beta}]\beta - C_{yp}\hat{p} + [2\mu_a - C_{yr}]\hat{r} - C_{L0}\varphi = C_{y\delta_r}\,\delta_r(\hat{t}_a) \quad (6.31a)$$

$$-C_{l\beta}\beta + [i_{xx}D_a - C_{lp}]\hat{p} - [i_{xz}D_a + C_{lr}]\hat{r}$$
$$= [C_{l\dot{\delta}_a}D_a + C_{l\delta_a}]\,\delta_a(\hat{t}_a) + C_{l\delta_r}\,\delta_r(\hat{t}_a) \quad (6.31b)$$

$$-C_{n\beta}\beta - [i_{xz}D_a + C_{np}]\hat{p} + [i_{zz}D_a - C_{nr}]\hat{r}$$
$$= C_{n\delta_a}\,\delta_a(\hat{t}_a) + [C_{n\dot{\delta}_r}D_a + C_{n\delta_r}]\,\delta_r(\hat{t}_a) \quad (6.31c)$$

$$D_a\varphi = \hat{p} + \hat{r}\tan\theta_0 \quad (6.31d)$$

$$D_a\psi = \hat{r}\sec\theta_0 \quad (6.31e)$$

Symbols in (6.30,31) are defined in Table 6.2, except for the derivatives of the aerodynamic-propulsive coefficients. Our next task is to discuss how the more significant of these derivatives may be roughly estimated from what we know about the forces and moments produced by the various parts of a cruising aircraft. Their precise determination must be left, of course, to that interactive process of testing and refined analysis to which we often allude in this book. As our review proceeds, we hope that you will seek a "feel" for both the physical meaning and the relative importance of each term in the two systems.

In anticipation of how they will be solved, let us mention that (6.30, 31) are ordinary differential equations in time, linear and with constant coefficients. Since it was for such situations that the Laplace-transform apparatus was evolved, we shall embrace it. Indeed, with zero initial conditions at $\hat{t}_{s,a} = 0$, transformation requires merely that we substitute the variable s for each of the D-operators.

6.3 ESTIMATION OF STABILITY DERIVATIVES: LONGITUDINAL

Chapters 6 through 8 of Etkin (1972) are the source for most of our insights in Sections 6.3 and 6.4. Appendix 1 and several chapters of Seckel (1964) also contain extensive derivative information, and such aerodynamic handbooks as the RAeS Data Sheets (see Anonymous) and the U.S. Air Force DATCOM (see Ellison and Malthan 1963) are valuable compendia.

We proceed through the derivatives in the order that the motion coordinates appear in (6.30, 31), introducing new concepts and notation as needed. With regard to the effects of propulsion systems, we make the simplifying assumption that thrust T acts through the CM and parallel to the x-direction of stability axes. But we warn that cases often arise in which this is unacceptable; for instance, on an airplane with turbine engines hung in pods beneath a low wing, T generates a substantial nose-up pitching moment during takeoff and slow flight. For consistency with aerodynamic usage, we adopt a *coefficient of thrust*

$$C_T \equiv \frac{T}{\dfrac{\rho_\infty}{2} v_c^2 S} \quad (6.32)$$

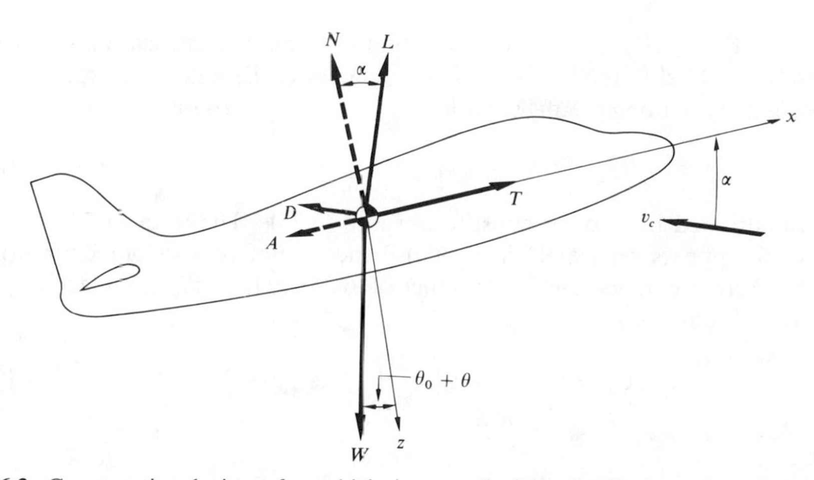

Fig. 6.3. Cross-sectional view of a vehicle in perturbed longitudinal motion, showing the instantaneous aerodynamic, propulsion, and gravitational forces.

Figure 6.3 shows a vehicle in longitudinal motion, with instantaneous values of v_c and α. The four principal forces act at the CM as indicated. Additionally, we depict in phantom the directions of normal force N and axial or *chord force* A as they might be related to the xz-system; observe that N and A are fixed to the vehicle and do not rotate with the relative wind as do L and D. The instantaneous X and Z can be expressed in the alternative forms

$$X = T - A = T + L \sin \alpha - D \cos \alpha \tag{6.33a}$$

$$Z = -N = -L \cos \alpha - D \sin \alpha \tag{6.33b}$$

When these are divided by the reference force from Table 6.1, and when $\alpha \ll 1$, we get

$$C_x = C_T - C_A = C_T + C_L\alpha - C_D \tag{6.34a}$$

$$C_z = -C_N = -[C_L + C_D\alpha] \tag{6.34b}$$

Thrust is, incidentally, independent of α under small perturbations, but may be a function of u.

1. The û-Derivatives

In order to explain the precise meaning attached to C_{xu} in the (6.30a) coefficient of $\hat{u}$, let us return to a dimensional examination of X_u in (6.16a). With $\alpha = 0$, differentiation of (6.33a) calls for

$$X_u u = \left[\left(\frac{\partial T}{\partial u}\right)_0 - \left(\frac{\partial D}{\partial u}\right)_0\right] u = \left[u_0\left(\frac{\partial T}{\partial u}\right)_0\right]\hat{u}$$

$$- \left[\frac{\partial\left[C_D \dfrac{\rho_\infty}{2} v_c^2 S\right]}{\partial u}\right]_0 u = \left[u_0\left(\frac{\partial T}{\partial u}\right)_0 - \frac{\rho_\infty}{2} u_0^2 S(C_{Du} + 2C_{D0})\right]\hat{u} \tag{6.35}$$

Now when Eqs. (3.10a,b) for rectilinear flight are made dimensionless and subscript zero is added to the coefficients, we can extract the following expression for force equilibrium along the flight path:

$$C_{D0} = C_{T0} - C_{L0} \tan \theta_0 \tag{6.36}$$

This is multiplied by two for substitution into (6.35). The term $(-2C_{L0} \tan \theta_0)$ from (6.36) appears separately in (6.30a), hence is not part of our definition of C_{xu}. We therefore divide the last member of (6.35) by $(\rho_\infty u_0^2 S/2)$, replace $2C_{D0}$ by $2C_{T0}$, and obtain

$$C_{xu} = \frac{1}{\frac{\rho_\infty}{2} u_0^2 S} u_0 \left(\frac{\partial T}{\partial u}\right)_0 - 2C_{T0} - C_{Du} \tag{6.37}$$

Examining the terms in (6.37), we note first that C_{D_u} refers to the change in drag coefficient with forward speed at fixed angle of attack, altitude, etc. As shown in Chapter 4, for a rigid vehicle Mach number M is the only parameter that has an appreciable effect on C_D. Hence we write

$$C_{Du} \equiv u_0 \left(\frac{\partial C_D}{\partial u}\right)_0 = \frac{u_0}{a_\infty}\left(\frac{\partial C_D}{\partial \mathrm{M}}\right)_0 = \mathrm{M}\left(\frac{\partial C_D}{\partial \mathrm{M}}\right)_0 \tag{6.38}$$

This derivative is usually most important transonically, in a range above the subsonic "drag break" M. At distinctly supersonic speeds we know that $C_D \sim (\mathrm{M}^2 - 1)^{-1/2}$, whence it can be reasoned that $\mathrm{M}(\partial C_D/\partial \mathrm{M})_0$ behaves asymptotically like $1/\mathrm{M}$.

For a rocket or a jet engine operating near its design point, analyses in Chapter 5 demonstrate that

$$\left(\frac{\partial T}{\partial u}\right)_0 \cong 0, \qquad \text{for rocket or gas-turbine propulsion} \tag{6.39}$$

On the other hand, many propeller-driven aircraft at fixed throttle tend to develop constant *thrust horsepower* (THP). It follows from

$$T[u_0 + u] \cong \text{constant} \tag{6.40a}$$

that

$$\frac{1}{\frac{\rho_\infty}{2} u_0^2 S} u_0 \left(\frac{\partial T}{\partial u}\right)_0 = -C_{T0}, \qquad \text{for constant THP} \tag{6.40b}$$

Etkin (1972, Section 7.8) adjusts (6.40b) for the influence of propeller efficiency, also furnishing more details on all the $\hat{u}$-derivatives.

A satisfactory working form for our purposes is*

$$C_{xu} = \frac{u_0}{\frac{\rho_\infty}{2} u_0^2 S}\left(\frac{\partial T}{\partial u}\right)_0 - 2C_{T0} - \mathrm{M}\left(\frac{\partial C_D}{\partial \mathrm{M}}\right)_0 \tag{6.41}$$

By reasoning similar to that underlying (6.38), we conclude that the other two derivatives involve only the effects of Mach number.

$$C_{zu} = -\mathrm{M}\left(\frac{\partial C_L}{\partial \mathrm{M}}\right)_0 \tag{6.42}$$

$$C_{mu} = \mathrm{M}\left(\frac{\partial C_m}{\partial \mathrm{M}}\right)_0 \tag{6.43}$$

We emphasize, however, that C_{zu} and C_{mu} are particularly susceptible to air-frame deformations. The variations in dynamic pressure due to u-changes, at fixed angle of attack, can give rise to a significant redistribution of air-loading due to flexibilities in wing and fuselage that are neglected in (6.42, 43). Even on a rigid vehicle, moreover, important transonic effects are manifested in $(\partial C_m/\partial \mathrm{M})_0$. The familiar rearward shift of AC near $\mathrm{M} = 1$ leads to negative values of this derivative. A substantial decrease in elevator setting δ_e is often required to maintain trim; especially on straight-winged aircraft, this is related to the "tucking under" tendency which was first experienced by pilots of certain World War II fighters.

2. The α-Derivatives and Elevator-Angle Derivatives

Recalling our neglect of the influence of angle of attack on thrust, we differentiate (6.34a) with respect to α and get

$$C_{x\alpha} \equiv \left.\left(\frac{\partial C_x}{\partial \alpha}\right)\right|_{\alpha=0} = \begin{cases} \left[C_{L0} - \left(\frac{\partial C_D}{\partial \alpha}\right)_0\right] \\ \text{or} \\ -\left(\frac{\partial C_A}{\partial \alpha}\right)_0 \equiv -C_{A\alpha} \end{cases} \tag{6.44}$$

For preliminary analyses, the dependence of C_D on C_L—hence on α—is usually approximated by Oswald's formula (4.168). (It was to avoid confusion with drag

---

* In what follows, we "box" that derivative formula which is deemed most suitable for preliminary aeronautical analyses. We cite Etkin (1959, 1972) and other texts for further valuable detail and recall that certain configurations (e.g., entry vehicles) always require special treatment.

at the reference condition that we chose a double subscript for the zero-lift $C_{D\,00}$.)
From (4.168) we obtain

$$\frac{\partial C_D}{\partial \alpha} = \frac{\partial C_D}{\partial C_L}\frac{\partial C_L}{\partial \alpha} = \frac{2C_{L0}}{\pi e} C_{L\alpha} \tag{6.45}$$

for substitution into (6.44) to produce the most useful form of the X-derivative.

$$\boxed{C_{x\alpha} = C_{L0}\left[1 - \frac{2C_{L\alpha}}{\pi e \, A\!\!R}\right]} \tag{6.46}$$

It follows from our Section 4.7 study of low-aspect-ratio wings [cf. (4.182)] that lift-curve slope $C_{L\alpha}$ has the quantity $(\pi \, A\!\!R/2)$ as a sort of upper bound. Since e is only a little less than unity, $C_{x\alpha}$ is normally a positive quantity.

Differentiation of (6.34b) yields

$$\boxed{C_{z\alpha} = \begin{cases} [-C_{L\alpha} + C_{D0}] \\ \text{or} \\ -C_{N\alpha} \end{cases}} \tag{6.47}$$

Since $C_{L\alpha}$ ranges between about 2 and 6, whereas C_{D0} is well below 0.1 on most designs, the last term in (6.47) may often be omitted. It is retained here, however, as a reminder that there are situations like ballistic entry in which it could easily dominate $C_{z\alpha}$.

The derivative $C_{m\alpha}$ provides us with an opportunity to introduce the elementary aerodynamics underlying what was known classically as "static stability and control." Although negative $C_{m\alpha}$ used to be deemed equivalent to positive longitudinal stability, Etkin (1972, page 199) observes that "positive pitch stiffness" is a better way to describe this condition. The discussion in Etkin's Section 9.3, however, reveals the importance of $C_{m\alpha} < 0$ in meeting the more meaningful criterion

$$\left.\frac{dC_m}{dC_L}\right|_{L=W} < 0 \tag{6.48}$$

for static or "speed stability." We shall return to this matter in Chapter 8.

A simplified approach to determining $C_{m\alpha}$ omits the contributions of thrust and drag, which often tend to cancel, as well as airloads induced by the action of propulsive slipstreams. In this approximation, it is convenient to separate pitching moments contributed by the wing and "body" (fuselage plus nacelles, if any) from those generated by the horizontal tail. The latter is nominally taken in an aft location, but the case of a canard can be covered by making *tail length* l_t negative, or the terms may be dropped entirely for a tailless configuration. The two parts of Fig. 6.4 illustrate how a reference line is positioned in the plane of symmetry for the purpose of locating various pitching-moment axes. This line

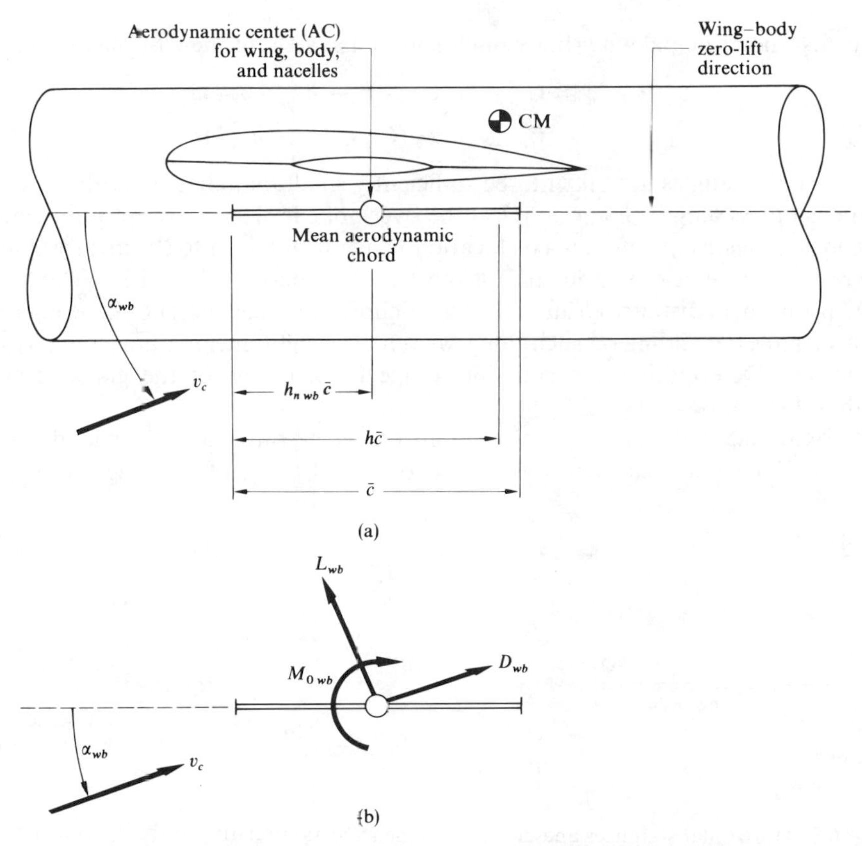

Fig. 6.4. The positioning of the reference mean aerodynamic chord in a flight vehicle. Part (b) locates the wing-body lift, drag, and pitching moment relative to this chord.

has the following properties:

a) Its length is equal to mean aerodynamic chord $\bar{c}$ [cf. (3.4)].

b) It is parallel to the zero-lift direction *for the wing and body with tail removed* (this definition is convenient because wind-tunnel tests are often made "tail-off" and "tail-on"; α_{wb} is the angle between this direction and the instantaneous relative wind).

c) It passes through an axis parallel to y which is the wing-body aerodynamic center (AC). That is, pitching moment M_{wb} about this axis is independent of angle-of-attack variations and equals its zero-lift value $M_{0\,wb}$.

d) Its fore-and-aft position, as in the upper figure, is fixed by that spanwise wing station in which the actual local chord $c(y)$ is equal to $\bar{c}$.

After nondimensionalization and in symbols defined by the lower part of

Fig. 6.4, the principal wing-body pitching moments about the CM may be written

$$\begin{aligned} C_{m\,wb} &\cong C_{m_0 wb} + C_{L\,wb}[h - h_{n\,wb}] \\ &= C_{m_0 wb} + \alpha_{wb}(C_{L\alpha})_{wb}[h - h_{n\,wb}] \end{aligned} \tag{6.49}$$

Observe that angles are taken to be sufficiently small that changes with α_{wb} in the dimensionless lengths h and $h_{n\,wb}$ can be overlooked. The subscript n introduces the idea of *neutral point*—an axis location which would lead to "neutral stability" (here for the vehicle without tail) if the CM happened to lie on it. The actual CM position, at distance $\bar{c}h$ aft of the mean-chord leading edge, should be regarded as a variable depending on such things as vehicle weight, fuel loading, and payload location. Determining a permissible range for h is one of the goals of static stability studies.

Figure 6.5 adds the horizontal stabilizer, the elevator, and their loads to the picture. The tail mean aerodynamic chord is also oriented along its zero-lift

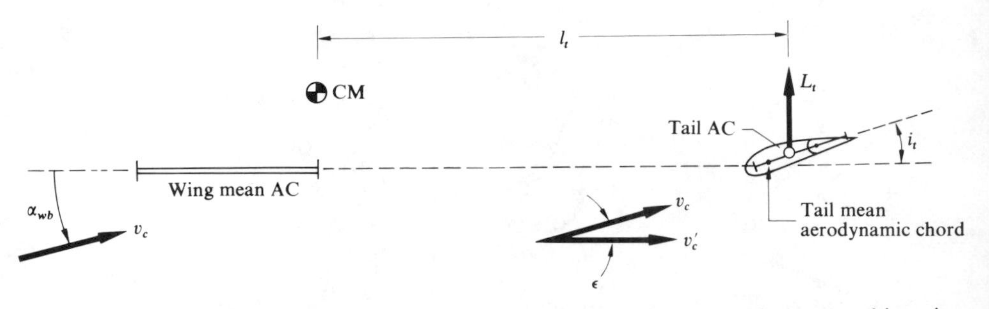

Fig. 6.5. Horizontal stabilizer and elevator, in their location relative to the approaching airstream and the wing aerodynamic chord. Note that tail lift is assumed always to act through the tail aerodynamic center.

direction and inclined at the *tail-setting angle* i_t below that of the wing. i_t is normally positive on an aft tail, and there is little or no camber—both reflections of the fact that tail lift acts downward in most flight conditions. Because of the downwash induced by the wing's wake near the xy-plane at stations between the tips and behind the trailing edge, we must also allow for an average *downwash angle* ϵ between the undisturbed airstream and the relative wind at the tail. The average magnitude of the speed there may also differ from v_c due to effects of slipstream or fuselage boundary layer, but we neglect this frequently small difference. l_t is the x-distance from vehicle CM to tail AC, where the tail lift is assumed to act. Note that, for a canard, ϵ is zero or slightly negative and $l_t < 0$.

In these circumstances the CM pitching moment due to stabilizer and elevator is roughly

$$M_t = -l_t L_t = -l_t \frac{\rho_\infty}{2} v_c^2 S_t C_{Lt} \tag{6.50}$$

where the tail lift coefficient C_{Lt} may be expressed as follows:

$$C_{Lt} = (C_{L\alpha})_t \alpha_{ZLt} + (C_{L\delta})_t \, \delta_e = (C_{L\alpha})_t[\alpha_{wb} - \epsilon - i_t] + (C_{L\delta})_t \, \delta_e \tag{6.51}$$

$(C_{L\alpha})_t$ obeys the conventional definition for any lifting surface, being based on the gross plan area S_t of the stabilizer. The lift-curve slope for the elevator is made dimensionless with this same area, so that it would be less than $(C_{L\alpha})_t$ for a partial-chord control and equal for an all-movable control.

When we insert (6.51) into (6.50) and determine a moment coefficient in the usual way, we get

$$C_{mt} \equiv \frac{M_t}{\dfrac{\rho_\infty}{2} v_c^2 S\bar{c}} = -V_H(C_{L\alpha})_t[\alpha_{wb} - \epsilon - i_t] - V_H(C_{L\delta})_t \, \delta_e \tag{6.52}$$

Here

$$V_H \equiv \frac{l_t S_t}{\bar{c}S} \tag{6.53}$$

is known as *volume ratio* for the horizontal stabilizer because of its numerator and denominator dimensions. It is a matter of observation that ϵ is directly proportional to wing-body lift, whence it follows that

$$\epsilon = \epsilon_0 + \frac{\partial \epsilon}{\partial \alpha} \alpha_{wb} \tag{6.54}$$

the constants ϵ_0 and $\partial\epsilon/\partial\alpha$ being functions of M and vehicle shape. More about downwash and numerical values of these constants can be found in any of the sources cited for aerodynamic data.

When we put (6.54) into (6.52) and combine the results with (6.49), we calculate for the total pitching moment

$$C_m = C_{m\,wb} + C_{mt} = C_{m_0 wb} + \alpha_{wb}\left\{(C_{L\alpha})_{wb}[h - h_{n\,wb}] - V_H(C_{L\alpha})_t\left[1 - \frac{\partial \epsilon}{\partial \alpha}\right]\right\}$$
$$+ V_H(C_{L\alpha})_t[\epsilon_0 + i_t] - V_H(C_{L\delta})_t \, \delta_e \tag{6.55}$$

From (6.55) we are already able to derive working forms for three derivatives. Since α_{wb} and the perturbation angle α vary in direct proportion, the following are self-evident.

$$\boxed{C_{m\alpha} = \frac{\partial C_m}{\partial \alpha_{wb}} = (C_{L\alpha})_{wb}[h - h_{n\,wb}] - V_H(C_{L\alpha})_t\left[1 - \frac{\partial \epsilon}{\partial \alpha}\right]} \tag{6.56}$$

$$C_{m\delta_e} = -V_H(C_{L\delta})_t \tag{6.57a}$$

$$C_{Z\delta_e} = -C_{L\delta_e} = -\frac{S_t}{S}(C_{L\delta})_t \tag{6.57b}$$

The value of h, from (6.56), which makes $C_{m\alpha} = 0$ is called the vehicle's *stick-fixed neutral point*, with the qualifying phrase appended as a reminder that the result would be different in the case of a reversible elevator which is "freed" by removing any restraint on the control column. h_n is written

$$h_n = h_{n\,wb} + V_{H\,n} \frac{(C_{L\alpha})_t}{(C_{L\alpha})_{wb}} \left[1 - \frac{\partial \epsilon}{\partial \alpha} \right] \tag{6.58}$$

generally an implicit formula, since $V_{H\,n}$ is volume ratio based on tail length l_{tn} measured from the new position of the CM. Often for conventional tails, however, it is sufficiently accurate to employ V_H based on some nominal l_t.

As shown, for instance, in Etkin (1959, Section 2.3), the following alternative forms may be used for C_m and its slope:

$$\boxed{C_m = C_{m0} + C_{m\alpha}\alpha_{ZL}} \tag{6.59a}$$

with

$$\alpha_{ZL} = \alpha_{wb} - \frac{(C_{L\alpha})_t}{C_{L\alpha}} \frac{S_t}{S} [\epsilon_0 + i_t] \tag{6.59b}$$

and

$$C_{m0} = C_{m0wb} + (C_{L\alpha})_t V_H [\epsilon_0 + i_t]$$
$$+ \frac{(C_{L\alpha})_{wb}}{C_{L\alpha}} (C_{L\alpha})_t \frac{S_t}{S} [\epsilon_0 + i_t] \left[h - h_{n\,wb} - V_H \frac{(C_{L\alpha})_t}{(C_{L\alpha})_{wb}} \left(1 - \frac{\partial \epsilon}{\partial \alpha} \right) \right] \tag{6.59c}$$

$$\boxed{C_{m\alpha} = C_{L\alpha}[h - h_n]} \tag{6.60}$$

where, finally, the overall vehicle lift-curve slope is

$$C_{L\alpha} = (C_{L\alpha})_{wb} \left[1 + \frac{(C_{L\alpha})_t}{(C_{L\alpha})_{wb}} \frac{S_t}{S} \left(1 - \frac{\partial \epsilon}{\partial \alpha} \right) \right] \tag{6.61}$$

All these will prove useful to us in Chapter 8. We shall see that having the actual CM ahead of the neutral point, $h_n > h$, is associated with the desired condition of positive pitch stiffness. The name *stick-fixed static margin* is applied to $[h_n - h]$, which is usually quoted as a percentage of the mean aerodynamic chord.

3. The q-Derivatives

The two main sources of lift and moment due to a pitching velocity about the y-axis are the wing and horizontal stabilizer. To estimate the former's contributions is an exercise in the sort of 3-D lifting-surface theory reviewed in Chapter 4. It is enough to state here that a small q gives any point (x, y) on planform area S a small upward velocity* $(-qx)$ relative to the atmosphere. This velocity, in

* We temporarily measure x aft from the CM, in order to be consistent with Chapter 4.

conjunction with the reference oncoming stream $v_c \cong u_0$, has the same effect as if the angle of attack at (x, y) had been increased locally in the amount qx/u_0. The identical result would be achieved by cambering the wing according to [cf. (4.26, 28)]

$$z_1(x) = -\frac{q}{2u_0} x^2 \tag{6.62}$$

Therefore, under the quasi-steady approximation, a suitable aerodynamic program may be applied to (6.62) and the total lift and pitching moment about the CM may be calculated. When reduced to dimensionless form and differentiated with respect to $\hat{q} \equiv q\bar{c}/2u_0$, these airloads furnish estimates for the wing portions of the derivatives

$$(C_{zq})_{wb} = -(C_{Lq})_{wb} \quad \text{and} \quad (C_{mq})_{wb}$$

These would, of course, be the totals for a tailless aircraft. On physical grounds, we anticipate that $(C_{mq})_{wb} < 0$ and that surfaces with large chords and/or appreciable sweep produce much greater *pitch damping* C_{mq} than those which have large Æ and $\Lambda \cong 0$.

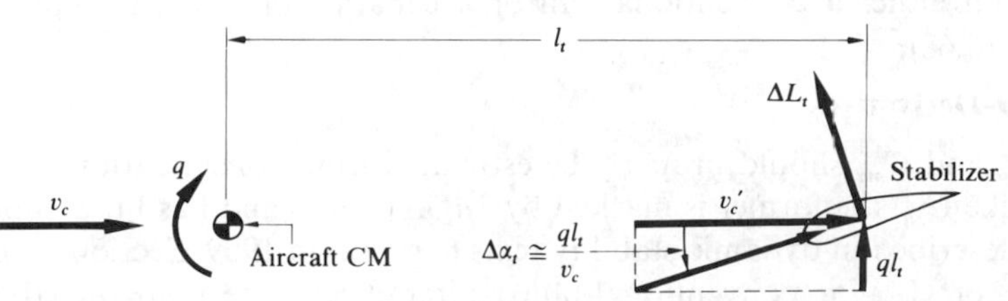

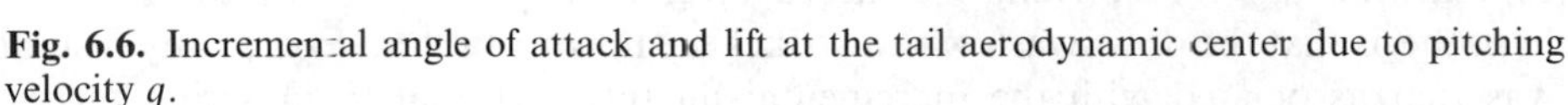

Fig. 6.6. Incremental angle of attack and lift at the tail aerodynamic center due to pitching velocity q.

Figure 6.6 depicts the way the angle of attack at the AC of an aft tail is augmented by pitching. Since the "camber" (6.62) is probably insignificant at such a distance behind the CM, we account only for

$$\Delta\alpha_t = \sin^{-1}\left(\frac{ql_t}{v_c}\right) \cong \frac{ql_t}{u_0} \tag{6.63}$$

The resulting lift increment

$$\Delta L_t = (C_{L\alpha})_t \frac{\rho_\infty}{2} v_c^2 S_t \left[\frac{ql_t}{u_0}\right] \tag{6.64}$$

is made dimensionless and differentiated with respect to $\hat{q}$ to obtain

$$(C_{Lq})_t = 2V_H(C_{L\alpha})_t \tag{6.65}$$

$\hat{q} = \frac{q\bar{c}}{2u_0}$ $\quad V_H = \frac{l_t S_t}{\bar{c} S}$ $\quad 2V_H\hat{q} = \frac{q l_t S_t}{S u_0}$

$\Delta L_t = (C_{Lq})_t \bar{q} S \hat{q}$

The pitching moment is due to ΔL_t acting with an arm $(-l_t)$, whence we compute

$$(C_{mq})_t = -2V_H \frac{l_t}{\bar{c}} (C_{L\alpha})_t \tag{6.66}$$

The q-derivatives are finally summarized.

$$C_{zq} = -(C_{Lq})_{wb} - 2V_H(C_{L\alpha})_t \tag{6.67}$$

$$C_{mq} = (C_{mq})_{wb} - 2V_H \frac{l_t}{\bar{c}} (C_{L\alpha})_t \tag{6.68}$$

Equation (6.30b) shows that C_{zq} occurs in combination with twice the density ratio μ_s. Since for aircraft the latter is a number in the hundreds, C_{zq} is usually inconsequential and leaves the "damping derivative" (6.68) as the principal aerodynamic effect of pitching. This is one of several derivatives, including $C_{z\alpha}$, C_{lp}, and C_{nr}, whose main role is to extract energy from the perturbed motions, damping out the natural oscillations which can be excited and opposing the build-up of angular velocity. We point out that tail length strongly controls the magnitude of $(-C_{mq})$ on either a conventional tail or a canard, since it contributes a factor $(l_t/\bar{c})^2$ to (6.68).

4. The $\dot{\alpha}$-Derivatives

Both $C_{z\dot{\alpha}}$ and $C_{m\dot{\alpha}}$ should properly be estimated from considerations of unsteady flow. Like C_{zq}, the former is masked by $2\mu_s$ in (6.30b) and has little importance.

As described in dynamic stability texts (e.g., Etkin 1959, Section 5.5), a rough estimate of $C_{m\dot{\alpha}}$ for conventional-tailed aircraft may be constructed from the hypothesis that the downwash which strikes the horizontal stabilizer at any instant was that associated with the incremental lift (proportional to α) acting on the wing at the time the corresponding wake elements left the wing. That is to say, the downwash "lags" its steady-state value by an interval l_t/u_0. With α increasing at the rate $\dot{\alpha}$, this approach would lead us to reduce the downwash angle appearing in equations like (6.52) and (6.55) by an amount

$$\Delta\epsilon = -\frac{\partial\epsilon}{\partial\alpha} \dot{\alpha} \frac{l_t}{u_0} \tag{6.69}$$

Now downwash causes a reduction in tail angle α_t, so the corresponding tail lift and CM pitching-moment effects would be

$$\Delta C_{Lt} = -(C_{L\alpha})_t \, \Delta\epsilon = (C_{L\alpha})_t \frac{\partial\epsilon}{\partial\alpha} \frac{l_t}{u_0} \dot{\alpha} \tag{6.70a}$$

$$\Delta C_{mt} = -V_H \, \Delta C_{Lt} = -2V_H \frac{l_t}{\bar{c}} (C_{L\alpha})_t \frac{\partial\epsilon}{\partial\alpha}\left(\frac{\bar{c}\dot{\alpha}}{2u_0}\right) \tag{6.70b}$$

When we differentiate (6.70b) with respect to $D_s\alpha \equiv (\bar{c}\dot{\alpha}/2u_0)$, we obtain the working approximation

$$\boxed{C_{m\dot{\alpha}} \cong -2V_H \frac{l_t}{\bar{c}} (C_{L\dot{\alpha}})_t \frac{\partial \epsilon}{\partial \alpha}} \tag{6.71}$$

6.4 ESTIMATION OF STABILITY DERIVATIVES: LATERAL

It is often remarked that theoretical analysis is less effective for predicting most lateral derivatives than for predicting longitudinal ones. On cruising aircraft this is because the relatively small vertical stabilizer is the only lifting surface directly activated in proportion to sideslip angle β or yaw rate r. Relatively intractable phenomena like fuselage side loads, incidence changes due to wing dihedral, and fuselage-wing interference are therefore able to dominate the lateral forces and moments. Accordingly—without apology, but with repeated reference to the necessity for accurate wind-tunnel data—we shall find ourselves including many terms about which little quantitative enlightenment can be offered here.

1. The β-Derivatives and Rudder-Angle Derivatives

Figure 6.7 is a top view of the vehicle in sideslip, with distance l_F indicated from the CM to fin AC and with the rudder rotated in the conventionally positive sense

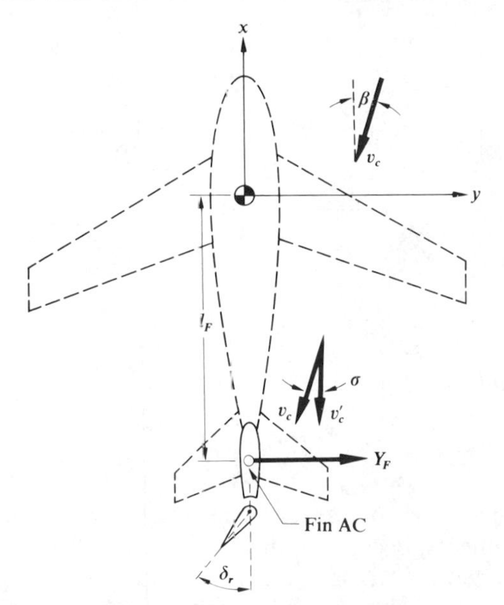

Fig. 6.7. Vertical stabilizer (fin) and rudder in their location relative to the approaching airstream when the vehicle has a positive sideslip β. Rest of the vehicle is shown in dashed outline.

for δ_r. As for the horizontal stabilizer, we assume no change in the airspeed magnitude v_c at the tail, but allow for a *sidewash angle* σ induced by the wake, interference, etc. Since most vertical area usually lies above the wake vortex sheet, whose centerline is shifted to the left when β is positive, one expects from the properties of vortex flows that $\partial\sigma/\partial\beta$ due to the wake would be negative. Appealing to symmetry and linearity, we assume that

$$\sigma = \frac{\partial\sigma}{\partial\beta}\beta \tag{6.72}$$

with $\partial\sigma/\partial\beta$ a constant.

The fin sideforce in the figure is evidently

$$Y_F = \frac{\rho_\infty}{2} v_c^2 S_F \left\{ (C_{L\alpha})_F \left[-1 + \frac{\partial\sigma}{\partial\beta} \right] \beta + (C_{L\delta})_F \, \delta_r \right\} \tag{6.73}$$

The symbolism here is self-explanatory, by analogy with corresponding quantities in the longitudinal case. The definition of S_F tends to be somewhat arbitrary, but on a conventional fin it is usually the sum of fixed and movable plan areas above the place at which the surface emerges from the rear fuselage.

Manipulations resembling those which yielded (6.56, 57) permit us to calculate from (6.73) the following four derivatives:

$$\boxed{C_{y\beta} = -\frac{S_F}{S}(C_{L\alpha})_F \left[1 - \frac{\partial\sigma}{\partial\beta} \right] + (C_{y\beta})_{wb}} \tag{6.74}$$

$$\boxed{C_{n\beta} = V_V (C_{L\alpha})_F \left[1 - \frac{\partial\sigma}{\partial\beta} \right] + (C_{n\beta})_{wb}} \tag{6.75}$$

$$\boxed{C_{y\delta_r} = \frac{S_F}{S}(C_{L\delta})_F} \tag{6.76a}$$

$$\boxed{C_{n\delta_r} = -V_V (C_{L\delta})_F} \tag{6.76b}$$

Here the *vertical tail volume ratio* is

$$V_V \equiv \frac{l_F}{b}\frac{S_F}{S} \tag{6.77}$$

Contributions to the β-derivatives from wing and body are indicated as they were for the pitching moments. Because of the fin moment-arm effectiveness, we expect $(C_{y\beta})_{wb}$ to be a considerably larger fraction of $C_{y\beta}$ than $(C_{n\beta})_{wb}$ is of the *yaw stiffness* or *weathercock* derivative $C_{n\beta}$. Large propellers are known to experience significant normal forces when set at incidence; because of the smallness of $|C_{y\beta}|$ compared to $C_{N\alpha}$, however, we expect that this propeller effect on $C_{y\beta}$, which is negative, will have to be accounted for much more often than in the case of the longitudinal derivative. Hence we refer to it here and cite the classical work of Ribner (1945) on this subject.

The *rolling moment due to sideslip*, $C_{l\beta}$, is an important cross-coupling derivative which affects lateral-directional oscillatory motions and over which the designer has considerable control, independent of other tradeoffs involved in configuring his aircraft. To see why this is so, we examine "semi-quantitatively" three phenomena that combine to fix its sign and magnitude. [The fin rolling moment (cf. Etkin 1959, Eq. 3.10,3) tends to be less important than these.]

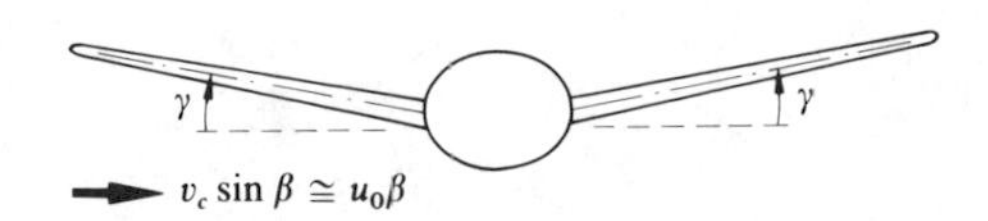

Fig. 6.8. Front view of fuselage and wing with positive dihedral angle γ, showing the lateral component of relative wind during sideslip.

a) *Dihedral.* Figure 6.8 shows the wing and fuselage in forward projection, with the *relative wind due to sideslip*, $v_c \sin\beta \cong u_0\beta$, seen coming from the pilot's right. In the presence of the indicated dihedral angle γ, there is an "upward" component of the relative wind $u_0\beta \sin\gamma$, which acts normal to the plane of the right wing. When associated with the main flow $\cong u_0$ parallel to the airplane centerline, this component causes the right-wing angle of attack to increase and that of the left wing to decrease, the amounts being

$$(\Delta\alpha)_\beta = \begin{cases} \beta \sin\gamma & \text{[right wing]} \\ -\beta \sin\gamma & \text{[left wing]} \end{cases} \tag{6.78}$$

Since these increments are antisymmetric, there is no change in total lift. There will, however, be a rolling moment, which might be predicted by the 3-D aerodynamic theory presented and referenced in Chapter 4. Since positive L_r depresses the right wing, we can write

$$C_{l\beta} = -K_1 \sin\gamma \tag{6.79}$$

where $K_1 > 0$ is a constant determined mainly by the vehicle shape and flight M.

b) *Vertical Wing Position.* Figure 6.9 sketches typical streamlines of the cross-

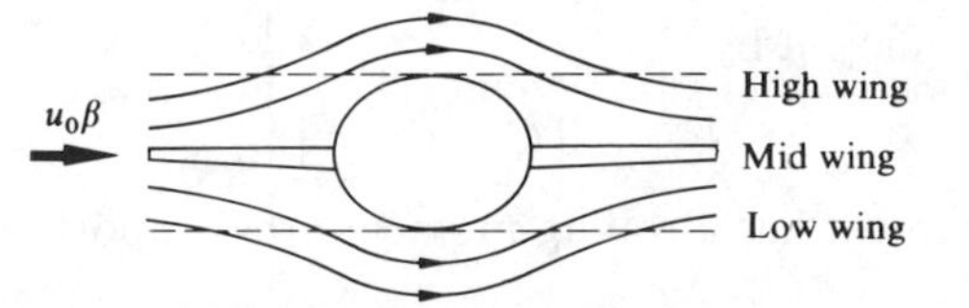

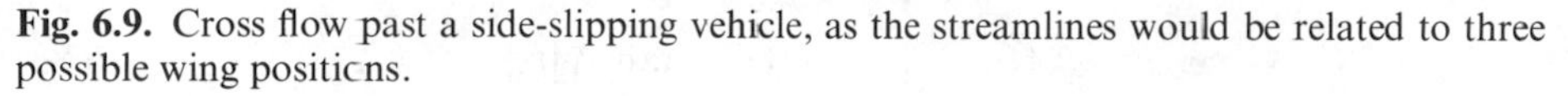

Fig. 6.9. Cross flow past a side-slipping vehicle, as the streamlines would be related to three possible wing positions.

flow due to sideslip as they might go past an elliptical-section fuselage. It also shows three possible vertical locations of the wing relative to the body. It is clear from a study of these pictures that a negative $C_{l\beta}$ will be experienced by the high wing due to crossflow effect. The sign of $C_{l\beta}$ is positive for the low wing, whereas the central position probably has little influence. Tests or interference methods must be employed to obtain any quantitative estimation.

c) *Sweep.* Figure 6.10 presents the plan view of a slipping wing with positive sweep angle Λ. Here a rolling moment arises because of the influence of sweep

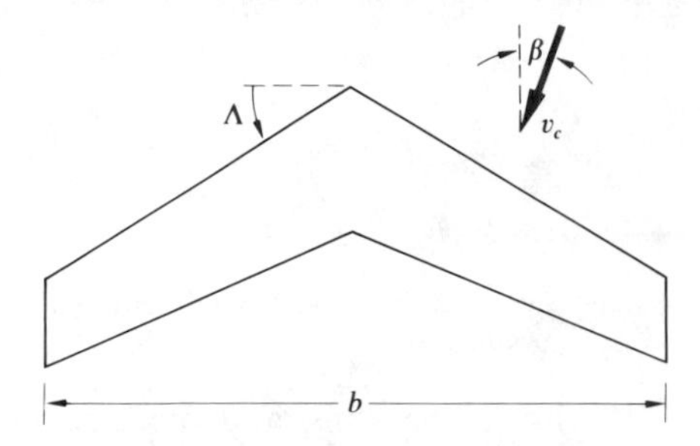

Fig. 6.10. Plan view of a side-slipping wing with sweep-back angle Λ.

on lift-curve slope discussed in Chapter 4. By 2-D considerations, an assigned problem demonstrated that $C_{L\alpha}$ varies as cos Λ, and therefore might be written

$$C_{L\alpha} \cong C_{L\alpha 0} \cos \Lambda \tag{6.80}$$

even in 3-D, where $C_{L\alpha 0}$ is the $\Lambda = 0$ value. Hence, if we assign half of area S to the right wing, its lift increase due to β would be

$$(\Delta L)_{\text{Right}} \cong \frac{\rho_\infty}{2} v_c^2 \frac{S}{2} C_{L\alpha 0}[\cos(\Lambda - \beta) - \cos \Lambda]\alpha_{ZL}$$

$$\cong \frac{\rho_\infty}{2} v_c^2 \frac{S}{2} C_{L\alpha 0}[\beta \sin \Lambda]\alpha_{ZL} \tag{6.81a}$$

By antisymmetry,

$$(\Delta L)_{\text{Left}} = -(\Delta L)_{\text{Right}} \tag{6.81b}$$

Here α_{ZL} is the zero-lift angle of attack, which would be approximately related to the trimmed lift coefficient by

$$C_{L0} \cong [C_{L\alpha 0} \cos \Lambda]\alpha_{ZL} \tag{6.82}$$

Since each lift increment has a rolling-moment arm proportional to $b/2$, it can be reasoned that

$$(C_l)_{\text{Sweep}} \sim -[C_{L0} \tan \Lambda]\beta \tag{6.83}$$

whence

$$(C_{l\beta})_{\text{Sweep}} = -K_2 C_{L0} \tan \Lambda \tag{6.84}$$

where $K_2 > 0$ is a second constant dependent on shape and M.

Inasmuch as other sweep laws can be found in the literature (Etkin 1959 gives $\sin 2\Lambda$), it is evident that (6.84) requires refinement. We see, however, that positive Λ and positive dihedral work together in causing negative $C_{l\beta}$. This is the sign that pilots regard as desirable, because it causes the wing toward which the vehicle is slipping to "be picked up" rather than "drop." We shall see in Chapter 7, however, that there must be some optimum intermediate magnitude for this derivative. When $|C_{l\beta}|$ is too large, an unpleasant amount of rolling is experienced during lateral-directional oscillation.

2. The *p*-Derivatives and Aileron-Angle Derivatives

Our reasoning with regard to sideslip derivatives enables us to set down at once some estimates of the fin's contribution to the sideforce and yawing moment due to rolling

$$(C_{yp})_{\text{Fin}} = -\frac{S_F}{S}(C_{L\alpha})_F\left[2\frac{Z_F}{b} - \frac{\partial\sigma}{\partial\hat{p}}\right] \tag{6.85}$$

$$(C_{np})_{\text{Fin}} = V_V(C_{L\alpha})_F\left[2\frac{Z_F}{b} - \frac{\partial\sigma}{\partial\hat{p}}\right] \tag{6.86}$$

The new symbol here is Z_F, the distance above the x-axis to the point on the fin at which an "effective angle of attack" due to rolling is to be calculated.

Equations (6.85) and (6.86) do not express the major airloads caused by the velocity p. These involve the rolling and yawing moments brought about by the antisymmetrical angle-of-attack distribution (py/u_0) induced across the wing. The associated incremental lift is, of course, odd in y. It can be computed by 3-D aerodynamic theory and then inserted into a formula like (4.150) or (4.151) to obtain the resulting C_l, hence the derivative C_{lp}. It should be clear that C_{lp} is always negative, except well into the stall regime, since the lift opposes the motion which induces it.

In a similar way, 3-D theory may be employed to estimate C_l and C_n due to the aileron deflection δ_a, as well as to any antisymmetrical rotation of the tail that may be employed to supplement roll control.

The determination of C_{lp} and $C_{l\delta_a}$ is relatively straightforward. C_{np} and $C_{n\delta_a}$ are rather more interesting, however, because they are consequences of differential chord forces between the two wings. By way of physical discussion of the cross-coupling derivative,

$$C_{np} = (C_{np})_{\text{Wing}} + (C_{np})_{\text{Fin}} \tag{6.87}$$

we remark that there are two competing effects of the wing incidence (py/u_0).

For example, for $p > 0$ at a station on the right wing, the local lift is increased and rotated forward, whereas there is also added drag due to lift. When components of these two changes are taken in the chordwise direction, one or the other may predominate. In subsonic flow, considerable "leading edge suction" usually develops, the net chord force is forward, and a negative value of C_{np} develops which tends to be proportional to C_{L0}. When the flight speed is supersonic, and especially when the wing leading edges are swept back less than the Mach lines, the incremental drag prevails and gives rise to $C_{np} > 0$.

In the wind tunnel, "rotary derivatives" like C_{np} and C_{lp} must be measured on a model which can be given a steady angular velocity about x—a procedure that is only occasionally made part of the developmental test program for a new design. Similarly, C_{nr} is found by oscillating the model about a vertical z-axis, while oscillation in pitch can be used to determine the combination $[C_{mq} + C_{m\dot{\alpha}}]$. Such tests have been made, but their role seems to be mainly for the verification of theory and for finding the influences on these derivatives of various configuration parameters.

3. The *r*-Derivatives

If we imagine the airplane in Fig. 6.7 to have a positive yawing rate, the right wing will be swinging backward and the fin AC will acquire a velocity $l_F r$ to the left. We add to the resulting fin angle of attack $(l_F r/u_0)$ a sidewash effect

$$(\sigma)_{\text{Yawing}} = \frac{\partial \sigma}{\partial \hat{r}} \hat{r} \equiv \frac{\partial \sigma}{\partial \hat{r}} \left(\frac{rb}{2u_0} \right) \tag{6.88}$$

It follows without difficulty that the sideforce and yawing-moment derivatives are

$$\boxed{C_{yr} = \frac{S_F}{S} (C_{L\alpha})_F \left[2 \frac{l_F}{b} + \frac{\partial \sigma}{\partial \hat{r}} \right] + (C_{yr})_{wb}} \tag{6.89}$$

$$\boxed{C_{nr} = -V_F (C_{L\alpha})_F \left[2 \frac{l_F}{b} + \frac{\partial \sigma}{\partial \hat{r}} \right] + (C_{nr})_{wb}} \tag{6.90}$$

Note that the (negative) "yaw damping" C_{nr} is related to l_F in the same way C_{mq} is to l_t.

If we further assume the fin sideforce due to yaw to act at a distance Z_F above the x-axis—observe that this need not equal the Z_F defined for (6.85, 86)—there will be a small rolling moment. The corresponding derivative might be written

$$\boxed{C_{lr} = \frac{Z_F S_F}{bS} (C_{L\alpha})_F \left[2 \frac{l_F}{b} + \frac{\partial \sigma}{\partial \hat{r}} \right] + (C_{lr})_{wb}} \tag{6.91}$$

for the complete airplane.

6.5 PROBLEMS

6A a) When a missile or acrobatic airplane starts a rapid rolling maneuver, the ailerons are suddenly thrown to a position δ_a. They produce a rolling moment proportional to $C_{l\delta}\delta_a$, where $C_{l\delta}$ is a (constant) "stability derivative" that depends on altitude, speed, airplane geometry, etc. Suppose that the resulting motion involves just the single degree of freedom $p(t)$, as if the airplane's x-axis were pivoted about a shaft. Show that the vehicle spins up in an exponential fashion to a final roll rate that is determined by the flight condition, by $C_{l\delta}$, and by the "damping-in-roll derivative" C_{lp}.

b) Discuss in some detail, including considerations of symmetry, how you would find $C_{l\delta}$ and C_{l_p} from subsonic lifting-surface theory or lifting-line theory (for the latter, see Problem 4M).

6B Simplified equations of motion for the longitudinal response of a horizontally cruising vehicle to relatively rapid elevator displacements $\delta_e(\hat{t})$ are as follows:

$$[2\mu D - C_{z\alpha}]\alpha - 2\mu\hat{q} = 0$$

$$-[C_{m\dot{\alpha}}D + C_{m\alpha}]\alpha + [i_{yy}D - C_{mq}]\hat{q} = C_{m\delta}\,\delta_e(\hat{t})$$

a) Discuss the physical assumptions underlying this approximation.
b) By methods of Laplace transform, find the response to a step elevator ixput $\delta_e = \delta_0 1(\hat{t})$. Initial conditions are zero.
c) Find formulas for the "steady-state" oscillatory response to sinusoidal elevator input of frequency ω. Discuss factors controlling the height of the resonance peak.

6C Consider small perturbations in the motion of a flight vehicle about a reference condition which consists of a banked, horizontal turn of radius R, flight speed u_0 and bank angle Φ_0. Examine the perturbed equations of motion to determine whether it is still possible to make a separation between "longitudinal" and "lateral" and to find which inertia terms and stability derivatives will be significantly changed from unaccelerated rectilinear flight. (Remember that the equilibrium flight path is curved.)

6D Specialize the general equations of motion for a rigid flight vehicle, moving in its plane of symmetry, to investigate the consequences of a sudden pull-up from trimmed flight by a glider pilot. At $t = 0$, he puts a step deflection in the elevator angle so that the angle of attack for zero pitching moment changes by an amount $\Delta\alpha$. You may assume (1) linear variations of C_L and C_m with α and Q, (2) $C_{m\alpha} < 0$, and (3) $C_D = C_{D0} + C_L^2/\pi e\mathcal{R}$.

First estimate the time history of Q and Θ from the pitch equation by taking a rectilinear flight path and approximating U with its initial value U_0. Then see what you can do toward determining the actual time histories of U and W from the other applicable equations of motion.

6E Explain in words how to mechanize the following stability augmentation systems, given that needed quantities can be accurately measured and processed electronically so as to actuate any of the three control surfaces in any reasonable way.

a) Heading stability (what important new "derivative" does this add to the equations of motion?).
b) Increased damping in yaw, the amount being programmed to vary with altitude.
c) Artificial longitudinal static stability, the amount to be under the pilot's control.
d) Favorable $C_{l\beta}$ for an airplane without sweep or wing dihedral.

6F Make a table with headings "Stability derivative," "Theory," "Wind tunnel." In its columns, enter brief but carefully thought-out statements about how you would estimate each of the derivatives listed below, using each of the two specified tools. (One or two formulas might help in the Theory column.) After each statement, give one percentage (for example, $\pm 25\%$) which is your best guess of the accuracy of that particular method.

a) Supersonic $C_{z\alpha}$ for a straight-tapered wing like X-15.
b) C_{mq} for a tailless delta.
c) The *aerodynamic* (not propulsion-system) portion of C_{xu} for high-speed operation of Boeing 747.
d) $C_{n\beta}$ and C_{nr} for a supersonic transport with a single large vertical tail.
e) $C_{l\beta}$ for a swept midwing of high Æ and no dihedral.
f) $C_{L\alpha}$ and $C_{y\beta}$ for a long slender booster in the form of a body of revolution without fins.
g) $C_{Du} \sim \partial C_D/\partial M$ for a transonic or supersonic transport.
h) $C_{m\alpha}$ for a conventional light airplane.

[*Note:* Many of these cases may also be used as a basis for quantitative estimations, by methods discussed in Chapters 4, 10, and Sections 6.3 and 6.4.]

6G (*Introduction to the analog computer*) Assume that you have available a collection of electronic devices, indicated by the symbols below and capable of performing the functions described. All voltages e are dc (actually slowly varying functions of time) relative to ground.

e_1, e_2, …, e_n → Σ → e_0

$$e_0 = -[e_1 + e_2 + \cdots + e_n]$$

A summer. (Note change of sign. Input voltages can also be multiplied by numerical constants like 0.1, 0.2, 0.5, . . . , 10.)

An integrator or summing integrator. (Note change of sign. Input voltages may also be multiplied by constants. When the apparatus is turned on at $t = 0$, an "initial condition potentiometer" can be set to give e_0 a specified initial value.)

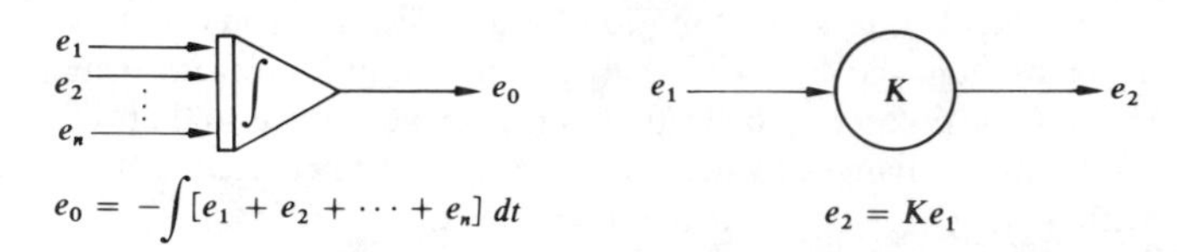

A coefficient potentiometer. (May be set to give K any value $0 < K \leq 1$. Easily adjusted during study of a problem.)

Show how you might wire these devices together to provide an electrical analogy to the solution of the equations of motion of the Problem 2E weathervane with rudder, or of the slightly perturbed airplane. You will need two voltages proportional to the dependent variables α, δ. Start the motion by applying some initial value $\alpha_0 = \alpha(0)$.

6H Now assume that to the equipment of Problem 6G the following nonlinear device is added:

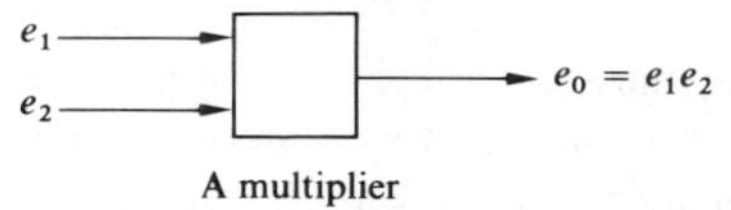

A multiplier

Discuss how you might study the problem of roll-pitch-yaw coupling of a flight vehicle whose CM is moving in a straight line so that only the rotational degrees of freedom are of interest. Consider *only* the simulation of the inertia terms (not aerodynamic) in the equations.

[*Note:* After studying Chapter 7, if analog equipment is available, you should try to learn about its operation, and set up and solve some problems like those discussed in Chapter 7.]

CHAPTER 7

SOLUTION OF THE SMALL-PERTURBATION EQUATIONS OF MOTION

7.1 A SIMPLIFIED LOOK AT LATERAL RESPONSE

In order to illustrate the analytic process whereby we can use linearized equations to study the motion of small-amplitude vehicles, let us address the lateral case first. The system (6.31) is further reduced by using aileron displacement $\delta_a(\hat{t})$ as the only source of disturbances and by assuming that the CM always proceeds along a horizontal straight-line path. The latter simplification has been found to represent quite well the short-period oscillations and lateral handling qualities of many designs. It is easily seen to be equivalent to the relation

$$\psi = -\beta \tag{7.1}$$

between the (small) angles of yaw and sideslip.

When we differentiate (7.1) and multiply it by u_0, we determine that

$$u_0[\dot{\beta} + \dot{\psi}] = \dot{v} + ru_0 \equiv a_y = 0 \tag{7.2}$$

Vanishing lateral acceleration a_y implies that we are neglecting the instantaneous sideforce. Although such an approximation must ultimately be justified by comparison with solutions of the complete set (6.31), we observe there are only two main contributors to Y: the sideslipping fuselage and the vertical stabilizer. The former is aerodynamically "inefficient," whereas the latter often has rather limited area in proportion, say, to the wing. Thus one might expect that the cumulative effects of sideforce during relatively rapid, periodic maneuvers would be inconsequential compared to those of the rolling and yawing moments.

Accordingly, we drop (6.31a), set $\theta_0 \cong 0$ and also omit the frequently insignificant aileron-velocity derivative $C_{l\dot{\delta}_a}$, obtaining from (6.31b, c)

$$-C_{l\beta} + [i_{xx}D_a - C_{lp}]\hat{p} - [i_{xz}D_a + C_{lr}]\hat{r} = C_{l\delta_a}\delta_a(\hat{t}_a) \tag{7.3a}$$

$$-C_{n\beta}\beta - [i_{xz}D_a + C_{np}]\hat{p} + [i_{zz}D_a - C_{nr}]\hat{r} = C_{n\delta_a}\delta_a(\hat{t}_a) \tag{7.3b}$$

From (7.1) and (6.31e), $\hat{r} = -D_a\beta$ in horizontal flight. We can therefore rewrite (7.3) in terms of two independent unknowns, which we choose to be β and the roll rate $\hat{p}$.

$$[i_{xz}D_a^2 + C_{lr}D_a - C_{l\beta}]\beta + [i_{xx}D_a - C_{lp}]\hat{p} = C_{l\delta_a}\delta_a(\hat{t}_a) \tag{7.4a}$$

$$-[i_{zz}D_a^2 - C_{nr}D_a + C_{n\beta}]\beta - [i_{xz}D_a + C_{np}]\hat{p} = C_{n\delta_a}\delta_a(\hat{t}_a) \tag{7.4b}$$

As previously discussed, this system (7.4) is linear with constant coefficients. The total order is three. Its various homogeneous and particular solutions can be worked out by methods elaborated in the field of automatic control theory—a classic text of this genre is Gardner and Barnes (1942), whereas Zadeh and deSouer (1963) exemplifies more modern formulations. Our approach here will be to examine first the steady-state response to sinusoidal forcing of arbitrary frequency. We shall then proceed to the transient responses known as *indicial* and *impulsive admittances*, to the case of a general input $\delta_a(\hat{t}_a)$, and to free motions of the vehicle as revealed by the homogeneous terms in these results.

1. Mechanical Admittance

This term is applied to the function relating the permanent sinusoidal response of a stable linear system to a sinusoidal input of circular frequency ω. For instance, we can pick

$$\delta_a = \delta_{a0} \cos \omega t = \delta_{a0} \cos k\hat{t}_a = \text{Re}\,\{\delta_{a0} e^{ik\hat{t}_a}\} \tag{7.5}$$

Here we have introduced a dimensionless *reduced frequency*

$$k \equiv \frac{\omega b}{2u_0} \tag{7.6}$$

Since the input period is $2\pi/\omega$, you may wish to verify that π/k equals the number of wingspan lengths traversed by the CM per cycle of the steady oscillation. k is normally a small number of order 10^{-1} to 10^{-3} for rigid-aircraft oscillations, a fact which, incidentally, helps to justify our use of quasi-steady aerodynamic theory.

Equation (7.5) also introduces a common device for facilitating the algebraic manipulation of such systems: replacement of a trigonometric expression in terms of the real part of a complex exponential. Because the problem is linear, the symbol Re $\{\ldots\}$ may be omitted from input and output quantities during intermediate steps, which then involve complex numbers. Only when a final result is obtained must we take its real part to determine the desired amplitude and phase relationships. With $\delta_0 \exp(ik\hat{t}_a)$ inserted for $\delta_a(\hat{t}_a)$, we thus calculate the permanent response from (7.4) by replacing β and $\hat{p}$ with the real parts of

$$\beta = \beta_0 \exp\left[i(k\hat{t}_a + \varphi_\beta)\right] \tag{7.7a}$$

$$\hat{p} = \hat{p}_0 \exp\left[i(k\hat{t}_a + \varphi_p)\right] \tag{7.7b}$$

Note that in (7.5) and (7.7) we are treating δ_0, β_0, and $\hat{p}_0$ as real. That is, the sinusoidal aileron displacement becomes the phase reference; φ_β and φ_p are, respectively and in terms of the senses in which these quantities are defined to be positive, the phase angles by which the sideslip angle and rolling velocity lead δ_a.

After these exponential substitutions into (7.4), each derivative operator D_a amounts to multiplication of its term by ik. Consequently, upon cancellation of a common factor exp $(ik\hat{t}_a)$, we recover an algebraic system. In matrix notation, it reads

$$\begin{bmatrix} [i_{xz}(ik)^2 + C_{l_r}ik - C_{l\beta}] & [i_{xx}ik - C_{lp}] \\ [-i_{zz}(ik)^2 + C_{nr}ik - C_{n\beta}] & [-i_{xz}ik - C_{np}] \end{bmatrix} \begin{Bmatrix} \beta_0 e^{i\varphi_\beta} \\ \hat{p}_0 e^{i\varphi_p} \end{Bmatrix} = \begin{Bmatrix} C_{l\delta_a} \\ C_{n\delta_a} \end{Bmatrix} \delta_{a0} \quad (7.8)$$

The determinantal solutions of (7.8) have, as their moduli and arguments, the *amplitude ratios* (β_0/δ_{a0}) and $(\hat{p}_0/\delta_{a0})$ and the corresponding phase leads. The two *complex* ratios are the aforementioned mechanical admittances. They depend on the various dimensionless system parameters and on the forcing frequency in the form of ik. Following Etkin (1959, 1972), we employ for them the symbol G, subscripted to show which input–output pair they relate.

$$G_{\beta\delta_a}(k) \equiv \frac{\beta_0 e^{i\varphi_\beta}}{\delta_{a0}} = \frac{\begin{vmatrix} C_{l\delta_a} & [i_{xx}ik - C_{lp}] \\ C_{n\delta_a} & [-i_{xz}ik - C_{np}] \end{vmatrix}}{D} \quad (7.9a)$$

$$G_{\hat{p}\delta_a}(k) \equiv \frac{\hat{p}_0 e^{i\varphi_p}}{\delta_{a0}} = \frac{\begin{vmatrix} [i_{xz}(ik)^2 + C_{l_r}ik - C_{l\beta}] & C_{l\delta_a} \\ [-i_{zz}(ik)^2 + C_{nr}ik - C_{n\beta}] & C_{n\delta_a} \end{vmatrix}}{D} \quad (7.9b)$$

In (7.9), the denominator D is simply the characteristic determinant of (7.8). Its roots, with k complex, will be examined later in connection with the normal modes of free motion.

$$D \equiv \begin{vmatrix} [i_{xz}(ik)^2 + C_{l_r}(ik) - C_{l_\beta}] & [i_{xx}ik - C_{lp}] \\ [-i_{zz}(ik)^2 + C_{nr}(ik) - C_{n_\beta}] & [-i_{xz}ik - C_{np}] \end{vmatrix} \quad (7.10)$$

A numerical example will help us understand the physical content of these formulas. For the B-70 airplane pictured in Fig. 1.7, we have assembled data for three representative conditions: subsonic cruising at intermediate altitudes, low supersonic operation during transitional accelerated or decelerated flight, and supersonic cruising at higher altitudes. Table 7.1 lists those geometrical

Table 7.1. Geometrical properties of XB-70-1 wing

S = 6297.8 ft^2
b = 105 ft
$\bar{c}$ = 78.5 ft
Æ = 1.751 (delta planform with approximately 70° leading-edge sweep)

quantities which remain constant throughout all the calculations of this chapter. Note that S denotes the maximum wing plan area, as seen when the angle δ_T of the foldable tips is zero.

Table 7.2 furnishes data, taken from Wolowicz *et al.* (1968), describing each of the three flight conditions employed for analyzing the lateral motion.

Table 7.2. Details of three XB-70-1 flight conditions used in calculations of lateral stability and response

M = 0.79	M = 1.2	M = 2.1
u_0 = 802.7 ft/sec	u_0 = 1161.6 ft/sec	u_0 = 2032.8 ft/sec
$h = 25 \times 10^3$ ft	$h = 40 \times 10^3$ ft	$h = 49.1 \times 10^3$ ft
$\rho_\infty = 1.066 \times 10^{-3}$ slug/ft³	$\rho_\infty = 5.87 \times 10^{-4}$ slug/ft³	$\rho_\infty = 3.81 \times 10^{-4}$ slug/ft³
$W = 378 \times 10^3$ lb	$W = 393 \times 10^3$ lb	$W = 383 \times 10^3$ lb
$I_{xx} = 1.98 \times 10^6$ slug-ft²	$I_{xx} = 1.95 \times 10^6$ slug-ft²	$I_{xx} = 1.83 \times 10^6$ slug-ft²
$I_{zz} = 22.0 \times 10^6$	$I_{zz} = 22.3 \times 10^6$	$I_{zz} = 21.4 \times 10^6$
$I_{xz} = -0.78 \times 10^6$	$I_{xz} = -0.85 \times 10^6$	$I_{xz} = -0.8 \times 10^6$
$\delta_T = 0$	$\delta_T = 25°$	$\delta_T = 65°$
$\alpha_{\text{Trim}} = 5.2°$	$\delta_{\text{Trim}} = 4.3°$	$\delta_{\text{Trim}} = 4.2°$
$\delta_{e\ \text{Trim}} = 2.5°$	$\delta_{e\ \text{Trim}} = 3.8°$	$\delta_{e\ \text{Trim}} = 4.9°$
$\theta_0 = 0$	$\theta_0 = 0$	$\theta_0 = 0$
$C_{LO} = 0.174$	$C_{LO} = 0.158$	$C_{LO} = 0.077$
CM = $0.218\bar{c}$	CM = $0.20\bar{c}$	CM = $0.203\bar{c}$

Table 7.3 presents all significant lateral derivatives, including those needed for (7.9, 10). Angles are taken in radian measure. Derivatives not listed, such as C_{yp} and $C_{l\dot{\delta}_a}$, may be assumed negligibly small. The given numbers account for the static influence of aeroelastic deformation under the particular conditions of flight, a phenomenon we discussed briefly in Section 2.7. Generally, the information in Table 7.3 was deduced from tables and graphs in Wolowicz *et al.* (1968), but I am indebted to Mr. John Wykes of North American Rockwell

Table 7.3. Lateral stability derivatives, deduced from flight test results on the XB-70-1

	M = 0.79	M = 1.2	M = 2.1
$C_{l\beta}$	−0.091	−0.057	0.011
$C_{n\beta}$	0.074	0.092	0.057
C_{lp}	−0.175	−0.24	−0.10
C_{np}	0	−0.13	−0.09
C_{lr}	−0.15	−0.47	0
C_{nr}	−0.10	−0.55	−0.38
$C_{y\beta}$	−0.315	−0.344	−0.309
C_{yr}	0.109	0.19	0.22
$C_{l\delta_a}$	0.0332	0.011	0.0052
$C_{n\delta_a}$	0.0023	−0.0023	0
$C_{y\delta_a}$	−0.018	−0.003	0.001

for filling a few gaps, notably C_{yr} and $C_{y\delta_a}$. All calculations and plots in this chapter were skillfully made by Messrs. James M. Summa and K. V. Krishna Rao.

Figures 7.1 and 7.2 give calculated results on the sinusoidal sideslip and bank-angle response to antisymmetrical elevon (i.e., "aileron") excitation. The quantities in Fig. 7.1 are precisely the amplitude and phase angle of $G_{\beta\delta_a}$ from (7.9a), the symbols being defined in (7.5) and (7.7a). The bank angle of Fig. 7.2 is the dimensionless time integral of roll rate $\hat{p}$ from (7.7b); it was chosen for presentation because its magnitude permits a direct comparison between angular displacements of bank and elevon.

These curves have many interesting features and merit careful study. Particularly clear from 7.1(a) are the resonances which occur in the vicinity of the natural frequency of the lateral-directional oscillation, known colloquially as *Dutch roll*. Thus at M = 0.79 we shall see that this frequency corresponds to $k = 0.06275$. The magnitudes of k tend to be quite typical of the lateral characteristics of many aircraft, but we should mention that resonant longitudinal k's found later in the chapter are unusually high because of the B-70's very large mean aerodynamic chord.

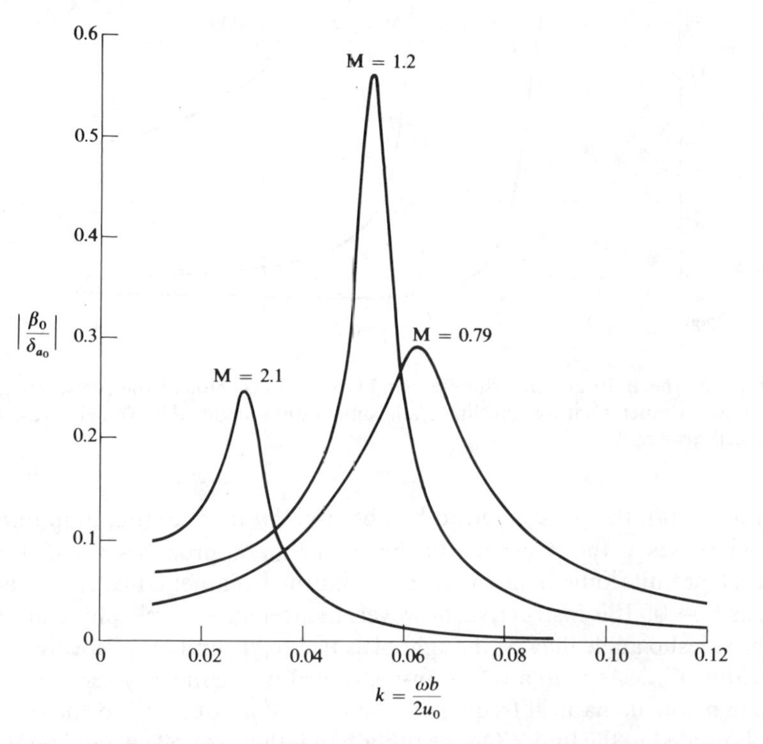

Fig. 7.1(a). For the B-60 at three different flight conditions, plot of the amplitude of the mechanical admittance relating sideslip to aileron rotation angle. The vehicle is assumed to follow a rectilinear path.

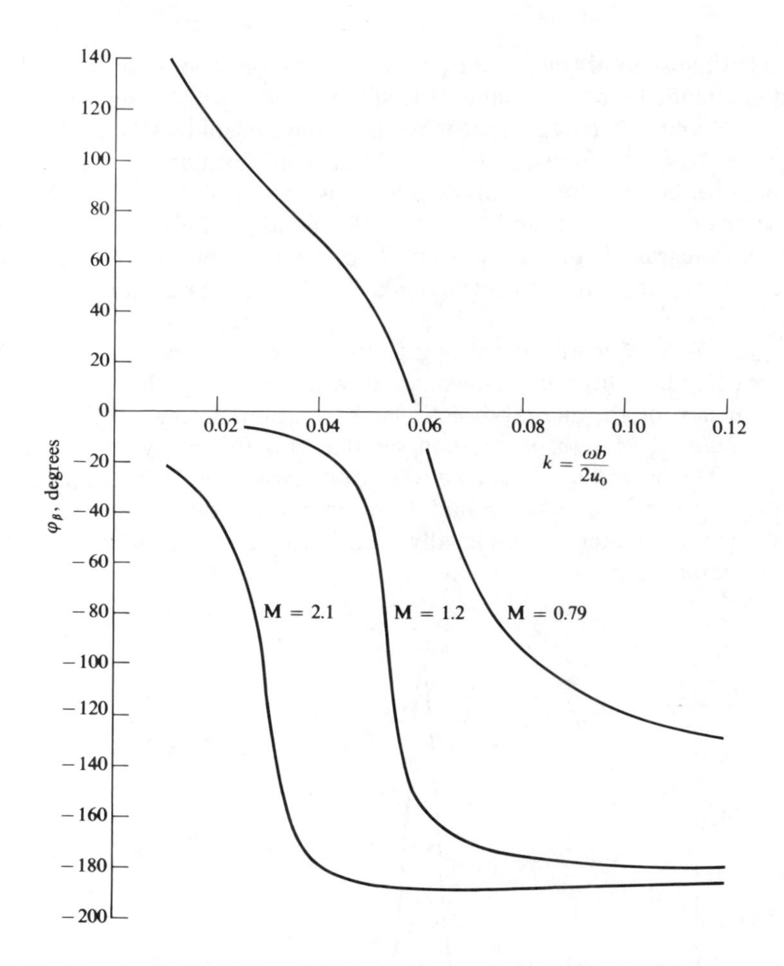

Fig. 7.1(b). For the B-70 at three different flight conditions, plot of the phase angle of the mechanical admittance relating sideslip to aileron rotation angle. The vehicle is assumed to follow a rectilinear path.

In Fig. 7.2(a), the peaks tend to be obscured by the fact that amplitude ratio $|\varphi_0/\delta_{a0}|$ increases without bound as the frequency approaches zero. Both this infinity and the finite limit for $|\beta_0/\delta_{a0}|$ can be confirmed mathematically by examining (7.9, 10) as $k \to 0$. It is instructive, however, to attempt a simple physical explanation. The sideslip angle may be thought of as tied to ground by a "spring" through the derivative $C_{n\beta}$. As with any spring-restrained mechanical system subjected to forcing far below its natural frequency, the ratio of β "output" to the δ_a "input" remains bounded in the limit. On the other hand, there is no "spring" on the bank angle, but only on the roll rate $\hat{p}$ through C_{lp}; this moment acts as a damper on φ. It follows that $G\hat{p}_{\delta a}$ stays finite as $k \to 0$, but that the bank angle response blows up.

It is also worth noting that the amplitudes of bank tend to be much larger than those of sideslip. This result is attributable to the elevon's greater effectiveness in exciting the roll degree of freedom, which may be seen from the relative magnitudes of $C_{l\delta_a}$ and $C_{n\delta_a}$ in Table 7.3.

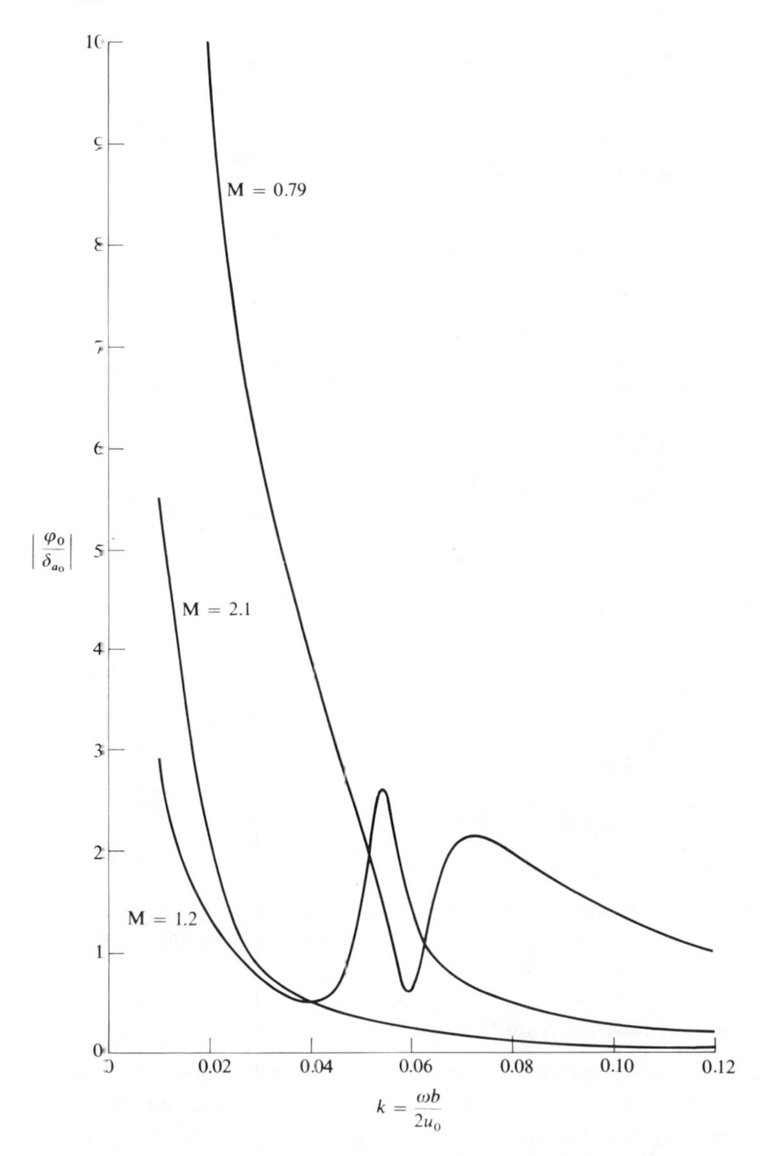

Fig. 7.2(a). For the B-70 at three different flight conditions, plot of the amplitude of the mechanical admittance relating bank angle to aileron rotation angle. The vehicle CM is assumed to follow a rectilinear path.

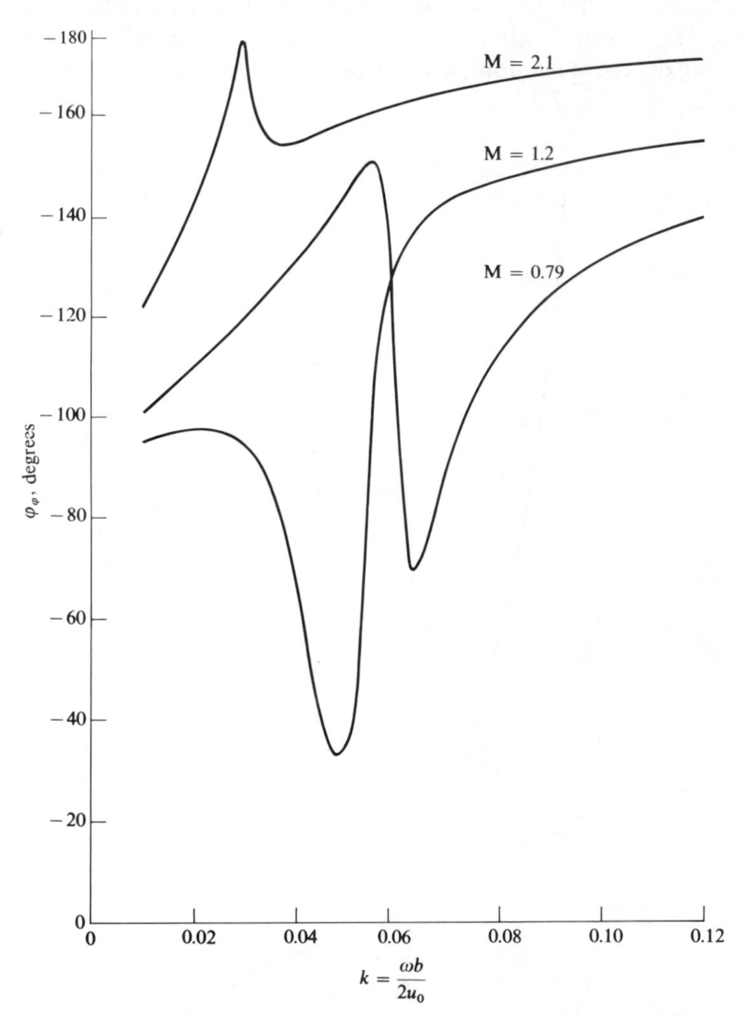

Fig. 7.2(b). For the B-70 at three different flight conditions, plot of the phase angle of the mechanical admittance relating bank angle to aileron rotation angle. The vehicle CM is assumed to follow a rectilinear path.

2. Impulsive and Indicial Admittances

There exists a duality between the mechanical admittance and the two elementary sorts of transient response which are customarily employed in the analysis of linear systems. Either may serve as a basis for examining the behavior when more complicated time-dependent inputs are introduced. Indeed, in this era of very high-speed digital computation, it is possible to pass back and forth in real time between mechanical and indicial admittances by means of a technique known as *fast Fourier transformation*.

The two basic and closely related indicial inputs are usually applied at $t = 0$, with zero initial conditions just prior to that instant. Working in terms of dimensionless lateral time $\hat{t}_a$, we define them as follows.

- *Unit impulse function*

$$\Delta(\hat{t}_a) = 0 \qquad \text{for } \hat{t}_a \neq 0 \tag{7.11a}$$

and such that

$$\int_{\substack{\text{through}\\ \hat{t}_a = 0}} \Delta(\hat{t}_a)\, d\hat{t}_a = 1 \tag{7.11b}$$

The impulse may be regarded as the limit of a sharply peaked, bell-shaped curve, under which unit area is enclosed, as the width is reduced to zero while the height becomes indefinitely large. When it acts as input to a single-degree-of-freedom mechanical system like a mass on a spring, the effect of Δ is equivalent to enforcing initial conditions of zero displacement and unit momentum at $\hat{t}_a = 0$. As explained in Chapter 2 of Lighthill (1958), this function and its derivatives are conveniently and rigorously expressible for mathematical purposes in terms of "generalized functions."

- *Step function*

$$1(\hat{t}_a) = \begin{cases} 1 & \text{for } \hat{t}_a \geq 0 \\ 0 & \text{for } \hat{t}_a < 0 \end{cases} \tag{7.12}$$

The unit step is merely the integral of $\Delta(\hat{t}_a)$. Although slightly more difficult to deal with analytically, this input often corresponds more closely to what can actually be applied to a flight vehicle. Like the mechanical admittance, the step-function response—or more realistically, the response to positive and negative steps separated by a known time interval—is thus capable of direct experimental determination.

Since Laplace transformation was developed in order to treat problems of the sort we are encountering, let us return to this technique, which we originally presented at (4.81). Here we are transforming on the single independent variable time. Hence we write for the transform of any function $f(\hat{t}_a)$, defined on $\hat{t}_a \geq 0$,

$$\mathcal{L}\{f(\hat{t}_a)\} \equiv \bar{f}(s) = \int_0^\infty e^{-s\hat{t}_a} f(\hat{t}_a)\, d\hat{t}_a \tag{7.13}$$

Key properties of this operation were reviewed following (4.81), and useful textbooks were cited. In the present context, it is elementary to show that the transforms of the impulse and step functions are, respectively,

$$\mathcal{L}\{\Delta(\hat{t}_a)\} = 1 \tag{7.14}$$

$$\mathcal{L}\{1(\hat{t}_a)\} = 1/s \tag{7.15}$$

In 7.14, we are implicitly using what is known as a "two-sided" transform whose lower limit lies at $0-$ and therefore captures the entire area under the unit impulse.

Returning to lateral-directional aileron forcing of the vehicle whose CM moves along a straight line, we carry over from control theory the concept of *system transfer function*. In general, this quantity is the Laplace transform of the response, with zero initial conditions, to a unit impulse applied at $t = 0$; for example, $G_{\beta\delta_a}(s)$ would be the transform of the sideslip angle following an aileron displacement

$$\delta_{a0}(\hat{t}_a) = \Delta(\hat{t}_a) \tag{7.16}$$

When (7.16) and (7.14) are substituted into the transformed equations of motion (7.4), we discover why the same symbol G_{ij} is chosen for both transfer function and mechanical admittance. Since Laplace transform of the derivatives D_a and D_a^2, under the undisturbed conditions at $\hat{t}_a = 0-$, merely calls for replacing them by s and s^2, the algebraic manipulations going from (7.4) through (7.8) to (7.9a) are seen to be identical with those required to calculate $G_{\beta\delta_a}(s)$. The only difference is that ik must everywhere be replaced by the complex transform variable s. Thus we can write, without further analysis,

$$G_{\beta\delta_a}(s) = \frac{\begin{vmatrix} C_{l\delta_a} & [i_{xx}s - C_{lp}] \\ C_{n\delta_a} & [-i_{xz}s - C_{np}] \end{vmatrix}}{\begin{vmatrix} [i_{xz}s^2 + C_{lr}s - C_{l\beta}] & [i_{xx}s - C_{lp}] \\ [i_{zz}s^2 + C_{nr}s - C_{n\beta}] & [-i_{xz}s - C_{np}] \end{vmatrix}} = \frac{\text{Linear polynomial in } s}{\text{Cubic polynomial in } s} \tag{7.17}$$

Similar expressions are readily constructed for $G_{p\delta_a}(s)$, $G_{\varphi\delta_a}(s)$, $G_{\beta\delta_r}(s)$, or any other desired transfer function. We remark that nonzero initial conditions can also be introduced as additive terms to relations like (7.17) by means of such transforms as

$$\mathscr{L}\{D_a\beta\} = s\bar{\beta}(s) - \beta(0) \tag{7.18a}$$

$$\mathscr{L}\{D_a^2\beta\} = s^2\bar{\beta}(s) - s\beta(0) - \dot{\beta}(0) \tag{7.18b}$$

Indicial admittance is an expression often used for the response, again with zero initial conditions, to a unit step input. Comparing (7.15) with (7.14), we find that the Laplace transform of any particular indicial admittance is just the corresponding transfer function divided by s. In the physical time domain, such division corresponds to an integration with respect to $\hat{t}_a$ and reflects the integral relationship between the unit impulse and the step. Let us adopt symbol A_{ij} for indicial admittance, with i and j representing the chosen output and input, respectively. We can then set down immediately formulas like

$$\bar{A}_{\beta\delta_a}(s) = \frac{G_{\beta\delta_a}(s)}{s} \tag{7.19a}$$

$$\bar{A}_{p\delta_a}(s) = \frac{G_{p\delta_a}(s)}{s} \tag{7.19b}$$

We leave as an exercise for the reader the details of inverting the various transfer functions and indicial admittances, which will yield time histories of responses to impulse and step inputs. Many such results can be found in Etkin (1959, 1972). Because the transforms are ratios of polynomials, it can be anticipated that the corresponding time functions will be sums of exponential terms and terms consisting of products between exponentials and sines and cosines of multiples of $\hat{t}_a$. The latter are, of course, oscillatory so that these $\hat{t}_a$-coefficients are dimensionless natural frequencies of the system.

Figures 7.3 and 7.4 plot indicial admittances for the B-70. Conditions of flight and input–output pairs correspond to the mechanical admittances of Figs.

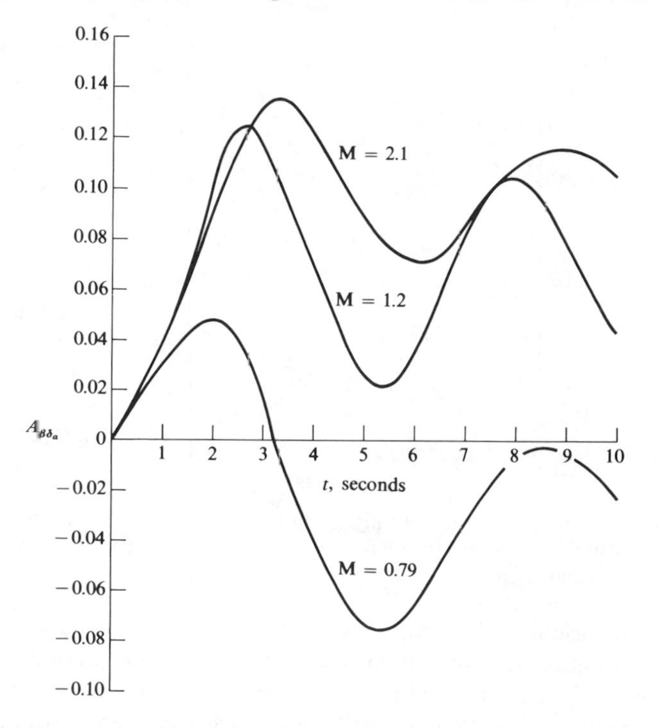

Fig. 7.3. For the B-70 at three different flight conditions, plot, versus physical time, of indicial admittance relating sideslip to aileron rotation angle. The vehicle CM is assumed to follow a rectilinear path.

7.1 and 7.2. To assist with visualizing these motions, the abscissa scales are converted to physical time through the relation $t = \hat{t}_a[b/2u_0]$. The sideslip resulting from a step elevon displacement is seen to be oscillatory, with frequencies and small dampings which agree roughly with what we might anticipate from the resonance peaks in Fig. 7.1. The $C_{n\beta}$ "spring" thus prevents an unbounded build-up of the angle β. By contrast, φ in Fig. 7.4 grows from zero with an

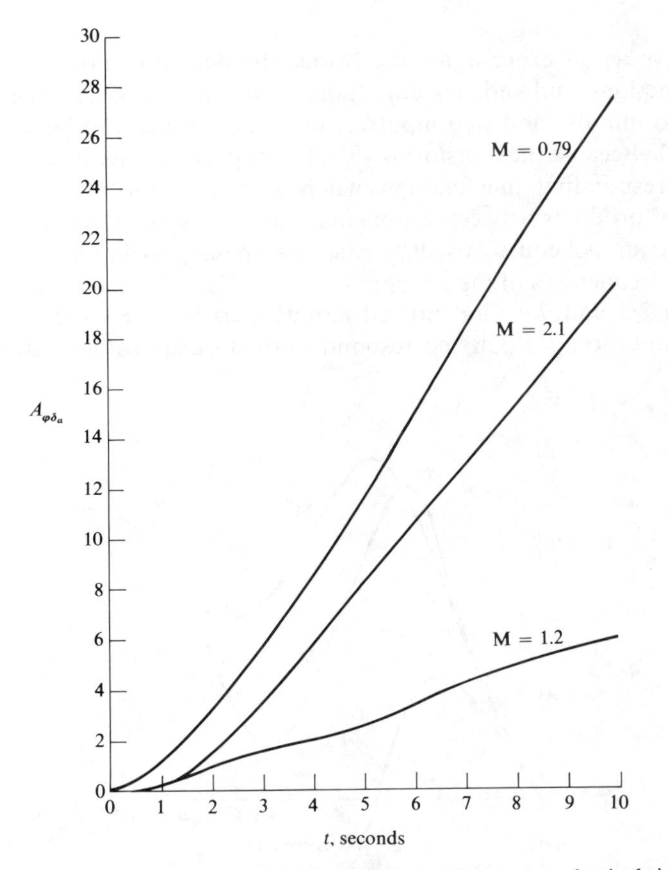

Fig. 7.4. For the B-70 at three different flight conditions, plot, versus physical time, of indicial admittance relating bank angle to aileron rotation angle. The vehicle CM is assumed to follow a rectilinear path.

exponential transient and then, except for the small oscillation at M = 1.2, approaches straight-line behavior consistent with constant roll rate. This behavior confirms our earlier diagnosis of the consequences of no "spring" on bank angle: The rolling velocity accelerates until an approximate equilibrium is attained between the damping moment due to C_{l_p} and the constant rolling moment due to the elevons. This steady spinning about the x-axis would obviously continue until modified by some compensatory change in δ_a. We comment that, even though $\varphi(t)$ grows to large values, these predictions of small perturbation theory may agree fairly well with reality, so long as $pb/2u_0$ remains small.

3. Response to a General Input

It is always possible to choose a time origin such that the input, for example, $\delta_a(\hat{t}_a)$, first begins to act at $\hat{t}_a = 0$. Moreover, no serious restriction is associated with picking zero initial conditions at $\hat{t}_a = 0-$; as mentioned above (7.18), Laplace transformation permits the incremental effects of a prior motion to be introduced straightforwardly. With these specifications, we transform (7.4) and are led immediately to the typical solution

$$\bar{\beta}(s) = G_{\beta\delta_a}(s)\bar{\delta}_a(s) \tag{7.20}$$

Equation (7.20) and its inverse clearly demonstrate the central importance of the system transfer function. We accomplish the inversion by means of the convolution theorem (see Churchill 1944, pages 36–38), which applies to the product of any two transforms $\bar{f}(s)$ and $\bar{g}(s)$:

$$\mathscr{L}^{-1}\{\bar{f}(s)\bar{g}(s)\} = \int_0^{\hat{t}_a} g(\hat{\tau}) f(\hat{t}_a - \hat{\tau})\, d\hat{\tau} \tag{7.21}$$

Here $\hat{\tau}$ is a dummy variable of integration. By an obvious interchange between $\hat{\tau}$ and $(\hat{t}_a - \hat{\tau})$, we can show that either f or g may be chosen to have the argument $(\hat{t}_a - \hat{\tau})$. The application of (7.21) to (7.20) yields our desired result.

$$\boxed{\beta(\hat{t}_a) = \int_0^{\hat{t}_a} \delta_a(\hat{\tau}) G_{\beta\delta_a}(\hat{t}_a - \hat{\tau})\, d\hat{\tau}} \tag{7.22}$$

You may also wish to derive a relationship between $\beta(\hat{t}_a)$ and the corresponding indicial admittance function. You can do this by suitable partial integrations on (7.22), taking account of the fact that $A_{\beta\delta_a}$ is the time integral of $G_{\beta\delta_a}$. Both (7.22) and the latter relation can be given physical explanations in terms of the superposition of responses to a succession of impulsive or step inputs applied at times $\hat{\tau}$ between zero and the present instant $\hat{t}_a$. The mathematical form of $G_{\beta\delta_a}$ is such that (7.22) can be integrated directly when δ_a is a simple enough time function. More generally, numerical integrations may be carried out on a digital computer. Alternatively, one can evaluate (7.22) by electrodynamic analogy when one applies a voltage representing the prescribed δ_a at the input terminal of an analog mechanization of the lateral equations of a vehicle.

7.2 LATERAL-DIRECTIONAL NORMAL MODES; COMPARISON WITH THE COMPLETE LATERAL EQUATIONS OF MOTION; ROOTS-LOCUS TECHNIQUE

With the exception of random inputs—which, despite their interest, fall beyond our scope—Section 7.1 gave a fairly comprehensive discussion of forced motion of a linear system. Let us next turn to the free or "characteristic" perturbed motions of aircraft and, through them, to an assessment of the straight-line approximation $\psi = -\beta$ underlying (7.4).

In the process of using (7.4) to analyze any transient response to a control input or specified initial conditions, one always finds that the solution contains a series of terms with exponential and sinusoidal functions of $\hat{t}_a$. The coefficients of $\hat{t}_a$ in these terms are determined from the roots of the denominator determinant of the Laplace-transformed equations. Thus in the present case one is working with the factors of

$$0 = \begin{vmatrix} [i_{xz}s^2 + C_{lr}s - C_{l\beta}] & [i_{xx}s - C_{lp}] \\ [-i_{zz}s^2 + C_{nr}s - C_{n\beta}] & [-i_{xz}s - C_{np}] \end{vmatrix}$$
$$\equiv As^3 + Bs^2 + Cs + E = 0 \tag{7.23}$$

In (7.23), the constant coefficients A through E can readily be expressed in terms of the stability derivatives and dimensionless inertias contained in the determinant. A key observation is that these coefficients depend only on inherent properties of the vehicle and are unrelated to quantities like $C_{l\delta_a}$, $C_{n\delta_a}$, $\beta(0)$, etc., which describe the inputs or initial conditions. The roots of (7.23) are therefore inherent "characteristics" of the vehicle and its flight regime. All transient responses are expected to involve, in different proportions, a superposition of the corresponding characteristic motions or "normal modes" of the system. Furthermore, it can be shown that "pure" motion can be set up in a single normal mode, associated either with one real root or one complex-conjugate root pair from (7.23). In the absence of subsequent disturbances, this motion will continue without exciting responses in any other mode.

We need a brief mathematical review of the sorts of factors that may be obtained from an equation like (7.23). Although it contains only a third-degree polynomial, we note that this degree differs from one linear system to another and generally equals the total order of the differential-equation set which governs the system. Uniformly, however, the coefficients A, etc., are real constants. As proved in algebra texts, it follows that two classes of roots are encountered.

1. Positive or Negative Real Roots

Let us consider $s = +\lambda$ or $-\lambda$, where $\lambda \geq 0$ is a real number. When we encounter such a root, the partial-fraction expansion of a transformed response such as (7.17) contains a term proportional to $[s \mp \lambda]^{-1}$. In view of the inverse

$$\mathscr{L}^{-1}\left[\frac{1}{s \mp \lambda}\right] = \exp\,[\pm\lambda\hat{t}_a] = \exp\left[\pm\left(\frac{2u_0\lambda}{b}\right)t\right] \tag{7.24}$$

we reason that the time variation of the normal mode associated with this root will involve a growing or decaying exponential.

When $s = +\lambda$, this yields an "unstable" divergent motion. The *time constant*—or number of seconds required for the amplitude of any coordinate to increase by a factor e—appears from (7.24) to be $(b/2u_0\lambda)$. Similarly, $s = -\lambda$ corresponds to an exponentially decaying motion, with time $(b/2u_0\lambda)$ to drop to $1/e$ of any given amplitude. Another common measure of this decay rate is the time for the amplitude to halve,

$$T_{1/2} = \frac{0.693b}{2u_0\lambda} \tag{7.25}$$

Two equal real roots $\pm\lambda$ give rise to a mode containing a linear factor $\hat{t}_a$, the time dependence being proportional to $[A + B\hat{t}_a] \exp [\pm\lambda\hat{t}_a]$. One or more zero roots are connected with constant displacements or velocities of the system coordinates. They usually signal the presence of "rigid-body degrees of freedom"—coordinates like bank angle φ—which are not restrained to ground by the equivalent of springs.

2. Complex Conjugate Root Pairs

These roots are always treated in pairs, $s = n \pm i\omega$, since the real coefficients in (7.23) ensure the appearance of equal positive and negative imaginary parts. The corresponding quantity to be inverted is $[(s - n)^2 + \omega^2]^{-1}$. It is a familiar fact from mechanical vibrations that such a transform is associated with an exponentially damped or divergent oscillation. The roots are usually rewritten

$$s = n \pm i\omega = -\zeta\omega_n \pm i\omega_n\sqrt{1 - \zeta^2} \tag{7.26}$$

where ω is the true or "damped" natural frequency, ω_n is the undamped natural frequency which this mode would have if $n = 0$, and ζ ($|\zeta| < 1$) is the critical damping ratio. Positive ζ bespeaks a stable or decaying motion, negative ζ an unstable one.

In the time domain, the general formula describing response in this mode is

$$f(\hat{t}_a) = \exp [-\zeta\omega_n\hat{t}_a][C_1 \cos \omega\hat{t}_a + C_2 \sin \omega\hat{t}_a] \tag{7.27}$$

with constants C_1 and C_2 to be determined from the initial conditions and/or the forcing input. Since ω is dimensionless, the period of this mode in seconds would be

$$T = \frac{2\pi}{\omega}\left(\frac{b}{2u_0}\right) \tag{7.28}$$

It is straightforward to work out the time constant, $T_{1/2}$ for a damped oscillation, or the time T_2 for the amplitude envelope to double in an unstable case.

We add a few more remarks about normal modes before returning to our example. First, the foregoing results imply one *sufficient* test for linear stability in a system—that is, a test which guarantees the absence of any free motions which are predicted by the small-perturbation approximation to grow without bound. The requirement is simply that:

- Re $\{s\} \leq 0$ for all roots of the characteristic equation.
- No multiple roots with zero imaginary parts.

Several procedures have been devised for ensuring that these requirements are met merely by working with the coefficients of the characteristic polynomial [cf. (7.23)]; thereby one can examine stability of a complicated system without having to calculate all the roots. Best known of these is Routh's criterion (Etkin 1959, pages 193–195). Inasmuch as there are now widely available computer programs which both expand and factor the characteristic determinant,* it usually makes sense to get the additional information about degree of stability that goes with full knowledge of these roots.

With each real root or conjugate pair there is connected a particular relationship between the amplitudes and phases of the system degrees of freedom (or components of the state vector). These relations describe what is called the "mode shape." They are found by assuming motion only in the particular mode, whereupon the time dependence can be eliminated from the homogeneous equations of motion and ratios calculated among the various amplitudes. For a real root, these ratios are positive or negative real numbers and show that the degrees of freedom are instantaneously in-phase or opposite. Conjugate roots lead, however, to complex ratios whose arguments give the phase differences among the various freedoms. As we shall see, complex vectors rotating at the frequency ω furnish a convenient visualization scheme.

We can base a simple illustration of how a mode shape is found on the lateral-directional equations (7.4). For anything that looks like an airplane (7.23) has one rapidly decaying negative real root. It is associated with the "roll subsidence" mode, wherein $\hat{p}(\hat{t}_a)$ dies out following a disturbance such as an aileron impulse. Accordingly, let us denote this root $-\lambda_1$ and assume that the sideslip and roll are described by

$$\beta = \beta_0 e^{-\lambda_1 \hat{t}_a} \tag{7.29a}$$

$$\hat{p} = \hat{p}_0 e^{-\lambda_1 \hat{t}_a} \tag{7.29b}$$

When (7.29) are substituted into the homogeneous (7.4), the exponential factor cancels and leaves an algebraic system,

$$[i_{xz}\lambda_1^2 - C_{lr}\lambda_1 - C_{l\beta}]\beta_0 - [i_{xx}\lambda_1 + C_{lp}]\hat{p}_0 = 0 \tag{7.30a}$$

$$-[i_{zz}\lambda_1^2 + C_{nr}\lambda_1 + C_{n\beta}]\beta_0 + [i_{xz}\lambda_1 - C_{np}]\hat{p}_0 = 0 \tag{7.30b}$$

Choosing (7.30a), we solve for the following real ratio between the amplitudes:

$$\frac{\beta_0}{\hat{p}_0} = \frac{i_{xx}\lambda_1 + C_{lp}}{i_{xz}\lambda_1^2 - C_{lr}\lambda_1 - C_{l\beta}} \tag{7.31}$$

[That the same solution would also be obtained from (7.30b) can be proved because $s = -\lambda_1$ makes the determinant in (7.23) vanish.] Although either β_0 or $\hat{p}_0$ individually would have to be determined from initial conditions, (7.31) indicates that they must appear in particular proportions if only this pure mode is excited. An interesting observation is that the roll-subsidence root is roughly estimable from

$$i_{xx}\lambda_1 + C_{lp} \cong 0 \tag{7.32}$$

Equation (7.31) therefore tells us that only a relatively small amount of sideslip is typically present in this mode.

---

* Roots loci of the sort discussed below are now drawn automatically.

Table 7.4 lists the approximate lateral-directional roots that are computed for the B-70 when appropriate data from Tables 7.1 through 7.3 are inserted into

Table 7.4. Roots of XB-70-1 characteristic determinant, for lateral motion under approximation that CM moves in a straight line

M	Lateral-directional oscillation	Roll subsidence
0.79	$-0.009067 \pm 0.06275i$	-0.07077
1.2	$-0.003068 \pm 0.05384i$	-0.06785
2.1	$-0.003771 \pm 0.02889i$	-0.01737

(7.23) and the factorization is carried out. For comparison, and to serve as a basis for further discussion, Table 7.5 gives the roots of the characteristic determinant of the complete set of lateral equations (6.31a,b,c) with $\theta_0 = 0$. It also adds information about mode shapes, periods, and decay times.

Table 7.5. Roots of full linearized lateral equations for XB-70-1 in horizontal flight

M	Lateral-directional oscillation	Roll subsidence mode	Spiral mode
0.79	$s = -0.004208 \pm 0.06883i$	$s = -0.08225$	$s = -0.002955$
	$(T = 5.97\text{ sec}, T_{1/2} = 10.7\text{ sec})$	$(T_{1/2} = 0.54\text{ sec})$	$(T_{1/2} = 1.52\text{ sec})$
	$\dfrac{\bar{\beta}_0}{\varphi_0} = 0.156e^{-i41.7^\circ}$	$\dfrac{\beta_0}{\varphi_0} = 0.0040$	$\dfrac{\beta_0}{\varphi_0} = 0.0013$
	$\dfrac{\bar{\psi}_0}{\varphi_0} = 0.134e^{i136.5^\circ}$	$\dfrac{\psi_0}{\varphi_0} = -0.0355$	$\dfrac{\psi_0}{\varphi_0} = -0.8818$
1.2	$s = -0.002080 \pm 0.05485i$	$s = -0.06957$	$s = -0.002978$
	$(T = 5.17\text{ sec}, T_{1/2} = 15.0\text{ sec})$	$(T_{1/2} = 0.448\text{ sec})$	$(T_{1/2} = 10.5\text{ sec})$
	$\dfrac{\bar{\beta}_0}{\varphi_0} = 0.2479e^{-i26.2^\circ}$	$\dfrac{\beta_0}{\varphi_0} = 0.0172$	$\dfrac{\beta_0}{\varphi_0} = 0.0017$
	$\dfrac{\bar{\psi}_0}{\varphi_0} = 0.2393e^{i155.2^\circ}$	$\dfrac{\psi_0}{\varphi_0} = -0.0014$	$\dfrac{\psi_0}{\varphi_0} = -0.4208$
2.1	$s = -0.004968 \pm 0.02891i$	$s = -0.01696$	$s = +0.3634 \times 10^{-3}$
	$(T = 5.61\text{ sec}, T_{1/2} = 3.59\text{ sec})$	$(T_{1/2} = 1.05\text{ sec})$	$(T_2 = 49.5\text{ sec})$
	$\dfrac{\bar{\beta}_0}{\varphi_0} = 0.5455e^{i167.1^\circ}$	$\dfrac{\beta_0}{\varphi_0} = 0.0169$	$\dfrac{\beta_0}{\varphi_0} = 0.0034$
	$\dfrac{\bar{\psi}_0}{\varphi_0} = 0.5435e^{-i17.6^\circ}$	$\dfrac{\psi_0}{\varphi_0} = -0.0087$	$\dfrac{\psi_0}{\varphi_0} = -1.098$

Let us address ourselves to the accuracy of the approximation in Section 7.1. To this end, we show in Fig. 7.5 for M = 2.1 the exact and approximate lateral-directional oscillatory mode shapes. Both are plotted as rotating and shrinking

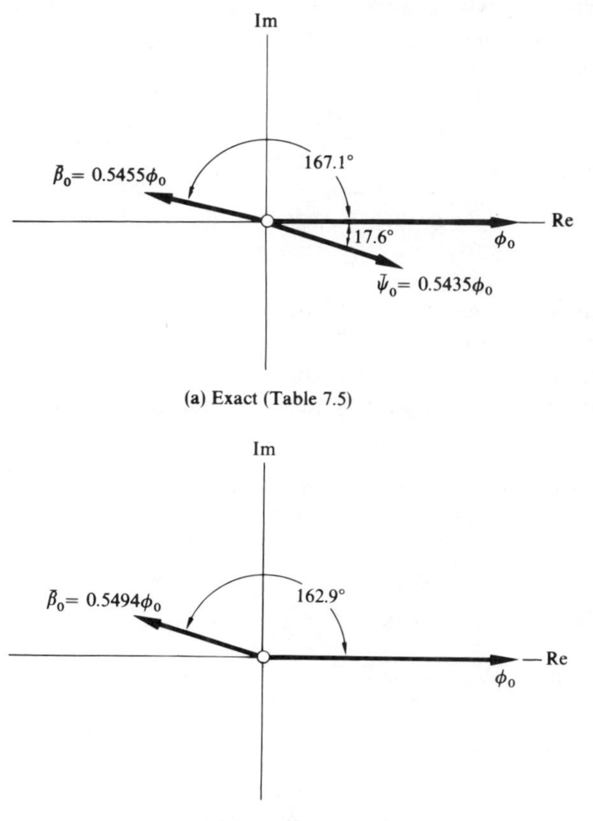

Fig. 7.5. Vector plots of the exact and approximate mode shapes for lateral-directional oscillation of the B-70 at M = 2.1 and altitude 49,000 ft. Approximate mode is based on the assumption of rectilinear flight.

complex vectors, whose horizontal projections might serve as actual time histories for the angles of bank, sideslip, and yaw. These vectors are caught at the instant when bank $\varphi(\hat{t}_a)$ has magnitude φ_0 and phase zero. $\beta(\hat{t}_a)$ is seen to be nearly opposite to $\varphi(\hat{t}_a)$ in phase, whereas the yaw angle lags slightly. (It is instructive to try to reproduce this motion with a small model, recalling that the period is about 5.6 sec and that about 0.64 of a period is required for the amplitudes to drop to half of their initial values.) In the case of Fig. 7.5, we see that the assumption of straight-line CM motion ($\psi = -\beta$) is quite satisfactory: The true magnitudes of these angles differ by two parts in 540, and their phase difference is 184.7°.

Furthermore, (7.4) estimate the period within 0.1 % and the damping within a less-acceptable 25 %. More detailed study of data from Tables 7.4 and 7.5 reveals that roll subsidence and the oscillatory mode are estimated with reasonable accuracy except at the lowest Mach number; the oscillation dampings leave something to be desired.

A few descriptive words are in order about the three normal modes as characterized in Table 7.5.

- The lateral-directional mode is a moderately damped vibration with period of the order of a few seconds. The number of semispans traversed during one cycle is $2\pi/\omega$; as with many aircraft, it falls in the vicinity of 100 to 200. Between one-half and three cycles are needed for a 50 % reduction in amplitude. It is worth mentioning that this damping tends to deteriorate with increasing altitude on most designs, but that deflecting the B-70's wingtips tends here to produce a compensatory augmentation of the damping derivative $|C_{nr}|$. During the oscillation, the CM translation is nearly rectilinear. There is more bank than sideslip or yaw, but the latter freedoms tend to participate more at higher M.
- Roll subsidence, sometimes called *roll convergence*, is the motion already discussed in connection with (7.31). The damping time constants are typically very short, and the other freedoms tend to contribute 0–4 % of what $\varphi(\hat{t}_a)$ does to the mode shape. Try comparing estimates from (7.32) with the exact roots in Table 7.5 and with the analysis of Problem 6A.
- The spiral mode goes by the names *convergence* or *divergence*, depending on whether it is stable or unstable. A slight instability is displayed by the B-70 only at the highest M we have chosen; many other vehicles have more rapid divergences however. Very little sideslip angle develops in this mode, since the "$C_{n\beta}$ spring" tends to align the vehicle with its path. Bank and yaw develop in roughly equal proportions, a right-wing-down attitude being associated with flight-path deviation to the left. Unstable spiral modes have been responsible for many accidents to inexperienced pilots under instrument conditions.

Let us close this section by reviewing and applying to the B-70 a valuable design tool known as the *roots-locus method*. [Refer to Seckel (1964, Chapter 2) for full details and for a set of rules whereby the sketching of loci is facilitated.] This approach, which has its greatest utility when gain constants are being selected for automatic control equipment, consists of varying a single parameter in the system and examining the effects of its changes on the characteristic-determinant roots, plotted in the complex *s*-plane. With a little experience and knowledge of how to interpret root positions, one may use this technique to optimize vehicle response and handling qualities within the limitations imposed by tradeoffs against other factors that may influence the configuration.

For illustrative purposes, let us study the construction of a locus for the weathercock stability derivative $C_{n\beta}$, as it affects the lateral-directional oscillatory mode. This could be done, of course, simply by inserting a series of values into the exact characteristic equation of (6.31) and factoring. There are, however, quicker ways of estimating important properties of loci, which help to speed up preliminary design. During the analysis, we adopt the assumptions of Section 7.1 and, for convenience, neglect the small product of inertia ($i_{xz} \cong 0$). These stipulations make it possible to expand (7.23) as follows:

$$i_{xx}i_{zz}s^3 - [i_{xx}C_{nr} + i_{zz}C_{lp}]s^2 + [-C_{lr}C_{np} + i_{xx}C_{n\beta} + C_{lp}C_{nr}] + [C_{l\beta}C_{np} - C_{n\beta}C_{lp}] = 0 \quad (7.33)$$

Attention is centered on $C_{n\beta}$ by rearranging (7.33) so that it appears as a separate term.

$$\frac{i_{xx}i_{zz}s^3 - [i_{xx}C_{nr} + i_{zz}C_{lp}]s^2 + [C_{lp}C_{nr} - C_{lr}C_{np}]s + C_{l\beta}C_{np}}{i_{xx}s - C_{lp}} + C_{n\beta} = 0 \quad (7.34)$$

Now $C_{n\beta} > 0$ is an absolute requirement, which suggests that we might profitably study what happens to the values of s computed from (7.34) as $C_{n\beta}$ ranges from 0 to $+\infty$. At $C_{n\beta} = 0$, there are evidently three finite roots, which arise from the factors of the numerator polynomial in the first term; in control texts, these are referred to as the *poles* of the locus. As $C_{n\beta} \to \infty$, two roots must increase without limit, but there is a third finite root at what is called a *zero* of the denominator polynomial

$$s = \frac{C_{lp}}{i_{xx}} = -\frac{|C_{lp}|}{i_{xx}} \quad (7.35)$$

[Compare this result with (7.32). It shows that, when $C_{n\beta}$ is very large, the x-axis is forced to line up with the relative wind and the roll-subsidence mode is decoupled from sideslip.]

According to Seckel's rules, which are elementary consequences of complex-variable theory, the roots of (7.34) must pass from the poles to the zeros, or to infinity, as $C_{n\beta}$ increases from 0. Here there are two infinities, and the rules therefore require that two loci go to $s = \pm i\infty$, parallel to the positive and negative imaginary axes.

With this brief background, we offer Fig. 7.6, a $C_{n\beta}$ locus plot for the B-70 at M = 2.1. Although these results are based on the full determinant of (6.31) rather than (7.34), the spiral mode remains distinct from the others, so that our statements above still apply qualitatively. Values of $C_{n\beta}$ are marked on the curve, the four poles falling, of course, at $C_{n\beta} = 0.0$. Of the two poles on the real axis, the unstable one is a spiral divergence, the one to the left a roll subsidence. The conjugate pair corresponds to a low-frequency, heavily damped lateral-directional oscillation. [Note that since $\zeta = -n/\sqrt{n^2 + \omega^2}$ for such a mode, this damping

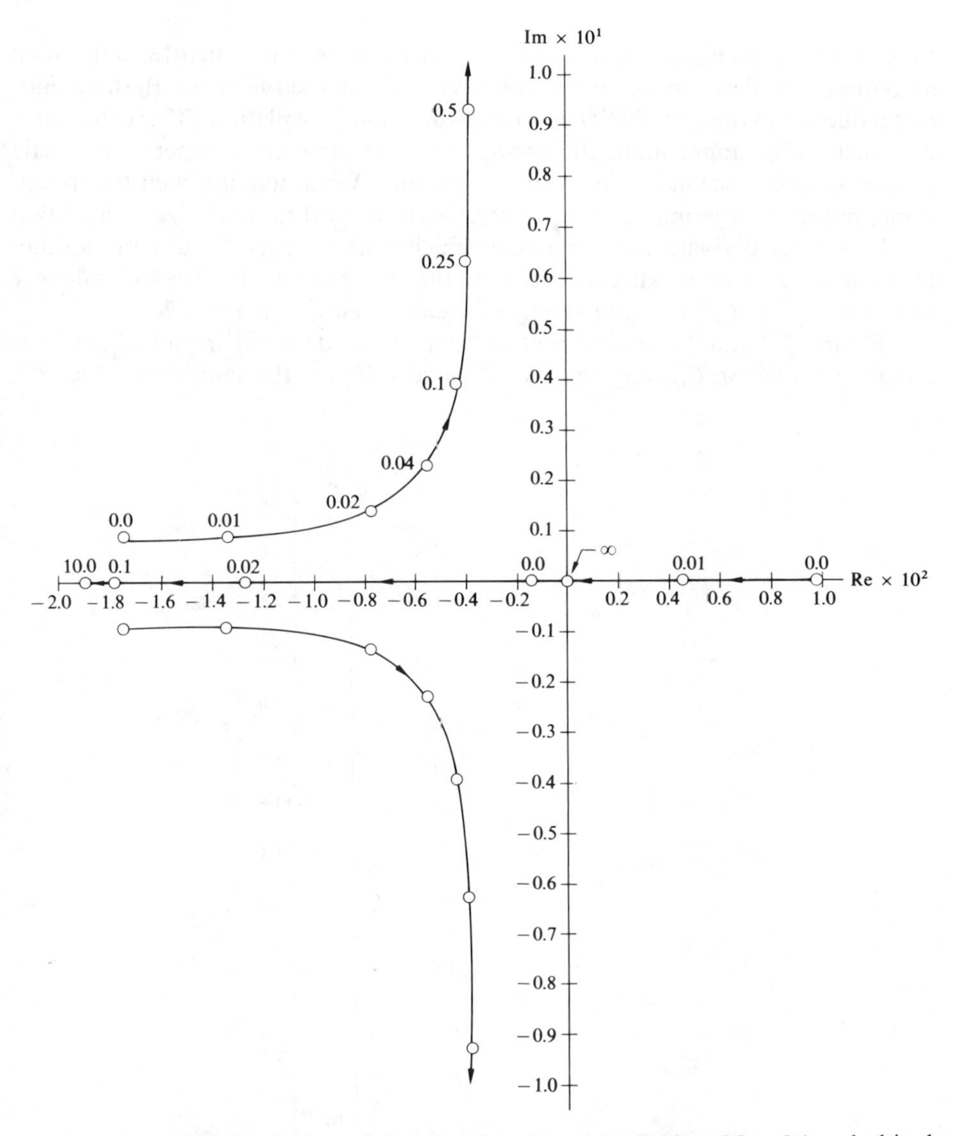

Fig. 7.6. Roots loci for the lateral-directional modes of the B-70 at M = 2.1 and altitude 49,000 ft. The parameter varied is directional stability $C_{n\beta}$, whose values are indicated on the loci. (Note difference in horizontal and vertical scales.)

ratio can be visualized as the sine of the angle between the imaginary axis and a line from the origin to the upper complex root. Such a visualization is unfortunately distorted in Fig. 7.6, in which the ordinate and abscissa scales differ by 10.]

As $C_{n\beta}$ is increased from 0 to its nominal value of 0.057, all roots are seen to approach the numbers given in Table 7.5. Further increases have relatively little effect on the nonoscillatory modes, which are known to involve only small amounts of sideslip. On the other hand they cause a marked growth in the frequency and reduction in the damping of the lateral-directional oscillation. This behavior is consistent with augmenting the spring, while leaving the damper and inertia unchanged, on a second-order vibratory system. When making such interpretations, however, it is important to recognize how hard it would be to modify a single stability derivative on an actual vehicle without affecting any of its other characteristics. For instance, additions to the fin area of the B-70 would influence not only $C_{n\beta}$ but C_{nr}, C_{np}, and several of the other entries in Table 7.3.

Figures 7.7 and 7.8 present roots-locus plots for damping in yaw C_{nr} and the coupling derivative $C_{l\beta}$, respectively. Ground rules are the same as for Fig. 7.6.

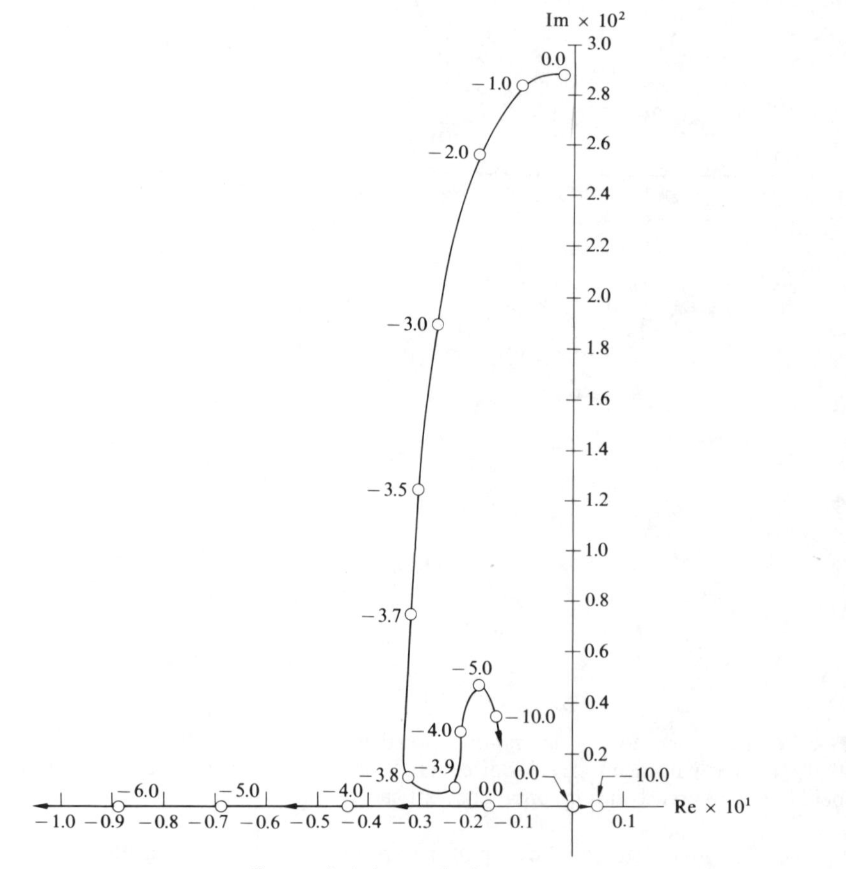

Fig. 7.7. Roots loci for the lateral-directional modes of the B-70 at M = 2.1 and altitude 49,000 ft. The parameter varied is yaw damping C_{nr}, whose values are indicated on the loci. (Note difference in horizontal and vertical scales.)

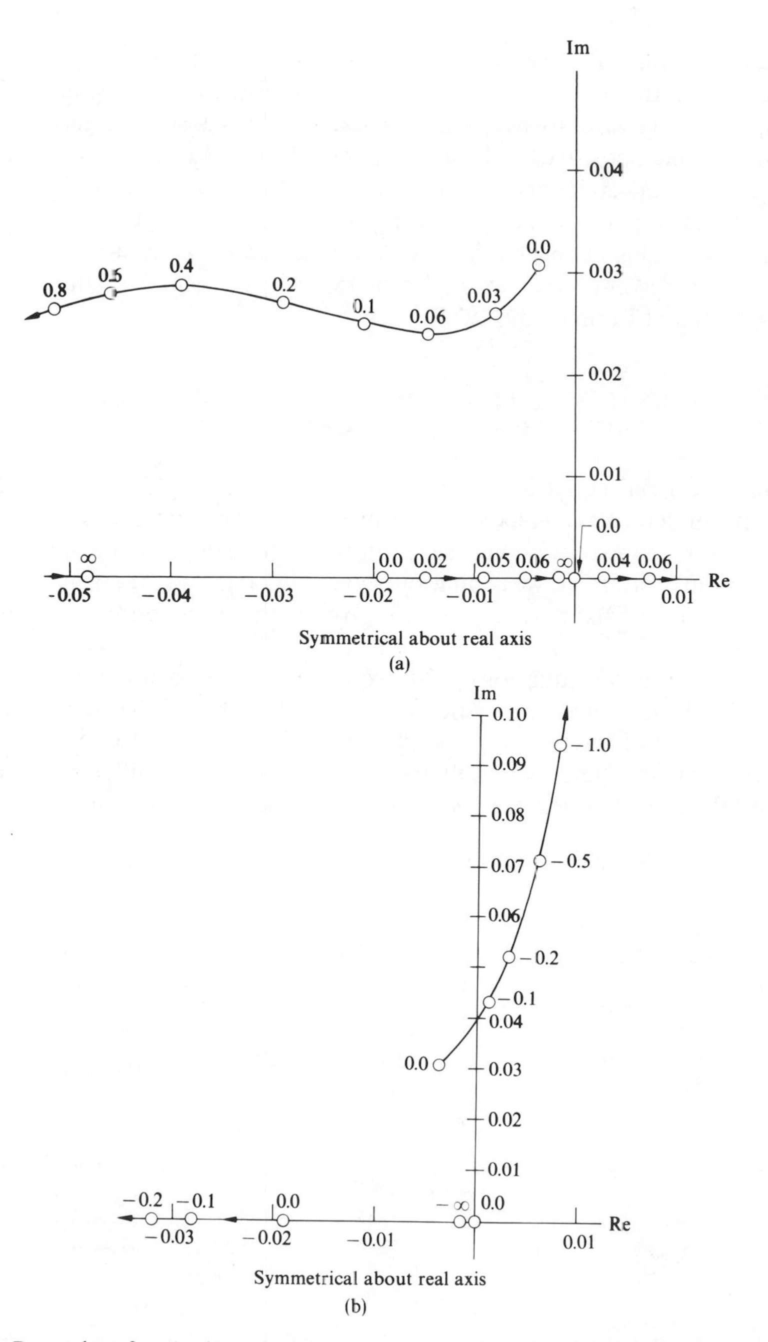

Fig. 7.8. Roots loc for the lateral-directional modes of the B-70 at M = 2.1 and altitude 49,000 ft. The parameter varied is the coupling derivative $C_{l\beta}$, with (a) positive values and (b) negative values. [Note difference in horizontal and vertical scales in (b).]

Because C_{nr} is inherently negative, only the range between 0 and $-\infty$ is covered in Fig. 7.7. On the other hand, $C_{l\beta}$ between $+\infty$ and $-\infty$ is covered in Fig. 7.8. Although negative values are regarded as desirable, this derivative may have either sign and is, in fact, positive for the unaugmented B-70 at M = 2.1. When studying the various loci, take note of nominal values from Table 7.3; substantial deviations away from these values are likely to be impractical. At the same time, it is useful to ask which favorable changes might be accomplished through automatic stability augmentation. Many other examples of loci will be found in sources such as Seckel (1964) and Etkin (1959, 1972).

7.3 LONGITUDINAL STABILITY AND RESPONSE; EXACT AND APPROXIMATE PROPERTIES OF THE NORMAL MODES

The full, linearized equations of motion in the vehicle's plane of symmetry are (6.30). Inasmuch as the calculation of mechanical and indicial admittances for longitudinal response has no unique features which distinguish it from the process exemplified in Section 7.1, we confine ourselves to examining the normal modes and certain useful approximations thereto. Again the B-70 serves for numerical illustrations.

The homogeneous equations (6.30), expressed in terms of unknowns $\hat{u}$, α, and θ, have total order 4. For aircraft which are "statically stable,"* the quartic characteristic polynomial of this system is nearly always found to have two complex conjugate root pairs. These two oscillatory modes have widely different frequencies, and are called *short-period* and *long-period* or *phugoid*. Let us study them on the

Table 7.6. Three XB-70-1 flight conditions used in calculations of longitudinal modes

M = 0.76	M = 1.21	M = 2.39
$u_0 = 802$ ft/sec	$u_0 = 1191$ ft/sec	$u_0 = 2314$ ft/sec
$h = 15{,}500$ ft	$h = 32{,}400$ ft	$h = 56{,}100$ ft
$\rho_\infty = 1.47 \times 10^{-3}$ slug/ft^3	$\rho_\infty = 8.27 \times 10^{-4}$ slug/ft^3	$\rho_\infty = 2.87 \times 10^{-4}$ slug/ft^3
$W = 480{,}425$ lb	$W = 423{,}474$ lb	$W = 385{,}155$ lb
$I_{yy} = 21{,}810{,}032$ slug-ft^2	$I_{yy} = 20{,}999{,}872$ slug-ft^2	$I_{yy} = 20{,}633{,}984$ slug-ft^2
Drag = 47,088 lb	Drag = 82,338 lb	Drag = 51,269 lb
$\delta_T = 0°$	$\delta_T = 25°$	$\delta_T = 65°$
$\alpha_{\text{Trim}} = 4.38°$	$\alpha_{\text{Trim}} = 3.54°$	$\alpha_{\text{Trim}} = 3.45°$
$\delta_{e\,\text{Trim}} = 3.21°$	$\delta_{e\,\text{Trim}} = 8.67°$	$\delta_{e\,\text{Trim}} = 4.72°$
$\delta_c = 2.45°$	$\delta_c = 1.52°$	$\delta_c = 2.52°$
$\alpha_c = 4.63°$	$\alpha_c = 3.55°$	$\alpha_c = 3.43°$
CM = $0.221\bar{c}$	CM = $0.228\bar{c}$	CM = $0.212\bar{c}$
$\theta_0 = 0$	$\theta_0 = 0$	$\theta_0 = 0$
$C_{L0} = 0.161$	$C_{L0} = 0.115$	$C_{L0} = 0.080$

* This expression is usually regarded as equivalent to $C_{m\alpha} < 0$; see, however, our discussion in Section 8.1.

Table 7.7. Longitudinal stability derivatives for XB-70-1, taken principally from Wykes and Lawrence (1971)

	M = 0.76	M = 1.21	M = 2.39
C_{xu}	−0.0534	−0.0411	−0.0309
$C_{x\alpha}$	+0.0879	+0.0639	+0.0464
C_{zu}	0	+0.058	+0.0595
$C_{z\alpha}$	−2.4206	−2.2434	−1.4814
$C_{z\dot{\alpha}}$	−1.509	−0.762	−0.073
C_{zq}	−3.195	−1.715	−0.335
C_{mu}	−0.0066	−0.0059	−0.0074
$C_{m\alpha}$	−0.1717	−0.2546	−0.1390
$C_{m\dot{\alpha}}$	−0.383	+0.127	+0.042
C_{mq}	−1.706	−1.392	−0.684
$C_{z\delta_e}$	−0.3244	−0.0596	−0.0366
$C_{m\delta_e}$	−0.1734	−0.0449	−0.0271

B-70, at the three Mach numbers employed by Wykes and Lawrence (1971). With aeroelastic effects accounted for as in the lateral case, Tables 7.6 and 7.7 give, respectively, data on the three flight conditions and a full set of stability derivatives. Again we are indebted to Wykes for supplementing his own report by furnishing C_{xu}, $C_{x\alpha}$, C_{zu}, C_{mu}, and the fact that $C_{m\dot{\delta}_e}$ may be neglected. In Table 7.6, two new symbols, δ_c and α_c, appear for the angle of the canard elevator and the angle of attack of the canard itself. These surfaces are not involved in control or trim of lateral motions.

With dimensionless properties of the B-70 developed from the tables by means of definitions taken from Chapter 6, we have expanded and factored the characteristic determinant of (6.30) at each value of M. The resulting roots, periods, decay times, and mode shapes are listed in Table 7.8.

We shall next take up each of these modes, looking at its important features and deriving approximate representations suggested by the mode shape.

1. Short-Period Oscillation

A complex vector plot of the short-period mode shape at M = 0.76 appears on the left side of Fig. 7.9, with pitch angle θ chosen for reference as in Table 7.8. We observe that very little change in speed occurs during these rapid, well-damped oscillations, since the amplitude of the dimensionless $\hat{u}$-perturbation remains less than $1\frac{1}{2}\%$ those of pitch and angle of attack. Thus we are led to try setting $\hat{u} \cong 0$ and, for consistency, to disregard the X-equation (6.30a).

A more radical approximation is hinted at by the facts that α and θ have nearly equal amplitudes, particularly at the higher M's, and that their phase difference does not exceed $\pi/8$. This reduction would additionally set $\alpha \cong \theta$, which amounts to specifying that the CM moves in a straight line at constant

Table 7.8. Roots of the full linearized longitudinal equations for the XB-70-1 in horizontal flight

M	Short-period oscillation	Long-period oscillation
0.76	$s = -0.04099 \pm 0.06122i$ $(T = 5.02$ sec, $T_{1/2} = 0.824$ sec$)$ $\dfrac{\bar{u}_0}{\theta_0} = 0.0133e^{i28.5°}$ $\dfrac{\bar{\alpha}_0}{\theta_0} = 1.114e^{i22.6°}$ [Approximate $s = -0.04096 \pm 0.06120i$]	$s = -0.2997 \times 10^{-3} \pm 0.002060i$ $(T = 149.2$ sec, $T_{1/2} = 112.6$ sec$)$ $\dfrac{\bar{u}_0}{\theta_0} = 0.9306e^{i101.4°}$ $\dfrac{\bar{\alpha}_0}{\theta_0} = 0.0564e^{-i79.1°}$ [Approximate $s = -0.324 \times 10^{-3} \pm 0.0028i$]
1.21	$s = -0.01815 \pm 0.06146i$ $(T = 3.37$ sec, $T_{1/2} = 1.253$ sec$)$ $\dfrac{\bar{u}_0}{\theta_0} = 0.0067e^{i55.2°}$ $\dfrac{\bar{\alpha}_0}{\theta_0} = 1.022e^{i15.7°}$ [Approximate $s = -0.01814 \pm 0.06146i$]	$s = -0.1528 \times 10^{-3} \pm 0.8617 \times 10^{-3}i$ $(T = 240.4$ sec, $T_{1/2} = 148.8$ sec$)$ $\dfrac{\bar{u}_0}{\theta_0} = 1.012e^{i101.8°}$ $\dfrac{\bar{\alpha}_0}{\theta_0} = 0.0282e^{i281.9°}$ [Approximate $s = -0.1592 \times 10^{-3} \pm 0.001085i$]
2.39	$s = -0.003885 \pm 0.02713i$ $(T = 3.93$ sec, $T_{1/2} = 3.02$ sec$)$ $\dfrac{\bar{u}_0}{\theta_0} = 0.0037e^{i69.6°}$ $\dfrac{\bar{\alpha}_0}{\theta_0} = 1.009e^{i9.16°}$ [Approximate $s = 0.003883 \pm 0.02713i$]	$s = -0.4295 \times 10^{-4} \pm 0.1135 \times 10^{-3}i$ $(T = 939.4$ sec, $T_{1/2} = 272.8$ sec$)$ $\dfrac{\bar{u}_0}{\theta_0} = 1.866e^{i116.0°}$ $\dfrac{\bar{\alpha}_0}{\theta_0} = 0.0999e^{-i63.98°}$ [Approximate $s = 0.4553 \times 10^{-4} \pm 0.2604 \times 10^{-3}i$]

speed—an equivalent to the rectilinear-flight assumption which successfully predicts at least the period of lateral-directional oscillation. We do not go this far here, however, because damping is not so well represented by $\alpha \cong \theta$. Moreover, it is easy enough to deal with the two unrelated degrees of freedom.

With $\hat{u} \cong 0$, $\theta_0 = 0$, and $\hat{q} = D_s\theta$, (6.30b,c) for free motions become

$$\{[2\mu_s - C_{z\dot{\alpha}}]D_s - C_{z\alpha}\}\alpha - [2\mu_s + C_{zq}]\hat{q} = 0 \tag{7.36a}$$

$$-[C_{m\dot{\alpha}}D_s + C_{m\alpha}]\alpha + [i_{yy}D_s - C_{mq}]\hat{q} = 0 \tag{7.36b}$$

The first equation can be solved for $\hat{q}$. When it is substituted into (7.36b), we obtain

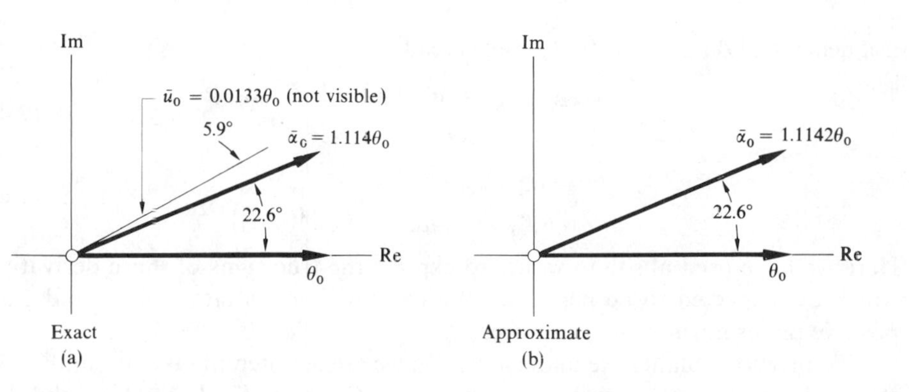

Fig. 7.9. Vector plots of the exact and approximate short-period longitudinal modes of the B-70 at M = 0.76 and altitude 15,500 ft. Approximate mode is based on neglect of perturbations in forward speed.

a second-order equation for α, which may be written in the following standard form in terms of the damping ratio and undamped natural frequency:

$$D_s^2\alpha + 2\zeta\omega_n D_s\alpha + \omega_n^2\alpha = 0 \tag{7.37}$$

Since the true dimensionless frequency is

$$\omega = \omega_n\sqrt{1 - \zeta^2} \tag{7.38}$$

we can deduce all required information about the free oscillation from the two constants in (7.37). Their relationship to the vehicle properties in (7.36) is

$$\omega_n \cong \left[\frac{C_{z\alpha}C_{mq} - C_{zq}C_{m\alpha} - 2\mu_s C_{m\alpha}}{i_{yy}[2\mu_s - C_{z\dot{\alpha}}]}\right]^{1/2} \tag{7.39}$$

and

$$\zeta \cong -\frac{[2\mu_s - C_{z\dot{\alpha}}]C_{mq} + i_{yy}C_{z\alpha} + [2\mu_s + C_{zq}]C_{m\dot{\alpha}}}{2\{i_{yy}[2\mu_s - C_{z\dot{\alpha}}][C_{z\alpha}C_{mq} - C_{zq}C_{m\alpha} - 2\mu_s C_{m\alpha}]\}^{1/2}} \tag{7.40}$$

The roots marked "approximate" in the short-period column of Table 7.8 and the right side of Fig. 7.9 are essentially consistent with (7.39)–(7.40). Inasmuch as all these approximations agree with their exact counterparts so closely that the difference could not be measured in a flight test, we conclude that our neglect of $\hat{u}(\hat{t}_s)$ is well justified.

A further simplification of (7.39)–(7.40) is usually acceptable, based on dropping $(-C_{z\dot{\alpha}})$ and C_{zq} by comparison with the much larger quantity $2\mu_s$ to which they are always added. Thus, for the B-70 at M = 0.76, we find $2\mu_s \cong 82.7$, with larger values at high speeds and altitudes. Table 7.7 then shows that the greatest error will be about 4%, at M = 0.76. The resulting reductions in the frequency and damping-ratio formulas read

$$\omega_n \cong \left[\frac{2\mu_s\,|C_{m\alpha}| + |C_{z\alpha}|\,|C_{mq}|}{2\mu_s i_{yy}}\right]^{1/2} \tag{7.41}$$

$$\zeta \cong \frac{2\mu_s[|C_{mq}| + |C_{m\dot{\alpha}}|] + i_{yy}\,|C_{z\alpha}|}{2\{2\mu_s i_{yy}[2\mu_s\,|C_{m\alpha}| + |C_{z\alpha}|\,|C_{mq}|]\}^{1/2}} \tag{7.42}$$

Here we have used absolute values to express the true signs of those derivatives which are expected to be negative. We thereby demonstrate that ω_n and ζ are positive under normal circumstances.

To promote qualitative understanding, we are tempted to take one additional step and omit the rather smaller product $|C_{z\alpha}|\,|C_{mq}|$ from (7.41, 42). If we did this, we would discover that the vehicle is analogous to a rotary spring-flywheel system, with the following properties:

- Rotational inertia proportional to i_{yy}
- Coefficient of damping proportional to $[|C_{mq}| + |C_{m\dot{\alpha}}|]$
- Torsional spring constant proportional to $C_{m\alpha} = C_{L\alpha}[h - h_n]$ [cf. (6.60) and note that this constant depends on the stick-fixed static margin].

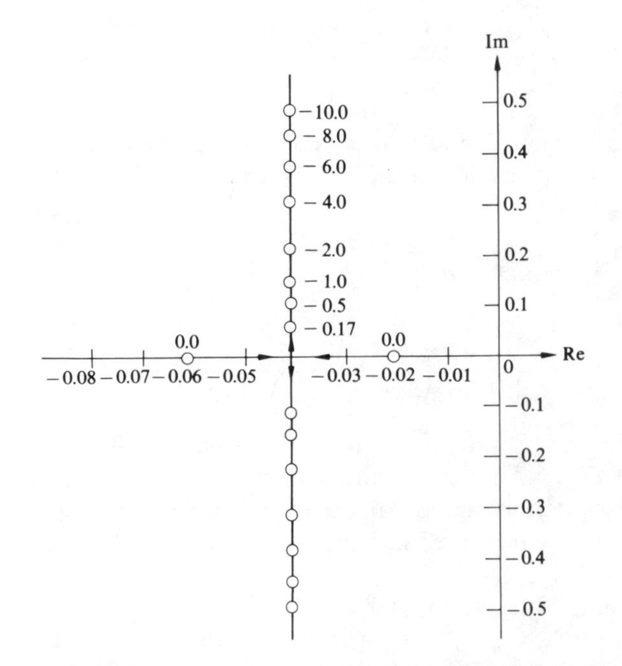

Fig. 7.10. Roots locus for the short-period mode of the B-70 at M = 0.76 and altitude 15,500 ft. Parameter varied is the static stability $C_{m\alpha}$, whose values are indicated on locus. (Note difference in horizontal and vertical scales.)

Although this idealization has some merit, it is oversimplified for most purposes, and always unnecessary. Indeed (7.39)–(7.40) demonstrate that both mass m and pitch inertia I_{yy} contribute to the overall vehicle inertia, through μ_s and i_{yy}. Furthermore, damping is supplied both by the moments C_{mq} and $C_{m\dot{\alpha}}$, which can extract energy from the pitching oscillation, and by $C_{z\alpha} \cong -C_{L\alpha}$, which represents a force opposing the vertical velocity of z-translation.

Figures 7.10 and 7.11 are short-period roots loci based on $\hat{u}(\hat{t}_a) \cong 0$ for the B-70 at M = 0.76. The scales in Fig. 7.11 are equal, so that damping ratio can here be found directly from the sin ζ construction mentioned in connection with Fig. 7.6. Both loci are roughly what one would calculate for a second-order

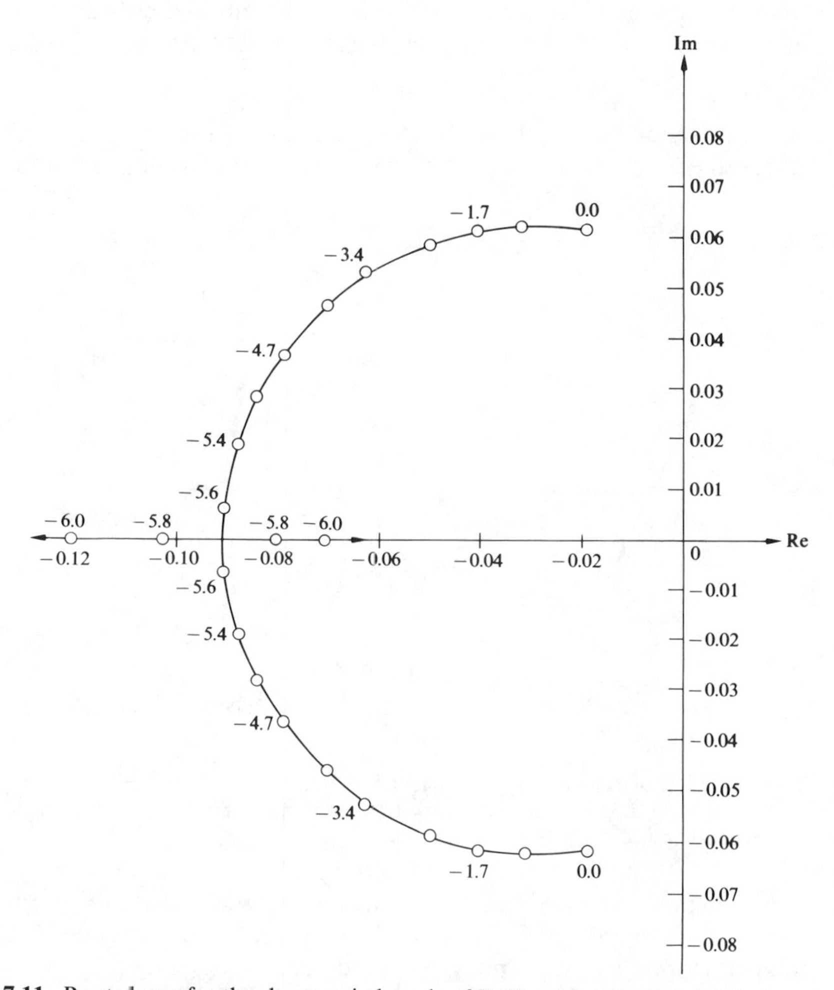

Fig. 7.11. Roots locus for the short-period mode of B-70 at M = 0.76 and altitude 15,500 ft. Parameter varied is the pitch damping C_{mq}, whose values are indicated on locus.

system with spring $\sim|C_{m\alpha}|$ and damper $\sim|C_{mq}|$. Figure 7.10 shows that decreases of $C_{m\alpha}$ below its nominal value have almost no effect on the real part of the root, but cause the imaginary part (very nearly, the frequency) to grow in rough proportion to $[|C_{m\alpha}| + \text{const.}]^{1/2}$. Driving the static stability toward zero causes the oscillatory root pair to approach the real axis and split into two exponentially convergent modes. A small amount of positive $C_{m\alpha}$ will quickly force the lesser-damped of these to become an unstable divergence, wholly unacceptable from the pilot's standpoint.

The influence of increasingly negative C_{mq} (Fig. 7.11) amounts to a progressive reduction in frequency and increase in damping. Around $C_{mq} = -5.7$ (some three times the nominal) this derivative brings the system to a critically damped condition, beyond which two rapidly decaying exponential modes are predicted.

2. Long-Period Oscillation

The long-period mode shape of the B-70 at M = 2.39 appears in part (a) of Fig. 7.12. Forward-speed perturbations and pitch are seen to be the main contributors to this

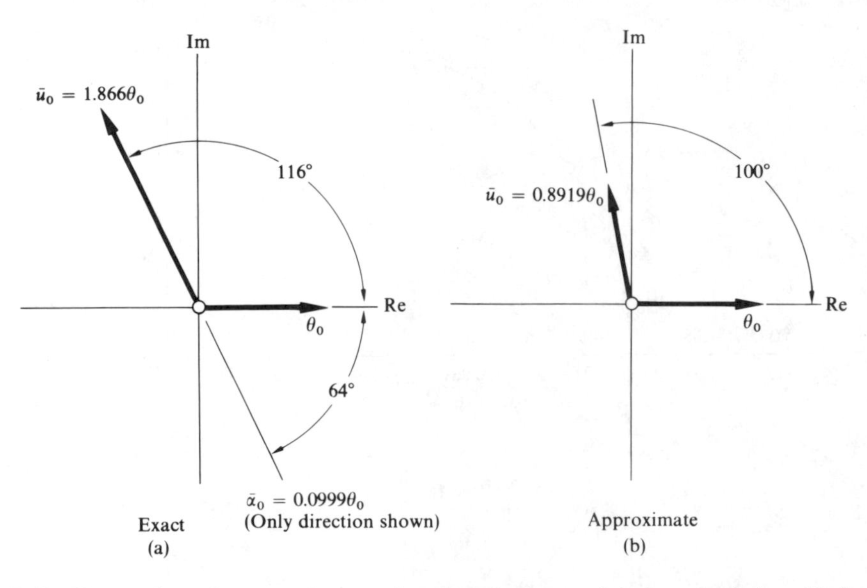

Fig. 7.12. Vector plots of exact and approximate mode shapes of the long-period oscillation of the B-70, flying at M = 2.39 and altitude 56,100 ft. Approximate mode is based on neglecting changes in angle of attack.

motion. Although α's amplitude is 10% of θ's, Table 7.8 shows that this percentage is smaller at other speeds and altitudes. As a general rule, therefore, a fairly satisfactory description of this "phugoid" mode can be developed by assuming that the airplane remains at its trim angle of attack ($\alpha \cong 0$) and moves like a "roller-coaster" on a sinusoidal track, cyclically exchanging potential energy of altitude with translational kinetic energy. Damping, if appreciable, is due to speed-induced changes in drag and thrust reflected in the derivative C_{xu}.

It is of interest that this energy-exchange idealization, extended to such large amplitudes that linearization is no longer acceptable [cf. Problem 7C, part (b)], has been identified as one of the earliest applications of nonlinear mechanics to vehicle motions. In consequence of the roughly $\pi/2$ phase angle between $\hat{u}$ and θ, one can show that the CM traces a sort of elliptical orbit when the oscillating vehicle is observed from a second vehicle, which flies in formation at constant $\mathbf{v}_{c0} = u_0\mathbf{i}$.

We turn to (6.30) for quantification of the $\alpha \cong 0$ approximation. Although angle of attack is constant, this does not imply that the z-translational equation (6.30b) is the one to be dropped. Rather, the condition of instantaneous trim means that $M_F \cong 0$, and the absence of pitching-moment perturbations calls for omitting (6.30c). Accordingly, we reduce (6.30a,b) for horizontal flight to

$$[2\mu_s D_s - C_{xu}]\hat{u} + C_{L0}\theta = 0 \tag{7.43a}$$

$$[2C_{L0} - C_{zu}]\hat{u} - [2\mu_s + C_{zq}]D_s\theta = 0 \tag{7.43b}$$

(Note that allowing for climbing or diving, $\theta_0 \neq 0$, does not complicate the analysis, but that in this case it would usually be necessary to consider variations of air density, thrust, etc., with altitude.)

Equations (7.43) can readily be manipulated into the standard form (7.37), with either $\hat{u}$ or θ as the dependent variable. The resulting undamped natural frequency and damping ratio are

$$\omega_n \cong \left[\frac{C_{L0}[2C_{L0} - C_{zu}]}{2\mu_s[2\mu_s + C_{zq}]}\right]^{1/2}$$

$$\cong \frac{\sqrt{C_{L0}[2C_{L0} - C_{zu}]}}{2\mu_s} \tag{7.44}$$

and

$$\zeta \cong -\frac{C_{xu}[2\mu_s + C_{zq}]}{2\{2\mu_s[2\mu_s + C_{zq}]C_{L0}[2C_{L0} - C_{zu}]\}^{1/2}}$$

$$\cong -\frac{C_{xu}}{2\sqrt{C_{L0}[2C_{L0} - C_{zu}]}} \tag{7.45}$$

In the final members of (7.44, 45), we have dropped C_{zq} relative to $2\mu_s$ for the same reasons as in (7.41, 42).

The approximate long-period mode shape in part (b) of Fig. 7.12 does include the small influence of C_{zq}, as do the estimated roots in the last column of Table 7.8. They show that the $\alpha \cong 0$ simplification yields an adequate qualitative representation of this mode, but that it is not as successful as $\hat{u} = 0$ for the short period.

We call particular attention to the period of over 15 min predicted for the B-70 at M = 2.39 and altitude 56,100 ft. This result does not correlate at all well with what is measured under these conditions. The reason is that, even with $\theta_0 = 0$, the added "spring" due to the vertical gradient of ρ_∞ in the atmosphere must be considered at such high speeds and altitudes. If you investigate this refinement in Problem 7C, part (a), you will discover that the period is markedly reduced. Use roots loci in examining the long-period effects of such parameters as C_{xu}, mass ratio, lift coefficient, and vehicle size.

We mention one final reduction of the formulas for frequency and damping ratio, which derives from the observation [cf. (6.42)] that derivative C_{zu} tends to be negligible at subcritical M and to be fairly small,* except in the transonic range, when compared with $2C_{L0}$. Under the same circumstances, and for level flight of a jet-powered aircraft, (6.41) shows that C_{xu} might be replaced by $(-2C_{T0}) = (-2C_{D0})$. Thus we are able to simplify (7.44) and (7.45) as follows.

$$\omega_n \cong \frac{C_{L0}}{\sqrt{2}\,\mu_s} \tag{7.46}$$

$$\zeta \cong -\frac{C_{xu}}{2\sqrt{2}\,C_{L0}} = \frac{1}{\sqrt{2}\,L/D} \tag{7.47}$$

Relation (7.47) may be interpreted by stating that highly efficient cruising aircraft with $L/D \gg 1$ are lightly damped in this mode.

A more illuminating estimate emerges when we convert the dimensionless ω_n from (7.46) into a physical period. If we simultaneously substitute for μ_s and level-flight C_{L0} in terms of weight, density, and the various geometrical and kinematic parameters, we can convert (7.46) to

$$T = \frac{2\pi}{\omega_n}\left(\frac{\bar{c}}{2u_0}\right) = \sqrt{2}\,\pi\,\frac{u_0}{g} \tag{7.48}$$

Equation (7.48) demonstrates a characteristic common to numerous flight vehicles: phugoid period is nearly independent of size, weight, and altitude, but is directly proportional to speed. The periods in Table 7.8, for M = 0.79 and 1.21 at least, follow this rule quite accurately. Another way of visualizing the content of (7.48) is to observe that the number of feet or chord lengths traveled during one cycle of this mode is proportional to speed of the vehicle squared.

7.4 HANDLING QUALITIES

Despite the complicated, adaptive, nonlinear mechanism which is a human pilot, it is a remarkable fact that close relationships, useful in design, have been established between the small-perturbation characteristics of aircraft and what pilots regard as desirable handling or flying qualities. Early research on this

* An exception is the B-70 at M = 2.39.

subject in the United States was mainly conducted by NACA [e.g., Gilruth 1943, Phillips 1948(b)]. Over the intervening years, a host of refinements have been developed to erstwhile simple criteria on damping and frequency of the longitudinal and lateral-directional short-period oscillations. A recent compendium appears in the relevant military specifications (Anonymous 1969) and references cited therein. Favorable ranges are given in these documents for a host of different parameters, these in turn being broken down by class of vehicle (jet transport, interceptor, etc.) and by flight regime. As an example, Stein and Henke (1971) developed from the specifications eleven distinct quantities whose values could be related just to the lateral handling qualities sought by pilots. Such items are included as ω_n and ζ for the oscillatory mode, the time constant of roll subsidence, maximum sideslip due to aileron application, steady roll rate per unit aileron displacement, and the like. All have in common, however, that they are predictable from the approximations underlying (6.31).

Although a few of his data are now obsolescent, we have found the discussion of handling qualities by Kolk (1961) a valuable introduction. The analyses and charts contained in his Chapter 8 retain considerable validity today as a basis for preliminary determination of what constitutes a "good-flying airplane." We therefore choose to reproduce some of his results by way of concluding our review of small-perturbation stability and response.

For motion in the plane of symmetry, there is a fair consensus that pilot evaluations can be correlated, for a given class of aircraft, with the damping ratio and the dimensional circular frequency $(2u_0\omega_n/\bar{c})$ of the short-period mode. Thus ζ in the vicinity 0.5–0.8 seems to be a universal desideratum; for fighters and other medium-weight vehicles, the favored frequencies center on 3–4 rad/sec. Figure 7.13, from Kolk (1961) is a typical plot. It summarizes results of flight testing, conducted by experienced pilots at Cornell Aeronautical Laboratory on the B-26, which is a light, propeller-driven bomber from World War II. The "best" qualities are associated with parameter combinations in the interior of the inner solid line. As one proceeds outward, the remaining areas are described as "good," "fair," and "poor." Some summaries of pilot comments are added on various regions of the chart.

Kolk (1961) makes an interesting analysis of the connection between the judgments expressed in Fig. 7.13 and the times required, following a step elevator input, for the airplane to achieve peak values of $\dot{\alpha}$ and α. For instance, the latter time can be approximated by

$$(T)_{\alpha\,\max} \cong \left(\frac{\bar{c}}{2u_0\omega_n}\right)\frac{\pi}{\sqrt{1-\zeta^2}} \tag{7.49}$$

As one alternative to this parameter, we suggest a characteristic called *control anticipation parameter* by Bihrle (1966). This quantity is the ratio, also after a

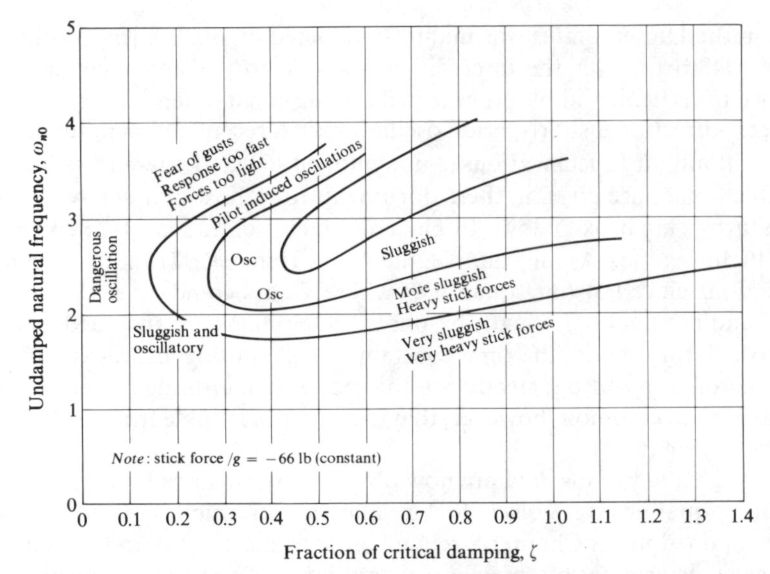

Fig. 7.13. Typical contours of handling qualities of the B-26, as rated by experienced pilots, shown on a plot of undamped natural frequency versus damping ratio of the short-period longitudinal mode. Best handling qualities fall inside the inner curve. [Adapted from Kolk (1961)].

step $\delta_e(t)$, between the initial pitch acceleration $\ddot{\theta}(0)$ and the incremental load factor ultimately produced by this step. The reader will find it challenging to derive a formula for this ratio from the forced-motion version of (7.36), as well as to study Bihrle's (1966) interpretation of its significance.

Those modes which have substantially longer time constants than the short period seem to have little effect on vehicle acceptance. For instance, it is immaterial to an experienced pilot whether there is some slight instability in either the phugoid or the spiral mode. In the process of normally controlling his vehicle, the operator counteracts any evidence of these motions long before they have time to build up. Moreover, a simple three-axis automatic pilot will stabilize both of them.

The other two lateral-directional modes are, however, primary determinants of handling qualities. Roll subsidence, as analyzed by a method like that in Problem 6A, has three important characteristics: its time constant, the initial roll acceleration $\dot{p}(0)$, and the final steady-state velocity $p(\infty)$, both taken per unit step of aileron deflection. Since desirable values of these parameters vary so much from one type to another (for example, $\dot{p}(0)/\delta_{a0}$ on a fighter should be several times that of a large transport), we offer no numerical data here.

As for the oscillatory mode, adequate damping is a necessity. What the pilot wishes, however, is also affected by the relative amplitudes of bank and sideslip ($|\bar{\beta}_0/\varphi_0|$ in Table 7.5), by the airspeed, and by the altitude. To be specific, a line of investigation detailed by Kolk (1961, Section 8.5) led to the conclusion that a "universal" boundary between satisfactory and unsatisfactory properties could be constructed, for a given aircraft, on a plot with the following ordinate and abscissa.

- The number of cycles required for an initial amplitude to decay to some specified fraction, as measured by its inverse $T/T_{1/2}$.
- The amplitude ratio between rolling and an "equivalent" sideslip velocity. With the sideslip velocity being $v = u_0\beta$, this equivalent value measures dynamic pressure rather than true speed; a suitable formula reads

$$v_e = u_0\beta\sqrt{\frac{\rho_\infty}{\rho_{\infty SL}}} \tag{7.50}$$

Figure 7.14 is a sketch of the sort of boundary of acceptability that results from such reasoning. We put no scales on the axes, because again the numerical

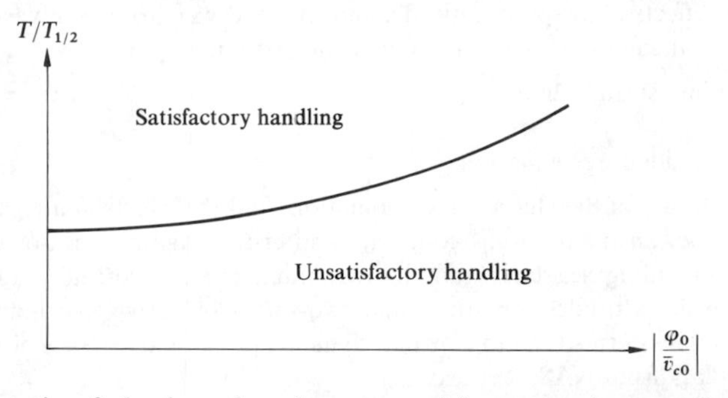

Fig. 7.14. Schematic of the boundary between satisfactory and unsatisfactory handling qualities based on characteristics of the lateral-directional oscillation. [Adapted from Kolk (1961)].

ranges differ considerably among various types. Nevertheless, it does appear to be a general conclusion that greater damping is sought from this mode when there is a larger amount of rolling present. This condition tends to be associated with high-altitude flight. But Table 7.5 indicates that the B-70, with its variable-geometry wing, has the exact opposite tendency. Incidentally, there have been many publications on B-70 handling qualities, of which Berry and Powers (1970) is a good example.

7.5 PROBLEMS*

7A You are familiar with the scheme for representing a simple harmonic motion like $A\cos(\omega t + \varphi)$ in terms of a complex vector $Ae^{i(\omega t+\varphi)}$ which rotates about the origin at angular velocity ω. In this representation, the time derivative is obtained by rotating the vector 90° counterclockwise (multiplication by i) and multiplying its length by ω.

a) Extend this scheme to damped or divergent oscillatory motion in a mode having damping ratio ζ and "undamped frequency" ω_n. Show that the head of the complex vector now traces

* Subscripts are often omitted from D, t, μ in the problems.

a logarithmic spiral as it rotates at angular velocity $\omega = \omega_n\sqrt{1-\zeta^2}$, and that time differentiation is accomplished by an amplification factor ω_n and rotation through the angle

$$\frac{\pi}{2} + \tan^{-1}\frac{\zeta}{\sqrt{1-\zeta^2}}$$

In this way, for example, each term in an equation of motion (at a particular instant) can be represented as a vector. If these vectors are laid head-to-tail, the balancing of the equation is revealed by the formation of a closed polygon. This scheme will be useful for comparing the magnitudes of various terms appearing in the equations of motion, as we shall soon discover.

b) For the B-70 in lateral-directional motion with $\beta = -\psi$, draw closed vector polygons showing the balance of terms in the two equations of pure oscillatory motion.

7B For any vehicle you wish to select, for which suitable data are given, make reasonable estimates of the effects of airspeed, altitude, and M (fixed weight) on the various terms in the *approximate* lateral equations of motion under the assumptions of

1) CM moving in a straight line
2) $\beta = -\psi$
3) Neglecting the side-force equation

By computer solution of the characteristic equation, find the effects of airspeed (two or three values per altitude) on the natural frequencies and critical damping ratios of the modes for four altitudes, including sea level and the maximum cruising altitude, and choosing two intermediate-density altitudes, one-third and two-thirds of the way to maximum. Let one set of airspeed values correspond to constant dynamic pressure over the whole altitude range. Here are some data from NASA reports.

	Subsonic bomber 1*	High-altitude fighter 1*	Small delta airplane, landing light 1*	Mach 3 supersonic transport 2†
u_0, ft/sec	700	776	100	2,920
b, ft	116	25	38	77
Altitude, ft	35,000	50,000	0	70,000
μ	63.7	364	23.7	456
i_{xx}	7.92	22.7	2.84	39.2
i_{zz}	18.33	227	6.43	346
i_{xz}	0	0	−2.84	0
$C_{y\beta}$	−0.61	−0.58	−0.286	−0.347
$C_{l\beta}$	−0.14	−0.18	−0.057	−0.093
C_{lp}	−0.44	−0.33	−0.02	−0.124
C_{lr}	0.149	0.23	0.60	0.102
$C_{n\beta}$	0.120	0.25	0.057	0.052
C_{np}	−0.0276	−0.050	−0.20	0.012
C_{nr}	−0.156	−0.069	−1.10	−0.453
		(Other derivatives zero)		

* B. B. Klawans (1956).
† W. D. Klopp, W. R. Witzke, and P. L. Raffo (1966).

7C Experience with large, high-speed aircraft like SST and B-70 caused a revival of interest in the long-period or phugoid mode of longitudinal oscillation. An excellent approximation to this motion can be obtained by assuming constant angle of attack ($\alpha = 0$) and studying eigenvalues of the coupled x- and z-equations.

a) For trimmed horizontal flight ($\theta_0 = 0$) and C_{zu} and C_{zq} negligible, first show that the dimensional natural frequency Ω_n is given by

$$\Omega_n = \sqrt{2}\,\frac{g}{u_0}$$

Then add a correction to the linearized equations which accounts for the variation of density with altitude, and show that the above is changed to

$$\Omega_n = \sqrt{g\left[\frac{2g}{u_0^2} - \frac{1}{\rho_\infty}\left(\frac{d\rho}{dh}\right)_0\right]}$$

Make a plot to show the range of flight Mach numbers in which this density effect might be significant (remember $d\rho/dh < 0$).

b) Consider the so-called nonlinear phugoid, in which $\alpha = 0$, but the assumption of small perturbations is discarded. If flight is horizontal and thrust is independent of speed, show that an appropriate set of equations might be

$$\dot{U} = -g \sin \Theta + C_{DO}\frac{\rho_\infty}{2}\frac{S[u_0^2 - U^2]}{m}$$

$$\dot{\Theta} = -\frac{g}{U}\cos \Theta + C_{LO}\frac{\rho_\infty}{2}\frac{US}{m}$$

Make a study of the possibility of solving these nonlinear equations. Show, for instance, that when the drag term is neglected the motion has constant energy; the oscillation then consists of periodic interchange between potential energy of altitude and kinetic energy $\frac{1}{2}mU^2$.

7D Criticize the merits of the following proposal: If effects of power, Reynolds and Mach numbers are unimportant, the phugoid motion of a large airplane can be conveniently studied by working with reduced-scale plastic models that are suitably trimmed, then allowed to glide underwater in a swimming pool.

7E Using a selected case involving the B-70 with the $\beta = -\psi$ approximation, solve for the free motion (controls centered) resulting from each of the following initial conditions.

a) $\beta(0) = 5°$, $\dot{\beta}(0) = 0$, $\hat{p}(0) = 0$
b) $\beta(0) = \dot{\beta}(0) = 0$, $\hat{p}(0) = 0.1$

Discuss how the choice of initial conditions determines the amount of response that occurs in the Dutch-roll and roll-subsidence modes, respectively.

7F Someone proposes using specified variations of propulsive thrust, while keeping the controls fixed, as a means of determining various admittance functions describing the dynamic response of a powered flight vehicle.

a) For the longitudinal case, adapt a set of *dimensional* small-perturbation equations which show clearly how you would go about predicting this response. (You may assume that the x-axis and the thrust line coincide.)

b) On a large airplane, it is difficult to create large-amplitude sinusoidal thrust variations at a rate faster than 0.1 cycle per second. Explain physically what limitations this puts on the thrust-variation scheme and why the approximation $\alpha = 0$ is probably now pretty good for describing the response. Show that the dimensionless equations of motion can now be written

$$\left[2\mu D + 2C_{DO} + M\frac{\partial C_D}{\partial M}\right]\hat{u} + C_{LO}\theta = \Delta C_T(t)$$

$$[2C_{LO} - C_{zu}]\hat{u} - [(2\mu + C_{zq})D - C_{LO}\tan\theta_0]\theta = 0$$

where ΔC_T is the variation of coefficient of thrust T from its trim value T_0.

c) Draw a diagram showing how these equations or those in part (d) below could be wired onto an analog computer. Also find analytically the mechanical admittance $G_{\theta C_T}(i\omega)$, showing how pitch angle responds to thrust-coefficient variations, and indicate how you would find indicial and impulsive admittances for the same input-output pair.

d) 

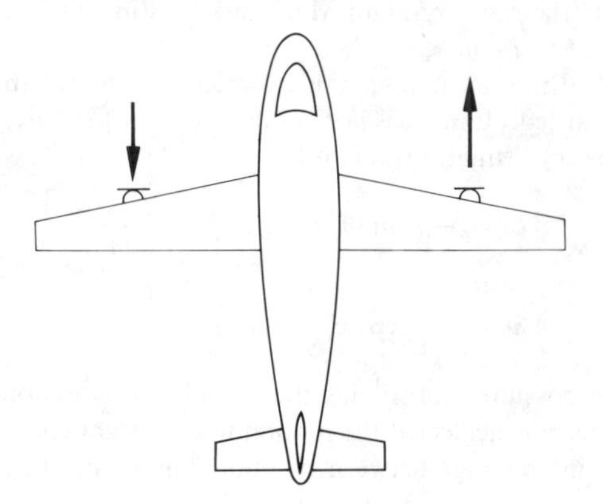

Discuss, with a couple of equations, the use of the same scheme for finding lateral response with antisymmetrical thrust variations on a twin-engine vehicle. What is the only degree of freedom that receives a substantial input?

7G For the aircraft of your choice, estimate the necessary longitudinal data to make a roots-locus study of the influences of C_{LO}, C_{xu}, and μ on the phugoid mode.

7H Choose a convenient time scale, and sketch an analog mechanization of longitudinal equations (6.30). Apply an initial condition to pitch rate q. Construct your numerical coefficients from any available data. Put numerical labels on all coefficient potentiometers, recalling that all settings must be < 1.00, but that inputs to amplifiers can be multiplied by 0.1, 10, or even 100. Make and report on a study of free short-period motion and of response to various inputs as key parameters are changed from one value to another. [If you have available only a small computer, work with the $\hat{u} = 0$ reduction of (6.30).]

7I a) We wish to study the *stability* of a spinning missile or rifle bullet (or perhaps a rapidly rolling fighter aircraft) as a follow-on to the work in Problem 2F. The equations of motion for such a body can be approximated as follows: The CM moves in a straight line so that $\mathbf{v}_c$ = const. (or u_0 = const.), and there is a constant rolling velocity P_0, so that there are now four degrees of freedom represented by Q, R, $V = \beta U_0$, and $W = \alpha U_0$. Note that we may assume that the F_x and L_r equations are automatically balanced by proper deflections of the control surfaces, so that we start with four equations of motion. We assume that the only forces and moments of any importance are

$$Y = -Y_\beta \beta, \qquad Z = -Z_\alpha \alpha$$
$$M_p = M_Q Q + M_\alpha \alpha$$
$$N = N_R R + N_\beta \beta$$

where Y_β, Z_α, . . . are constant coefficients which depend on geometry and flight conditions. For the bullet, explain in words and sketches why the following are true:

$$Z_\alpha = Y_\beta > 0, \qquad M_Q = N_R < 0$$
$$M_\alpha = (-N_\beta) > 0, \qquad I_{yy} = I_{zz} \cong I$$

Make all the above substitutions in the equations of motion. Show how the definitions

$$\dot{\xi} = \alpha + i\beta, \qquad \eta = Q - iR$$

permit these equations to be combined in the forms:

$$\dot{\xi} + \frac{Z_\alpha}{mu_0}\xi - iP_0\xi - \eta = 0$$

$$\dot{\eta} - \frac{M_Q}{I}\eta - i\left[1 - \frac{I_{xx}}{I}\right]P_0\eta - \frac{M_\alpha}{I}\xi = 0$$

where $i = \sqrt{-1}$. By inserting $\xi = \xi_0 e^{\lambda t}$, $\eta = \eta_0 e^{\lambda t}$, see if you can arrive at the key test for the bullet's flight to be stable.

b) Study on the digital computer the effect of changing dimensionless roll rate $P_0 l/2u_0$ (where l is reference length) on the dynamic stability of a spinning missile or rifle bullet. Nondimensionalize the equations of motion and examine the natural frequencies and the damping *or* divergence characteristics for the following two cases.

1) A conventional artillery shell, which is statically unstable, and which has the following characteristics: $C_{L\alpha} = 2$, $C_{mq} = -5$, $C_{m\alpha} = 1$, $I_{xx}/I = 0.1$, $u_0/l = 1000\ \text{sec}^{-1}$.
2) A statically stable cruciform missile (see figure), which has the following characteristics: $C_{L\alpha} = 3$, $C_{mq} = -20$, $C_{m\alpha} = -0.5$, $I_{xx}/I = 0.2$, $u_0/l = 1000\ \text{sec}^{-1}$.

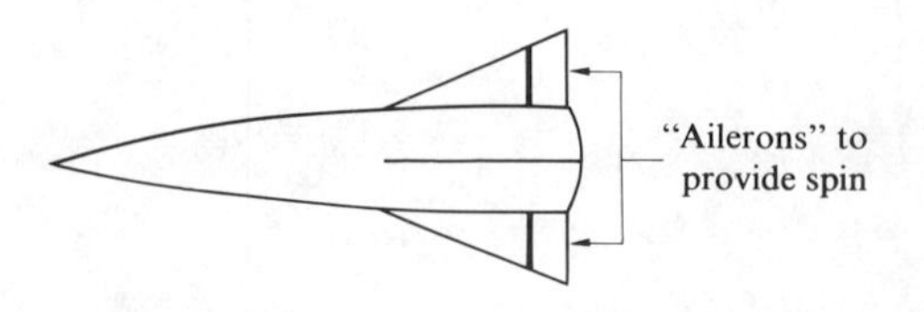

Remember that $C_{z\alpha} = C_{y\beta}\ (= C_{L\alpha})$, $C_{mq} = C_{nr}$, and $C_{m\alpha} = -C_{n\beta}$.
Use what you know about rifles to estimate reasonable spin rates and use smaller values for the missile.

7J *Speed stability* (cf. Etkin 1972, Section 11.5) is a term usually applied to a property of longitudinal motion, whereby, with the CM controlled to move along a straight line ($\theta_0 =$ const., $\alpha = \theta$), small $\hat{u}$-disturbances tend to die out. The problem can be studied with (6.30) by neglecting the pitching-moment equation (6.30c), using (6.30d) to eliminate $\hat{q}$ from (6.30b), and examining the *free* motion described by what remains. It is also useful to assume that C_{zq}, $C_{z\dot{\alpha}} \ll 2\mu_s$ and neglect $\partial C_L/\partial M$, $\partial C_D/\partial M$. This last approximation is consistent with the fact that speed stability is most important during a landing approach.

Investigate the effects on stability of the parameters which are left, notably speed as expressed by C_{L0} and thrust-drag characteristics as they appear in C_{xu}. Confirm Etkin's result that, when $\theta \cong 0$ and T is independent of $\hat{u}$, the question of stability rests on the sign of

$$\frac{C_{L0}}{\mu_s}\left[\frac{\dfrac{\partial C_D}{\partial \alpha}}{C_{L\alpha}} - \frac{D}{L}\right]$$

What is the role of flaps and drag spoilers in helping stability?

7K In the general area of small-perturbation dynamic stability and response of flight vehicles, make up and solve an original problem about something you find particularly interesting. Try to choose some fairly realistic and meaningful situation. Start by writing a brief, accurate, complete statement of what you plan to do. Use care in stating your assumptions or referencing sections in a suitable textbook.

CHAPTER 8

STATIC STABILITY, TRIM, STATIC PERFORMANCE AND RELATED SUBJECTS

8.1 IMPACT OF STABILITY REQUIREMENTS ON DESIGN AND LONGITUDINAL CONTROL

The condition

$$C_m = 0 \tag{8.1}$$

is, of course, a requirement for longitudinal equilibrium in undisturbed flight. In Chapter 7 we discussed a less universal, but nevertheless commonly accepted, criterion

$$C_{m\alpha} < 0 \tag{8.2}$$

for static stability. Relation (8.2) is intimately associated with the need for satisfactory longitudinal handling qualities as they are measured by the short-period frequency, but its imposition also relates to other vehicle characteristics desired by the pilot. For example, let us say that a vehicle encounters a gust which causes the angle of attack to be altered from the trimmed α_{ZL_0}. If the control column is not displaced nor the speed changed significantly from v_{c0}, then negative $C_{m\alpha}$ ensures a restoring tendency that drives this angle back toward equilibrium.

Figure 8.1 gives pitching-moment curves for a vehicle that can meet both (8.1) and (8.2). Note that this configuration has a conventionally located horizontal stabilizer and that, since δ_e is positive when the elevator trailing-edge is down, negative changes in δ_e apply positive increments to C_m. Zero elevator deflection is here taken to be where the profile drag and drag due to lift for a given C_{Lt} are near their minima; hence the design point, $C_L = C_{L_{01}}$, is arranged to be along the line $\delta_e = 0$. We mention that this situation is attainable only at one loading condition, because a change in location of CM rotates all these curves around fixed points on the ordinate axis where $C_L = 0$.

It is evident that the zero-lift moment must be positive when $\delta_e = 0$,

$$C_{m0} > 0 \tag{8.3}$$

As may be seen from the two-dimensional analyses in Chapter 4, one way to satisfy (8.3) is to camber the wing "negatively," i.e., make it concave upward. Certainly for an aircraft that does most of its flying subsonically this is undesirable, however, since positive camber promotes lower drag around the positive design C_L. Hence the best method, when a separate horizontal stabilizer is used, is to adjust its incidence i_t so that tail lift at $C_L = 0$ provides the desired C_{m0}. A swept-back

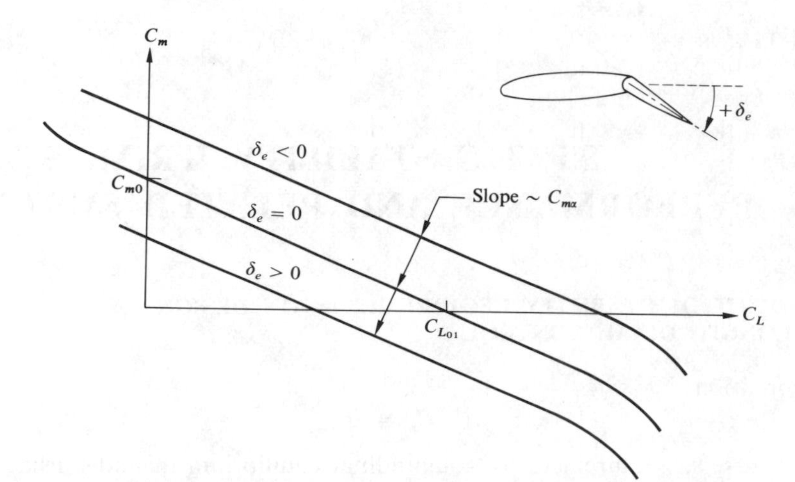

Fig. 8.1. Typical variations of moment coefficient with lift coefficient for a vehicle with positive static stability. The nonlinearity near the ends of the curves is due to stalling.

wing with its tips twisted nose-downward may also achieve adequate positive C_{m0}, but excessive twisting is unacceptable from the induced-drag standpoint.

Figure 8.2 depicts conventional and canard arrangements at $C_L = 0$, with i_t set so that $C_{Lt} < 0$, as required by the former, and conversely for the latter. In the light of these facts, the designer is faced with some particularly interesting tradeoffs when it comes to selecting the tail volume V_H and deciding whether to use a canard. One key question involves how the job of supporting the weight is to be apportioned between the wing and stabilizer. If both are developing positive lift—for instance, if $C_{L\,wb}$ and C_{Lt} are approximately equal—there is a tendency to minimize the individual lift coefficients in a given flight condition, and therefore to reduce the total drag due to lift. On the conventional arrangement the tail setting shown in Fig. 8.2, plus the effects of downwash, make this almost impossible.

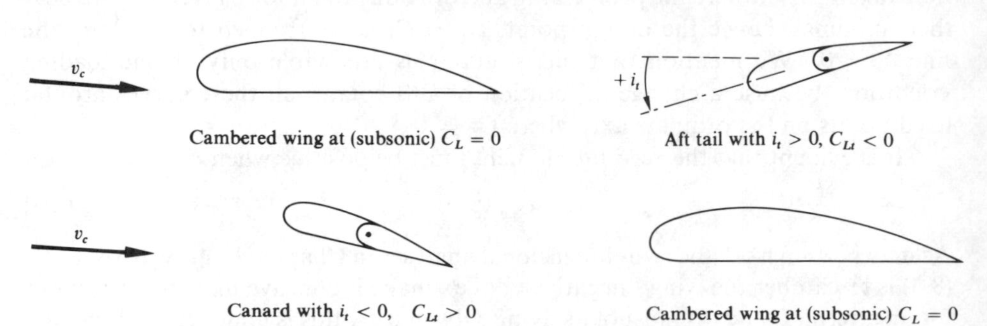

Fig. 8.2. Conventional and canard wing-tail arrangements, shown at angle of attack corresponding to zero airplane C_L.

Indeed, the tail load is downward a good part of the time, even at the high C_L's during landing, when trailing-edge flaps add a large negative increment to C_{m0}. As mentioned relative to the B-70 in Chapter 1, however, the canard not only carries positive lift but the "trim drag" penalties of going to off-design flight conditions are not so serious with the canard as with the aft tail (cf. treatments of this subject in such papers as Graham and Ryan 1960 and McKinney and Dollyhigh 1971).

The canard's great disadvantage occurs at the stall, where its tendency to exceed $C_{L\,max}$ before the wing can result in a violent, uncontrollable downward pitching condition called the *hammerhead*. By contrast, a conventional stabilizer stays within the unstalled range while the flow is breaking away from its wing, so that the pilot retains his ability to adjust the pitching moment and recover smoothly from the maneuver. (An exception may be encountered with swept-back wings and T-tails; with these, the vehicle may pitch up and become locked into a deeply stalled situation; for this reason, designs like the C-141 embody devices to prevent the attainment of dangerously high angles of attack.)

In only a few cases to date have these tradeoffs led to adopting a pure canard arrangement—the original Wright biplane,* the B-70, and certain configurations proposed for the U.S. SST being among them. The Swedish Viggen fighter is a unique example, on which vortex interference between a high delta canard and delta wing are said both to enhance longitudinal control and to reduce trim drag. Forward trimming surfaces will be seen more frequently in the future, we believe, because the importance of their minimization of supersonic drag, the avoidance of jet-exhaust damage to the stabilizer, and other benefits work strongly in their favor.

We shall now discuss some questions connected with ensuring enough "tail power" for trim and therefore with sizing the stabilizer and elevator. Turning to Section 6.3, Part 2, for the necessary formulas, we take the elevator pitching moment from (6.57a) and add it to the overall vehicle C_m, (6.59a), to obtain

$$C_m = C_{m0} + C_{m\alpha}\alpha_{ZL} + C_{m\delta_e}\delta_e \tag{8.4}$$

Operation within the linear ranges of the vehicle's angle of attack and that of a distinct trimming surface is assumed. From (6.59c), (6.60), and (6.57a), we reproduce the coefficients appearing in (8.4).

$$C_{m0} = C_{m0_{wb}} + (C_{L\alpha})_t V_{Hn}[\epsilon_0 + i_t] \tag{8.5}$$

$$C_{m\alpha} = C_{L\alpha}[h - h_n] \tag{8.6}$$

$$C_{m\delta_e} = -(C_{L\delta})_t V_H \tag{8.7}$$

* Obviously the Wright brothers, although they were outstanding engineers, did not have the benefit of the refined analytical considerations that can now be applied to such a decision.

All these quantities were discussed in Section 6.3, notably the stick-fixed neutral point coordinate

$$h_n = h_{n\,wb} + V_{Hn}\frac{(C_{L\alpha})_t}{(C_{L\alpha})_{wb}}\left[1 - \frac{\partial\epsilon}{\partial\alpha}\right] \tag{8.8}$$

The tail volume is

$$V_H \equiv \frac{l_t S_t}{\bar{c}S} \tag{8.9}$$

negative for a canard and positive for an aft tail. V_{Hn} is the value of V_H when the CM falls on the neutral point; except for very short tail lengths, V_{Hn} is not substantially different from the actual operational values of this "volume ratio."

As an aircraft accelerates, climbs, turns, flares, and performs other maneuvers, the pilot constantly needs to adjust the trimmed α_{ZL}. An unconventional scheme for doing this would be to change the CM position, since this hardly affects C_{m0} but drastically alters $C_{m\alpha}$ in direct proportion to h. On the U.S. SST, this method was actually planned during the subsonic–supersonic transition by transferring fuel between forward and aft tanks. The obvious agency for trim is, however, the elevator or elevon.

The required setting, $\delta_e = \delta_{\text{Trim}}$, is determined as follows: Let the lift coefficient be written

$$C_{L\alpha}\alpha_{ZL} + C_{L\delta_e}\delta_{\text{Trim}} = C_{L0} \tag{8.10}$$

where $C_{L\delta_e}$ comes from (6.57b). Substituting δ_{Trim} into (8.4) and equating C_m to zero, we obtain

$$C_{m\alpha}\alpha_{ZL} + C_{m\delta_e}\delta_{\text{Trim}} = -C_{m0} \tag{8.11}$$

Equations (8.10) and (8.11) are simultaneous equations for α_{ZL} and δ_{Trim}, in terms of the known C_{m0} and of C_{L0}, which can be calculated from the flight condition by inserting known data into the dimensionless form of (3.10b) to get (e.g., for a rectilinear trajectory)

$$C_{L0} = \frac{mg\cos\theta_0}{\frac{\rho_\infty}{2}v_c^2 S} \tag{8.12}$$

By eliminating α_{ZL} between (8.10) and (8.11), we find that

$$\boxed{\delta_{\text{Trim}} = \frac{C_{m0}C_{L\alpha} + C_{L0}C_{m\alpha}}{-C_{m\delta_e}C_{L\alpha} + C_{m\alpha}C_{L\delta_e}}} \tag{8.13a}$$

or

$$\delta_{\text{Trim}} = \frac{C_{m0}C_{L\alpha} - C_{L0}\,|C_{m\alpha}|}{|C_{m\delta_e}|\,C_{L\alpha} - |C_{m\alpha}|\,C_{L\delta_e}}, \qquad \text{for } C_{m\alpha} < 0 \text{ and } C_{m\delta_e} < 0 \tag{8.13b}$$

In order to highlight the roles of the various parameters, but only for a statically stable vehicle with conventional stabilizer, we have replaced the two negative derivatives with their absolute values in (8.13b). For this case the denominator is numerically positive, because $C_{L\alpha} > C_{L\delta_e}$, whereas the magnitude of $C_{m\delta_e}$ tends to be two or more times that of $C_{m\alpha}$.

By comparing (8.12) and (8.13), we can construct a curve of δ_{Trim} versus true speed v_c or versus dynamic pressure. Figure 8.3 is typical of the former, showing

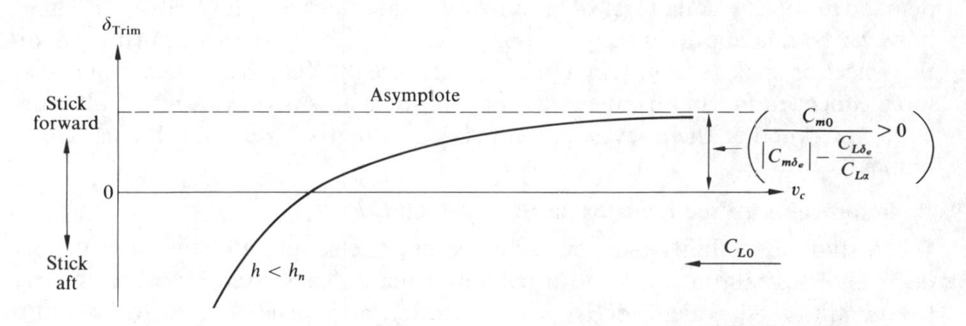

Fig. 8.3. Variation of elevator position for trim with flight speed, for a vehicle with representative positive static margin.

the expected requirement for large backward stick positions in slow flight and the positive asymptotic value of δ_{Trim} associated with very high speed and/or vertical flight ($\theta_0 \to \pm 90°$). It is also apparent that the pilot must push forward on the control column to maintain constant altitude (or a straight-line trajectory) when he advances the throttle to increase v_c. Such behavior, which is again seen to be a consequence of negative $C_{m\alpha}$, must normally characterize any aircraft. If $C_{m\alpha} = 0$, either with or without stability augmentation, δ_{Trim} would evidently be a constant; although the approach to such a condition makes for high maneuverability, it is dangerous except in the hands of a skilled operator.

Calculating elevator angle to trim is one of a series of similar analyses that must be made relative to longitudinal and lateral control. These analyses have, as a principal aim, to establish the limits between which the vehicle CM may safely lie. This whole subject is systematically and exhaustively addressed in texts like Etkin (1959, Chapters 2 and 3, and 1972, Chapter 6). Let us confine ourselves here to a qualitative discussion of the more important questions.

1. Elevator Hinge Moment, Stick Force, and Stick Force to Trim

As mentioned just after (2.49), the deflected elevator experiences an aerodynamic moment H_{δ_e} about its hingeline. For small enough angles, this is related to other parameters by*

$$H_{\delta_e} = \frac{\rho_\infty}{2} v_c^2 S_{\delta_e} \{C_{h\delta_e}\delta_e + (C_{h\alpha})_t \alpha_{ZL_t}\} \tag{8.14}$$

* Both a constant term and a term due to the tab system, if any, must often be added within the braces of (8.14). The former would be necessary, for instance, in the case of an elevon mounted on a cambered wing, where there is no symmetry to ensure zero hinge moment at $\alpha_{ZL} = 0$.

Here $S_{\delta e}$ is customarily defined as the plan area behind the hingeline, but would be total area on an all-movable control. The *hinge-moment derivatives* $C_{h\delta_e}$, $(C_{h\alpha})_t$ depend on M and tail or wing-elevon geometry, but they are particularly sensitive to "aerodynamic balance," which is connected with the fraction and shape of movable area that extends forward of the hinge.

On a powered elevator (8.14) is useful mainly for finding the maximum outputs demanded of its actuators. On directly connected or "reversible" designs, however, $H_{\delta e}$ is approximately proportional to the force the pilot must exert on the wheel or stick to maintain a given δ_e. Hence (8.14) helps to determine what force is needed for any trimmed flight condition; how it varies with v_c, altitude, etc.; and whether these levels are always consistent with normal human capabilities.

2. Influence of a Free Elevator on Static Stability

When studying stability of a reversible design, one should allow for all possible degrees of restraint at the control column, from a "stick-fixed" to a completely free condition equivalent to $H_{\delta e} = 0$. A free elevator usually tends to float into the relative wind, thus reducing the tail lift and CM pitching moment that would otherwise oppose changes in the vehicle angle of attack. Consequently a stiffness derivative less negative than $C_{m\alpha}$ and a neutral-point location ahead of h_n (called $C'_{m\alpha}$ and h'_n by Etkin 1959) are associated with the "free stick." A stricter limitation is placed thereby on permissible aft locations of the CM, because, unless $h < h'_n$, an angle-of-attack disturbance would tend to augment itself.

3. The Trim Tab

Most reversible elevators are furnished with a small trailing-edge flap known as a *trim tab*, whose setting is irreversibly adjustable in the cockpit. Its only significant effect is to make changes in $H_{\delta e}$, and it adds in (8.14) a term roughly proportional to its angle δ_{Tab}. The pilot is thus able to relieve himself of the inconvenience of continually pushing or pulling on the stick or column in order to hold a particular flight condition. He simply adjusts the tab so that $H_{\delta e} = 0$ when $\delta_e = \delta_{\text{Trim}}$.

Frequently, on larger aircraft, all reversible primary controls are provided with tabs. When restrained with springs or geared to the elevator rotation, similar devices are also employed to modify the inherent hinge-moment characteristics. Indeed, such balance tabs may be essential to keep stick forces within bounds throughout the operating envelope.

4. Stick-Force Gradient

This term is applied to the rate of change with speed v_c of the force required to trim. The gradient is affected by the setting of a trim tab, by the behavior of any other tab system which is included in the design, and by the location of the CM.

5. Maneuverability; Elevator Angle to Trim in Curvilinear Flight

The term "maneuver" comprehends almost any intentional deviation from a rectilinear trajectory, so time-dependent linear and angular accelerations are often involved. There are, however, several basic maneuvers wherein the angular accelerations are essentially zero except during entry and exit, and $C_m = 0$ is still necessary for pitch equilibrium. Examples are steady level or climbing turns and gradual pullouts. In such cases there is a nearly constant pitching velocity q, which, through the derivative C_{mq}, affects the problem of trimming moments about the CM.

Gates (1942) seems to have been the first to point out the significance of these observations relative to permissible CM range and similar questions. Etkin (1972) gives detailed analyses equivalent to Gates's; for turning flight, see his page 426, and for a constant-g pull-up, pages 238–243. The latter analysis leads to the discovery of yet another neutral point, called the *stick-fixed maneuver point*, whose coordinate is approximately

$$h_m = h_n - \frac{C_{mq}}{2\mu_s} \tag{8.15}$$

The quantity h_m has the property that the elevator rotation per unit increment of normal acceleration developed in the pull-up is proportional to $(h_m - h)$. Since C_{mq} is negative [cf. (6.66, 68)], we see that $h_m > h_n$. For aircraft with separate horizontal stabilizers at subsonic cruising altitudes, a typical difference would be 5–10% of $\bar{c}$, and the maneuver point is therefore not the factor which determines aft CM limit. There are cases, such as tailless vehicles and flight at very high altitude, in which μ_s may become a large number, when the "maneuver margin" $[h_m - h_n]$ can be much smaller. There may then be a CM restriction to avoid excessive sensitivity of the normal load factor to δ_e changes.

6. Stick Force per g

As with the subject of trim in rectilinear flight, when we analyze steady maneuvers, we must pay special attention to the directly connected elevator. Because of the influence of q on the hinge moment $H_{\delta e}$, it turns out that the stick force needed to hold a given number of g's during pull-up or steady turning differs from the force for that same elevator setting in the absence of such acceleration. This force is roughly proportional to the distance by which the CM lies forward of a *stick-free maneuver point* h'_m. Were the margin between these points negative, an undesirable situation might arise in which the pilot had to push forward to prevent the "tightening up" of a turning maneuver. Since $h'_n < h'_m$, however, the stick-free neutral point usually constitutes a barrier to such difficulties.

Etkin (1959, pages 63–65) discusses the influence of the main vehicle parameters on these phenomena. His Section 3.3 gives details on the use of springs attached to the control column and of unbalanced masses called "bob-weights" to make needed adjustments in various stick-force characteristics.

7. Trim for Landing and Permissible CM Limits

All considerations outlined above relate to rearward positions for the CM. They may be summarized by stating that either h_n or h'_n defines a station ahead of which the CM should lie by at least a few percent of $\bar{c}$. Accurate test data must always be available when this limit is finally established, and the combined influences of flaps, aeroelasticity, and the propulsive system should be carefully accounted for.

If we examine forward CM limits, the assurance of adequate elevator power to pull back the nose during landing flare is often the determining factor on conventional aircraft. This is mainly because of the flaps, which add to the wing camber, thereby decreasing C_{m0} and requiring greater downward force on the stabilizer for trim. The aerodynamic effects of proximity to the ground must be considered, as should contributions from engine thrust at all throttle settings between idle and full. Full power is associated with the "go-around" maneuver, which is typically executed just before the wheels have touched. If the thrust line is below the CM, the pitching moment due to T will evidently push the nose up and thus assist a conventional elevator in its trimming function. On the other hand, engines in a high position may call for even greater negative δ_e.

In addition to the need for adequate elevator lift for landing and go-around, Etkin (1959, Section 3.8) lists three other characteristics that may restrict forward CM when the control is reversible.

- The stick force per g in a pull-up or steady turn should not be excessive.
- The stick-force gradient around the most critical trim condition should not be excessive.
- The stick force required to land, from trimmed flight at the landing approach speed, should fall within the pilot's capabilities.

8. Automatic Stability Derivative Augmentation

At the other extreme of design sophistication from manual control are those modern aircraft which make extensive use of automatic equipment to augment both static and dynamic stability. The miniaturization of electronic components and the digital processing of data in large, central on-board computers have recently increased the possibilities of this approach almost beyond human imagination. The realization of these potentials is being delayed, however, by controversy between designers and operators over whether any requirements should be placed on the "open-loop" vehicle (e.g., whether the flying qualities should meet certain criteria even if all augmentation systems should be knocked out of action) and whether pilots will accept "fly-by-wire." The latter concept involves eliminating all mechanical connections between the control column and the aerodynamic-surface actuator valves, communicating commands entirely by means of highly redundant and fail-safe electrical signals.

The cause of stability augmentation has been advanced by such programs as the Cornell Aeronautical Laboratory's In-Flight Simulators (see Reynolds and Pruner 1972). The vehicles in this series were furnished with special controls and electronic means for changing many important stability derivatives over wide ranges. Not only is such a system useful for training and for investigating the properties of new designs before first flight, but it gives the individual pilot a chance to experiment rapidly and discover what combination of longitudinal and lateral characteristics optimizes the handling qualities that he perceives in a class of aircraft.

Recent studies have also focused on the *control-configured vehicle* (CCV). This notion stems from a revised design philosophy, according to which the entire process makes the greatest use of whatever reliable stability-augmentation technology has to offer (see Holloway, Burris, Johannes 1970). It is, of course, not possible to dispense completely with the empennage or the movable controls, because there are always certain minimum requirements on the aerodynamic moments that must be available for trimming and maneuvering. When these have been met, however, it still turns out that desirable values of such quantities as $C_{m\alpha}$, C_{mq}, $C_{n\beta}$, C_{nr}, C_{np}, and $C_{l\beta}$ can be achieved more efficiently by automatic means than by open-loop sizing and shaping of the aerodynamic surfaces. The consequent design improvements are only just beginning to be understood. For instance, one recent preliminary study suggested that the gross takeoff weight of a large strategic aircraft might be reduced by nearly 20%, without affecting the missions it is to fly, by maximum realistic application of the CCV scheme.

9. Lateral-Directional Static Stability and Control

Desirable magnitudes for the lateral derivatives must be deduced almost wholly from such dynamic-stability considerations as the effect of lateral oscillation properties on handling qualities. One obvious requirement is that the "weathercock stability" should be positive,*

$$C_{n\beta} > 0 \tag{8.16}$$

and indeed that this derivative should exceed some minimum value at all attitudes and flight conditions. The vertical stabilizer is located and sized mainly by this $C_{n\beta}$ criterion; when there is need for much high-α maneuvering, as on the F-15, some ventral fin area often proves necessary.

Lateral trim is a trivial matter in normal operation with the wings level, because ideally all moments about the CM are balanced when both $\delta_a = 0$ and $\delta_r = 0$. On aircraft with more than one engine, however, the rudder size must be selected so that all yawing moments can be safely trimmed even under the worst conditions, with a single engine shut down. This is why such large rudders can be seen on vehicles, like the Boeing 737, where there are only two engines and each has a substantial moment arm about the z-axis of yaw. They also tend to demonstrate lively takeoff and climbout performance, since the need for safe operation

* Note the analogy with (8.2), the difference in sign resulting from opposite definitions of the senses of α and β. You can improve your physical understanding of these motion coordinates by observing that α and θ, as seen from the right side, are analogous to β and $(-\psi)$ seen from above. We recall that θ and ψ are so defined that no change in the airload system results from a steady alteration in either of these angles.

on one engine tends to make the ratio of maximum thrust to weight greater than on aircraft with three or more separate propulsive units.

There is no inherent aerodynamic way to achieve open-loop static stability about a given bank angle Φ, although an autopilot or "automatic wing leveler" may be provided to activate the ailerons and relieve the pilot of this task.

Both the ailerons and rudder must be given static deflections for trim during either a sideslip or rectilinear flight with $\Phi \neq 0$. The former maneuver is used for landing approach into a crosswind, and small-plane pilots also find it convenient as a means for rapidly reducing altitude. When the vehicle is sideslipping, the rudder is set at a fairly large angle so as to produce constant $\mathbf{v}_c$ at a nonzero β. Some δ_a is then needed to trim the rolling moment due to sideslip, and the Y-force due to gravity must be canceled if, as is often the case, the wing toward which the slip occurs is allowed to drop a few degrees.

A quantitative discussion of all these matters will be found in Sections 3.9 and 3.10 of Etkin (1959).

8.2 STATIC PERFORMANCE

In Section 3.2, we chose the adjective "static" to characterize the class of performance problems which can be adequately analyzed under the assumption that $\mathbf{v}_c$ = constant. That is, the motion is not only taken to be rectilinear, but no dynamic interchange whatever is permitted between the kinetic and potential energies. We believe that such problems should, most logically, be regarded as a subset of the more interesting and comprehensive field of dynamic performance, to which most of the rest of this book is dedicated. We acknowledge their inherent importance, however—as well as the reverence in which they are held by many aeronautical engineers—by devoting a preliminary section to their treatment. It is a matter of convenience also to include here cases like takeoff roll, in which the CM is accelerated but moves along a straight line, and turns when it moves on a fixed circle or spiral.

We presume that sufficient knowledge is available on the thrust capability of the propulsion system, on the weights, and on the aerodynamic forces. We discussed in Chapter 5 how T can be determined as a function of throttle setting, speed, and altitude in some standardized atmosphere. Drag or "thrust required" also depends on speed and altitude, but for turning flight C_L must first be computed from the normal load factor. C_D and D then follow from relations such as (4.168) and (3.5a).

For rectilinear cruising, climbing, or diving, the equilibrium equations are (3.10), here rewritten with $mg = W$ and $\Theta = \theta_0$:

$$T - D - W \sin \theta_0 = 0 \tag{8.17a}$$

$$L - W \cos \theta_0 = 0 \tag{8.17b}$$

When the vehicle is rolling along a horizontal runway during takeoff or landing, only the longitudinal acceleration in an x-direction parallel to the runway is significant. Accordingly we modify (3.9a) to get

$$m\dot{U} = T - D - \mu[W - L] \tag{8.18}$$

Here μ is a coefficient of rolling friction, which relates the resisting force on the wheels to the net vertical force supported by the struts. For landing, maximum deceleration is desired. Hence the throttle is pulled back so that $T \cong 0$, the nose is often pushed forward to make L as small as possible, and brakes are applied to increase μ. Obviously, negative T should be inserted into (8.18) to account for any thrust-reverser operation, and D should include contributions from flaps, spoilers, aerodynamic brakes, drag parachutes, and the like.

By contrast, we assign T its maximum value during takeoff. When an aircraft with tricycle gear takes off, normal practice is to leave the nose wheel on the runway until "rotation" for liftoff; the resulting angle of attack, in the presence of a ground plane, then yields C_L and C_D. There may be another option, however, for improving the performance: Pull off the nose gear immediately and try to program α_{ZL} so that the $(-D)$ and $(+\mu L)$ terms in (8.18) cooperate to minimize the ground run up to a given speed.

One good example of steady turning performance is a spiral climb at fixed speed and load factor. The circumstances are described analytically by (2.39a,b,c). If the x-axis is tangent to the flight path and there is no sideslip, we have

$$Y = V = W = \dot{U} = \dot{V} = \dot{W} = 0 \tag{8.19}$$

U, Φ, Θ should be assigned their fixed values v_{c0}, φ_0, θ_0. If there is a constant rate of turn Ω, taken positive about a vertically downward axis, we can componentize by means of Euler angles to obtain

$$P = -\Omega \sin \theta_0 \tag{8.20a}$$

$$Q = \Omega \sin \varphi_0 \cos \theta_0 \tag{8.20b}$$

$$R = \Omega \cos \varphi_0 \cos \theta_0 \tag{8.20c}$$

Finally we replace X and Z by $(T - D)$ and $(-L)$, respectively, to derive

$$T - D - W \sin \theta_0 = 0 \tag{8.21a}$$

$$W \sin \varphi_0 \cos \theta_0 = m v_{c0} \Omega \cos \varphi_0 \cos \theta_0 \tag{8.21b}$$

$$L - W \cos \varphi_0 \cos \theta_0 = m v_{c0} \Omega \sin \varphi_0 \cos \theta_0 \tag{8.21c}$$

From (8.21b) we see that the "coordinated" bank angle for a given speed and Ω is

$$\varphi_0 = \tan^{-1} \left[\frac{v_{c0} \Omega}{g} \right] \tag{8.22}$$

The normal load factor n_z experienced by the pilot would be due to the z-component of the resultant force on the airplane, which lies in the plane of symmetry. We calculate it from (8.21c) and (8.22) as

$$|n_z| \equiv \frac{L}{W} = \sec \varphi_0 \cos \theta_0 = \sqrt{1 + \left(\frac{v_{c0}\Omega}{g}\right)^2} \cos \theta_0 \tag{8.23}$$

We might use (8.23) to find the lift necessary to turn at a given rate, while achieving a given angle of climb at some v_{c0}; with the drag corresponding to this lift, (8.21a) would then yield the thrust required to perform the maneuver. Alternatively, the system (8.21, 23) can be implicitly solved for the maximum possible rate of climb, at full throttle and with a specified n_z or Ω. Etkin (1972, Section 10.4) analyzes this situation only with $\theta_0 \ll 1$, but also discusses the control requirements for angular trim.

With this analytical background, we list the topics addressed under the heading of "classical performance." Investigation of most of these is left to assigned problems or to your own initiative. At the very least, given the necessary data on a particular vehicle, you should ask yourself how you would go about computing the desired information.

a) Length of takeoff run, either to liftoff or to the moment of clearing a fixed obstacle such as a 35- or 50-ft barrier.

b) Length of landing run, either from touchdown or after clearing some specified obstacle at the runway threshhold.

c) Maximum airspeed as a function of altitude, either in level flight, in a prescribed turn, or in climb at a specified angle θ_0 or rate $v_c \sin \theta_0$.

d) Maximum rate of climb as a function of altitude, either in rectilinear flight or in a prescribed turn.

e) Ceiling—often defined statically as that altitude at which maximum rate of climb falls below some particular value such as 100 ft/min.

f) Time to climb from sea level to a specified altitude or to the ceiling.

g) Range, which is the maximum distance a vehicle can fly through the air with a given fuel load, given rules of operation and specified allowances for delay or emergency.

h) Radius of action, which has to do with the maximum distance to which a vehicle can fly and return to its starting point, with or without credit for aerial refueling.

i) Endurance, which is the longest time a vehicle can remain in continuous flight with a given fuel load, given rules of operation and specified allowances for emergency.

Let us conclude this chapter by examining a particular performance question. It has to do with estimating range under a set of ground rules typifying the cruise operation of a subsonic jet transport. Our example has the merit of underlining the importance of both high L/D and high specific impulse I_{sp} to the attainment of long range.

We assume that a turbojet-powered aircraft has already climbed to its initial cruising altitude in the constant-temperature stratosphere. Our objective is to determine how far it will travel through the air while burning a given mass of fuel. Let it further be assumed that flight occurs at constant M and constant lift coefficient C_{L0}, whence the ratio L/D is fixed and may be held close to its maximum. Since the weight goes down as fuel is used, this means that $L = W$ equilibrium can be maintained only by permitting a very gradual climb ($\theta_0 \ll 1$); we can determine the density altitude corresponding to the instantaneous weight from

$$\rho_\infty = \frac{2W}{C_{L0} v_c^2 S} \tag{8.24}$$

At fixed throttle, the engines are working in the "kinematic similarity" mode described by Chapter 5. Thus $T/\dot{m}$ and I_{sp} remain constant, while the thrust T itself drops off in direct proportion to ρ_∞. Since C_D = constant, this means that the balance between thrust and drag is preserved independent of altitude.

The mass of fuel consumed during any time interval dt is

$$dm_f = \frac{1}{I_{sp}} T\,dt = \frac{L}{I_{sp}}\left(\frac{D}{L}\right) dt = \frac{W\,dt}{I_{sp}(L/D)} \tag{8.25}$$

Furthermore $-dm_f = +(dW/g)$, where dW is the change in vehicle weight during dt; hence (8.25) may be rearranged into the following form:

$$dt = -\left(\frac{I_{sp}}{g}\right)\frac{L}{D}\frac{dW}{W} \tag{8.26}$$

Multiplying (8.26) by $v_c \equiv \mathrm{M}a_\infty$ yields a quantity $v_c\,dt$, which can be integrated from the starting time t_0 to the end of cruise flight t_1. This definite integral is the desired range R,

$$R = \int_{t_0}^{t_1} v_c\,dt = -\mathrm{M}\left(\frac{I_{sp}}{g}\right)\frac{L}{D}\,a_\infty \int_{W_0}^{W_1} \frac{dW}{W} \tag{8.27}$$

whence

$$\boxed{R = \left(\mathrm{M}\,\frac{L}{D}\right)\left(\frac{I_{sp}}{g}\right)a_\infty \ln\frac{W_0}{W_1} = \left(\mathrm{M}\,\frac{L}{D}\right)\left(\frac{I_{sp}}{g}\right)a_\infty \ln\left(\frac{W_1 + \Delta W_f}{W_1}\right)} \tag{8.28}$$

Here W_1 is the vehicle weight at the end of cruise, during which a weight ΔW_f of fuel has been burned.

The useful formula (8.28) is sometimes called the *Breguet range equation*, although Breguet's original analysis referred to steady flight of an airplane driven by a combination of reciprocating engines and propellers, for which the rate of fuel consumption is proportional to HP rather than T. In any event, (8.28) demonstrates the key importance of high I_{sp} (low SFC) and high values of the product ($\mathrm{M}L/D$) to the achievement of long-distance unrefueled operation. In Chapter 1 we anticipated these conclusions during our discussion of subsonic versus supersonic transportation. It was suggested, for instance, that the economic feasibility of an SST partly depends on its ability to make good an ($\mathrm{M}L/D$) equal to or greater than those for the best subsonic turbojet airliners. By the middle 1960's there remained few doubts that this test could be met.

8.3 PROBLEMS

8A Allowing for the influence of steady pitch rate Q on the balance of pitching moments about the CM, derive a formula for the elevator angle required to trim an airplane in a pullout at a specified number of g's. From this result, verify the correctness of (8.15).

8B A scheme recently developed for enabling an airplane to move on a curved flight path without banking the wings ($\Phi = 0$) is as follows: A steady sideslip angle (with associated Y-force) is produced by deflecting the rudder. The resulting rudder rolling moment is trimmed with ailerons, whereas the rudder yawing moment is trimmed with the differential drag forces due to differentially splitting the trailing-edge ailerons, thus making them function as drag brakes with large moment arms in yaw. Assuming you have any required aerodynamic information, take a typical configuration and analyze this maneuver.

8C It is often said that a canard surface is preferable to a conventional aft horizontal stabilizer from the standpoint of "off-design trim drag," because the trimming loads tend to be downward on the aft surface, thus forcing the airplane wing to carry more lift than necessary and therefore to have more induced drag than necessary. Investigate this possibility analytically by comparing two subsonic aircraft, whose only significant difference is that one has l_t positive, the other negative. S_t is large enough compared to S that tail contributions both to total airplane C_L and to C_D must be accounted for. Assume the total drag to be composed of a constant zero-lift drag C_{DO} plus the usual sort of induced-drag contributions from the wing and the tail.

To be specific, *design* each airplane so as to have L/D a *maximum* at its highest flight dynamic pressure (i.e., lowest C_L). As a suggestion, see if this can be done by having $C_{Lt} = 0$ at this condition and putting the wing at its "ideal angle of attack." Then let q_∞ go down (and C_L go up), and study in each case the variation of trimmed airplane drag with C_L.

8D Study analytically the performance of a jet-powered aircraft flying in the stratosphere under the following approximations (whose basis you should understand from previous work).

$$\text{Thrust } T = T_s \frac{\rho}{\rho_s}, \qquad \text{Density } \rho = \rho_s e^{-h/H}, \qquad L = W, \qquad C_D = C_{DO} + \frac{C_L^2}{\pi e A\!\!R}$$

Here h is altitude measured from base of the stratosphere, at which altitude quantities are identified by subscript s. Also H, W, C_{DO}, and $\pi e A\!\!R$ are all constants, and T at full throttle is independent of speed for a given h.

Find the following as functions of h.

1) Maximum rate of climb for constant velocity in direction of flight path, or $v_c[(T - D)/W]$—"specific excess power."
2) Maximum rate of climb.
3) Minimum *time* to climb between two altitudes.

8E Consider a simplified analysis of the spinning motion of a flight vehicle by assuming that the spin is steady such that the rotational motions involved amount to simple, steady turning about a vertical axis, with the airplane nosed down and banked with respect to the horizontal (see figure).

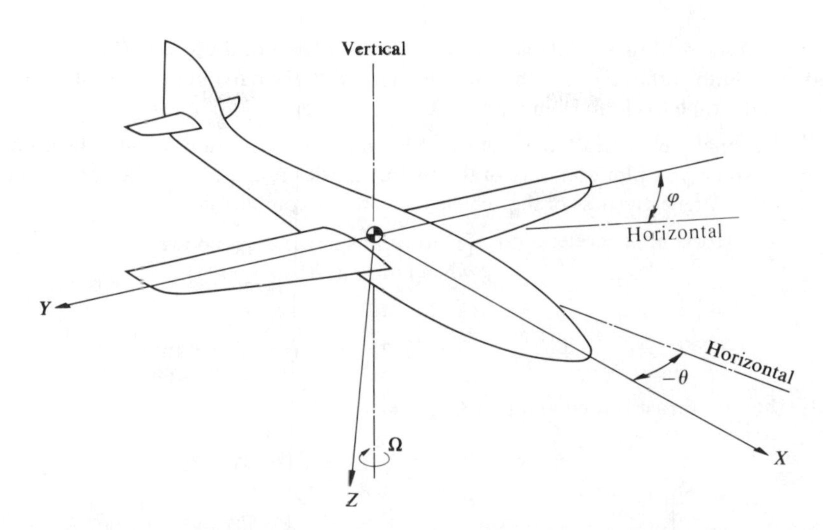

Given the body-fixed axis system taken through the center of gravity of the aircraft, the rate of rotation about the vertical, Ω, a constant; and θ a constant negative angle and φ a positive or negative one, show that the general governing differential equations reduce to a set of nonlinear algebraic equations for this simplified case.

A simple particular case is a spin in which the wings are held horizontal. What do the above set of equations reduce to for this case, if in addition the axis system coincides with the principal directions?

On the basis of your foregoing results, discuss why the most effective means of recovery for a "typical aircraft" ($I_z > I_x$) would be a simultaneous application of rudder and elevator. In what direction should they be applied?

8F You are given the following data for a small jet aircraft: wing loading $= W/S = 110$ lb/ft^2, wing area $= S = 200$ ft^2, aspect ratio $= \mathcal{R} = 4$, span efficiency factor $= e = 0.9$. You are

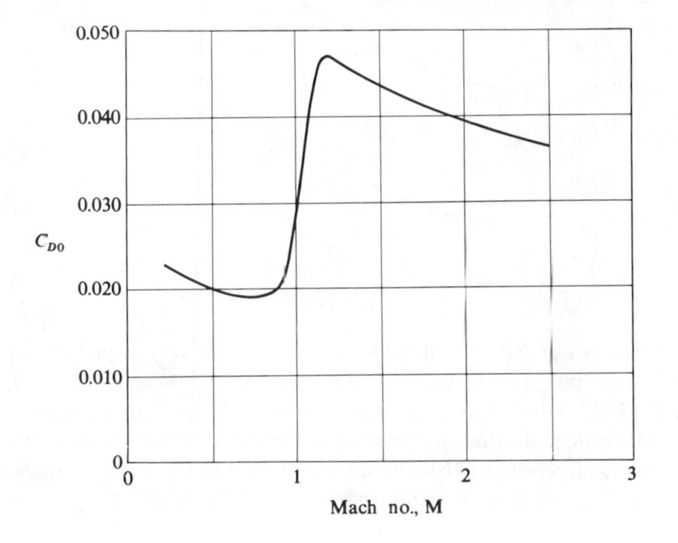

also given curves of thrust available and C_{D0} versus M. [Recall: $C_D = C_{D0} + C_L^2/\pi e \mathcal{R}$.] Estimate the climb performance of this aircraft, along with the maximum speed. Discuss how to minimize the time to climb from sea level to 45,000 ft, ending up in level flight at M = 1.

8G For the small jet aircraft of Problem 8F, prepare an altitude-versus-Mach-number diagram on which you plot contours of the two quantities regarded as measures of "energy maneuverability" (effectiveness of the vehicle in air-to-air combat)

$$\frac{T - D}{W} = \frac{\text{Horizontal acceleration}}{g}$$ [Known as specific excess power when multiplied by speed v_c]

and

$$\frac{L_{\max}}{W} = \sqrt{1 + \frac{U_0^2}{gR}},$$ where R = radius of a level turn.

Consider that full thrust is used and that $C_{L\max} = 1$.

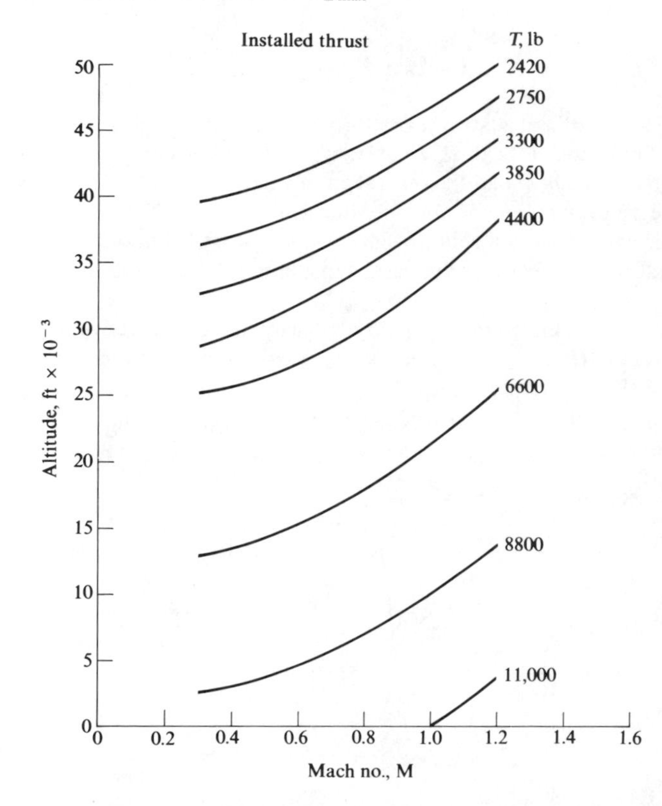

8H Specific excess power (SEP) for a fighter airplane is also very important, as it affects the ability to turn and "pull g." Consider steady climbing turns at altitude 10,000 ft for an airplane like the F-15, which can be assumed to have SEP = 0 in a 5-g turn at M = 0.8. Assume for full-throttle flight that the subsonic maximum thrust is independent of speed and that C_{D0} and e are independent of M. Find rate of climb (or descent) as a function of g and M, allowing for a $C_{L\max}$ of 1.5 and an allowable limit load factor of 7g.

8I a) Consider an aircraft with $W = 70{,}000$ lb and $S = 1551$ ft^2. Given standard sea level conditions, compute the landing distance after clearance of a 50-ft obstacle with power off and flaps fully extended. The nose wheel is held off the ground for as long as possible ("soft-field landing") so you can assume for the entire maneuver that $C_L = 2.0$, $C_D = 0.2$, and $\mu = 0.4$ (μ is the coefficient of coulomb friction between wheels and ground after touchdown). Neglect changes in aerodynamic characteristics due to ground proximity, wind, etc.

b) Discuss takeoff over a 50-ft obstacle for the same vehicle.

CHAPTER 9

DYNAMIC PERFORMANCE: BOOST FROM NONROTATING AND ROTATING PLANETS; NUMERICAL INTEGRATION OF ORDINARY DIFFERENTIAL EQUATIONS

9.1 INTRODUCTION

In Section 2.6 we discussed the state-vector formulation of equations of motion for flight vehicles. We also identified the n-dimensional vector $\{\mathscr{X}\}$ of time-dependent state variables, and mentioned and exemplified the frequent appearance of an m-dimensional vector $\{\mathscr{U}\}$ of specified controls. A key feature of this approach is that n is chosen so as to cast the equations in the form [cf. (2.41) or (2.45)] of a first-order, nonlinear system solved explicitly for the first derivatives of the state variables. This scheme has special relevance to studies of dynamic performance, such as the determination of boost and entry trajectories, because of the high efficiency of numerical integration methods that are applicable to (2.41) and (2.45).

Usually but not always, dynamic performance parallels its static counterpart in viewing the vehicle as a mass point and focusing on the path traced by its CM. Thus, if we return for the moment to the airplane moving over a flat earth, but restricted to a fixed plane of symmetry and with its wings level, two velocity variables (for example, U, W) would be adequate to characterize its motion through the atmosphere, whereas two position coordinates (for example, x', z') fix the CM instantaneously in the earth's frame of reference. The resulting four-dimensional state vector is not, however, the most convenient to work with. This is because the directions for U and W are taken in a frame determined by the airplane's angular orientation, which is not of primary interest—a disadvantage manifested by the appearance of the angle $\Theta(t)$ in every member of the system of equations we would derive by appropriate reductions of (2.42) and (2.29).

The customary way of avoiding this difficulty is to characterize the motion by the resultant speed $v_c(t)$ and by the angle which the flight path makes above the local horizontal. The latter is designated $\gamma(t)$. For the case at hand, it is easily seen that these quantities are related to the variables of Chapters 2 and 6 by

$$v_c = \sqrt{U^2 + W^2} \tag{9.1a}$$

$$\gamma = \Theta - \alpha \tag{9.1b}$$

α being defined by (6.20).

You may find it helpful to derive flat-earth equations of motion for the components of

$$\{\mathscr{X}\} = \begin{Bmatrix} v_c \\ \gamma \\ x' \\ z' \end{Bmatrix} \tag{9.2}$$

If aerodynamic loads are computed under the quasi-steady approximation, the corresponding control vector might be

$$\{\mathscr{U}\} = \begin{Bmatrix} \alpha_{ZL} \\ T \end{Bmatrix} \tag{9.3}$$

Thus to regard the angle of attack and thrust as preassigned functions of time is, of course, tantamount to neglecting the rotary inertias of both the pitching airplane and the propulsion system. But, on the time scale of most performance phenomena, this simplification is quite well justified.

Let us now introduce the subject of booster operation by studying trajectories which are confined to a single plane passing through the center of a spherical, nonrotating planet. Figure 9.1 illustrates the situation we envisage and defines a number of quantities often employed to describe such motions. We take acceleration components tangent and normal to the path, expressing Newton's law as follows:

$$a_T = \dot{v}_c = F_T/m \tag{9.4a}$$

$$a_N = v_c[\dot{\gamma} - \dot{v}] = F_N/m \tag{9.4b}$$

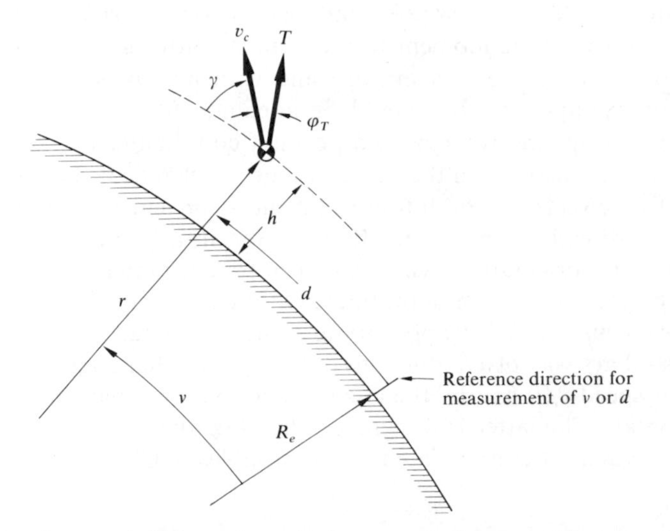

Fig. 9.1. Illustrating speed, flight-path angle, thrust, and other quantities related to boost in a fixed vertical plane. The radii r and R_e are measured from the center of a nonrotating spherical planet.

Here r and ν are polar coordinates of the CM, with origin at the planet's center. Sphericity is accounted for, under the senses we have chosen for γ and ν, by noting how the inertial rate of change of the direction of vector $\mathbf{v}_c$ is the difference between $\dot{\gamma}$ and $\dot{\nu}$. From the figures, we see that the derivative $\dot{\nu}$ is related to the horizontal component of $\mathbf{v}_c$,

$$\dot{\nu} = \frac{v_c \cos \gamma}{r} \tag{9.5}$$

The rate of change of r, or of altitude h, is evidently equal to the vertical component $v_c \sin \gamma$.

The same four forces act on the booster as on any other atmospheric vehicle. We allow for the inverse-square behavior of gravitation by writing the local acceleration g directed toward the center, as

$$g = \frac{g_e R_e^2}{r^2} \equiv \frac{k}{r^2} \tag{9.6}$$

Here R_e is the radius of the planet's surface, at which g has the constant value g_e. Lift L and drag D are defined according to standard conventions. As before, we assume that they are related to v_c, h, and the geometry of the vehicle.

Initially we suppose that the rocket thrust is controllable both in magnitude and direction e.g., one can control the thrust by throttling the propellant flow and by rotating one or more engines about gimballed mounts. We temporarily avoid any reference to the vehicle's angular orientation by selecting, as controls, T itself and the angle φ_T between T and the velocity vector.* The mass of a booster decreases rapidly as propellant is consumed, so that m itself here becomes a state variable. If we take the specific impulse to be nearly independent of thrust level, then (5.16), with $\dot{m}_f = -\dot{m}$, provides the required law of mass variation.

With these preliminaries, we are in a position to construct a determinate set of equations. To this end, we substitute the aforementioned forces into (9.4) and add (5.16) plus the kinematic relations for $\dot{r}$ and $\dot{\nu}$.

$$\dot{v}_c = \frac{T \cos \varphi_T}{m} - \frac{D}{m} - \frac{k \sin \gamma}{r^2} \tag{9.7a}$$

$$\dot{\gamma} = \frac{v_c \cos \gamma}{r} + \frac{T \sin \varphi_T}{v_c m} + \frac{L}{v_c m} - \frac{k \cos \gamma}{v_c r^2} \tag{9.7b}$$

$$\dot{r} \equiv \dot{h} = v_c \sin \gamma \tag{9.7c}$$

$$\dot{\nu} = \frac{v_c \cos \gamma}{r} \tag{9.7d}$$

$$\dot{m} = -\frac{T}{I_{sp}} \tag{9.7e}$$

* To ensure maintenance of the assumed planar trajectory, φ_T must remain in the plane of motion. Also, no out-of-plane aerodynamic-force components are permitted.

The form of (9.7) corresponds precisely to (2.45), with

$$\{\mathscr{X}\} = \begin{Bmatrix} v_c \\ \gamma \\ r \\ \nu \\ m \end{Bmatrix} \tag{9.8}$$

$$\{\mathscr{U}\} = \begin{Bmatrix} T \\ \varphi_T \end{Bmatrix} \tag{9.9}$$

When (9.7) are applied, observe that r and h may be used interchangeably, since

$$r = R_e + h \tag{9.10}$$

Moreover, ν appears only in (9.7d) and is therefore decoupled from other state variables when the planet has negligible rotation. Normally one requires information on booster position over the planet's surface. After the other four members of (9.7) are solved, this position can be calculated by finding $\nu(t)$ from (9.7d) and combining it with data on the launch-site location and the orientation of the plane of boost. For instance, if boost were toward the east, ν would be the angle of east longitude measured from launch, whereas it plays the role of latitude for a northerly trajectory.

We add another comment relative to staging. Since boost even to earth orbit usually involves two or three stages, it is necessary to account for known discontinuities in m, T, D, etc., during the integration of (9.7). This is done by computing the trajectory in segments, reestablishing the initial conditions $\{\mathscr{X}(t_k)\}$ for each segment at a series of instants t_k corresponding to first-stage engine cutoff, second-stage ignition, and the like. Often the moment of cutoff is not fixed by a certain preassigned time interval after liftoff, but by when the mass m falls to the value associated with the exhaustion of usable propellants in that stage.

9.2 NUMERICAL INTEGRATION OF ORDINARY DIFFERENTIAL EQUATIONS

Our development has reached the point at which some discussion of the numerical-integration process will help prepare the way for the topics covered later in this chapter and in Chapters 11 and 12. Among the many references on this subject we have found Henrici (1962) and Hildebrand (1956, especially his Chapter 6) very useful. Also cited is the compendium by Ralston and Wilf (1960), which treats a wide variety of other engineering applications of the digital computer.

As we have seen and will see further, a host of important maneuver and trajectory calculations can be reduced to solving the following system:

$$\{\dot{\mathscr{X}}\} = \{f(\{\mathscr{X}\}, \{\mathscr{U}\}; t)\} \tag{9.11}$$

with initial conditions

$$\{\mathscr{X}(t_1)\} = \{\mathscr{X}_1\} \tag{9.12}$$

specified and with the $m \times 1$ control vector $\{\mathscr{U}\}$ given as a function of t (or of $\{\mathscr{X}\}$ and t) in an interval $t_1 \le t \le t_F$. In order to avoid repeated use of complex symbols, we first concentrate attention on the special case in which $n = 1 = m$. $\{\mathscr{X}\}$ and $\{\mathscr{U}\}$ may then be replaced by the scalars $x(t)$ and $u(t)$ [or $u(x, t)$]. The transition to more than one dimension is quite straightforward, as can be studied in the case of "one-step" methods by proceeding from Chapters 1–2 to Chapter 3 of Henrici (1962).

Perhaps the most direct approach to (9.11, 12) is by Euler's method of linear extrapolation. This is based on the assumption that, over any small interval of length $h = t_{n+1} - t_n$, the first derivative of x differs negligibly from its estimated value $f(x_n, t_n)$ at the beginning. In particular, let us break the range from t_1 to the final time t_F into equal intervals

$$h = t_2 - t_1 = t_3 - t_2 = \cdots = t_{n+1} - t_n = \cdots \tag{9.13}$$

If x_n represents our estimate of x at t_n, we then start at t_1 and go successively over each step, the discrete scalar equivalent of (9.11) over any typical step being

$$\boxed{x_{n+1} = x_n + hf(x_n, t_n)} \tag{9.14}$$

Note that the control $u(t_n)$ is not written explicitly in (9.14). Since u is a known function of other quantities in the examples in which it appears, however, its inclusion is a trivial matter.

Figure 9.2 illustrates Euler's method, showing how the piecewise-continuous approximation to $x(t)$ may be expected gradually to diverge away from its exact counterpart. The favorable influences on computational accuracy both of adopting the smallest possible step size h and of working with a function f that does not change too rapidly with x or t should be evident from this sketch.

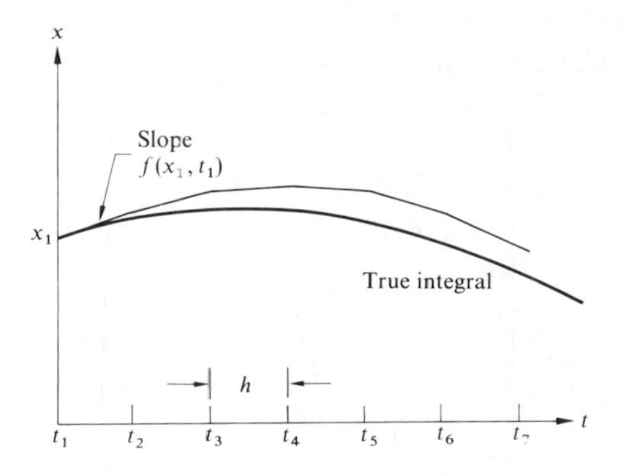

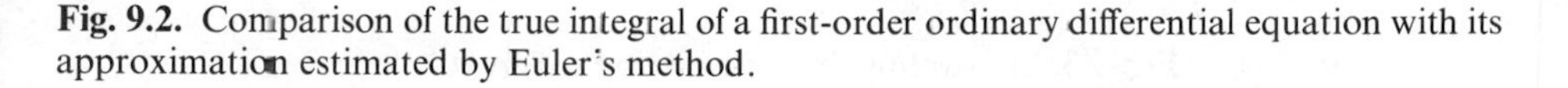
Fig. 9.2. Comparison of the true integral of a first-order ordinary differential equation with its approximation estimated by Euler's method.

Although many improvements and generalizations of (9.14) have been proposed over the years, it still retains its utility in some practical situations. We regard it as very interesting that circumstances can be defined (cf. Henrici 1962, pages 15–26) under which the Euler solution converges uniformly to the exact $x(t)$ as h is systematically reduced to zero. Such convergence is assured, for instance in the scalar case, when the following two specifications are met:

1) $f(x, t)$ is a unique and continuous function of both variables in the ranges $t_1 \leq t \leq t_F$, $-\infty < x < \infty$.
2) A *Lipschitz condition* is satisfied, according to which there is some constant L such that, for any t in the range and any two finite x and x^*,

$$|f(x, t) - f(x^*, t)| < L\,|x - x^*| \tag{9.15}$$

As suggested in Fig. 9.3, (9.15) means that a plot of f versus x for any t must always fall between two straight lines, drawn with slopes $\pm L$ through any point $[x^*, f(x^*, t)]$ on the curve.

We remark that condition (1) can be immediately generalized to vectors $\{f\}$ and $\{\mathscr{X}\}$, but that the Lipschitz condition calls for the introduction of a new concept known as the *norm of a vector*. For instance, the norm employed by Henrici (1962, Chapter 3) is simply the sum of magnitudes of the components:

$$\|\{\mathscr{X}\}\| = \sum_{j=1}^{n} |x_j| \tag{9.16}$$

Inequality (9.15) is not often violated in nature, but it does imply that the derivative $\partial f/\partial x$ (and $\partial f/\partial u$, where appropriate) is everywhere bounded. Certain idealized control laws are therefore potential sources of difficulty. You may wish to examine, as an illustration, the consequences of employing a discrete-position or "bang-bang" controller. If such a device were used to exert forces on a point mass constrained to move frictionlessly along a line, the dimensionless equation of motion would take the form

$$\ddot{x} = u = \pm 1 \tag{9.17}$$

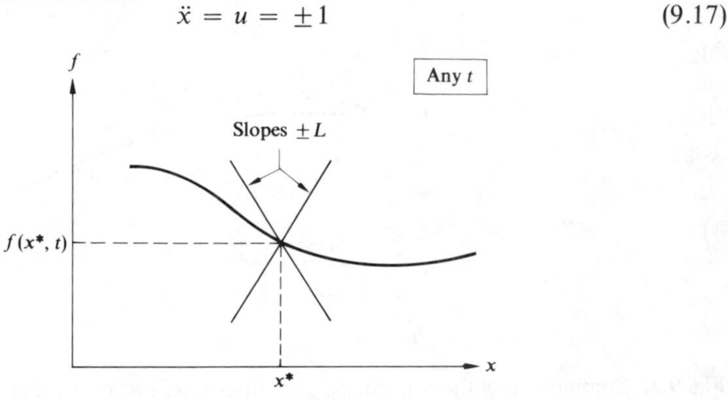

Fig. 9.3. Illustrating the Lipschitz condition (9.15).

Here u might be arranged to drive x to zero by changing sign along a *switching boundary*, e.g., by means of

$$u = -\text{sign}\,(\dot{x} + kx) \tag{9.18}$$

with $k > 0$ a gain constant. Defining the state vector

$$\{\mathscr{X}\} \equiv \begin{Bmatrix} x \\ \dot{x} \end{Bmatrix} \tag{9.19}$$

you will discover (1) that "trajectories" of the system in the plane* of $\dot{x}$ versus x have discontinuous slopes at the switching boundary, and (2) that, when $k\,|\dot{x}| < 1$ at a point of boundary crossing, it becomes impossible to determine the behavior by straightforward integration of (9.11). This failure is associated mathematically with the infinite derivatives of $\{f\}$ at the boundary.

There are two sources of computational inaccuracy when a method like Euler's is applied to (9.11, 12). *Roundoff error* occurs because nearly all quantities involved are actually irrational, but are rounded off to a finite number of places for digital processing. The estimation of such an error is a complicated exercise in statistics (cf. Henrici 1962, pages 50–57). For "nonpathological" systems, however, one may expect the percentage roundoff error to increase in rough proportion to the total number of arithmetic steps and to be inversely related to the total number of figures carried.

Discretization error accumulates as a result of finite step size, here because the continuously varying derivative $\dot{x}$ is approximated by a constant over each discrete interval h. An interesting feature of the aforementioned theorem on convergence of Euler's method is its demonstration that the discretization error goes uniformly to zero as $h \to 0$. For instance, the proof in Henrici (1962) proceeds by successively halving the time step according to

$$h = \frac{t_F - t_1}{2^p}, \qquad p = 1, 2, \ldots \tag{9.20}$$

Let us next discuss the family of more elaborate and accurate single-step methods, of which Euler's is the archetype. All these have in common that, for predicting x_{n+1} or $\{\mathscr{X}(t_{n+1})\}$, they start with data only at the instant t_n at which this step begins. They are typified by a sequence associated with the names of Runge and Kutta. Leaving the subjects of proof and error estimation until later, we summarize the first four of these as follows.

1) *First-Order Runge–Kutta Method*

This scheme is identical with Euler's. The nth step is described by (9.14).

2) *Second-Order Runge–Kutta Method*

$$x_{n+1} = x_n + \tfrac{1}{2}[k_1 + k_2] \tag{9.21a}$$

* This plane is called the *state space* or *phase plane* for this state-two system.

where

$$k_1 = hf(x_n, t_n) \tag{9.21b}$$

$$k_2 = hf(x_n + k_1, t_n + h) \tag{9.21c}$$

In qualitative terms, the idea is to estimate an average value for the first derivative of x over the interval $[t_{n+1} - t_n]$. Since this derivative is not known initially at t_{n+1}, the value needed for the average is approximated by the Euler procedure.

3) *Third-Order Runge–Kutta Method*

$$x_{n+1} = x_n + \tfrac{1}{6}[l_1 + 4l_2 + l_3] \tag{9.22a}$$

where

$$l_1 = hf(x_n, t_n) \tag{9.22b}$$

$$l_2 = hf(x_n + \tfrac{1}{2}l_1, t_n + \tfrac{1}{2}h) \tag{9.22c}$$

$$l_3 = hf(x_n + 2l_2 - l_1, t_n + h) \tag{9.22d}$$

4) *Fourth-Order Runge–Kutta Method*

$$x_{n+1} = x_n + \tfrac{1}{6}[m_1 + 2m_2 + 2m_3 + m_4] \tag{9.23a}$$

where

$$m_1 = hf(x_n, t_n) \tag{9.23b}$$

$$m_2 = hf(x_n + \tfrac{1}{2}m_1, t_n + \tfrac{1}{2}h) \tag{9.23c}$$

$$m_3 = hf(x_n + \tfrac{1}{2}m_2, t_n + \tfrac{1}{2}h) \tag{9.23d}$$

$$m_4 = hf(x_n + m_3, t_n + h) \tag{9.23e}$$

Equations (9.23) are very often used in practice. They have a systematic, simple form and require no special arrangements to initiate the computation at $t = t_1$. Moreover, the discretization error is quite attractive, having an absolute value $O(h^5)$ per step when all quantities are suitably nondimensionalized.

The derivation of (9.21, 23) and similar formulas can be approached methodically through series expansions of the exact relation

$$x_{n+1} = x_n + \int_{t_n}^{t_n+h} f(x(\tau), \tau)\, d\tau \equiv x_n + F(h) \tag{9.24}$$

Here τ is a dummy time variable. Again we work with the scalar reduction of (9.11) and omit explicit reference to the control vector, reminding you of the ease with which you can generalize our development. We adopt the notation

$$f_n \equiv f(x_n, t_n) \tag{9.25a}$$

$$f_{nx} \equiv \left.\frac{\partial f}{\partial x}\right|_{x_n, t_n} \tag{9.25b}$$

$$f_{nt} \equiv \left.\frac{\partial f}{\partial t}\right|_{x_n, t_n} \tag{9.25c}$$

[When the state vector is multi-dimensional, (9.25a) and (9.25c) are replaced by column vectors, whereas (9.25b) becomes an $n \times n$ matrix, each of whose columns contains the partial derivatives of the set of functions f with respect to one of the state variables.]

Our approach is to compare two different approximate representations of the function $F(h)$ in (9.24). The first of these is straightforward Maclaurin expansion about the point $h = 0$, where $x = x_n$, $t = t_n$.

$$F(h) = F(0) + hF'(0) + \frac{h^2}{2!}F''(0) + \cdots = 0 + hF'(0) + \frac{h^2}{2!}F''(0) + O(h^3) \quad (9.26)$$

The lead term here vanishes, of course, because $h = 0$ reduces the integration interval in (9.24) to zero. By Leibnitz' rule for differentiating a definite integral,

$$\frac{dF}{dh} \equiv \frac{d}{dh}\int_{t_n}^{t_n+h} f(x(\tau), \tau)\, d\tau = f(x(t_n + h), t_n + h) \quad (9.27a)$$

whence, with (9.25a),

$$F'(0) = f_n \quad (9.27b)$$

Another differentiation of (9.27a) gives

$$\frac{d^2F}{dh^2} = \frac{\partial f}{\partial x}\frac{\partial x(t_n + h)}{\partial h} + \frac{\partial f}{\partial t}\frac{\partial(t_n + h)}{\partial h} \quad (9.28)$$

However,

$$\frac{\partial x(t_n + h)}{\partial h} = \dot{x}(t_n + h) = f(t_n + h) \quad (9.29a)$$

and

$$\frac{\partial(t_n + h)}{\partial h} = 1 \quad (9.29b)$$

since t_n is a fixed constant. Combining (9.29) with (9.28) and letting $h \to 0$, we obtain

$$F''(0) = f_{nx}f_n + f_{nt} \quad (9.30)$$

The $F(h)$ expansion therefore reads

$$F(h) = hf_n + \frac{h^2}{2!}[f_{nx}f_n + f_{nt}] + O(h^3) \quad (9.31)$$

We remark that (9.24) and (9.31) together furnish an acceptable numerical integration formula, with absolute error $O(h^3)$, but that using it requires computing first derivatives of the function f. In order to avoid this considerable inconvenience, we introduce a second way of estimating the change in x over the interval. The idea is to determine four numerical constants $\alpha, \beta, \gamma, \delta$ in such a manner that the formula

$$x_{n+1} \cong x_n + \alpha hf_n + \beta hf(x_n + \delta hf_n, t_n + \gamma h) \quad (9.32)$$

is, in some sense, a "best" approximation. In words, is there a way of averaging the first derivative of x so as to improve efficiently on Euler's specification of this derivative at the beginning of the interval?

Now (9.32) can also be subjected to a Maclaurin expansion in h, whereby we calculate

$$x_{n+1} \cong x_n + \alpha h f_n + \beta h\left\{f(x_n, t_n) + \delta h f_n \left[\frac{\partial f}{\partial x}\right]\Bigg|_{x_n,t_n} + \gamma h \left[\frac{\partial f}{\partial t}\right]\Bigg|_{x_n,t_n} + O(h^2)\right\}$$

$$= x_n + h[\alpha + \beta] f_n + h^2[\beta\delta f_n f_{nx} + \beta\gamma f_{nt}] + O(h^3) \tag{9.33}$$

Although by no means unique, an obvious means of fixing the constants is to require that (9.33) and (9.31) agree up through $O(h^2)$. Thus we obtain the following three equations:

$$\alpha + \beta = 1 \tag{9.34a}$$

$$\beta\delta = \tfrac{1}{2} \tag{9.34b}$$

$$\beta\gamma = \tfrac{1}{2} \tag{9.34c}$$

Suppose that we arbitrarily pick $\gamma = 1$, so that the last two terms in (9.32) just span the time interval. Equations (9.34) then yield $\delta = 1$ and $\alpha = \beta = \frac{1}{2}$. The resulting explicit form of (9.32) reads

$$x_{n+1} \cong x_n + \tfrac{1}{2}h[f_n + f(x_n + hf_n, t_n + h)] \tag{9.35}$$

with error $O(h^3)$. Equation (9.35) is identical with the second-order Runge–Kutta method (9.21).

The higher-order Runge–Kutta formulas can be developed, in a parallel fashion, by going to more terms in (9.25, 31) and suitably generalizing (9.32). Since it continues to be necessary to choose certain constants arbitrarily when matching the two estimates of $[x_{n+1} - x_n]$, there is no formal way of proving "optimality" for the methods so derived. They do, however, involve discretization errors of increasing orders in h, and they turn out to be as satisfactory in practice as any competing one-step schemes.

Reference is made to Chapter 2 of Henrici (1962) for comprehensive information on the one-step approach. There are, incidentally, alternative names for (9.21) through (9.23) and a variety of others carried to the same degree in h. Henrici's discussion of errors is particularly thorough and illuminating.

The generic term *multi-step* identifies integration methods which incorporate data from times before t_n during the projection from this instant to t_{n+1}. When applied to a problem like (9.11, 12), any such scheme must fall back on a single-step process to get started from $t = t_1$, since usually one has no knowledge of prior states. The fourth-order Runge–Kutta, (9.23), is often selected for this purpose.

Inasmuch as the whole subject is in a rapid state of evolution, we would be foolhardy either to attempt a listing of multi-step schemes or to point out one as superior. Accordingly, we confine ourselves to summarizing two which have been very helpful in our own research. They have the parochial advantage of forming the basis for an efficient program available at the Stanford University Computation Center. There is a message: The "best" method for any particular engineering analysis is frequently the most accurate and reliable which is already in the library of the local computer.

We generalize the notation (9.25a) by using $f_{n-1}, f_{n-2}, \ldots$ to abbreviate the first derivative of x, previously determined at times t_{n-1}, t_{n-2}, etc. Again we employ the scalar reduction of (9.11) for simplicity. In these terms, our first example is the four-step Adams–Bashforth method (Henrici 1962, pages 192–194), which reads

$$x_{n+1} = x_n + \frac{h}{24}[55f_n - 59f_{n-1} + 37f_{n-2} - 9f_{n-3}] \tag{9.36}$$

After one performs single-step computations up to $t = t_4$, (9.36) is capable of carrying the solution onward to a remarkably high t_F in an accurate and stable manner. Its convergence as $h \to 0$ to the unique solution can be proved under the same conditions as Euler's, the absolute error being $O(h^5)$.

From the standpoint of the precision obtainable for a given number of arithmetic operations, however, (9.36) can be improved on by associating it with an Adams–Moulton procedure into a so-called predictor-corrector scheme. The isolated Adams–Moulton formula resembles (9.36), except that the coefficients are different and $f_{n+1} \equiv f(x_{n+1}, t_{n+1})$ appears in the brackets on the right. It is therefore "implicit," since the value of x_{n+1} would have to be found by some sort of auxiliary calculation. But if a first estimate $P_{n+1} \cong x_{n+1}$ is "predicted" from (9.36) and then used to provide f_{n+1} in a "correction" step, the implicit aspect of Adams–Moulton can be avoided. We are thus led to a method which, in its four-step version, proceeds as follows.

$$P_{n+1} = x_n + \frac{h}{24}[55f_n - 59f_{n-1} + 37f_{n-2} - 9f_{n-3}] \tag{9.37a}$$

$$x_{n+1} \equiv C_{n+1} = x_n + \frac{h}{24}[9f(P_{n+1}, t_{n+1}) + 19f_n - 5f_{n-1} + f_{n-2}] \tag{9.37b}$$

One by-product of the predictor-corrector approach is that the absolute error during any time step is related to the difference between the prediction and the correction. In the case of Adams–Bashforth–Moulton, Henrici (1962) gives

$$(\text{Absolute error})_n \cong \frac{19}{270}|C_{n+1} - P_{n+1}| \tag{9.38}$$

The order of magnitude is generally h^5.

The concept underlying these and most other multi-step methods calls for evaluating the integral in (9.24) by passing an interpolating polynomial through a certain number of the known values of f prior to the time t_n or t_{n+1}. The name

Lagrange polynomial interpolation is given to this curve-fitting process when the points at which the polynomial is required to fit the data are spaced equally in the independent variable. Both Henrici (1962) and Hildebrand (1956) furnish many details, and we therefore omit here the somewhat lengthy development.

As an example, one can derive (9.36) or (9.37a) by passing a third-degree polynomial in t through the four points $f(t_n), f(t_{n-1}), f(t_{n-2}), f(t_{n-3})$. The result is unique, because such a polynomial contains exactly four constants, which are found by solution of the four linear, simultaneous equations that enforce the fitting. With t replaced by τ, the polynomial is inserted into (9.24). Elementary integrations then yield the desired formula. This procedure is properly described as *extrapolation*, because an approximation computed over the interval t_{n-3} to t_n is employed for integration from t_n to t_{n+1}. When f is a reasonably well-behaved function and h is sufficiently small, however, excellent accuracy is obtained. You will have no trouble comprehending the generalization to N-step methods, which is accomplished by fitting an $(N - 1)$st-degree polynomial to the preceding N values of f, or in seeing how implicit methods like Adams–Moulton follow from shifting the range of interpolation by one step to the right on the time scale.

9.3 SIMPLIFIED TREATMENT OF BOOST FROM A NONROTATING PLANET

We seek further insight into the performance of boosters by starting at the level of approximation embodied in (9.7). Until recently the liquid and solid rockets used for missiles and space operation had no provisions for throttling. During passage upward through the atmosphere, considerations from Section 5.3 show that we should still expect some gradual increase of T and I_{sp} with altitude, but it is often acceptable to employ averaged values for these quantities over the trajectory of a particular stage. With T, I_{sp} assumed constants, we see from (9.7e) that $\dot{m} < 0$ is constant also. Hence we can eliminate vehicle mass $m(t)$ from the state vector by writing

$$m = m_0 + \int_0^t \dot{m}\, dt = m_0\left[1 + \frac{\dot{m}}{m_0} t\right] = m_0\left[1 - \frac{T}{m_0 I_{sp}} t\right] \tag{9.39}$$

where m_0 is mass at time $t = 0$ when the engines are started. Any transient buildup of thrust is presumed to take place very rapidly. In actuality, many first stages are clamped to the launch pad until T becomes high enough to permit a a safe liftoff; m_0 would then be the liftoff mass.

In the present case, the ratio T/m appearing in (9.7a,b) can be expressed as

$$\boxed{\frac{T}{m} = \frac{T}{m_0\left[1 + \dfrac{\dot{m}}{m_0} t\right]} = \frac{g_e(T/W_0)}{1 - \dfrac{g_e T}{I_{sp} W_0} t}} \tag{9.40}$$

This quantity might be termed the *inertial acceleration*, for it represents that rate of increase of booster speed that would occur in field-free space. Figure 9.4 shows the hyperbolic growth of T/m with time and identifies a typical instant at which thrust might be cut off because of exhaustion of the propellant. The approach to infinite T/m at a time $t_\omega = I_{sp}W_0/g_eT$ corresponds to the physical impossibility of a vehicle composed entirely of propellant, whose mass is all ejected through its own engines and ultimately drops to zero.

The ratio T/W_0 of thrust to initial weight is an important characteristic of any stage. For the first stages of boosters employed to date in the manned space program, its value has ranged from 1.25 to 1.3. This is no accident, since T/W_0 has a lower limit set by the need for adequate $\dot{v}_c$ just after liftoff, but yet the acceleration at the end of the burn, when m has dropped to a small fraction of m_0, must not be excessive. The apparent acceleration perceived by the crew just before engine cutoff has been restricted to 4–$5g_e$. On the Saturn V, this first-stage peak was reduced by shutting down the center engine several seconds before the other four.

We remark further that a second parameter appearing in (9.40) is I_{sp}/g_e. This ratio is precisely the specific impulse in "seconds" discussed after (5.16), and is the principal figure of merit for the propulsion system.

Continuing the simplification process in our quest for understanding, we next adopt the approximations for (9.7) that $\varphi_T \cong 0$, $L \cong 0$, and the drag term in (9.7a) may be neglected. The last of these is certainly questionable, and we shall return to it for *a posteriori* justification. The first two, on the other hand, are reasonable, in the light of how booster control systems operate. Typically, an idealized angular-rate controller acts on measurements of the three components of absolute angular velocity of the vehicle. It calls for moments to be exerted about the CM by rotating the thrust line relative to the longitudinal axis. In the absence of disturbances and on a nonrotating planet, the angles of roll and yaw would be held very close to fixed initial values, in an effort to maintain the pitch

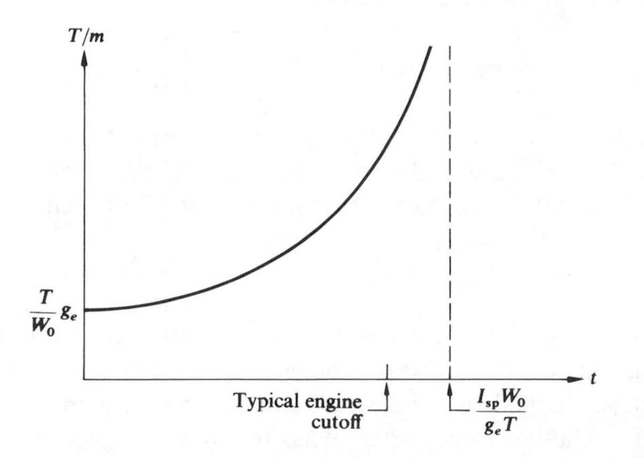

Fig. 9.4. Variation of T/m for a rocket engine with constant specific impulse.

plane coincident with the plane of the trajectory. A pre-programmed, slow rate of pitching would be commanded, so as to bring γ from 90° at liftoff over to 0° at the instant of injection when h and v_c reach the correct values (e.g., at perigee) for the desired orbit.

Any perturbations to the angles of attack or sideslip occurring during flight in the lower atmosphere would immediately be sensed by lateral accelerometers and driven to zero, because the structure is able to withstand only very limited lift or sideforce. Such disturbances arise mainly from the rate of change of the horizontal wind component with altitude ("wind shear"). When these shears are strong, the vehicle is typically permitted a temporary small drift away from its pre-programmed course so as to minimize the associated airloads.

The foregoing considerations add up to the conclusions that, during most of the time along nearly every boost trajectory, the relative wind is practically aligned with the longitudinal axis and the thrust vector passes almost through the CM. The vanishing of lift and of φ_T (the angle between T and $\mathbf{v}_c$) is particularly valid when assumed, on the average, over many seconds of flight.

Under the circumstances thus defined, we may adopt (9.40) and reduce the three most important of the motion equations (9.7) to the following forms:

$$\dot{v}_c = \frac{g_e(T/W_0)}{1 - \dfrac{g_e T}{I_{sp} W_0} t} - \frac{k \sin \gamma}{r^2} \tag{9.41a}$$

$$\dot{\gamma} = \frac{v_c \cos \gamma}{r} - \frac{k \cos \gamma}{v_c r^2} \tag{9.41b}$$

$$\dot{r} \equiv \dot{h} = v_c \sin \gamma \tag{9.41c}$$

If the position above the planetary surface needs to be calculated, (9.7d) still applies. Alternatively, we may define a distance d, measured over the spherical surface from the point of launch, such that

$$\dot{d} = \frac{R_e}{r} v_c \cos \gamma = \left(\frac{R_e}{R_e + h}\right) v_c \cos \gamma \tag{9.42}$$

We recognize the first term on the right of (9.41b) as the rate of change of the angle ν in Fig. 9.1. Equation (9.41b) reveals an interesting property of such trajectories, namely, that the acceleration normal to the flight path [cf. (9.4b)] is

$$a_N \equiv v_c[\dot{\gamma} - \dot{\nu}] = -\frac{k \cos \gamma}{r^2} = -g \cos \gamma \tag{9.43}$$

a_N is evidently a consequence only of the component of weight in the N-direction. During such a "gravity turn," the booster is allowed to fall over slowly toward the horizontal, while its thrust is devoted entirely to producing acceleration along the flight path. The net force experienced by crew and payload therefore turns out to be parallel to $\mathbf{v}_c$ and to the longitudinal axis.

When initial conditions are prescribed for integrating (9.41), it is convenient to choose an instant slightly after liftoff, when v_c is already finite and a denominator zero in (9.41b) is avoided. For a vertical sounding rocket we would make $\gamma(0)$ exactly 90°, and this value would hold all along the trajectory, since $\dot{\gamma} = 0$ throughout. For boost to orbit, however, the gravity turn would have to be started with an initial $\gamma < 90°$. For example, some experimentation with (9.41) has yielded the following results, which you may wish to reproduce: After setting $T/W_0 = 1.3$, $I_{sp} = 300g_e$ ft/sec, $R_e = 20.89 \times 10^6$ ft, $v_c(0) = 100$ ft/sec, and $\gamma(0) = 89.866°$, one can achieve quite a satisfactory single-stage boost to a circular near-earth orbit by integration over about 220 sec. In this connection, we observe that a circular orbit is characterized by $\gamma = 0$; the speed can be found from (9.41b) as $\sqrt{gr}$. Unfortunately, the final weight turns out to be less than 5% of W_0. Not only is this beyond present technological capability for building such a vehicle, but the constant-thrust acceleration just before injection would be excessive for many payloads, including man.

Somewhat more realism can be added to this sort of calculation by estimating a drag term $(-D/m)$ for the right-hand side of (9.41a). Its relationship to other elements of the state vector is developed by writing

$$-\frac{D}{m} = -\frac{\left[C_D(\mathrm{M})S\,\dfrac{\rho_\infty(h)}{2}\,v_c^2\right]}{m_0\left[1 - \dfrac{g_e T}{I_{sp}W_0}\,t\right]} \tag{9.44a}$$

Here Mach number $\mathrm{M} = v_c/a_\infty$ can be computed from the dependence of speed of sound on altitude h. Similarly, a density-altitude law is required; for our purposes, the curve fit of the lower atmosphere suggested by Chapman (1958)

$$\rho_\infty(h) \cong \rho_0 e^{-h/H} \tag{9.44b}$$

(for earth, $\rho_0 = 0.0027$ slugs/ft^3 and the "scale height" $H = 23{,}500$ ft) might be a suitable choice. More information will be furnished in Chapter 10 about drag coefficients and common definitions of the reference area S.

Let us conclude this preliminary treatment by making what is known as a "Δv budget" to determine the significance of various influences on booster performance. This is done by study of terms in the time integral of (9.41a), to which we add the drag. Some rearrangement yields

$$\frac{T}{m} = \dot{v}_c + \frac{k \sin \gamma}{r^2} + \frac{D}{m} \tag{9.45}$$

Dealing with these one by one, let us carry out the following integrations from $v_c(0) = 0$ to $t = t_{\text{Orbit}}$ at injection.

1) *Thrust and Characteristic Velocity*

The assumption of constant specific impulse is retained.

$$\int_0^{t_{\text{Orbit}}} \frac{T}{m}\,dt = -I_{\text{sp}}\int_0^{t_{\text{Orbit}}} \frac{\dot{m}\,dt}{m} = I_{\text{sp}}\int_{m_{\text{Orbit}}}^{m_0} \frac{dm}{m}$$

$$= I_{\text{sp}} \ln \frac{m_0}{m_F} \equiv \Delta v_{\text{Char.}} \tag{9.46}$$

The quantity thus calculated [cf. the range formula (8.28)] is called *characteristic velocity*. An integral of the inertial acceleration, it measures the inherent booster capability in terms of the velocity increment that could be achieved in field-free space.

2) *Actual Change in Speed*

$$\int_0^{t_{\text{Orbit}}} \dot{v}_c\,dt = v_{\text{Orbit}} \tag{9.47}$$

If we were looking just at one stage of a multistage vehicle, this integral would, of course, represent the actual change of speed during that segment of the trajectory.

3) *Effect of Gravity*

The integrated value of $(k \sin \gamma/r^2)$ depends on flight-path details. Both an upper limit and an order-of-magnitude estimate can be obtained, however, by imagining that the boost is vertical under constant gravity $g \cong g_e$.

$$\Delta v_{\text{Gravity}} = \int_0^{t_{\text{Orbit}}} \frac{k \sin \gamma}{r^2}\,dt < \int_0^{t_{\text{Orbit}}} g_e\,dt = g_e t_{\text{Orbit}} \tag{9.48}$$

4) *Loss Due to Drag*

$$\Delta v_{\text{Drag loss}} = \int_0^{t_{\text{Orbit}}} \frac{D}{m}\,dt = g_e \int_0^{t_{\text{Orbit}}} \frac{(D/W_0)\,dt}{\left[1 - \dfrac{g_e T}{I_{\text{sp}} W_0} t\right]} \tag{9.49}$$

can be estimated only roughly, unless complete information is available on the trajectory, atmospheric properties, and configuration. The important fact is that, for ballistic boosters, it is a remarkably small number.

The "budget" consists of adding the last three terms above to equal the first. Thus we can see how the capability represented by $\Delta v_{\text{Char.}}$ is divided among the tasks of creating kinetic energy of orbital speed, increasing potential energy of height in the gravity field, and overcoming air resistance. The Δv-equation is written, along with numerical estimates below each term that might be appropriate for vertical launch and single-stage boost to a near-earth orbit,

$$\begin{array}{ccccccc} \Delta v_{\text{Char.}} & = & v_{\text{Orbit}} & + & \Delta v_{\text{Gravity}} & + & \Delta v_{\text{Drag loss}} \\ (30\text{–}34 \times 10^3 & & (25 \times 10^3 & & (6\text{–}8 \times 10^3 & & (200 \text{ ft/sec}) \\ \text{ft/sec}) & & \text{ft/sec}) & & \text{ft/sec}) & & \end{array} \tag{9.50}$$

The results are presented in this way to reemphasize the fact that the effect of drag on performance is almost negligible. There are many other ways, however, of judging the efficiency of the operation. For instance, a much less rosy picture emerges if we examine the orbital energy (i.e., kinetic plus potential referred to a state of rest at the planet's surface). For the relatively small mass which remains after injection of the vehicle into orbit, this energy turns out to be a tiny fraction of the total contained in the heating value of all propellants consumed during the flight.

The modest loss due to drag is one of two major attractions ascribed to ballistic boost along an initially vertical trajectory, the other being the low fraction of structural weight attainable in a vehicle without lifting surfaces which is always flown close to zero angle of attack. Boosters of the space-shuttle type must pay a high structural cost for the wings that enable them to maneuver during reentry and cruise in the atmosphere, and the first-generation shuttle is still going to be launched vertically in order to minimize loads and $\Delta v_{\text{Drag loss}}$. The outstanding advantage accrues from routine reuse of the expensive orbiter stage. It remains to be seen whether we shall ever arrive at the often-heralded ideal of airplanes which operate directly to orbit from conventional runways. Not only will they resemble cruisers in having structures that weigh 20–30% of takeoff gross weight, but their long times of flight at lower altitudes will cause the integral (9.49) to absorb a much greater portion of the available characteristic velocity.

9.4 AN ELEMENTARY LOOK AT STAGING

We have already mentioned how multiple staging can be accounted for in the calculation of boost trajectories. Our aim in this section is to clarify the selection and sizing of stages by recourse to a simplified approach that goes back to early theoretical research on space travel. In essence, the ideas were published by Malina and Summerfield (1947).

Consider an n-stage booster with payload weight P and the distribution of individual gross weights W_i illustrated in Fig. 9.5. Let us ask how these stages should be proportioned in order to maximize performance. Dimensionless notation is introduced by using P as the reference weight and defining

$$w_i \equiv \frac{W_i}{P} \tag{9.51}$$

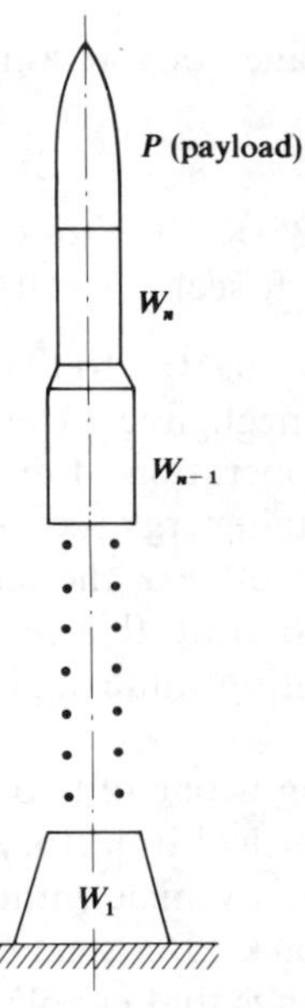

Fig. 9.5. An idealized n-stage booster, carrying a payload P. The individual weights of fully fueled stages are designated W_i.

The total launch weight is thus Ps_1, where

$$s_1 = 1 + \sum_{j=1}^{n} w_j \tag{9.52}$$

The subscript on s decreases as we proceed down the "stack,"

$$s_i = 1 + \sum_{j=i}^{n} w_j \tag{9.53}$$

being the dimensionless fueled weight of everything from the top through stage i.

Let us describe some important quantities which characterize each stage. The efficiency with which the structure, propulsion system, controls, etc., are designed is measured by a parameter $\beta_i < 1$, which is defined for the isolated ith stage by

$$\beta_i \equiv \left[\frac{\text{Gross weight} - \text{Weight of usable propellant}}{\text{Gross weight}}\right]\Bigg|_{i\text{th stage only}} \tag{9.54}$$

By combining (9.53) and (9.54), we can write

$$\mu_i \equiv \left[\frac{\text{Total stack weight before } i\text{th burn}}{\begin{array}{c}\text{Total stack weight after } i\text{th burn}\\ \text{but before } i\text{th stage separation}\end{array}}\right]$$

$$= \frac{s_{i+1} + w_i}{s_{i+1} + \beta_i w_i} \tag{9.55}$$

The significance of μ_i is that, by the same reasoning that led to (9.46), we know the characteristic velocity $\Delta v_{\text{Char.}}$ for the ith burn to be

$$\Delta v_i = I_{\text{sp}_i} \ln \mu_i \tag{9.56}$$

Here I_{sp_i} is the specific impulse (averaged exhaust speed) of the ith-stage propulsion system.

Experience has shown that the cost of a given class of aerospace vehicles is roughly proportional to gross weight. This suggests that we should seek a minimum value of s_1, while holding fixed both P and a suitable index of the performance required from the booster. Since drag losses and gravity effects are not substantially influenced by moderate changes in stage proportions, it is logical to specify a constant overall characteristic velocity

$$\boxed{v_f = \sum_{i=1}^{n} \Delta v_i = \sum_{i=1}^{n} I_{\text{sp}_i} \ln \mu_i} \tag{9.57}$$

Available technology also constrains both I_{sp_i} and β_i. Some representative target values for the former are listed in Table 5.1. The latter have fallen in a range somewhat above 10% for the majority of liquid-fueled stages of recent large boosters. The S-II stage of Saturn V is cited as an especially successful design, because its insulation was placed outside the aluminum alloy from which the cryogenic oxygen and hydrogen tanks were constructed. The increased strength of this alloy at low temperature made it possible for a β_2 below 10% to be achieved.

With v_f, β_i and I_{sp_i} prescribed, we seek that set of s_i (or, more precisely, μ_i) that minimizes s_1. To this end, a convenient expression for s_1 can be obtained from the identity

$$s_1 \equiv \frac{s_1}{s_2} \cdot \frac{s_2}{s_3} \cdots \frac{s_{n-1}}{s_n} s_n = \prod_{i=1}^{n} \frac{s_i}{s_{i+1}} \tag{9.58}$$

Here the $\prod$ (capital π) is a notation, analogous to $\sum$, often used to abbreviate the product of a series of indexed quantities; $s_{n+1} = 1$, since this refers to the dimensionless payload.

To compute the ratios in (9.58), we first solve (9.55) for

$$s_{i+1} = \frac{w_i[1 - \beta_i \mu_i]}{\mu_i - 1} \tag{9.59}$$

We add w_i to (9.59) and get

$$s_i = s_{i+1} + w_i = \frac{w_i \mu_i [1 - \beta_i]}{\mu_i - 1} \tag{9.60}$$

When we divide (9.60) by (9.59), w_i then cancels:

$$\frac{s_i}{s_{i+1}} = \frac{\mu_i[1 - \beta_i]}{1 - \beta_i \mu_i} \tag{9.61}$$

Substitution into (9.58) yields

$$\boxed{s_1 = \prod_{i=1}^{n} \frac{\mu_i[1 - \beta_i]}{1 - \beta_i\mu_i}} \tag{9.62}$$

We are faced with a classical problem of algebraic or parameter optimization. n scalars μ_i are to be determined so as to attain a minimum of s_1 or (more conveniently) of

$$\ln s_1 = \sum_{i=1}^{n} \{\ln \mu_i - \ln [1 - \beta_i\mu_i] + \ln [1 - \beta_i]\} \tag{9.63}$$

while the sum in the right-hand member of (9.57) is constrained to be fixed. In the absence of this "constraint," it is well known that n equations among the optimal μ_i would result from the necessary requirement that all partial derivatives of $\ln s_1$ with respect to μ_i must vanish. Additionally, the μ_i so calculated are sufficient for a minimum if the matrix of second partial derivatives $\partial^2(\ln s_1)/\partial u_i\, \partial u_j$ is positive semidefinite (see, e.g., Bryson and Ho 1969, Section 1.1).

The most efficient way of imposing the constraint during minimization is to "adjoin" it to $\ln s_1$ by means of a *Lagrange multiplier* λ, as discussed in Section 10.2 of Halfman (1962) or Section 1.2 of Bryson and Ho (1969). Specifically, if we rewrite (9.57) by introducing the function

$$\varphi(\mu_i) \equiv \sum_{i=1}^{n} I_{\text{sp}_i} \ln \mu_i - v_f = 0 \tag{9.64}$$

it is required that the partials with respect to μ_i of the quantity

$$\begin{aligned} F &\equiv \ln s_1 + \lambda\varphi(\mu_i) \\ &= \sum_{i=1}^{n} \{[1 + \lambda I_{\text{sp}_i}] \ln \mu_i - \ln [1 - \beta_i\mu_i] + \ln [1 - \beta_i] - \lambda v_f\} \end{aligned} \tag{9.65}$$

all vanish at the minimum. Thus we obtain

$$\boxed{\frac{\partial F}{\partial \mu_i} = \frac{1 + \lambda I_{\text{sp}_i}}{\mu_i} + \frac{\beta_i}{1 - \mu_i\beta_i} = 0, \qquad \text{for } i = 1, 2, \ldots, n} \tag{9.66}$$

We can solve (9.66) explicitly,

$$\mu_i = \frac{1 + \lambda I_{\text{sp}_i}}{\lambda\beta_i I_{\text{sp}_i}} \tag{9.67}$$

One additional relation for the elimination of λ in terms of the known v_f is provided by (9.57) and (9.67),

$$v_f = \sum_{i=1}^{n} I_{\text{sp}_i} \ln \left[\frac{1 + \lambda I_{\text{sp}_i}}{\lambda\beta_i I_{\text{sp}_i}}\right] \tag{9.68}$$

It is illuminating to examine special cases of the solution (9.67, 68). Let us consider two which are qualitatively applicable to practical systems.

1) *Specific Impulse and Weight Fraction the Same for Each Stage*

If we replace I_{sp_i} and β_i, respectively, with constant overall numbers I_{sp} and β, (9.67) shows μ to be invariant from one stage to another.

$$\mu_i \equiv \mu = \frac{1 + \lambda I_{\mathrm{sp}}}{\lambda \beta I_{\mathrm{sp}}} \tag{9.69}$$

From (9.61), the ratio of total weights above successive stages is also constant,

$$\frac{s_i}{s_{i+1}} = \frac{\mu[1 - \beta]}{1 - \mu\beta} \tag{9.70}$$

whence these ratios proceed in geometric series. (9.57) gives us

$$v_f = nI_{\mathrm{sp}} \ln \mu \tag{9.71a}$$

from which we compute the explicit relation

$$\mu = \exp\left[v_f/nI_{\mathrm{sp}}\right] \tag{9.71b}$$

for substitution into (9.70).

For example, a typical three-stage boost to earth orbit might have $v_f \cong 3I_{\mathrm{sp}}$. Therefore $\mu \cong e = 2.718$. With a 10% weight fraction, we get

$$\frac{s_i}{s_{i+1}} \cong \frac{e[1 - 0.1]}{1 - 0.1e} = 3.36 \tag{9.72}$$

A comparison with the Saturn V–Apollo vehicle is perhaps not quite fair, because the first-stage propellants have much lower I_{sp} than the others and the third stage is not exhausted at orbital injection. Nevertheless, it is interesting that rough approximations to the first two ratios are (with the weights in pounds)

$$\frac{s_1}{s_2} \cong \frac{6.2 \times 10^6}{[6.2 - 4.7] \times 10^6} \cong 4.1 \tag{9.73a}$$

and

$$\frac{s_2}{s_3} \cong \frac{1.5 \times 10^6}{[1.5 - 1.03] \times 10^6} \cong 3.2 \tag{9.73b}$$

2) *Two Stages Only*

A suitable procedure in this case is to set $i = 1$ and 2 in (9.67), solve each for λ, and equate

$$\lambda^{-1} = I_{\mathrm{sp}_1}[1 - \beta_1\mu_1] = I_{\mathrm{sp}_2}[1 - \beta_2\mu_2] \tag{9.74}$$

This linear relation between the two μ_i's can be solved simultaneously with

$$v_f = I_{\mathrm{sp}_1} \ln \mu_1 + I_{\mathrm{sp}_2} \ln \mu_2 \tag{9.75}$$

Numerical examples and comparisons with data on existing systems are suggested as an exercise for the reader.

9.5 EQUATIONS OF BOOST FROM A ROTATING PLANET

Our last topic brings us closer to the situation facing the operators of an actual space-transportation system. We shall try to indicate ways of accounting for such phenomena as a steady angular velocity of the planet and small deviations from perfectly spherical geopotential surfaces. Even at that, it will be necessary to specialize our analysis in certain ways, and you will begin to appreciate how complex is the task of precisely applying orbital mechanics in the presence of all significant sources of disturbance.

The principal goal of an investigation such as this is to generate reference trajectories with initial conditions fixed by the place, time, and direction of launch and final conditions corresponding to the desired orbit. Only when such trajectories are known can guidance laws be developed which will cause a vehicle to follow them during actual flights. After we have put the equations of motion into a suitable form, a brief indication will be furnished of the trial-and-correction process involved in trajectory generation.

We assume that a nonrotating reference frame (x_I, y_I, z_I) at the planet's center is inertial. In the absence of the influences of satellites like the moon,* the error in so doing is $O(10^{-8})$ at earth's distance from the sun, as can be worked out by an analysis similar to that on pages 58–60 of Halfman (1962). Let us also suppose that liftoff occurs at $t = 0$ from a site on the equator, and that the x-axes of two planet-centered Cartesian frames both pass through this site at this instant. As illustrated in the two parts of Fig. 9.6, we employ altogether four coordinate systems when proceeding from (x_I, y_I, z_I) to one connected with the moving vehicle. They are as follows.

1) *Inertial System* (x_I, y_I, z_I)

Origin at the planetary center; z_I upward through the north pole and parallel to the (constant) absolute angular velocity

$$\boldsymbol{\omega}^{E-I} = \omega_E \mathbf{k}_I \tag{9.76}$$

of the planet. x_I outward through the equatorial launch site at $t = 0$.

2) *Planet-Fixed Central System* (x_E, y_E, z_E)

Origin at the planetary center; $z_E = z_I$ upward through the north pole; x_E outward through the equatorial launch site so that $x_E = x_I$ at $t = 0$. The angles Λ and λ are, respectively, east longitude and north latitude measured from the

* When the moon is accounted for in earth-based trajectories, the proper inertial frame from which to start the analysis is a nonrotating one, with origin at the CM of the earth–moon system.

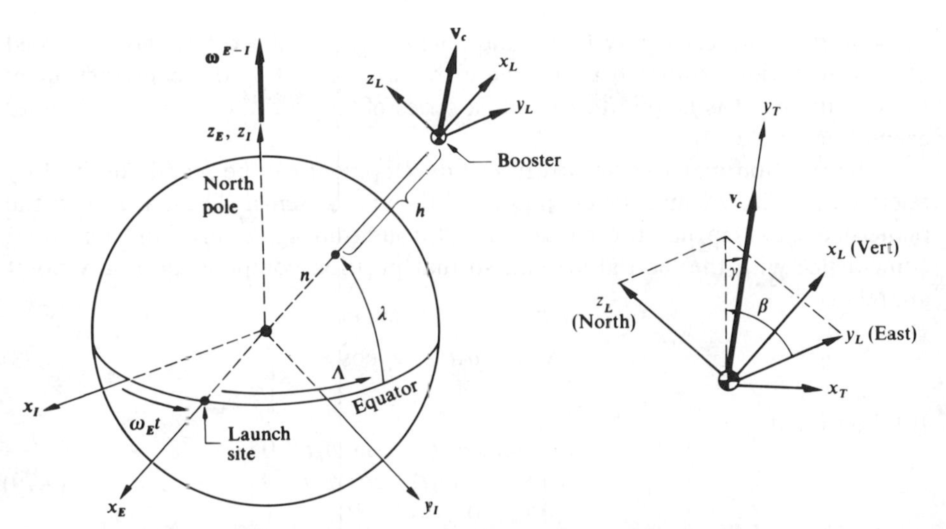

Fig. 9.6. The four coordinate systems involved in analyzing boost from a rotating planet. The sketch on the right shows the instantaneous relative velocity vector and the angles β, γ which locate it in the locally level system.

launch site. These angles are used to specify the direction of radius vector $\mathbf{r}$ from the planetary center to the booster CM.

3) *Locally Level System* (x_L, y_L, z_L)

Origin at the booster CM; x_L along the local radial direction, positive vertically upward. y_L and z_L point, respectively, toward local east and generally northward in the plane containing x_L and the axis $z_E = z_I$ of planetary rotation.

4) *Trajectory-Tangent System* (x_T, y_T, z_T)

Origin at the booster CM; y_T parallel to the booster velocity vector $\mathbf{v}_c$ (observed from the rotating planet). x_T normal to y_T in the locally vertical plane containing x_L and x_T. The direction of $\mathbf{v}_c$ in the locally level system is determined by an azimuth angle β, measured positively about x_L from the y_L-direction, and an elevation angle γ measured upward from the local horizontal. Note that γ is equivalent to the flight-path angle of Section 9.1.

We shall find it necessary repeatedly to relate the components of various vectors, as projected on the coordinate directions of the four systems. When so doing, the concept of rotation matrix introduced in Section 2.4 proves valuable, as abbreviated by the equation*

$$\{r_2\} = [(\theta_{\text{rot}})_{21}]\{r_1\} \tag{9.77}$$

Here the two column matrices might contain, for instance, the xyz-components of

---

* Specialists in automatic control and orbital mechanics usually omit the brackets and write rotations as boldface letters, with subscripts to identify the axes involved. For obvious reasons, they often replace "cos" and "sin" by just "C" and "S." Although tempted, we eschew these practices in our introductory presentation.

the same position vector in systems 2 and 1, whereas the 3×3 rotation is composed of nine direction cosines between these two systems. For the construction of these matrices, it is helpful to cascade a series of single rotations in the manner exemplified by (2.32).

The rotation matrices needed in our development can be readily derived by reference to Fig. 9.6 and the definitions of the axis systems. Thus, at time t, the planet-fixed system has turned eastward through an angle $\omega_E t$ from its initial coincidence with the inertial system, so that position components of any point are related by

$$\left.\begin{aligned} x_E &= x_I \cos \omega_E t + y_I \sin \omega_E t \\ y_E &= -x_I \sin \omega_E t + y_I \cos \omega_E t \\ z_E &= z_I \end{aligned}\right\} \tag{9.78}$$

It follows that

$$[(\theta_{\text{rot}})_{EI}] = \begin{bmatrix} \cos \omega_E t & \sin \omega_E t & 0 \\ (-\sin \omega_E t) & \cos \omega_E t & 0 \\ 0 & 0 & 1 \end{bmatrix} \tag{9.79}$$

We might use (9.79) as our model for any single rotation. Observe that the principal diagonal will always consist of two cosines of the relevant angle and a "one," the latter being in the location corresponding to the axis about which rotation occurs. This "one" is accompanied by zeros off the diagonal in its row and column. The remaining elements are sines of the angle; one of these has a minus sign, whose position depends on whether the rotation is counterclockwise (i.e., positive, as here) or clockwise.

To get to the locally level from the planet-fixed directions, we can first specify a positive rotation Λ about z_E. This brings y into the correct orientation. Hence a positive rotation λ about this intermediate y-direction completes the job. In matrix notation,

$$\begin{aligned} [(\theta_{\text{rot}})_{LE}] &= [(\theta_{\text{rot}})_{\lambda \text{ about } y}][(\theta_{\text{rot}})_{\Lambda \text{ about } z}] \\ &= \begin{bmatrix} \cos \lambda & 0 & \sin \lambda \\ 0 & 1 & 0 \\ (-\sin \lambda) & 0 & \cos \lambda \end{bmatrix} \begin{bmatrix} \cos \Lambda & \sin \Lambda & 0 \\ (-\sin \Lambda) & \cos \Lambda & 0 \\ 0 & 0 & 1 \end{bmatrix} \\ &= \begin{bmatrix} \cos \lambda \cos \Lambda & \cos \lambda \sin \Lambda & \sin \lambda \\ (-\sin \Lambda) & \cos \Lambda & 0 \\ (-\sin \lambda \cos \Lambda) & (-\sin \lambda \sin \Lambda) & \cos \lambda \end{bmatrix} \end{aligned} \tag{9.80}$$

Similarly,

$$\begin{aligned} [(\theta_{\text{rot}})_{TL}] &= [(\theta_{\text{rot}})_{-\gamma \text{ about } z}][(\theta_{\text{rot}})_{\beta \text{ about } x}] \\ &= \begin{bmatrix} \cos \gamma & (-\sin \gamma) & 0 \\ \sin \gamma & \cos \gamma & 0 \\ 0 & 0 & 1 \end{bmatrix} \begin{bmatrix} 1 & 0 & 0 \\ 0 & \cos \beta & \sin \beta \\ 0 & (-\sin \beta) & \cos \beta \end{bmatrix} \\ &= \begin{bmatrix} \cos \gamma & (-\sin \gamma \cos \beta) & (-\sin \gamma \sin \beta) \\ \sin \gamma & \cos \gamma \cos \beta & \cos \gamma \sin \beta \\ 0 & (-\sin \beta) & \cos \beta \end{bmatrix} \end{aligned} \tag{9.81}$$

We remark that the reason for orienting y_T along the trajectory is that our choice for azimuth and elevation rotations naturally carries y_L into this direction. To make x_T parallel to $\mathbf{v}_c$ would be in better accord with practice in the dynamic stability field and elsewhere; such a change can be accomplished by some trivial substitutions after the equations have been derived.

Turning to vehicle kinematics, let us adopt the symbols $\mathbf{u}_c$ and $\mathbf{v}_c$ for the CM velocity vector as seen, respectively, by an inertial observer and one who partakes of the planetary rotation. They are related by the Coriolis law (2.10),

$$\mathbf{u}_c = \mathbf{v}_c + \boldsymbol{\omega}^{E-I} \times \mathbf{r} \tag{9.82}$$

Equation (2.10) introduced the useful operations $d(\ldots)/dt$ and $\delta(\ldots)/\delta t$, which refer to the rates of change of a vector perceived by the two observers. In those terms, the two velocities may be written and componentized as follows:

$$\mathbf{u}_c = \frac{d\mathbf{r}}{dt} = \dot{x}_I \mathbf{i}_I + \dot{y}_I \mathbf{j}_I + \dot{z}_I \mathbf{k}_I \tag{9.83}$$

$$\mathbf{v}_c = \frac{\delta_E \mathbf{r}}{\delta_E t} = \dot{x}_E \mathbf{i}_E + \dot{y}_E \mathbf{j}_E + \dot{z}_E \mathbf{k}_E \tag{9.84}$$

(Observe that no ambiguity results from using the "dot" for derivatives of components. Such derivatives represent rates of change of scalars, which are the projections onto a uniquely specified set of directions of a unique vector.)

The fundamental vector equation of motion for the booster CM is clearly

$$\frac{d^2\mathbf{r}}{dt^2} \equiv \frac{d\mathbf{u}_c}{dt} = \frac{\mathbf{F}}{m} \tag{9.85}$$

The resultant external force $\mathbf{F}$ will later be divided into contributions from weight, thrust, lift, and drag.

Our objective is to transform (9.85) so as to express the time derivatives of v_c, γ, and β—quantities associated with the components of $\delta_L \mathbf{v}_c/\delta_L t$ in the trajectory-tangent system. Using appropriate subscripts on $\delta(\ldots)/\delta t$ to indicate which rotating observer is intended, and superscripts like $(E - I)$ in (9.76) to clarify which relative angular velocity is meant, we examine the vector transformation of (9.85). Thus the inertial derivative of (9.82), with $\boldsymbol{\omega}^{E-I}$ constant, is

$$\frac{d\mathbf{u}_c}{dt} = \frac{d\mathbf{v}_c}{dt} + \boldsymbol{\omega}^{E-I} \times \frac{d\mathbf{r}}{dt} = \frac{d\mathbf{v}_c}{dt} + \boldsymbol{\omega}^{E-I} \times \mathbf{u}_c \tag{9.86}$$

The Coriolis law takes us from the inertial to the locally level system, giving

$$\frac{d\mathbf{v}_c}{dt} = \frac{\delta_L \mathbf{v}_c}{\delta_L t} + \boldsymbol{\omega}^{L-I} \times \mathbf{v}_c \tag{9.87}$$

Substituting (9.87) and (9.82) into (9.86), we obtain

$$\frac{d\mathbf{u}_c}{dt} = \frac{\delta_L \mathbf{v}_c}{\delta_L t} + \boldsymbol{\omega}^{L-I} \times \mathbf{v}_c + \boldsymbol{\omega}^{E-I} \times \mathbf{v}_c + \boldsymbol{\omega}^{E-I} \times [\boldsymbol{\omega}^{E-I} \times \mathbf{r}]$$

$$= \frac{\delta_L \mathbf{v}_c}{\delta_L t} + [\boldsymbol{\omega}^{L-E} + 2\boldsymbol{\omega}^{E-I}] \times \mathbf{v}_c + \boldsymbol{\omega}^{E-I} \times [\boldsymbol{\omega}^{E-I} \times \mathbf{r}] \tag{9.88}$$

In combination with (9.85), (9.88) yields

$$\frac{\delta_L \mathbf{v}_c}{\delta_L t} = -[\boldsymbol{\omega}^{L-E} + 2\boldsymbol{\omega}^{E-I}] \times \mathbf{v}_c - \boldsymbol{\omega}^{E-I} \times [\boldsymbol{\omega}^{E-I} \times \mathbf{r}] + \frac{\mathbf{F}}{m} \tag{9.89}$$

We are ultimately going to take components of (9.89) in the T-system. This process involves working successively on the various terms. As a start, observe that the components of $\mathbf{v}_c$ in the L-system can be written in the following two ways (see Fig. 9.7):

$$\{\mathbf{v}_c\}_L = \begin{Bmatrix} v_c \sin \gamma \\ v_c \cos \gamma \cos \beta \\ v_c \cos \gamma \sin \beta \end{Bmatrix} = \begin{Bmatrix} \dot{r} \\ \dot{\Lambda} r \cos \lambda \\ \dot{\lambda} r \end{Bmatrix} \tag{9.90}$$

(9.90) furnishes expressions, which will later prove useful, for the rates of change of radial distance, longitude, and latitude,

$$\dot{r} = v_c \sin \gamma \tag{9.91a}$$

$$\dot{\Lambda} = \frac{v_c \cos \gamma \cos \beta}{r \cos \lambda} \tag{9.91b}$$

$$\dot{\lambda} = \frac{v_c \cos \gamma \sin \beta}{r} \tag{9.91c}$$

Next consider the double cross product in (9.89), which we componentize in the L-system and then transform by means of (9.81). $\boldsymbol{\omega}^{E-I} \times \mathbf{r}$, according to its definition, must be a vector pointing in the positive y_L-direction with magnitude $\omega_E r \sin [(\pi/2) - \lambda] \equiv \omega_E r \cos \lambda$,

$$\{\boldsymbol{\omega}^{E-I} \times \mathbf{r}\}_L = \begin{Bmatrix} 0 \\ \omega_E r \cos \lambda \\ 0 \end{Bmatrix} \tag{9.92}$$

Moreover, this vector is already normal to $\boldsymbol{\omega}^{E-I}$, so that the double product has a magnitude $\omega_E^2 r \cos \lambda$ and lies in the $x_L z_L$-plane with components

$$\{\boldsymbol{\omega}^{E-I} \times [\boldsymbol{\omega}^{E-I} \times \mathbf{r}]\}_L = \omega_E^2 r \begin{Bmatrix} -\cos^2 \lambda \\ 0 \\ \sin \lambda \cos \lambda \end{Bmatrix} \tag{9.93}$$

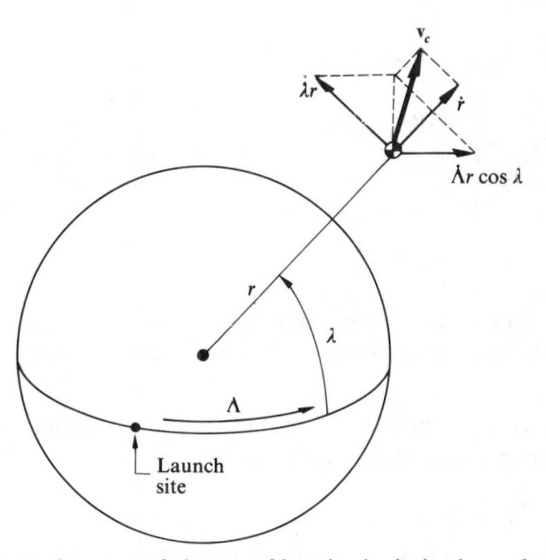

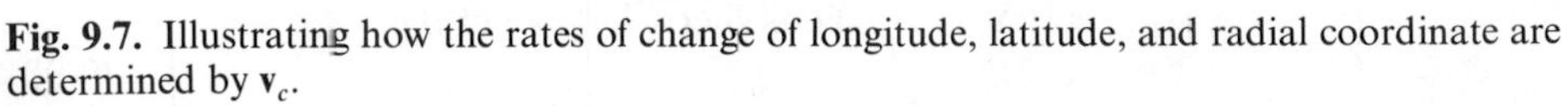
Fig. 9.7. Illustrating how the rates of change of longitude, latitude, and radial coordinate are determined by $\mathbf{v}_c$.

Application of (9.81) yields

$$\{\boldsymbol{\omega}^{E-I} \times [\boldsymbol{\omega}^{E-I} \times \mathbf{r}]\}_T = [(\theta_{\text{rot}})_{TL}]\{\boldsymbol{\omega}^{E-I} \times [\boldsymbol{\omega}^{E-I} \times \mathbf{r}]\}_L$$

$$= \omega_E^2 r \begin{Bmatrix} -\cos\gamma\cos^2\lambda - \sin\gamma\sin\beta\sin\lambda\cos\lambda \\ -\sin\gamma\cos^2\lambda + \cos\gamma\sin\beta\sin\lambda\cos\lambda \\ \cos\beta\sin\lambda\cos\lambda \end{Bmatrix} \tag{9.94}$$

In order to componentize the first right-hand term of (9.89) into the T-system, refer to Fig. 9.8 for the following:

$$\{\boldsymbol{\omega}^{L-E} + 2\boldsymbol{\omega}^{E-I}\}_L = \begin{Bmatrix} \dot{\Lambda}\sin\lambda \\ -\dot{\lambda} \\ \dot{\Lambda}\cos\lambda \end{Bmatrix} + 2\begin{Bmatrix} \omega_E \sin\lambda \\ 0 \\ \omega_E\cos\lambda \end{Bmatrix} = \begin{Bmatrix} [\dot{\Lambda} + 2\omega_E]\sin\lambda \\ -\dot{\lambda} \\ [\dot{\Lambda} + 2\omega_E]\cos\lambda \end{Bmatrix} \tag{9.95}$$

After another application of (9.81), we obtain

$$\{\boldsymbol{\omega}^{L-E} + 2\boldsymbol{\omega}^{E-I}\}_T = [(\theta_{\text{rot}})_{TL}] \begin{Bmatrix} [\dot{\Lambda} + 2\omega_E]\sin\lambda \\ -\dot{\lambda} \\ [\dot{\Lambda} + 2\omega_E]\cos\lambda \end{Bmatrix}$$

$$= \begin{Bmatrix} [\dot{\Lambda} - 2\omega_E][\cos\gamma\sin\lambda - \sin\gamma\sin\beta\cos\lambda] + \dot{\lambda}\sin\gamma\cos\beta \\ [\dot{\Lambda} - 2\omega_E][\sin\gamma\sin\lambda + \cos\gamma\sin\beta\cos\lambda] - \dot{\lambda}\cos\gamma\cos\beta \\ [\dot{\Lambda} + 2\omega_E]\cos\beta\cos\lambda + \dot{\lambda}\sin\beta \end{Bmatrix} \tag{9.96}$$

For the purpose of crossing this vector into

$$\{\mathbf{v}_c\}_T = \begin{Bmatrix} 0 \\ v_c \\ 0 \end{Bmatrix} \tag{9.97}$$

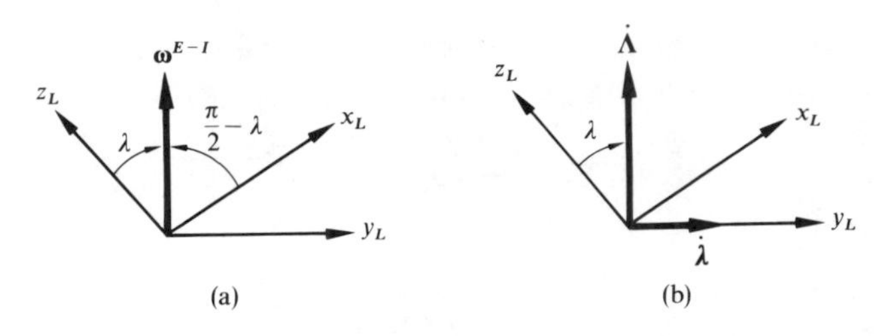

Fig. 9.8. Illustrating how the planetary rotation vector (a) and the rates of change of longitude and latitude (b) would be componentized in the locally level system.

we employ (2.14) as a systematic formula for the components. After some algebraic manipulation, we can make the result read

$$\{[\boldsymbol{\omega}^{L-E} + 2\boldsymbol{\omega}^{E-I}] \times \mathbf{v}_c\}_T = \begin{Bmatrix} -v_c[(\dot{\Lambda} + 2\omega_E)\cos\beta\cos\lambda + \dot{\lambda}\sin\beta] \\ 0 \\ v_c[(\dot{\Lambda} + 2\omega_E)(\cos\gamma\sin\lambda - \sin\gamma\sin\beta\cos\lambda) + \dot{\lambda}\sin\gamma\cos\beta] \end{Bmatrix} \tag{9.98}$$

The left-hand side of (9.89) requires another use of the Coriolis law,

$$\frac{\delta_L \mathbf{v}_c}{\delta_L t} = \frac{\delta_T \mathbf{v}_c}{\delta_T t} - \boldsymbol{\omega}^{T-L} \times \mathbf{v}_c \tag{9.99}$$

It is clear that

$$\{\mathbf{v}_c\}_T = \begin{Bmatrix} 0 \\ v_c \\ 0 \end{Bmatrix}, \qquad \left\{\frac{\delta_T \mathbf{v}_c}{\delta_T t}\right\}_T = \begin{Bmatrix} 0 \\ \dot{v}_c \\ 0 \end{Bmatrix} \tag{9.100a,b}$$

Moreover, $\boldsymbol{\omega}^{T-L}$ represents the effect of the angular rates $\dot{\gamma}$ and $\dot{\beta}$, whence

$$\{\boldsymbol{\omega}^{T-L}\}_L = \begin{Bmatrix} \dot{\beta} \\ \dot{\gamma}\sin\beta \\ -\dot{\gamma}\cos\beta \end{Bmatrix} \tag{9.101}$$

and

$$\{\boldsymbol{\omega}^{T-L}\}_T = [(\theta_{\text{rot}})_{TL}]\{\boldsymbol{\omega}^{T-L}\}_L = \begin{Bmatrix} \dot{\beta}\cos\gamma \\ \dot{\beta}\sin\gamma \\ -\dot{\gamma} \end{Bmatrix} \tag{9.102}$$

[Equation (9.102) is readily confirmed by direct examination.] Equations (9.100) and (9.102) can be substituted into (9.99), along with another use of (2.14), to produce

$$\left\{\frac{\delta_L \mathbf{v}_c}{\delta_L t}\right\}_T = \begin{Bmatrix} v_c\dot{\gamma} \\ \dot{v}_c \\ -v_c\dot{\beta}\cos\gamma \end{Bmatrix} \tag{9.103}$$

which is also fairly obvious physically.

The components of $\mathbf{F}/m$ must be studied in more detail. The four contributions to this force mentioned following (9.85) are temporarily characterized by vectors as follows:

$$\frac{\mathbf{F}}{m} = \mathbf{g} + \frac{\mathbf{T}}{m} + \frac{\mathbf{L}}{m} + \frac{\mathbf{D}}{m} \tag{9.104}$$

If the gravitational acceleration $\mathbf{g}$ were to be based on the spherical-planet assumption of Sections 9.1 and 9.3, its T-components would evidently be $(-k\cos\gamma/r^2)$, $(-k\sin\gamma/r^2)$, and 0. No celestial object is a perfect sphere, however, so we adjust these by adding small corrections e_x, e_y, e_z. The latter are functions of r, Λ, λ, γ, β which can be calculated from the gradient of the planet's potential per unit mass; for instance, information about the earth might be taken from a source such as Kaula (1966). Hence,

$$\{\mathbf{g}\}_T = \begin{Bmatrix} \dfrac{-k\cos\gamma}{r^2} + e_x \\[2ex] \dfrac{-k\sin\gamma}{r^2} + e_y \\[2ex] e_z \end{Bmatrix} \tag{9.105}$$

Thrust, lift, and drag will here be written in the general forms

$$\left\{\frac{\mathbf{T}}{m}\right\}_T = \begin{Bmatrix} T_x/m \\ T_y/m \\ T_z/m \end{Bmatrix} \qquad \left\{\frac{\mathbf{L}}{m}\right\}_T = \begin{Bmatrix} L_x/m \\ 0 \\ L_z/m \end{Bmatrix} \qquad \left\{\frac{\mathbf{D}}{m}\right\}_T = \begin{Bmatrix} 0 \\ -D/m \\ 0 \end{Bmatrix} \tag{9.106a,b,c}$$

These will be discussed further after the final equations of motion have been derived.

When we substitute (9.94), (9.98), (9.103), (9.105), and (9.106) into (9.89), we obtain a penultimate set of relations among v_c, γ, β, and other state variables.

$$\begin{Bmatrix} v_c\dot{\gamma} \\ \dot{v}_c \\ v_c\dot{\beta}\cos\gamma \end{Bmatrix} = v_c \begin{Bmatrix} [(\dot{\Lambda} + 2\omega_E)\cos\beta\cos\lambda + \dot{\lambda}\sin\beta] \\ 0 \\ -[(\dot{\Lambda} + 2\omega_E)(\cos\gamma\sin\lambda - \sin\gamma\sin\beta\cos\lambda) + \dot{\lambda}\sin\gamma\cos\beta] \end{Bmatrix}$$
$$+ \omega_E^2 r\cos\lambda \begin{Bmatrix} \cos\gamma\cos\lambda + \sin\gamma\sin\beta\sin\lambda \\ \sin\gamma\cos\lambda - \cos\gamma\sin\beta\sin\lambda \\ -\cos\beta\sin\lambda \end{Bmatrix}$$
$$+ \begin{Bmatrix} \dfrac{-k\cos\gamma}{r^2} + e_x \\[2ex] \dfrac{-k\sin\gamma}{r^2} + e_y \\[2ex] e_z \end{Bmatrix} + \begin{Bmatrix} T_x/m \\ T_y/m \\ T_z/m \end{Bmatrix} + \begin{Bmatrix} L_x/m \\ 0 \\ L_z/m \end{Bmatrix} + \begin{Bmatrix} 0 \\ -D/m \\ 0 \end{Bmatrix} \tag{9.107}$$

In order to place (9.107) in state-vector form, we still need to eliminate $\dot{\Lambda}$ and $\dot{\lambda}$ from the right. This can be accomplished by appropriate additions of the second and third members of (9.91), with multiplicative factors, as follows:

$$\dot{\Lambda}\cos\beta\cos\lambda + \dot{\lambda}\sin\beta = \frac{v_c\cos\gamma}{r}\left[\frac{\cos^2\beta\cos\lambda}{\cos\lambda} + \sin^2\beta\right] = \frac{v_c\cos\gamma}{r} \tag{9.108a}$$

Similarly, after some manipulation,

$$-\dot{\Lambda}(\cos\gamma\sin\lambda - \sin\gamma\sin\beta\cos\lambda)\dot{\lambda}\sin\gamma\cos\beta = -\left(\frac{v_c\cos\gamma}{r}\right)\frac{\cos\gamma\cos\beta\sin\lambda}{\cos\lambda} \tag{9.108b}$$

We insert (9.108) into (9.107), then divide the first and third members of the latter, respectively, by v_c and $v_c\cos\gamma$. Explicit formulas for the first time derivatives are the result; we rearrange the order to conform with (9.7).

$$\dot{v}_c = \omega_E^2 r\cos\lambda[\sin\gamma\cos\lambda - \cos\gamma\sin\beta\sin\lambda] - \frac{k\sin\gamma}{r^2} + e_y + \frac{T_y}{m} - \frac{D}{m} \tag{9.109a}$$

$$\dot{\gamma} = \frac{v_c\cos\gamma}{r} + 2\omega_E\cos\beta\cos\lambda + \frac{\omega_E^2 r\cos\lambda}{v_c} \times [\cos\gamma\cos\lambda + \sin\gamma\sin\beta\sin\lambda] - \frac{k\cos\gamma}{v_c r^2} + \frac{e_x}{v_c} + \frac{T_x}{mv_c} + \frac{L_x}{mv_c} \tag{9.109b}$$

$$\dot{\beta} = -\left(\frac{v_c\cos\gamma}{r}\right)\cos\beta\tan\lambda - 2\omega_E[\sin\lambda - \tan\gamma\sin\beta\cos\lambda] - \frac{\omega_E^2 r\cos\beta\sin\lambda\cos\lambda}{v_c\cos\gamma} + \frac{e_z}{v_c\cos\gamma} + \frac{T_z + L_z}{mv_c\cos\gamma} \tag{9.109c}$$

There is some justification for regarding (9.109) and (9.91) as determinate state-vector equations for a six-state system with the variables

$$\{\mathscr{X}\} \equiv \begin{Bmatrix} v_c \\ \gamma \\ \beta \\ r \\ \Lambda \\ \lambda \end{Bmatrix} \tag{9.110}$$

although a seventh component governed by something equivalent to (9.7e) would have to be added if the magnitude of the thrust were a controlled quantity. [These can, of course, be brought into agreement with (9.7) by a series of simplifications, including spherical earth, $\omega_E = 0$, $\beta = 0$, etc.]

Let us assume, however, that $m(t)$ is known from a relation resembling (9.39) plus data on the initial weight of each stage. Given size and shape of the vehicle, one can closely estimate D from v_c and $\rho_\infty(r) \equiv \rho_\infty(h)$. For the same reasons as in the case of the nonrotating planet, the two components of lift might well be assigned zero averaged values. Were they not small enough to permit this approximation, vehicle rotational dynamics and aerodynamic angles of attack and sideslip would have to be brought into the picture, at the expense of enlarging the state vector and adding complexity which would impair the basic understanding we seek.

The foregoing considerations leave direction of thrust as the primary control mechanism during analyses based on (9.109)–(9.91). In such cases, establishing a typical trajectory might proceed generally as follows. Following liftoff, there would be a short period of stabilized vertical flight (say 20 sec), during which all ground obstructions would be cleared and the booster would be rolled about its longitudinal axis to bring the pitch plane into coincidence with the desired inertial plane for both the boost flight path and the orbit. Because $\gamma = 90°$ and $\cos\gamma = 0$, (9.109c) would be singular during the vertical segment. This is no problem, however, since the initial value of β need not enter until tipping over begins. The required equation of motion is (9.109a), which for the equatorial launch ($\lambda = 0$) might be simplified to

$$\dot{v}_c = \omega_E^2 R_e - g_e + \frac{T_y}{m} - \frac{D}{m} \tag{9.111}$$

The important terms on the right of (9.111) are obviously the second and third, since there is little drag at low speed and $\omega_E^2 R_e$ is just the centrifugal correction that gives rise to an apparent reduction in gravity at the equator.

Next a control law takes over which causes the trajectory gradually to tilt away from the vertical, but to remain in a plane, fixed in inertial space, passing through the axis x_I and inclined at the desired initial azimuth angle. In the absence of throttling, the components of the control vector $\{\mathscr{U}\}$ may be thought of as two small angles φ, ψ (Fig. 9.9) defining the direction of $\mathbf{T}$ relative to velocity $\mathbf{v}_c$ in the trajectory-tangent coordinate system.

$$\begin{Bmatrix} T_x \\ T_y \\ T_z \end{Bmatrix} = \begin{Bmatrix} T\sin\varphi \\ T\cos\varphi\cos\psi \\ T\cos\varphi\sin\psi \end{Bmatrix} \tag{9.112}$$

Note that, unless there are large wind-shear disturbances, $\mathbf{v}_c$ is controlled to be along the booster axis, whence φ, ψ are determined by the gimbal angles of the

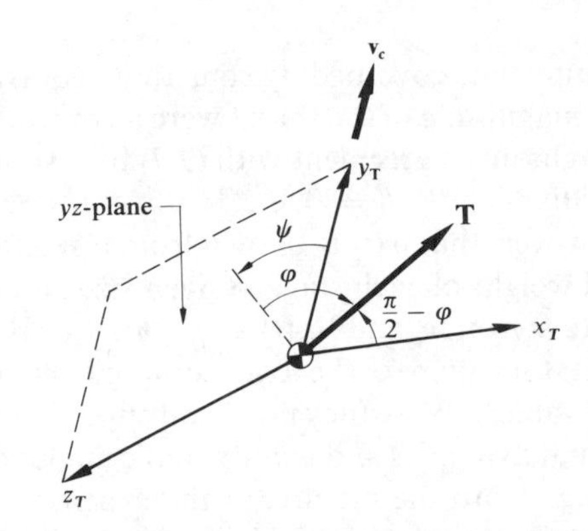

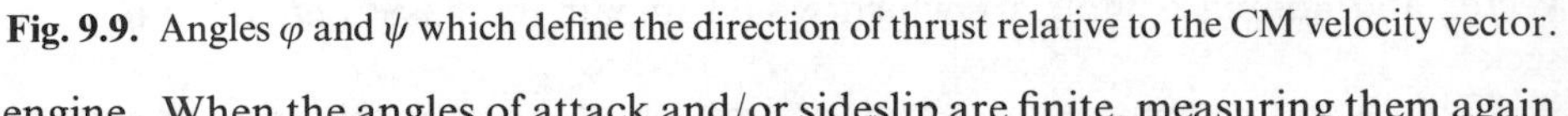
Fig. 9.9. Angles φ and ψ which define the direction of thrust relative to the CM velocity vector.

engine. When the angles of attack and/or sideslip are finite, measuring them again enables us to compute φ, ψ.

Often φ, ψ are programmed so as to constrain to zero the absolute angular rate out of the pitch plane and to keep the pitch rate $[\dot{\gamma} - v_c \cos \gamma/r]$ at a constant value over each of a series of segments. With no guidance errors or aerodynamic disturbances, **T** remains in the chosen inertial plane of the orbit. If the "pitch rate segments" have been properly selected, final-stage burnout occurs at the prescribed orbital speed, altitude, and flight-path angle. For example, a circular orbit calls for $\gamma = 0$ and $u_c = \sqrt{gr_0}$ when $r = r_0$ is attained.

We round out our treatment of the subject with a few more words on the generation of reference trajectories and control laws. We can gain some generality and save space by working with the state vector equation. Recall that it is

$$\{\dot{\mathscr{X}}\} = \{f(\{\mathscr{X}\}, \{\mathscr{U}\}; t)\} \tag{2.45}$$

The foregoing systems are all examples of (2.45).

A preliminary calculation must always be made with (2.45), whose result is a trajectory corresponding to a first estimate of the control law $\{\mathscr{U}(t)\}$ and of initial conditions

$$\{\mathscr{X}(0)\} \equiv \{\mathscr{X}_0\} \tag{9.113}$$

The time history of the state vector along this trajectory might be written

$$\{\mathscr{X}(t)\} = \{\Phi_{\mathscr{U}}(t; \{\mathscr{X}_0\})\} \tag{9.114}$$

with subscript $\mathscr{U}$ and argument $\{\mathscr{X}_0\}$ indicating the dependence of Φ on the manner in which these quantities are chosen. At the time $t = t_F$ of final-stage burnout, which we presume known, the conditions $\{\Phi_{\mathscr{U}}(t_F;\{\mathscr{X}_0\})\}$ will not generally correspond to the desired state at orbital injection. Provided they are close enough, however, there are several systematic processes for adjusting $\{\mathscr{U}\}$ and/or $\{\mathscr{X}_0\}$ so as to produce the necessary corrections.

Perhaps the simplest approach, which is conceptually very straightforward when only $\{\mathscr{X}_0\}$ are to be adjusted, is to calculate the "sensitivities" of the final conditions $\{\mathscr{X}(t_F)\}$ to these initial conditions. Let us employ the "variation" symbol δ to denote the small change in any quantity of interest; for example, $\{\delta\mathscr{X}_0\}$ would be an $n \times 1$ column of arbitrary adjustments made to the original initial conditions. A "sensitivity matrix" $[S]$ can then be defined by the equation

$$\{\delta\mathscr{X}(t_F)\} = [S]\{\delta\mathscr{X}_0\} \tag{9.115}$$

The quantity in the ith row and jth column of this matrix represents the amount by which $x_i(t_F)$ would change due to a unit change in the initial condition $x_j(0)$, with all others held fixed. (9.115) is obviously linearized, in the sense that the members of $\{\delta\mathscr{X}_0\}$ must be small enough to permit neglect of higher-order terms on the right.

The columns of $[S]$ might be found by modifying the initial conditions, one at a time, away from those of the preliminary trajectory, then solving (2.45) n times so as numerically to find the corresponding final-condition changes. Given $[S]$, one could in principle use the desired final state to get

$$\delta x_{i_D}(t_F) = x_{i\text{Desired}}(t_F) - \Phi_{\mathscr{U}_i}(t_F; \mathscr{X}_0) \qquad \text{for } i = 1, 2, \ldots, n \tag{9.116}$$

A correction

$$\{\delta\mathscr{X}_0\} = [S]^{-1}\{\delta\mathscr{X}_D(t_F)\} \tag{9.117}$$

to the preliminary initial conditions would then produce the sought-after reference trajectory. Even a small change δt_F in the burnout time could be included, by an obvious modification of the process. Furthermore, (9.117) proves satisfactory in practice for determining such results as the initial conditions for single-stage boost to orbit from a nonrotating earth mentioned following (9.41, 43). Unfortunately, more complicated problems do not prove so tractable, and only simple changes in the control law (e.g., a step in φ or ψ at a prescribed instant) can be readily accounted for.

A commonly used refinement on the direct numerical computation of sensitivities is to apply variational calculus to the control law and the time history of the entire flight. In this connection, symbols like $\{\delta\mathscr{X}(t)\}$ and $\{\delta\mathscr{U}(t)\}$ are understood to denote small functional changes, respectively, to $\{\Phi_{\mathscr{U}}(t; \{\mathscr{X}_0\})\}$ and the preliminary control, defined over the range $0 \le t \le t_F$. The most general $\{\delta\mathscr{X}(t)\}$ may be thought of as due to varied initial conditions, to a varied control, to errors in measurement, and to unanticipated external disturbances. We assume, however, that all these are small enough to make possible linear superposition of their effects, and here we consider only the first two. The independent variable, time t, is not varied; hence any $\delta(\ldots)$ refers to a modification in the value of a function occurring at a particular time.

Symbolically, the varied equation of motion (2.45) reads

$$\{\delta\dot{\mathscr{X}}\} = \left[\frac{\partial f}{\partial \mathscr{X}}\right]\{\delta\mathscr{X}\} + \left[\frac{\partial f}{\partial \mathscr{U}}\right]\{\delta\mathscr{U}\} \tag{9.118}$$

The first matrix on the right is an $n \times n$ array of partials of the n functions f with respect to the n state variables, whereas the second matrix is $n \times m$ and contains similar derivatives with respect to the control-vector elements.

It is important that analytic expressions can often be obtained for these partials. For instance, examine the highly simplified system (9.41), wherein the state vector has been reduced to three variables v_c, γ, and r. In the absence of aerodynamic forces, we have

$$\{f\} = \begin{Bmatrix} \dfrac{g_e \dfrac{T}{W_0}}{1 - \dfrac{g_e T}{I_{\text{sp}} W_0} t} - \dfrac{k \sin \gamma}{r^2} \\ \dfrac{v_c \cos \gamma}{r} - \dfrac{k \cos \gamma}{v_c r^2} \\ v_c \sin \gamma \end{Bmatrix} \tag{9.119}$$

It follows that

$$\frac{\partial f_1}{\partial x_1} = \frac{\partial}{\partial v_c}\left[\frac{g_e \dfrac{T}{W_0}}{1 - \dfrac{g_e T}{I_{\text{sp}} W_0} t} - \frac{k \sin \gamma}{r^2}\right] = 0 \tag{9.120a}$$

$$\frac{\partial f_2}{\partial x_1} = \frac{\partial}{\partial v_c}\left[\frac{v_c \cos \gamma}{r} - \frac{k \cos \gamma}{v_c r^2}\right] = \frac{\cos \gamma}{r} + \frac{k \cos \gamma}{(v_c r)^2} \tag{9.120b}$$

etc., so that the entire construction of $[\partial f/\partial \mathscr{X}]$ is manifest. A control angle $\varphi(t)$ between v_c and the thrust might also be used to generalize (9.119), in which case the elements of $[\partial f/\partial \mathscr{U}]$ would be a column of partials of f_i with respect to φ.

Given these functional "sensitivity matrices" and the preliminary trajectory (9.114), we calculate each member of each matrix *as it depends on t along that trajectory*. In consequence, we can write (9.118) in the form

$$\{\delta\dot{\mathscr{X}}\} = [F(t)]\{\delta\mathscr{X}\} + [D(t)]\{\delta\mathscr{U}\} \tag{9.121}$$

the members of $[F(t)]$ and $[D(t)]$ being known time functions. Procedures for trajectory adjustment and optimization are frequently based on manipulation of "unit solutions" of the system (9.121). These are linear equations with variable coefficients, which fact suggests that each integral be separated into a complementary function plus a particular solution,

$$\{\delta\mathscr{X}(t)\} = \{\delta\mathscr{X}_c(t)\} + \{\delta\mathscr{X}_p(t)\} \tag{9.122}$$

There should, in general, be n independent columns of solutions to the homogeneous system

$$\{\delta\dot{\mathscr{X}}_c\} = [F(t)]\{\delta\mathscr{X}_c\} \tag{9.123}$$

They may be defined, in an especially convenient way for correcting initial conditions, by successively prescribing these conditions in the manner used for the matrix $[S]$. That is, $\{\delta\mathscr{X}_c^1(t)\}$ would correspond to

$$\{\delta\mathscr{X}_c^1(0)\} = \begin{Bmatrix} 1 \\ 0 \\ 0 \\ \cdot \\ \cdot \\ \cdot \\ 0 \end{Bmatrix} \tag{9.124}$$

and so forth. In general, we would calculate a matrix $[\mathscr{Y}(t, 0)]$ of complementary functions from

$$\frac{d}{dt}[\mathscr{Y}(t, 0)] = [F(t)][\mathscr{Y}(t, 0)] \tag{9.125a}$$

with

$$[\mathscr{Y}(0, 0)] = \begin{bmatrix} 1 & 0 & \cdots & 0 \\ 0 & 1 & & \\ \cdot & & \cdot & \\ \cdot & & \cdot & \\ \cdot & & & \cdot \\ 0 & & & 1 \end{bmatrix} \equiv I \tag{9.125b}$$

the $n \times n$ unit matrix.

With $[\mathscr{Y}(t, 0)]$ available, it can be proved that the entire solution to (9.121) is

$$\{\delta\mathscr{X}(t)\} = [\mathscr{Y}(t, 0)]\{\delta\mathscr{X}_0\} + \int_0^t [\mathscr{Y}(t, \tau)][D(\tau)]\{\delta\mathscr{U}(\tau)\}\, d\tau \tag{9.126}$$

Here $[\mathscr{Y}(t, \tau)]$ refers to a case in which the unit initial conditions are applied at time τ rather than 0; it can be determined without difficulty once the elements for $\tau = 0$ are known. Note that (9.115) is a specialization of (9.126) for $t = t_F$ and without alterations in the preliminary control law.

Equation (9.126) is the basis of efficient schemes for trajectory adjustment, a comprehensive description of which may be found in such references as Battin (1964), Chapter 9. For example, if the control is to be corrected so as to cause the final conditions at $t = t_F$ to match desired values, one can proceed as follows: there exists a procedure for "backward integration" from this time by which $\{\delta\mathscr{U}(t)\}$ can be estimated so as to correct the final conditions systematically toward their desired values. Recent developments in the field revolve around the analysis of this situation when random errors of measurement and external disturbances are present to complicate the task. Refer to Bryson (1972) for an account and citations of the definitive literature.

9.6 PROBLEMS

9A Discuss the circumstances under which (9.109), and associated boost equations for a rotating planet, reduce to (9.7) when $\omega_E = 0$.

9B Specialize the equations of motion for boost over a rotating planet to determine the variations with time of latitude and longitude for a satellite in circular orbit over the earth. For a typical case, make a plot of the track of the subsatellite point over the earth's surface. For the *very* special case of a circular orbit over the equator, verify that

$$v_c = \pm gr - \omega_E r$$

9C Derive the Runge–Kutta formula for X_{n+1} to *third order* or *fourth order* in h for general $(n \times 1)$-column vectors X and $f(X, t) = \dot{X}$.

9D For vertical firing of a constant-thrust sounding rocket over a nonrotating planet, determine burnout and maximum altitudes as a function of propellant mass fraction for the case in which drag losses are negligible.

Study the effects of drag in an exponential atmosphere in which the density is approximated by

$$\rho = \rho_0 e^{-h/H}$$

A sounding rocket is fired vertically from the spherical rotating earth ($\gamma = 90°$) with constant thrust from a point on the equator. Show by elementary physical considerations that the equation of motion for the velocity is

$$\dot{v}_c = -g - \frac{I_{sp}\dot{m}}{m} - \frac{D}{m} + (\omega_E)^2 r, \qquad \omega_E = \text{earth's rotational velocity}$$

If D is negligible and changes in r and g can be overlooked, find v_c as a function of $m(t)$ and t. Determine the components of the inertial velocity $\mathbf{u}_c$ as seen in the I-frame.

9E Suppose you wish to launch a sounding rocket under the following conditions:

Total required $\Delta v = 20{,}000$ ft/sec
Total weight on the pad = 3500 lb
$I_{sp} = 14{,}000$ ft/sec, for all stages
Propellant loading fraction: $\dfrac{W_{\text{Propellant}}}{W_{\text{Gross}}} = \dfrac{15}{17}$, for all stages

Compare the maximum payload that can be carried by

a) A single stage
b) Two stages of equal stage gross weight
c) An optimized pair of stages

Estimate the absolute maximum payload that could theoretically be obtained with an "infinite" number of stages.

9F Using the local computer, incorporate the equations of boost on a nonrotating planet into a numerical integration program. By varying the initial velocity and the initial angle at which the booster is fired (γ approximately 90°), place the vehicle described below into an orbit with a velocity of 25,000 ft/sec and an altitude of 100 miles. The following information is all you will need: initial $T/W = 1.2$, initial $W/S = 7000$, $I_{sp} = 260$, coefficient of drag = 0.26, burn time = 151 sec, $\rho_0 = 0.0027$ for an exponential fit to the atmosphere, radius of earth = 20,890,000 ft. [*Note:* Input initial velocity in terms of feet/second (about 100 ft/sec). Input initial gamma in terms of radians (about 1.48 radians).]

9G Study the design of that portion of the control system which keeps an aerodynamically unstable booster pointed along its flight path (the "balancing-broomstick" problem). To simplify, work just in the pitch plane. Assume that the desired pitch angle $\theta_0(t)$ is specified and that the control system is provided with one rate gyro that measures instantaneous $\dot{\theta}$. The constant thrust T can be pointed at an angle $\varphi_T \ll 1$ relative to vehicle centerline in an effort to maintain $\theta = \theta_0$. Study what kind of linear operations (e.g., one or more integrations) have to be applied to the error $\epsilon = \dot{\theta} - \dot{\theta}_0$ in order to get satisfactory performance. As functions of the mass m, moment of inertia I, and CM distance l, see what you can learn about the gain constants in your control system that ensure stability. [*Note*: Concentrate on the equation of motion for the θ-degree of freedom.]

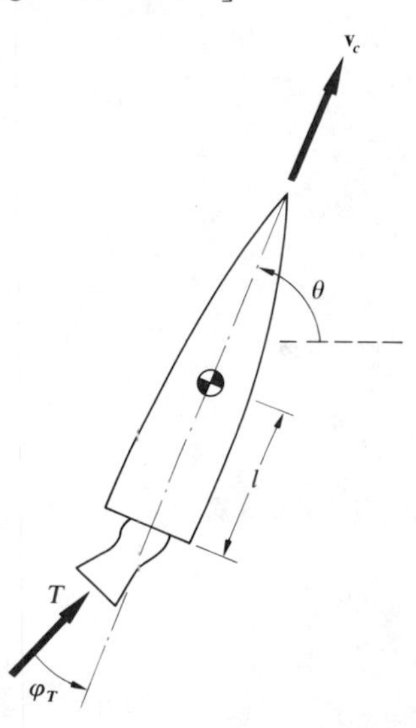

9H Verify the equations (9.10, 17–18) from Etkin (1972) for longitudinal motion of a hypersonic vehicle over the equator of a rotating planet. See if you can check, for the general case rather than a particular numerical example, his interesting result that the period of the phugoid mode of oscillation approaches the period of a circular orbit as orbital speed is approached.

CHAPTER 10

AERODYNAMIC TERMS FOR EQUATIONS OF MOTION; SLENDER AND BLUNT BODIES

10.1 SOME INFORMATION ON BOOSTER AIRLOADS

We have put off discussing the aerodynamics of bodies until the subject arose naturally in connection with the analysis of boost and entry trajectories. Most of this chapter will be devoted to summarizing useful predictive methods, but we shall make the transition from Chapter 9 by introducing in Section 10.1 some facts about flow over pointed, streamlined bodies of revolution.

The definitions of coefficients of drag, lift, and pitching moment given in (3.5) remain applicable, with two modifications. First, the wing area S is commonly replaced, for an axisymmetric shape, by either the maximum cross-sectional area S_{max} or the area S_B of the cut-off base. Second, the characteristic length in M_p must be revised; maximum or base diameter sometimes serves this purpose, but overall length l may also be used. Especially when test data are involved, we cannot caution too strongly about determining which area and length are implicit in numerical values cited for C_D, C_L, C_m, etc.

The axial and normal forces mentioned in Section 3.1 appear frequently among body data. They are typically defined in terms of the following coefficients:

$$A \equiv \frac{\rho_\infty}{2} v_c^2 S_B C_A \tag{10.1a}$$

$$N \equiv \frac{\rho_\infty}{2} v_c^2 S_B C_N \tag{10.1b}$$

A and N are always directed, respectively, downstream along the axis of revolution and perpendicular thereto in the plane of the angle of attack. Since $C_{m0} = 0$, a preferable alternative to plotting moment coefficient is to present the *center of pressure*, CP. This is the point of action of the force N, often quoted as x_{CP}/l, where x is the distance aft along the axis from the nose. Two measures of aerodynamic nonlinearity are (1) changes in the slope of the curve of C_N or C_L versus α_{ZL} and (2) movement of the location of the CP as α_{ZL} is increased.

We have mentioned that the time history of events along the flight path of a ballistic booster is subject to rather tight constraints, and this is especially true for the first stage of a booster which is man-rated. In consequence, the buildup and falloff of dynamic pressure tend to follow a "universal" curve. Figure 10.1 shows a representative variation with Mach number and suggests that the range

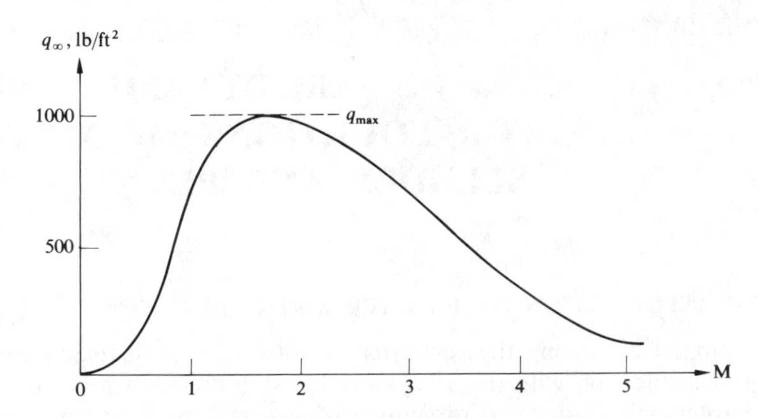

Fig. 10.1. Qualitative variation of flight dynamic pressure with Mach number along a typical boost trajectory.

wherein airloads are of particular importance runs from moderate subsonic to M = 3 or 4. Maximum dynamic pressure is normally encountered at low supersonic speed, but unfortunately the nearby transonic regime concerns the designer even more than q_{max}. Not only are large static loads to be expected here, but close attention must be paid to boundary-layer noise and to violent unsteady effects such as oscillatory interaction between shocks and intermittent separated regions, occurring near corners like those at which the transition is accommodated from one stage to another of different diameter. Because most boosters have large fineness ratio, pointed noses, and limited α_{ZL}, subsonic and supersonic linearized theory may be applied to them with some confidence. However, a crucial exception occurs transonically. Here, as we have stressed repeatedly, careful wind-tunnel testing and flight experience are still the only reliable sources of information.

Let us now examine the various aerodynamic quantities, along with some observational aspects of the flow fields which are associated with them. As for drag, we noted in Chapter 9 its overall insignificance to manned booster performance. Nevertheless, drag during the early portion of first-stage flight must definitely be considered, and it is a far-from-negligible factor in the operation of a sounding rocket or a rapidly accelerated device like an antiballistic missile. $C_D = C_A$ at zero lift is of primary interest, especially since C_A tends, for many configurations, to be almost independent of α_{ZL} in a range between, say $\pm 5°$.

When M < 1, the fluid motion past a multistage vehicle at $\alpha_{ZL} = 0$ involves a smooth pattern of streamlines like that depicted in Fig. 10.2. The boundary-layer thickness is here exaggerated over what it would be at the high Reynolds numbers Re near q_{max}. For the moment, we omit engines and suppose that the flat base is exposed to an average pressure p_B. It is then logical to think of total drag as composed of three parts, as follows:

$$C_D = C_{Df} + C_{Dp} + C_{DB} \tag{10.2}$$

Estimates of these terms can be taken from references like Hoerner (1958) or Nielsen (1960). As a point of departure, subsonic C_D including the base drag C_{DB} falls in a range 0.25–0.30 for many streamlined bodies with $S_{max} = S_B$.

We have previously discussed *friction drag* C_{Df}, which is the downstream resultant of all shear forces experienced by the forebody, and is therefore related to the wetted area and to some averaged Re. A first guess at its magnitude might be to take the drag on the outer surface of a circular cylinder of the same length l and maximum diameter d,

$$C_{Df} \cong 4C_f \frac{l}{d} \tag{10.3}$$

Here C_f would be the skin-friction coefficient averaged over the area of a flat plate or slightly curved shape at the same length-Reynolds number Re_l as the body. For Re_l in the range 10^6–10^7, one familiar formula reads

$$C_f \cong \frac{0.043}{[Re_l]^{1/6}} \tag{10.4}$$

Nielsen (1960, Chapter 9) suggests that, when the maximum boundary layer displacement thickness δ^* is such that $(\delta^*/l) \leq 0.02$, flat-plate results like (10.4) are fairly accurate.

The *pressure drag* C_{Dp}, as for a wing, is due to the summation of axial components of $[p - p_\infty]$ over the forebody area. This tends to be rather small at $M < 1$, 5–20% of the total C_D being a typical range. It is due entirely to displacement effects of the boundary layer, because the same paradox mentioned in Section 4.3 applies to bodies and shows—at least when both ends are pointed†—that the inviscid $C_{Dp} = 0$.

The definition of base drag,

$$C_{DB} = \frac{[p_\infty - p_B]S_B}{\frac{\rho_\infty}{2} v_c^2 S_{ref}} \tag{10.5}$$

† This result is proved for very general shapes by slender-body theory in Nielsen (1960, Section 3-14). Since the disturbance caused by a cut-off base can propagate upstream and violate the small-disturbance hypothesis, the only rigorous theoretical representation at subsonic M is to replace the actual configuration with a "semi-infinite body," extending downstream as a circular cylinder of area S_B to infinity. By considerations like those in a very interesting note of Jones and Van Dyke (1958), we expect the overall C_{Dp} of such a pointed, smooth semi-infinite body to be zero. Some small drag might be predicted, however, on just the forward portion which is supposed to approximate the actual truncated shape.

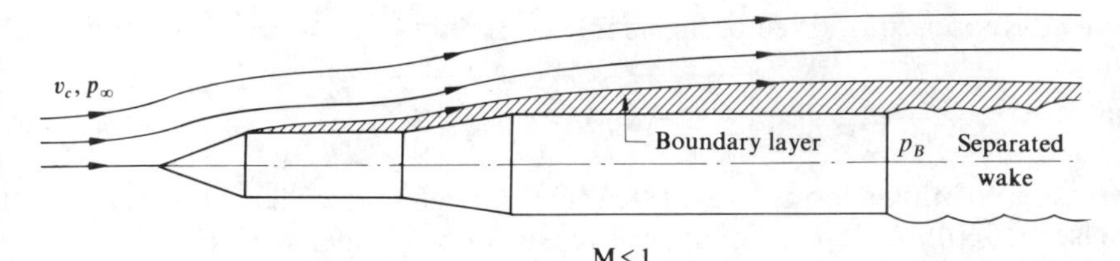

Fig. 10.2. Sketch of streamline pattern and boundary layer evolution on a staged booster in subsonic flight at $\alpha_{ZL} = 0$. Thickness of the boundary layer is exaggerated.

where $S_{ref} = S_{max}$ or S_B, takes account of the practice of subtracting p_∞ over the rest of the surface when calculating pressure drag. Since $p_B < p_\infty$, this quantity is positive; Nielsen (1960) gives many relevant references. A useful first guess for high Re and $S_{ref} = S_B$ is Hoerner (1958), page 3-19.

$$C_{DB} \cong \frac{0.029}{\sqrt{C_{Df}}} \tag{10.6}$$

C_{Df} being taken from (10.3) or some other source. For a booster stage with the engines on, a common practice is to omit C_{DB} and allow for pressure disturbances over the base when quoting net values for thrust. It is vital to be aware, however, that recirculating flow caused by the operation of one or several rockets may produce large fluctuations in pressure and severe heating in the base region.

As a simple illustration, let us apply the foregoing estimates to some data given by Stoney (1958) and reproduced in Fig. 10.3. The reference area is S_{max} for all bodies involved here, but we study the cylindrical-based model No. 98, with after-body length equal to five times the diameter. Although the Mach-number range for these C_D's is centered transonically, it extends low enough to suggest a subcritical value of 0.29–0.3 for the chosen shape. The Re_l is of the order 10^6, and the effect of the fins is both small at zero lift and partially compensatory for the overestimate of body wetted area in (10.3). Hence we compute from (10.4)

$$C_f \cong \frac{0.043}{(10^6)^{1/6}} = 0.0043 \tag{10.7a}$$

With $l/d = 12.13$, (10.3) gives

$$C_{Df} \cong 0.21 \tag{10.7b}$$

and (10.6) yields

$$C_{DB} \cong 0.063 \tag{10.7c}$$

By subtraction, the pressure drag is in the vicinity of 0.02–0.03 at low M.

Figures 10.3 and 10.4 [the latter taken from Spencer and Fox (1966)] demonstrate that slender bodies share with wings a tendency to high drag peaks near M = 1. Although C_{DB} is included in all these data, the body's plan area is the reference for Fig. 10.4. Also the forebodies with $n < 1.0$ are blunt-nosed. The principal source of drag rise in these cases is C_{Dp}, which is passing from its relatively insignificant subsonic value to the substantial "wave-making" drag of supersonic flight. It is interesting that the Fig. 10.4 data generally have maxima near M = 1.1 at about twice the low subsonic C_D. On these smooth shapes, there is much less transonic variation either in C_{DB} or in the direct and indirect effects of the boundary layer. On a staged or stepped body of revolution, however, shocks would be forming ahead of abrupt changes in section; their interaction with the boundary layer would cause separated zones in concave corners and appreciably affect all airloads.

Figure 10.5 sketches the conditions that establish themselves at distinctly supersonic M and $\alpha_{ZL} = 0$. The three terms in (10.2) still characterize the drag, except that $C_{Dp} \cong C_{DW}$ now equals or outweighs the contributions of skin friction. Figure 10.4 suggests a rapid falling-off of total C_D as M increases. But Spencer and Fox (1966) also demonstrate the interesting fact that, when C_{DB} is subtracted from the data on noses with $n \geq 0.5$, the forebody pressure plus friction drag tends to settle down almost to a constant between M = 3 and 10. The strong influence of Mach number on base drag is indicated by (10.5) if we just assume a vacuum there ($p_B \cong 0$); the formula then gives $C_{DB} \sim M^{-2}$.

Before leaving our examination of smooth, unstepped bodies, we call attention to the subject of *boat-tailing*, which refers to the tapering of the aft end so that $S_B < S_{max}$ (cf. the lower nine models pictured in Fig. 10.3). On a booster or rocket

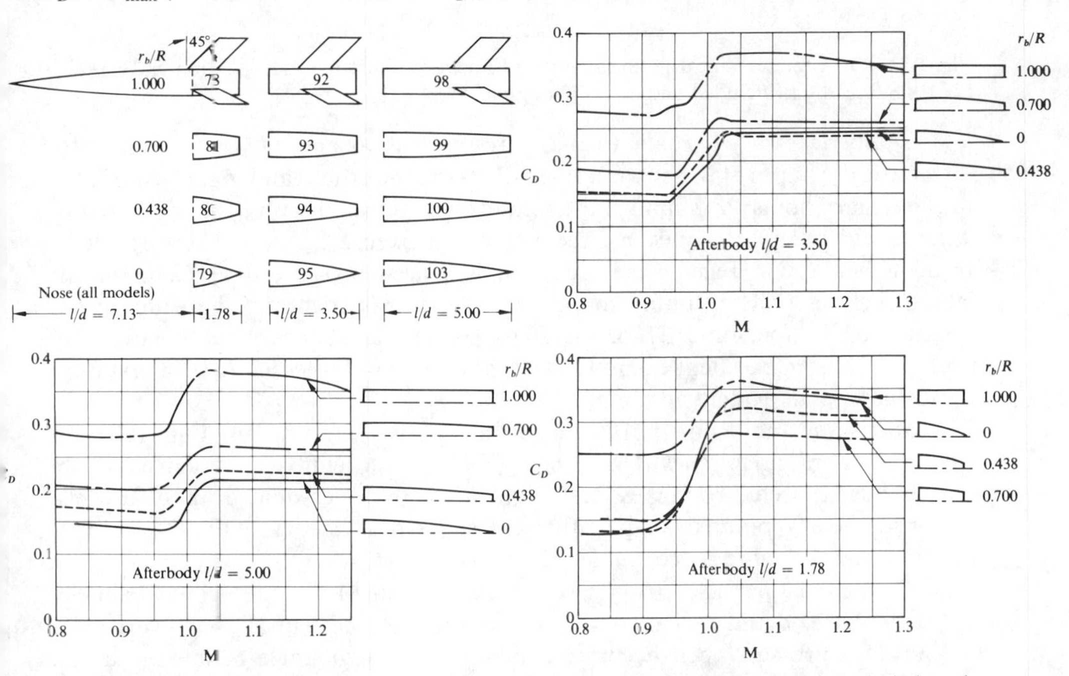

Fig. 10.3. Data for zero-lift drag on the illustrated family of bodies. Base drag is included, and S_{ref} is maximum cross-sectional area.

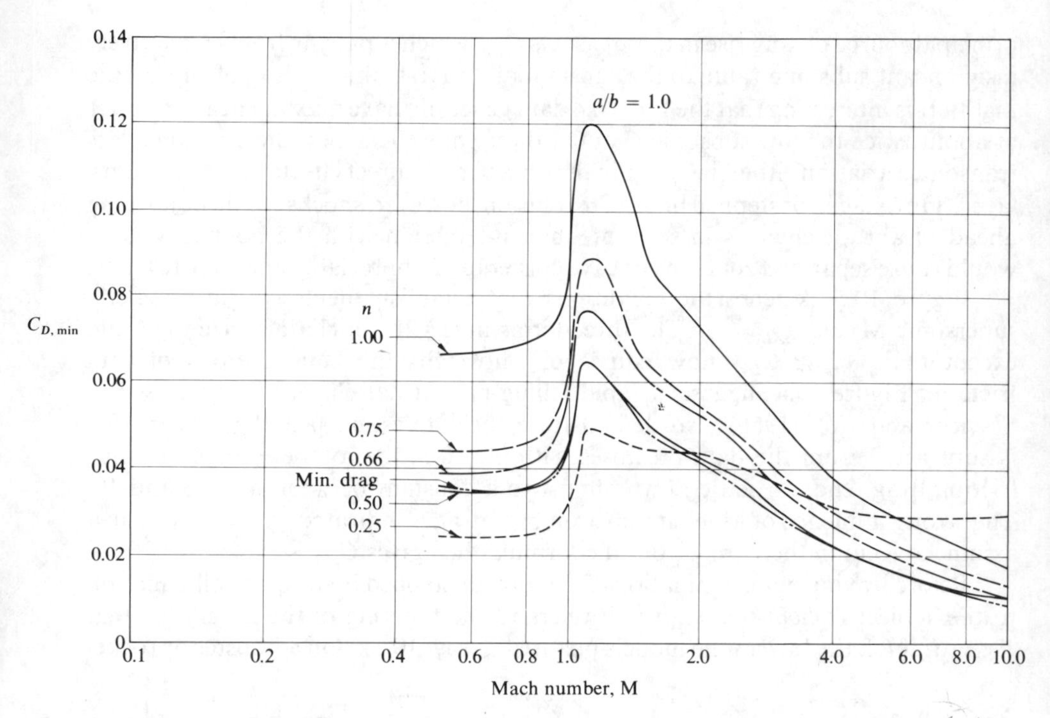

Fig. 10.4. Data for zero-lift drag for a family of noses with circular cross section and $r_0(x) \sim (x/l)^n$. Base drag is included and S_{ref} is plan area.

it is usually necessary to make the base area as large as possible to accommodate the nozzles. Shapes like shells and rifle bullets, on the other hand, may be modified in this regard so as to minimize the total drag associated with some design constraint, which might require a fixed volume, a fixed S_{max}, etc. There results a tradeoff between reduced base drag and increased pressure drag from the aft lateral surfaces. The optimum configuration often has an intermediate amount of boat-tailing. Supersonic C_D of the Fig. 10.3 family of models with afterbody $(l/d) = 1.78$ is an excellent example, the case of base radius equal to 0.7 maximum radius having an evident advantage.

Morrisette and Romeo (1965) provide extensive results (cf. Fig. 10.6) on staged boosters and missiles at M = 6. Again with C_{DB} removed, one can conclude that 0.2 is a useful rough estimate for the high supersonic drag of a well-designed, nearly pointed shape. Inviscid supersonic theory, with C_{Df} added, proves generally quite successful for these cases at M = 6.

When α_{ZL} is increased from zero, for instance at M > 1, the conical shocks of Fig. 10.5 are no longer coaxial with the body, but intensify on the windward side. The surface pressure then becomes a function of polar angle θ, measured circumferentially around any section. If, say, $\theta = 0$ corresponds to the windward meridian, the stronger shock there gives rise to the highest pressure along this line. The analytical determination, from given $p(x, \theta)$, of normal force and moment might proceed by the following steps: Suppose $d(x)$ is the body diameter as a function of axial distance x from the nose. First, a coefficient of normal force per unit x is defined by

$$C_n = \frac{1}{\frac{\rho_\infty}{2} v_c^2 \, d(x)} \int_0^{2\pi} [p - p_\infty][\cos \theta] \frac{d(x)}{2} \, d\theta \tag{10.8}$$

Integrations lengthwise then yield

$$C_N = \frac{1}{S_{\text{ref}}} \int_0^l C_n \, d(x) \, dx \tag{10.9}$$

(recall that $C_L \cong C_N$ at small angles) and

$$C_m = \frac{-1}{l S_{\text{ref}}} \int_0^l C_n [x - x_0] \, d(x) \, dx \tag{10.10}$$

or

$$\frac{x_{\text{CP}}}{l} \equiv \frac{1}{l} \frac{\int_0^l C_n x \, d(x) \, dx}{\int_0^l C_n \, d(x) \, dx} \tag{10.11}$$

In (10.10), we employ l as the reference length for moment coefficient; C_m is calculated for a pitch axis at $x = x_0$. Especially when there are no fins, it often turns out that the CP is far forward, one consequence being the (open-loop) static instability which characterizes boosters. The C_m curves of Fig. 10.6, which refer to the axes identified by small crosses, show this behavior clearly.

Several examples of the effects of angle of attack on the supersonic airloads of two-stage boosters are reproduced from Morrisette and Romeo (1965) and Blackwell (1966) in Figs. 10.6 and 10.7, respectively. Their results are generally self-explanatory, and merit your careful study. Thus the data at the top of Fig. 10.6

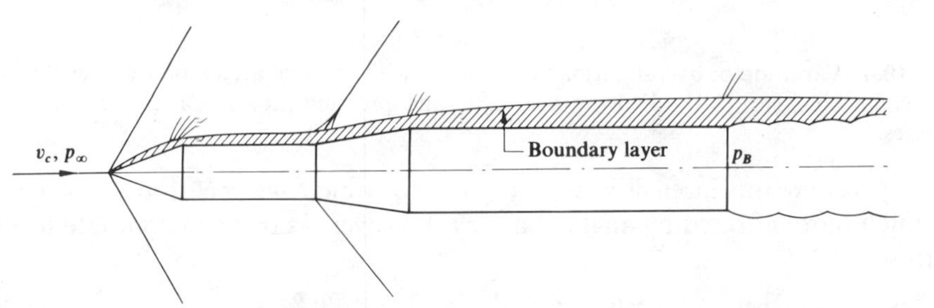

Fig. 10.5. Sketch of streamline pattern and evolution of boundary layer on a staged booster in supersonic flight at $\alpha_{ZL} = 0$. Positions of shock waves and expansion fans are indicated; thickness of boundary layer is exaggerated.

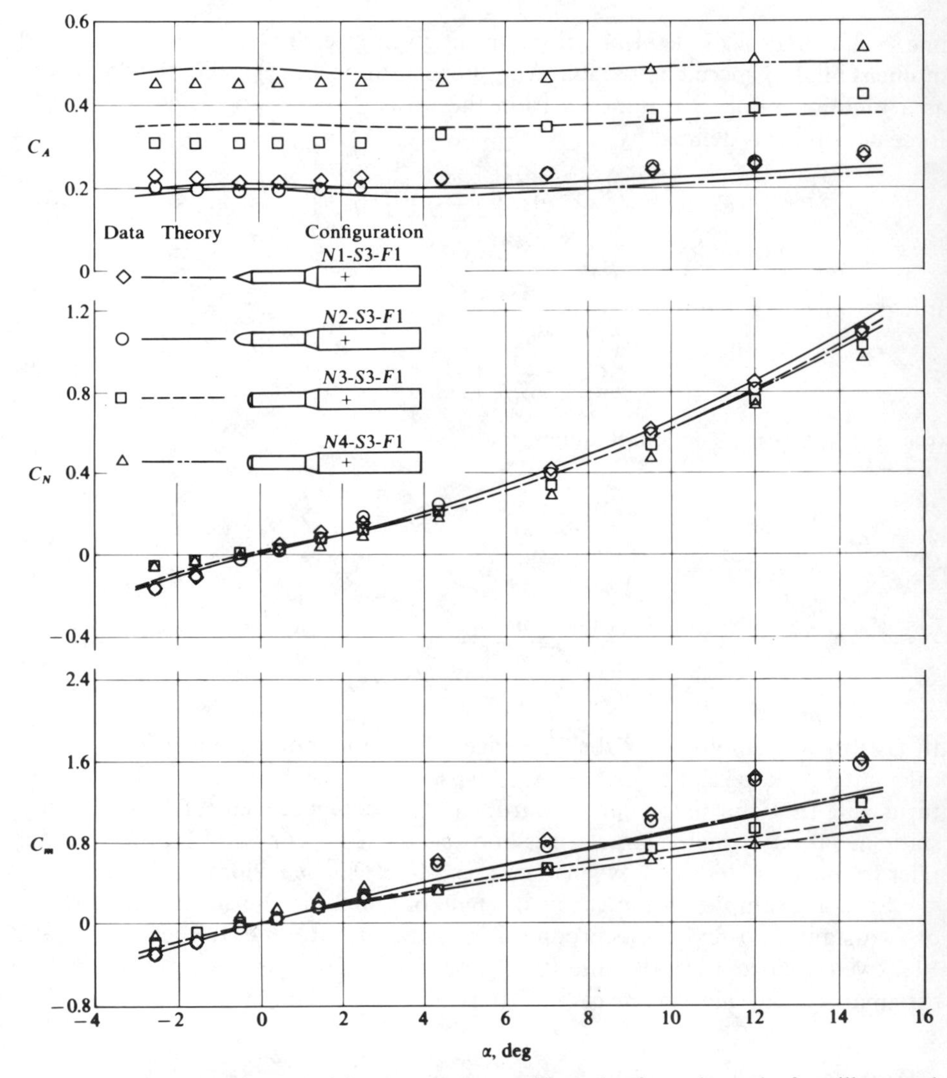

Fig. 10.6. Variation of overall airload coefficients with angle of attack on the four illustrated two-stage boosters. The Mach number is 6, and pitching moment axes are indicated by crosses.

clearly reveal the insensitivity of C_A to α_{ZL} which we mentioned earlier; C_D is much more affected by angle of attack, however, as one can compute from the formula

$$C_D = C_A \cos \alpha_{ZL} + C_N \sin \alpha_{ZL} \tag{10.12}$$

The C_N-curves of Fig. 10.6 typify a characteristic of slender bodies which is opposite to that of larger-Æ wings over the practical range of α_{ZL}: There is a

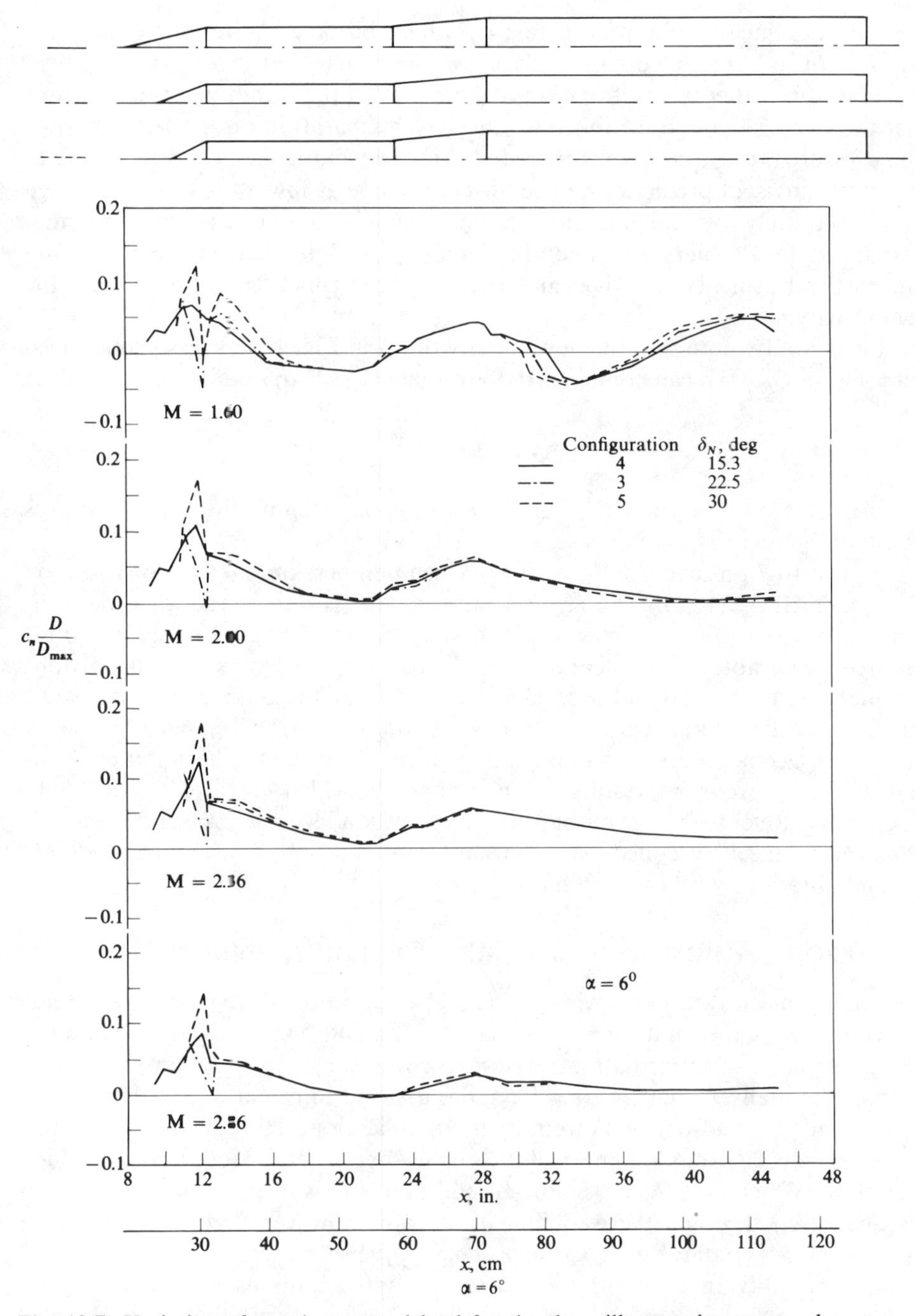

Fig. 10.7. Variation of running normal load for the three illustrated two-stage boosters at four Mach numbers.

positive, nonlinear buildup of C_N or C_L which becomes quite apparent above $\alpha_{ZL} = 4°$ or so. We shall discuss the causes in more detail during our development of slender-body theory. Suffice it to say here that, at the higher α_{ZL}, the crossflow past the body begins to exhibit boundary-layer separation on the leeward side. The pressure there is consequently reduced below what it should be according to linearized, inviscid predictions. The lift-curve slope at low α_{ZL} can be estimated quite accurately by the crossflow momentum considerations which we first introduced in Problem 4P, and this simple model is even capable of semi-empirical refinements (cf. Allen and Perkins 1951) which can approximate the nonlinearity.

Incidentally, note that the nonlinearity of C_m in Fig. 10.6 is less pronounced than that of C_N. We can combine (10.9) through (10.11) to obtain

$$x_{CP} = x_0 - \frac{C_m}{C_N} \tag{10.13}$$

and use (10.13) to interpret our last observation as meaning that the CP moves aft along the body as α_{ZL} increases.

Figure 10.7 presents data at various Mach numbers on the axial distribution of C_n, defined by (10.8). The angle of attack of 6° is quite large for a booster, and above where nonlinear effects have first appeared. For all the configurations pictured here, note the tendency for C_n to build up over the conical portions in which the cross-sectional area increases, either at the nose or the interstage transition. The lateral load then tapers off toward zero, except in the case of M = 1.60, as one proceeds aft along the cylindrical segments. This latter observation will later prove interesting as an instance in which classical slender-body theory fails. Refer to these same papers (Morrisette and Romeo 1965 and Blackwell 1966) for an excellent collection of measurements on both $C_n(x)$ and total loads for configurations with two and more stages.

10.2 AERODYNAMIC THEORY FOR SLENDER, POINTED BODIES

We shall concentrate our review of small-perturbation theory on streamlined axisymmetric shapes in a supersonic main stream, although some of our findings will be capable of immediate extension to cover subsonic and transonic flows. Among the many textual sources for more detailed information, Nielsen (1960) still prevails beyond ten years from its first publication. His Chapter 3 amounts to a complete, rigorous summary of the slender-body method for all M's at which it applies. We can make few significant additions to his reference lists, if those in his Chapters 3 through 10 are combined. We add, however, that the short article by Sears (1954) contains many clear insights on the subject.

Our simplifying assumptions are essentially those discussed in Sections 4.1 and 4.2 above. Where appropriate, we shall retain equation numbers from Chapter 4. Consider first the case of $\alpha_{ZL} = 0$, the situation under analysis being illustrated

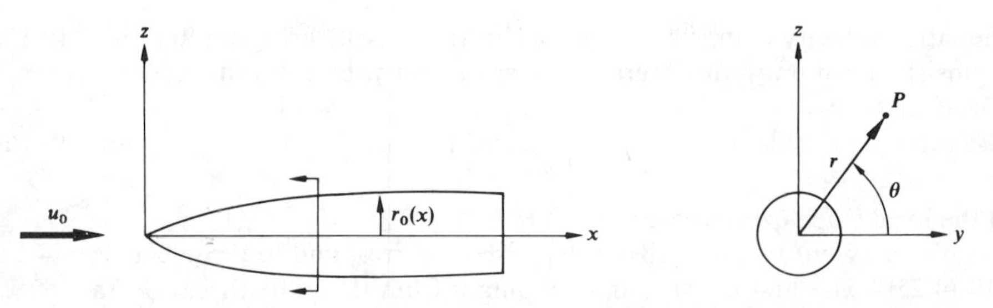

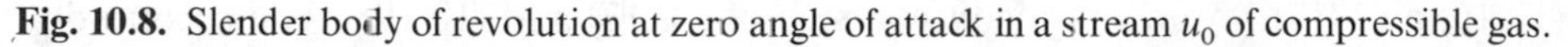
Fig. 10.8. Slender body of revolution at zero angle of attack in a stream u_0 of compressible gas.

in Fig. 10.8. Irrctationality is a consequence of assuming small disturbances and weak waves in an initially uniform flow of inviscid, perfect gas. We therefore retain the perturbation potential φ as the principal unknown, separating out the free stream according to

$$\Phi = u_0 x + \varphi \tag{4.17}$$

φ is governed by the partial differential equation

$$\nabla^2\varphi - \mathrm{M}^2 \frac{\partial^2\varphi}{\partial x^2} = 0 \tag{4.18}$$

with $\mathrm{M} > 1$ here. It is obviously advantageous to work in the cylindrical polar coordinates x, r, θ of Fig. 10.8. Since the corresponding Laplace operator is

$$\nabla^2 \equiv \frac{\partial^2}{\partial x^2} + \frac{\partial^2}{\partial r^2} + \frac{1}{r}\frac{\partial}{\partial r} + \frac{1}{r^2}\frac{\partial^2}{\partial \theta^2} \tag{10.14}$$

and since we expect φ to be independent of θ in the axisymmetric streamline pattern, the appropriate form of (4.18) is

$$\boxed{(\mathrm{M}^2 - 1)\frac{\partial^2\varphi}{\partial x^2} - \frac{\partial^2\varphi}{\partial r^2} - \frac{1}{r}\frac{\partial\varphi}{\partial r} = 0} \tag{10.15}$$

The exact flow-tangency boundary condition, equivalent to (4.22) for wings, requires the streamline in contact with the surface at $r = r_0(x)$ to have the same slope as the body meridian there. That is,

$$\frac{1}{(u_0 + u)}\frac{\partial\varphi(x, r_0)}{\partial r} = \frac{dr_0}{dx} \tag{10.16}$$

In cylindricals, the hypothesis (4.2) of small perturbations reads

$$|u|, |v_r|, |v_\theta| \ll u_0 \tag{10.17}$$

where

$$v_r = \frac{\partial\varphi}{\partial r} \quad \text{and} \quad v_\theta = \frac{1}{r}\frac{\partial\varphi}{\partial\theta} \tag{10.18a,b}$$

The latter velocity component vanishes in the present analysis. Relation (10.17) is consistent not only with overall slenderness but with a condition

$$\left|\frac{dr_0}{dx}\right| \ll 1 \tag{10.19}$$

on the local slope everywhere.

We carry out two linearizing steps which correspond to those between (4.22) and (4.25). The first consists of dropping u from the parentheses on the left of (10.16) and is an immediate consequence of (10.17) and (10.19). The second demands more careful explanation, because Maclaurin-series expansion of v_r about its value at the axis $r = 0$ [cf. (4.24)] runs afoul of a singularity. As will be verified *a posteriori*, v_r behaves like r^{-1} near the axis in the sought-after solution of (10.15). We therefore rewrite the approximate (10.16)

$$\left.\frac{\partial \varphi}{\partial r}\right|_{r=r_0} = u_0 \frac{dr_0}{dx} \tag{10.20}$$

and multiply the right side by r_0 and the left by r, which is to be equated to r_0. The right member can then be modified as follows:

$$u_0 r_0 \frac{dr_0}{dx} = \frac{u_0}{2\pi}\frac{d}{dx}(\pi r_0^2) \equiv \frac{u_0}{2\pi}\frac{dS}{dx} \tag{10.21}$$

where $S(x)$ is the cross-sectional area distribution. On the left, the product $r\,\partial\varphi/\partial r$ is essentially independent of r in the range $0 \leq r \leq r_0$. The convenient limit $r \to 0$ can thus be taken, and we get for the principal boundary condition

$$\boxed{\operatorname*{Lim}_{r\to 0}\left\{r\frac{\partial \varphi}{\partial r}\right\} = \frac{u_0}{2\pi}\frac{dS}{dx}, \qquad \text{for } 0 \leq x \leq l} \tag{10.22}$$

The other condition of vanishing of disturbances at infinity is met by our choice of singular solutions of (10.15).

Recall from Section 4.8 the usefulness of supersonic sources (4.186) for treating the problem of wing thickness. If placed on the axis $r = 0$, this same solution has just the symmetry needed to represent perturbations due to our body at $\alpha_{ZL} = 0$. Accordingly, we rewrite (4.186) for a unit-strength source centered at $x = x_1$ on the axis, noting that $(y^2 + z^2)$ can be replaced by r^2.

$$\begin{aligned}\varphi_S(x, r) &= -\frac{1}{2\pi\sqrt{(x - x_1)^2 - (\mathrm{M}^2 - 1)r^2}} \\ &\equiv -\frac{\partial}{\partial x}\left[\cosh^{-1}\left(\frac{x - x_1}{r\sqrt{\mathrm{M}^2 - 1}}\right)\right]\end{aligned} \tag{10.23}$$

We remind ourselves that $\varphi_S = 0$, identically, outside the downstream Mach cone from $x = x_1$, whose location is found by requiring $x \geq x_1$ and equating to zero the quantity under the radical in (10.23). The final member of (10.23) can readily be checked by means of the derivative formula for the inverse hyperbolic cosine; it is sometimes useful for calculating $\partial\varphi/\partial r$ and other manipulations.

By analogy with the development of (4.187), we center a line of sources along $r = 0$ between $x = 0$ and l. Let their running strength—undetermined for the moment—be designated $f(x_1)$ per unit axial distance:

$$\varphi(x, r) = -\frac{1}{2\pi}\int_0^{(x-\sqrt{M^2-1}\,r)} \frac{f(x_1)\,dx_1}{\sqrt{(x - x_1)^2 - (M^2 - 1)r^2}} \tag{10.24}$$

The upper limit in (10.24) has been reduced below l, because the law of forbidden signals insists that no disturbances get to (x, r) from $x_1 > (x - \sqrt{M^2 - 1}\,r)$. It will become clear that the limit should be l for points behind the base at which $x > (l + \sqrt{M^2 - 1}\,r)$.

In order to apply (10.22) to (10.24), we must take advantage of r's smallness during the limiting process. In particular, we use the fact that

$$\sqrt{(x - x_1)^2 - (M^2 - 1)r^2} \cong x - x_1 \tag{10.25}$$

except in the vicinity of the upper limit. The latter region can be isolated, so long as $f(x)$ is a continuous function, by substituting

$$f(x_1) \equiv f(x) + [f(x_1) - f(x)] \tag{10.26}$$

into (10.24) and evaluating an elementary integral to obtain

$$\begin{aligned}
\varphi(x, r) &= -\frac{f(x)}{2\pi}\int_0^{(x-\sqrt{M^2-1}\,r)} \frac{dx_1}{\sqrt{(x - x_1)^2 - (M^2 - 1)r^2}} \\
&\quad - \frac{1}{2\pi}\int_0^{(x-\sqrt{M^2-1}\,r)} \frac{f(x_1) - f(x)}{\sqrt{(x - x_1)^2 - (M^2 - 1)r^2}}\,dx_1 \\
&= -\frac{f(x)}{2\pi}\{-\ln(r\sqrt{M^2 - 1}) + \ln[x + \sqrt{x^2 - (M^2 - 1)r^2}]\} \\
&\quad - \frac{1}{2\pi}\int_0^{(x-\sqrt{M^2-1}\,r)} \frac{f(x_1) - f(x)}{\sqrt{(x - x_1)^2 - (M^2 - 1)r^2}}\,dx_1
\end{aligned} \tag{10.27}$$

For small enough r, the difference of logarithms in braces here may be approximated by $-\ln[r\sqrt{M^2 - 1}/2x]$. Moreover, because the integrand numerator vanishes at $x = x_1$, no violence is done by using (10.25) and approximating the upper limit by x in the last integral. Thus we recover

$$\varphi(x, r) \cong \frac{f(x)}{2\pi} \ln \left[\frac{r\sqrt{M^2 - 1}}{2x} \right] - \frac{1}{2\pi} \int_0^x \frac{f(x_1) - f(x)}{x - x_1} dx_1 \qquad (10.28)$$

[The last step here is described in more detail in Ashley and Landahl (1965, Section 6-3) and rigorously confirmed by the alternative development in Chapter 3 of Nielsen (1960).]

Not only is our previous claim that $\partial\varphi/\partial r$ behaves like r^{-1} near the axis obvious from the logarithmic term in (10.28), but the $\ln r$ portion of this potential can be identified with a two-dimensional, incompressible flow whose complex potential function (cf. Section 4.3 and Problem 4B) is

$$F(\xi) = \frac{f}{2\pi} \ln \xi \qquad (10.29a)$$

Here $\xi = y + iz = re^{i\theta}$. Accordingly, the real part of (10.29a),

$$\text{Re}\{F(\xi)\} \equiv \varphi_{2\text{-D}}(r) = \frac{f}{2\pi} \ln r \qquad (10.29b)$$

describes a 2-D source with streamlines directed radially away and a volume f of fluid, per unit distance along the body's axis, emanating outward per unit time.

Substitution of (10.28) into (10.22) is now straightforward,

$$\lim_{r \to 0} \left\{ r \frac{\partial \varphi}{\partial r} \right\} = \lim_{r \to 0} \left\{ r \frac{f(x)}{2\pi r} \right\} = \frac{f(x)}{2\pi} = \frac{u_0}{2\pi} \frac{dS}{dx} \qquad (10.30a)$$

whence for the dummy argument

$$f(x_1) = u_0 \frac{dS(x_1)}{dx_1} \equiv u_0 S'(x_1) \qquad (10.30b)$$

Finally, the original potential (10.24) can be written, both for any r and (through 10.26) for r in the vicinity of the body itself.

$$\begin{aligned} \varphi(x, r) &= -\frac{u_0}{2\pi} \int_0^{(x - \sqrt{M^2 - 1}\, r)} \frac{S'(x_1)\, dx_1}{\sqrt{(x - x_1)^2 - (M^2 - 1)r^2}} \\ &\cong \frac{u_0 S'(x)}{2\pi} \ln \left[\frac{r\sqrt{M^2 - 1}}{2x} \right] \\ &\quad - \frac{u_0}{2\pi} \int_0^x \frac{S'(x_1) - S'(x)}{x - x_1} dx_1 \qquad \text{for small } r,\ M > 1 \end{aligned} \qquad (10.31)$$

By way of interpreting the latter form in its range of applicability near $r = r_0(x)$, we observe that (10.31) may be recast as follows:*

$$\varphi(x, r) \cong \frac{u_0}{2\pi} S'(x) \ln \frac{r}{l} + g(x) \tag{10.32a}$$

where

$$g(x) = \frac{u_0}{2\pi} S'(x) \ln \left[\frac{\sqrt{M^2 - 1}}{2x/l}\right] - \frac{u_0}{2\pi} \int_0^x \frac{S'(x_1) - S'(x)}{x - x_1} dx_1$$

$$= \frac{u_0}{2\pi} S'(x) \ln \left[\frac{\sqrt{M^2 - 1}}{2}\right] - \frac{u_0}{2\pi} \int_0^x S''(x_1) \ln \left[\frac{x - x_1}{l}\right] dx_1 \tag{10.32b}$$

[The final term in (10.32b) is calculated by a partial integration, wherein both continuous body curvature and the pointed-nose condition $S'(0) = 2\pi r'_0(0) r_0(0) = 0$ are assumed. This version proves useful when one is predicting drag.]

Clearly the first term in (10.32a) is the source line discussed in connection with (10.29), and it is governed mathematically by the Laplace equation

$$\nabla^2 \varphi = \frac{\partial^2 \varphi}{\partial r^2} + \frac{1}{r} \frac{\partial \varphi}{\partial r} = 0 \tag{10.33}$$

which would hold for a purely 2-D, incompressible crossflow. Among many others, Munk (1924) and Harder and Klunker (1957) arrived at slender-body theory generally by assuming, in advance, that the solution to near field or inner flow could be approximated in this way. Indeed, someone once compared the motion, in a crossflow plane normal to the x-axis and fixed in the remote fluid at rest, as resembling a spear being thrown through a sheet of foam rubber. The viewpoint is similar to that taken in Problem 4P and to the interpretation of low-Æ wing theory at the end of Section 4.7. If one sets out initially to solve (10.33) for $\varphi = \varphi(x, r)$, subject to (10.22), one arrives immediately at the $\ln(r/l)$ term of (10.32a). The possibility must also be allowed for, however, that an additive function of x alone will satisfy both the differential equation and the axisymmetry requirement. In the present case, the $\ln(r/l)$ is a 2-D source whose strength is fixed by the observation that, as seen in the crossflow, the body of revolution looks like an expanding circle. We can reason that the source's emanation, by continuity considerations, must equal the rate $u_0 S'(x)$ at which the body is pushing fluid volume outward.

In our example, the additive function of x can be identified with $g(x)$ from (10.32b). We were here able to find it by working directly with the full equation

---

* From here onward in the development, we abandon the questionable practice of working with logarithms of dimensioned quantities. Although no difficulty is encountered during general discussion and interpretation of results, errors can arise from trying improperly to compute flow fields, pressures, and loads with such logarithms. Note that it does not matter whether we write $\ln r$ or $\ln(r/l)$ in formulas like (10.29b), since an arbitrary constant may be added to φ itself

(10.15), but even in more difficult cases the method of inner and outer expansions permits $g(x)$ to be calculated by matching, as described in Ashley and Landahl (1965, Chapter 6). We also note that the source portion of our inner flow is independent of Mach number and therefore can be expected to hold for subsonic and even transonic flight. Only $g(x)$ is a function of M. We shall see that an interesting consequence of this behavior is that drag is the only resultant airload on a slender body which is predicted to vary with M.

Perhaps the most fascinating deduction from the slender-body approach is called the *equivalence rule*, announced for transonic speeds by Oswatitsch and Keune (1955), but essentially given for $M > 1$ in a paper by Ward (1949). It applies for any smooth, pointed streamlined shape and even when small amounts of lift are being developed. We state the general results without proof.

- Remote from the body surface, the lowest-order flow field is axisymmetric and identical with what would be produced, at the specified Mach number, by an "equivalent" body of revolution at $\alpha_{ZL} = 0$ having the same cross-sectional area distribution $S(x)$ as the actual shape. [Our first form of $\varphi(x, r)$ is precisely the result needed for $M > 1$.]
- Near the surface the flow differs from that around the equivalent body by a correction, which is a 2-D incompressible crossflow that causes the proper tangency condition to be met at the actual body.

Some instances of these corrective crossflows will be given below. Looking at (10.32), we note that $g(x)$ is the same at this M for all bodies with a given $S(x)$; it is the $\ln (r/l)$ term that has to be replaced in order to accommodate different boundary conditions.

Before leaving our axisymmetric result (10.31), we add that there are two ways of determining aerodynamic forces. Most direct would seem to be the integration of appropriate components of surface pressure. This method runs into the difficulty, however, that the C_p formula (4.16) may not be approximated by (4.19), as in the case of a wing. The reason is that, near the body, we can estimate from (10.32) that

$$\frac{1}{u_0}\frac{\partial \varphi}{\partial x} = O(\tau^2 \ln \tau) \tag{10.34a}$$

whereas

$$\frac{1}{u_0}\frac{\partial \varphi}{\partial r} = O(\tau) \tag{10.34b}$$

Here τ is the body fineness ratio; we add that, when $\varphi = \varphi(x, r, \theta)$, the velocity component v_θ also obeys (10.34b). Now the $\ln \tau$ in (10.34a) is not usually different enough from unity to justify dropping the squares of the crossflow velocities, relative to $\partial\varphi/\partial x$, when the small-disturbance terms in (4.16) are expanded. It follows that the Bernoulli equation generally suitable for slender-body flows reads

$$C_p = -\frac{2}{u_0}\frac{\partial\varphi}{\partial x} - \frac{1}{u_0^2}\left[\left(\frac{\partial\varphi}{\partial r}\right)^2 + \left(\frac{1}{r}\frac{\partial\varphi}{\partial\theta}\right)^2\right] \tag{10.35}$$

Workable expressions for the pressures at the surface can, of course, be developed from (10.35) and (10.32). But if only resultant airloads are needed—forebody pressure drag C_{Dp} in the axisymmetric case—it proves easier to use the approach of computing the corresponding component of momentum flux out of a control volume, surrounding the body and having its lateral boundaries in the outer flow. We have already discussed this scheme in connection with (4.197) for wings. When used together with the equivalence rule, it yields formulas mainly in terms of the area $S(x)$.

Thus Ashley and Landahl (1965, Section 6-6) treat the general supersonic slender body. Their control-volume boundary consists of a circular cylinder about the x-axis with generators parallel to the flight direction, cut off front and back by planes intersecting the nose and base, respectively. One useful form of their result is the following:

$$\begin{aligned} C_{Dp}S_{\text{ref}} = &-\frac{1}{\pi}\int_0^x S''(x)\int_0^x S''(x_1)\ln\left[\frac{x-x_1}{l}\right]dx_1\,dx \\ &+\frac{1}{\pi}S'(l)\int_0^l S''(x_1)\ln\left[1-\frac{x_1}{l}\right]dx_1 \\ &-\frac{1}{2\pi}[S'(l)]^2\ln\left[\frac{\sqrt{M^2-1}}{2}\right] - \oint_{C_B}\varphi_2\frac{\partial\varphi_2}{\partial n}\,ds \end{aligned} \tag{10.36}$$

The final term in (10.36) is a line integral, taken counterclockwise around the base contour (cf. Fig. 10.9) and involving the 2-D portion of the inner-flow potential. n is a normal coordinate directed outward from this contour, and φ_2 is the quantity that must be added to $g(x)$ to construct the complete perturbation; for instance, it would be the first term on the right of (10.32a) for the axisymmetric body flow. Clearly this term vanishes if the back end is cut off cylindrically, for then $\partial\varphi_2/\partial n = 0$. Note also that base drag [cf. (10.5)] is not included in (10.36) but must be added, along with friction drag, as in (10.2).

All terms but the first in (10.36) are zero when the base is either cylindrical [$S'(l) = 0$] or pointed [$S(l) = 0 = S'(l)$]. Such cases are very common in practice, boat-tailing being the principal exception. Among other interesting properties, it is evident that the zero-lift drag of any such body is predicted to equal that of its equivalent body of revolution. This observation, together with the fact that $\ln[\sqrt{M^2 - 1}/2]$ has now disappeared from (10.36) and $C_{Dp}S_{\text{ref}}$ is therefore independent of Mach number, provides theoretical justification for the earliest form of the *transonic area rule* [cf. Whitcomb (1956)]. The application of this rule may be thought of as consisting of two steps:

- Variational calculus or some other optimizing technique is used to find that $S(x)$ which minimizes C_{Dp} under suitable physical constraints. The total volume and S_{max} might, for example, be held constant.
- The cross-sectional area distribution of the actual vehicle, which would include contributions from the wing, empennage, engine pods, stores, etc., is made to agree with the optimum $S(x)$ as closely as possible.

Since optimal bodies of revolution tend to be smoothly contoured, one consequence of "area-ruling" is that the fuselage cross section is drawn in or "waisted" at stations at which the wing contributes a substantial amount of $S(x)$. This phenomenon can be seen in the F-102 airplane (Fig. 1.6), which was the first American fighter to benefit from Whitcomb's (1956) discovery. It is worth adding that a more refined "supersonic area rule" [cf. Ashley and Landahl (1965, Chapter 10)]—more precisely, a methodology for minimization of supersonic pressure drag—has been applied to many more recent aircraft whose design points are at $M > 1$.

To complete our discussion of axisymmetric flow, we set down some results for $M < 1$. The differential equation and principal boundary condition are unchanged from (10.15) and (10.22), except for an important sign reversal in the coefficient of $\partial^2\varphi/\partial x^2$. The unit subsonic source, analogous to (10.23) and similarly located at $x = x_1$ on the axis, has a potential

$$\varphi_s(x, r) = -\frac{1}{4\pi\sqrt{(x - x_1)^2 + (1 - M^2)r^2}} \tag{10.37}$$

When these are placed along the body centerline between $x = 0$ and l, it is not hard to derive the following solution as a counterpart to (10.31, 32).

$$\begin{aligned}\varphi(x, r) &= -\frac{u_0}{4\pi}\int_0^l \frac{S'(x_1)\,dx_1}{\sqrt{(x - x_1)^2 + (1 - M^2)r^2}} \\ &\cong \frac{u_0 S'(x)}{2\pi}\ln\frac{r}{l} + \frac{u_0 S'(x)}{4\pi}\ln\left[\frac{(1 - M^2)l^2}{4x(l - x)}\right] \\ &\quad - \frac{u_0}{4\pi}\int_0^l \frac{S'(x_1) - S'(x)}{|x - x_1|}\,dx_1 \qquad \text{for small } r,\ M < 1\end{aligned} \tag{10.38}$$

The near-field crossflow potential, as anticipated, is identical with that for $M > 1$. $g(x)$ is given by the last two terms of (10.38).

We have remarked earlier that (10.38) is rigorously consistent with the slender-body assumptions only when both ends are pointed. A high-fineness-ratio shape with a cylindrically cut-off base is, however, quite well approximated if the cross

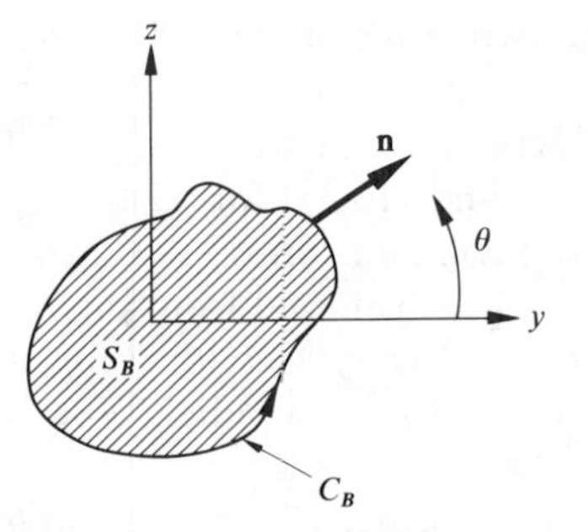

Fig. 10.9. Rear view of the base of an arbitrary slender body, illustrating the bounding contour C_B and the unit normal vector.

section is imagined to extend to $x = \infty$ as a circular cylinder of area S_B. (10.38) would still represent the potential correctly because $S'(x) = 0$ for $x \geq l$. The actual truncation is then accounted for by including C_{DB} in the total drag. Either for this case or for the doubly pointed body, a calculation similar to that underlying (10.36) yields a theoretically zero forebody pressure drag, in conformity with the d'Alembert paradox.

When the body of revolution in Fig. 10.8 is given a positive incidence α_{ZL} in the xz-plane, there results the situation shown in Fig. 10.10. It is mathematically convenient to retain x as the body axial coordinate, so a small free-stream component of velocity

$$v_z = u_0 \sin \alpha_{ZL} \cong u_0 \alpha_{ZL} \tag{10.39}$$

is thereby added at infinity. The right side of the figure suggests the streamline pattern of crossflow past a typical station, as it would appear to an observer fixed with respect to the body.

Let the complete potential for the motion under study be written ($z = r \sin \theta$)

$$\Phi = u_0 x + u_0 \alpha_{ZL} z + \varphi(x, r) + \varphi_\alpha(x, r, \theta) \tag{10.40}$$

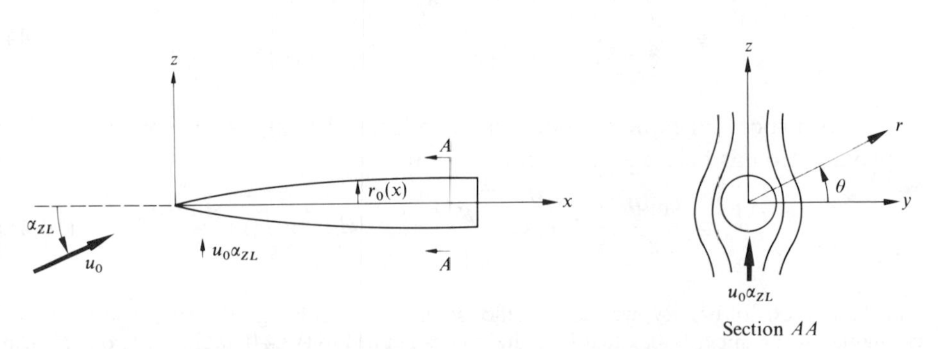

Fig. 10.10. Slender body of revolution at finite angle of attack in a stream u_0 of compressible gas. Sketch on the right illustrates a typical section with crossflow streamlines as seen by an observer on the body.

With $\cos \alpha_{ZL} \cong 1$, the first two terms on the right describe the free stream. $\varphi(x, r)$ is the axisymmetric disturbance potential, derived above, which accounts for the rate of change of cross-sectional area by satisfying (10.22). It continues to be given by (10.38) for $M < 1$ and (10.31) for $M > 1$. The incremental effects of angle of attack are therefore contained in φ_α.

A boundary condition for φ_α is found from the requirement that, aside from the effects of dr_0/dx which are already accounted for, v_r must vanish at $r = r_0(x)$. Due to the altered free stream, there is a radial velocity

$$v_{r_0} = u_0 \alpha_{ZL} \frac{\partial z}{\partial r} = u_0 \alpha_{ZL} \sin \theta \tag{10.41}$$

which has to be canceled by φ_α. Hence we obtain

$$\frac{\partial \varphi_\alpha[x, r_0(x), \theta]}{\partial r} + u_0 \alpha_{ZL} \sin \theta = 0 \tag{10.42}$$

The condition (10.42) is conveniently solved by distributing along x a line of supersonic doublets with their axes directed parallel to z. A useful form of the doublet potential, here centered at the origin and with unit strength, is

$$\varphi_d = \begin{cases} \dfrac{\partial}{\partial x}\left[\dfrac{x \sin \theta}{r\sqrt{x^2 - (M^2 - 1)r^2}}\right], & x > \sqrt{M^2 - 1}\, r \\ 0 \qquad , & x < \sqrt{M^2 - 1}\, r \end{cases} \tag{10.43}$$

You can readily confirm that (10.43) satisfies (4.18), with (10.14), and that its disturbances die out as x and $r \to \infty$. Let the doublets in the vicinity of $x = x_1$ have strength $h(x_1)$ per unit axial distance. Accounting for forbidden signals, we can then set down a trial expression* for φ_α,

$$\varphi_\alpha = \frac{\sin \theta}{r} \frac{\partial}{\partial x} \int_0^{(x - \sqrt{M^2 - 1}\, r)} \frac{h(x_1)[x - x_1]\, dx_1}{\sqrt{(x - x_1)^2 - (M^2 - 1)r^2}} \tag{10.44}$$

In the process of enforcing boundary condition (10.42), we employ the small-r_0 approximation as before and write, for $r \to 0$,

$$\varphi_\alpha \cong \frac{\sin \theta}{r} \frac{\partial}{\partial x} \int_0^x h(x_1) \frac{[x - x_1]}{\sqrt{(x - x_1)^2}}\, dx_1 = h(x) \frac{\sin \theta}{r} \tag{10.45}$$

---

* In the interests of brevity, we here sacrifice some rigor. Although the integrand of (10.44) is singular at the upper limit when Leibnitz' rule is applied to (10.44), taking $\partial/\partial x$ outside can be justified by reference to the "finite part" of a singular integral (e.g., Heaslet and Lomax, 1954). Equivalent but more elaborate derivations, such as that in Nielsen (1960), verify that our results are correct.

Therefore

$$\frac{\partial \varphi_\alpha(x, r_0(x), \theta)}{\partial r} \cong -h(x)\frac{\sin\theta}{r_0^2} \tag{10.46a}$$

and (10.42) yields

$$h(x) = u_0\alpha_{ZL}r_0^2(x) \tag{10.46b}$$

Substituting (10.46b) into (10.44) and (10.45), we determine the following forms for the general solution and its inner-flow approximation.

$$\varphi_\alpha(x, r, \theta) = \frac{u_0\alpha_{ZL}\sin\theta}{r}\int_0^{(x-\sqrt{M^2-1}\,r)} \frac{r_0^2(x_1)[x - x_1]}{\sqrt{(x - x_1)^2 - (M^2 - 1)r^2}}\,dx_1$$

$$\cong u_0\alpha_{ZL}\frac{r_0^2(x)\sin\theta}{r}, \qquad \text{for small } r,\ M > 1 \tag{10.47}$$

The last member of (10.47) is, as in the case of the ln (r/l) in (10.32a), a solution of Laplace's equation for incompressible crossflow. It is known as a 2-D doublet, with its axis parallel to z and "strength" proportional to $u_0\alpha_{ZL}r_0^2(x)$. In combination with free-stream potential $u_0\alpha_{ZL}z$, it describes steady liquid motion, of velocity $v_z = u_0\alpha_{ZL}$ at infinity, past a circular cylinder of radius $r_0(x)$. Alternatively, φ_α alone (at small r) is the instantaneous potential for the same cylinder, moving downward at velocity $u_0\alpha_{ZL}$ through liquid at rest. We could also have arrived at this result by reasoning similar to that following (10.33). Moreover, we would have to reject any additive function of x, because the potential we seek to satisfy (10.42) should be purely an odd function of z (or θ), and a function independent of z is automatically even. It follows that $g(x)$ from (10.32b) is sufficient and that, by adding $[(u_0\alpha_{ZL}r_0^2(x)\sin\theta/r) + u_0\alpha_{ZL}z]$ to (10.32), we get the complete supersonic potential of the inner flow, including angle of attack.

We can further conclude that, at transonic and subsonic Mach numbers, the last member of (10.47) still applies. Only $g(x)$ is changed in the $M < 1$ inner flow, (10.38) providing the appropriate formula. Since the crossflow is identical throughout the range, we may already anticipate that all lateral airloads are theoretically unaffected by changes in M.

As with the calculation of drag, we have a choice for the axisymmetric configuration between finding $C_n(x)$, C_N, C_m, etc., by integration of surface pressures from Bernoulli's equation (10.35) or resorting to momentum considerations. Again we select the latter, because we can thereby introduce a method applicable to more general slender shapes. We now operate in the near field. Here the flow which produces lateral loads is essentially 2-D incompressible, and from the standpoint of estimating its loading on the body we can assume this same type of flow to extend to $r = \infty$. We turn aside from the mainstream of the development to prove the needed results about momentum.

Consider the 2-D motion, generated in an unbounded mass of incompressible liquid at rest at infinity, by a rigid solid of cross-sectional area S bounded by a contour C. As shown in Fig. 10.9, the flow due to this solid's translational velocity is described by an instantaneous potential $\varphi(y, z)$. We assert that the vector quantity, whose rate of change gives the force exerted by the solid on the liquid (cf. Problem 4P), is

$$\mathcal{M} = -\rho_\infty \oint_C \varphi \mathbf{n}\, ds \tag{10.48}$$

The contour integral in (10.48) is taken counterclockwise around C, as in the drag formula (10.36), and $\mathbf{n}$ is an outward unit normal vector. $\mathcal{M}$ can be regarded as like fluid momentum per unit x-distance, although the total momentum is actually indeterminate up to an additive constant, which may be unbounded (see discussion in Lamb 1945 or Ashley and Landahl 1965, Section 2-4).

An elementary way of deriving (10.48) is to make use of a physical interpretation of the incompressible velocity potential, according to which $-\rho_\infty\varphi(\mathbf{r}, t)$ is assimilated to the distribution of sudden, large pressure impulses $P(\mathbf{r}, t)$ that would be required to produce this particular flow from rest if they acted over a very short interval from time $(t - \Delta t)$ to t. This connection can be demonstrated by integrating the constant-density Bernoulli equation (4.13) over such an interval. We have $\varphi = 0$ prior to $(t - \Delta t)$, and $u_0 = 0$. If Δt is sufficiently small, the integrals of the moderately sized quantities p_∞ and $\rho_\infty q^2/2$ are negligible. Thus we obtain

$$P = \int_{t-\Delta t}^{t} p\, dt = -\rho_\infty \int_{t-\Delta t}^{t} \frac{\partial \varphi}{\partial t}\, dt = -\rho_\infty \varphi \tag{10.49}$$

Both the hypothetical impulse P and the potential itself are, of course, functions of position throughout the field. In particular, the values of P along the contour C represent the force per unit area that would have to be exerted impulsively by the body surface onto the liquid in order to bring the motion up from rest. The corresponding total impulse would be, per unit length

$$\iint_{\substack{\text{Body}\\\text{surface}}} P\mathbf{n}\, dS_{\text{Body}} = \oint_C P\mathbf{n}\, ds = -\rho_\infty \oint_C \varphi\mathbf{n}\, ds \tag{10.50}$$

The last member of (10.50) is evidently the same as $\mathcal{M}$ in (10.48). We complete our proof that the actual force is equal to $\mathcal{M}$'s rate of change in a continuous, nonimpulsive, but unsteady flow as follows: Consider an infinitesimal time step dt. Since the impulses required to generate the flows immediately before and after dt are $\mathcal{M}$ and $\mathcal{M} + d\mathcal{M}$, then

$$d\mathcal{M} = \frac{d\mathcal{M}}{dt}\, dt \tag{10.51}$$

by itself constitutes the infinitesimal impulse per unit length that the body would have to exert over dt to produce just the actual small change. By the definition of impulse, the associated force is

$$\mathbf{F} = \frac{d\mathscr{M}}{dt} = -\rho_\infty \frac{d}{dt} \oint_C \varphi \mathbf{n}\, ds \tag{10.52}$$

(More elegant demonstrations of the foregoing will be found in Milne-Thompson 1962, Lamb 1945, and Ashley and Landahl 1965, Section 2-4.)

Let us now return to the crossflow potential in (10.47), for which the body S at longitudinal station x is just the cylinder of radius $r_0(x)$. φ for insertion into (10.52) is φ_α alone, since this term characterizes motion relative to resting fluid at infinity. (Because of its symmetry in θ, $\varphi(x, r)$, as expected, adds nothing to the "momentum" integral.) By specializing Fig. 10.9 to the cylinder, we easily see that

$$\mathbf{n}\, ds = [\mathbf{j} \cos\theta + \mathbf{k} \sin\theta] r_0(x)\, d\theta, \tag{10.53}$$

where $\mathbf{j}, \mathbf{k}$ are the customary y, z-unit vectors. When φ_α from (10.47) and (10.53) are substituted into (10.48), we calculate that

$$\begin{aligned} \mathscr{M} &= -\rho_\infty \int_{\theta=0}^{2\pi} \left[u_0 \alpha_{ZL} \frac{r_0^2 \sin\theta}{r_0} \right] [\mathbf{j} \cos\theta + \mathbf{k} \sin\theta] r_0\, dx \\ &= -[\rho_\infty u_0 \alpha_{ZL} r_0^2] \pi \mathbf{k} = -\rho_\infty u_0 \alpha_{ZL} S(x) \mathbf{k} \end{aligned} \tag{10.54}$$

$S(x) = \pi r_0^2(x)$ is, as before, the cross-sectional area of the body.

As we could have anticipated, (10.54) shows that the running "momentum," calculated with respect to the remote fluid, is in the negative z-direction. It is proportional to the relative downward velocity $u_0 \alpha_{ZL}$ of the cross section and to the mass (or "virtual mass") $\rho_\infty S(x)$ of fluid displaced at station x. The normal force $n(x)$, per unit length, and *applied by the fluid to the body*, equals the negative time derivative of the z-component of $\mathscr{M}$. Since again this is the rate of change seen by the stationary observer, we reason that

$$\frac{d}{dt}(\ldots) = u_0 \frac{d}{dx}(\ldots) \tag{10.55}$$

is determined by the speed u_0 with which the body is translating longitudinally past the fluid slab at station x. Finally,

$$n(x) = -u_0 \frac{d}{dx}[-\rho_\infty u_0 \alpha_{ZL} S(x)] = \rho_\infty u_0^2 \alpha_{ZL} S'(x) \tag{10.56}$$

From $n(x)$ we can readily compute normal force, pitching moment, etc. For instance, if base area is adopted as the reference,

$$C_N = \frac{1}{q_\infty S_B} \int_0^l n(x)\, dx = 2\alpha_{ZL} \tag{10.57}$$

We differentiate (10.57) to get a very interesting formula [cf. (4.182)],

$$\boxed{C_{N\alpha} \cong C_{L\alpha} = 2} \tag{10.58}$$

As is obvious from momentum considerations, total lift is fixed entirely by circumstances in which the fluid passes off the back end, the lift-curve slope based on the area there being exactly two.

Similar results for two other cross-sectional shapes are covered by Problem 4P, and many more are discussed in such references as Nielsen (1960). It can, incidentally, be proved that the line of action of the resultant lateral force lies halfway between the normals to the direction of the free stream and the body axis. Consequently there is a drag due to lift

$$C_{Di} = C_N \frac{\alpha_{ZL}}{2} = \alpha_{ZL}^2 \tag{10.59}$$

the last member being true if $S_{\text{ref}} = S_B$.

There have been many attempts to refine the appealingly simple slender-body model and to account for the nonlinearity and separated flow which are observed when α_{ZL} exceeds a few degrees. One of the first and most successful was Allen's scheme (Allen and Perkins 1951). He adds to the normal force from (10.56) the drag of a circular cylinder which is suddenly accelerated from rest up to speed $u_0\alpha_{ZL}$, thereafter leaving behind a transient wake that causes pressures on the leeward side lower than those predicted by inviscid theory. More recently, sophisticated efforts were made to trace the concentrated vortices which trail backward from the vertex of a pointed configuration at higher α_{ZL} and large Reynolds numbers. An account of some of these will be found in Nielsen (1960, Chapter 4).

10.3 IMPROVED METHODS FOR SUPERSONIC BODIES OF REVOLUTION: HYPERSONIC FLOW

Despite its usefulness and simplicity, slender-body theory fails in some respects as a practical design tool. One of these is its inability to account rigorously for situations, like conical transition sections, in which there is a discontinuity in the surface slope dr_0/dx. Another is its prediction that there is no lateral loading $n(x)$ due to angle of attack on cylindrical sections where $S'(x) = 0$ in (10.56). Figure 10.7 contains several instances of the obvious incorrectness of this result, notably in the region immediately behind a step from positive to zero cross-sectional slope.

We choose to discuss quantitatively only one of the several available methods for remedying these weaknesses: a supersonic theory which is based on the original analysis by Lighthill (1948). In particular, our presentation summarizes some work of Kacprzynski (1965) and Kacprzynski and Landahl (1966). The key idea is to satisfy boundary conditions at the actual body surface rather than in the limit

$r \to 0$. The concentrated sources and doublets along the axes are replaced, in effect, by rings of similar singularities at $r = r_0(x)$.

Experience indicates that, when $M > 1$ but the *hypersonic parameter* $M\tau$ is still reasonably small (strictly, $M\tau \ll 1$), the fully linearized potential differential equation remains an excellent description of physics in the near field of flow. We consider the axisymmetric problem first and are, therefore, seeking to solve (10.15). Laplace transformation on x proves a powerful aid, for the same reasons set forth in Section 4.5 on supersonic airfoils. As a preliminary step, which will be helpful later for treating general shapes, let us calculate the potential $\varphi_1(x, r;\ x_0, r_0)$ which corresponds to a unit step at $x = x_0$ in dr_0/dx for a body with radius $r = r_0$. That is, we apply the boundary condition

$$\frac{1}{u_0}\frac{\partial \varphi_1}{\partial r}\bigg|_{r=r_0} = 1(x - x_0) \tag{10.60}$$

We remark that such a step would correspond to the beginning of a conical section, along which the surface radius would, in actuality, be steadily increasing. It will cause no difficulty to enforce (10.60) at $r = r_0$ for all x, however, since the body will later be divided into a series of infinitesimal sections and separate terms like φ_1 will be superimposed for each.

The transform is defined by [cf. (4.80)]

$$\mathscr{L}\{\varphi_1(x, r)\} \equiv \bar{\varphi}_1(s, r) = \int_0^\infty e^{-sx}\varphi_1(x, r)\, dx \tag{10.61}$$

If $x_0 > 0$, the potential and its derivatives must vanish for $x = 0$ and all r. It is, therefore, straightforward to transform (10.15) and (10.60), with the following results:

$$\frac{d^2\bar{\varphi}_1}{dr^2} + \frac{1}{r}\frac{d\bar{\varphi}_1}{dr} - (M^2 - 1)s^2\bar{\varphi}_1 = 0 \tag{10.62}$$

$$\frac{1}{u_0}\frac{d\bar{\varphi}_1}{dr}\bigg|_{r=r_0} = \frac{e^{-sx_0}}{s} \tag{10.63}$$

One general solution of (10.62) consists of a combination of the two "modified" Bessel functions of order zero and imaginary argument (see, e.g., Morse and Feshbach 1953, pages 1296 and 1323),

$$\bar{\varphi}_1 = C_1(s)I_0(\sqrt{M^2 - 1}\, sr) + C_2(s)K_0(\sqrt{M^2 - 1}\, sr) \tag{10.64}$$

Here the coefficients C_1 and C_2 depend parametrically on the Laplace variable s. The familiar requirement of vanishing disturbances as $r \to \infty$, coupled with the exponentially growing behavior of I_0 as its argument becomes large, forces us to drop this term by setting $C_1(s) = 0$. $C_2(s)$ is determined from (10.63),

$$\frac{1}{u_0}\frac{d\bar{\varphi}_1}{dr}\bigg|_{r=r_0} = \frac{C_2(s)}{u_0}\sqrt{M^2 - 1}\, sK_0'(\sqrt{M^2 - 1}\, sr_0) = \frac{e^{-sx_0}}{s} \tag{10.65}$$

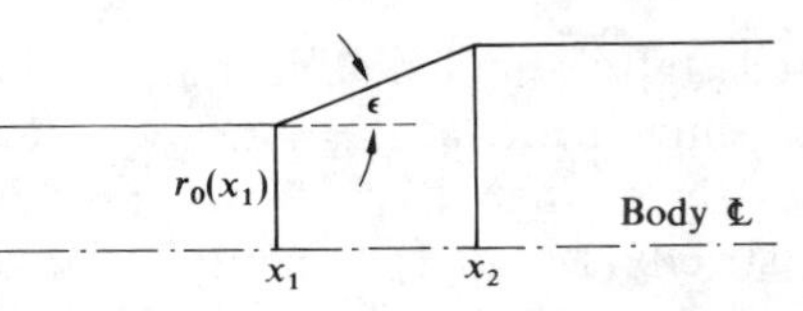

Fig. 10.11. Short sloping segment of an otherwise cylindrical body at zero angle of attack in supersonic flow.

Thus we compute the transformed potential

$$\frac{\bar{\varphi}_1(s, r; x_0, r_0)}{u_0} = \frac{e^{-sx_0}K_0(\sqrt{M^2 - 1}\, sr)}{\sqrt{M^2 - 1}\, s^2 K_0'(\sqrt{M^2 - 1}\, sr_0)} \tag{10.66}$$

As a simple illustration of how φ_1 might be applied, Fig. 10.11 pictures a portion $x_1 \leq x \leq x_2$ of an otherwise cylindrical shape in which the cone angle is ϵ. We assume this section to be short enough that the initial radius $r_0(x_1)$ is adequate for applying the boundary condition. Clearly this case can be analyzed by the application of successive positive and negative steps ϵu_0 in $\partial\varphi/\partial r$, the first starting at $x = x_1$ and the second at $x = x_2$. Since our assumption about r_0 means that each boundary condition has a form like (10.60), we are able immediately to identify the transform of the desired potential from (10.66) as

$$\frac{\Delta\bar{\varphi}}{u_0} = \frac{\epsilon[e^{-sx_1} - e^{-sx_2}]K_0(\sqrt{M^2 - 1}\, sr)}{\sqrt{M^2 - 1}\, s^2 K_0'(\sqrt{M^2 - 1}\, sr_0)} \tag{10.67}$$

One can construct the formal inverse transform of (10.67) by observing that it is a convolution between two factors. The first, $[e^{-sx_1} - e^{-sx_2}]/s$, inverts to a positive unit step followed by a negative unit step; the range of integration is therefore narrowed to $x_1 \to x_2$, since this quantity vanishes elsewhere. The second factor is just the inverse of the ratio K_0/sK_0'. In general, it would depend on the four quantities x, r, r_0, and $\sqrt{M^2 - 1}$. Kacprzynski (1965) has used the method of matched expansions to show, however, that the result is quite insensitive to small variations in r and that, when one is computing surface pressures or the flow field in the vicinity of the body, it is sufficiently accurate to replace r by $r_0(x_1)$ in (10.67). If, therefore, we define

$$F_{01}(x) \equiv \mathscr{L}^{-1}\left\{-\frac{K_0(s)}{sK_0'(s)}\right\} \tag{10.68}$$

we find that the sought-after "ramp" potential is, for $r \cong r_0$

$$\frac{\Delta\varphi(x, r_0)}{u_0} = -\frac{1}{\sqrt{M^2 - 1}} \int_{x_1}^{x_2} \epsilon F_{01}\left(\frac{x - x_0}{\sqrt{M^2 - 1}\, r_0}\right) dx_0 \tag{10.69}$$

The form of F_{01}'s argument, as it appears in (10.69), follows from two facts: This is the factor in the convolution which is supposed to be made a function of

$(x - x_0)$; and the "dual" in the physical space of the transform variable $s[\sqrt{M^2 - 1}\, r]$ is x divided by the coefficient $\sqrt{M^2 - 1}\, r$.

No closed-form expression for $F_{01}(x)$ is known. But Kacprzynski (1965) describes how it can be efficiently computed by numerical integration of a representation based on the general inversion formula,

$$\frac{1}{2\pi i}\int_{\gamma - i\infty}^{\gamma + i\infty} \frac{e^{sx}}{s}\frac{K_0(s\sqrt{M^2 - 1}\, r_0)}{K'_0(s\sqrt{M^2 - 1}\, r_0)}\, ds$$

$$= \frac{1}{2\pi i}\int_{(\sqrt{M^2-1}\, r_0\gamma - i\infty)}^{(\sqrt{M^2-1}\, r_0\gamma + i\infty)} \frac{\exp\left[S\dfrac{x}{\sqrt{M^2 - 1}\, r_0}\right]}{S}\frac{K_0(S)}{K'_0(S)}\, dS \qquad (10.70)$$

In (10.70), $S \equiv s\sqrt{M^2 - 1}\, r_0$ is a modified integration variable which demonstrates our statement about the dual arguments. γ is a positive constant which places the vertical contour to the right of all integrand singularities in the complex s-plane (cf. Churchill 1944, page 157), an acceptable choice being $\gamma = 0+$ here.

From (10.69) we can derive an especially convenient formula for the perturbation of axial velocity,

$$\frac{1}{u_0}\frac{\partial\, \Delta\varphi}{\partial x} = \frac{\epsilon}{\sqrt{M^2 - 1}}\int_{x_1}^{x_2} \frac{\partial}{\partial x_0} F_{01}\left(\frac{x - x_0}{\sqrt{M^2 - 1}\, r_0}\right) dx_0$$

$$= \frac{\epsilon}{\sqrt{M^2 - 1}}\left\{F_{01}\left(\frac{x - x_2}{\sqrt{M^2 - 1}\, r_0}\right) - F_{01}\left(\frac{x - x_1}{\sqrt{M^2 - 1}\, r_0}\right)\right\} \qquad (10.71)$$

During the integration here, we have used the fact that $\partial/\partial x$ may be replaced by $(-\partial/\partial x_0)$ when applied to a function of $(x - x_0)$ only.

The generalization of this solution for the $x_1 - x_2$ ramp to an arbitrary slender body of revolution can be illustrated by working with the near-surface velocity $\partial\varphi/\partial x$. We may conceive of the body as composed of a series of ramps of infinitesimal length dx_0, each with its local radius $r_0(x)$ and slope $\epsilon = dr_0/dx$. As terms resembling (10.71) are added for each successive ramp, we find quantities like

$$F_{01}\left(\frac{x - x_2}{\sqrt{M^2 - 1}\, r_0}\right)$$

canceling from one ramp to the next, unless there is a small change in slope $\Delta\epsilon$ between them. Expressed as a summation over n short body segments, the resulting potential would read approximately

$$\frac{1}{u_0}\frac{\partial\varphi}{\partial x} \cong -\frac{1}{\sqrt{M^2 - 1}}\sum_{i=0}^{n} \Delta\epsilon_i F_{01}\left(\frac{x - x_i}{\sqrt{M^2 - 1}\, r_0(x_i)}\right) \qquad (10.72)$$

Here $i = 0$ corresponds to the pointed nose, whereas by the law of forbidden signals $i = n$ would have to be at surface station x at which the velocity is being calculated. Now we can write

$$\Delta\epsilon_i = \frac{d}{dx_0}\frac{dr_0}{dx_0}\bigg|_{x_0 = x_{0i}} \Delta x_0 = \frac{1}{2\pi r_0(x_{0i})}\frac{d^2S}{dx_0^2}\bigg|_{x_0 = x_{0i}} \Delta x_0 \tag{10.73}$$

With (10.73) substituted into (10.72), it is clear that a limit ($\Delta x_0 \to 0, n \to \infty$) makes it possible for the sum to be replaced with an integral.

$$\frac{1}{u_0}\frac{\partial\varphi}{\partial x} = -\frac{1}{2\pi\sqrt{M^2 - 1}}\int_0^x F_{01}\left(\frac{x - x_0}{\sqrt{M^2 - 1}\,r_0(x_0)}\right)\left[\frac{d^2S/dx_0^2}{r_0(x_0)}\right] dx_0$$

$$= -\frac{1}{2\pi\sqrt{M^2 - 1}}\int_0^x F_{01}\left(\frac{x - x_0}{\sqrt{M^2 - 1}\,r_0(x_0)}\right)\frac{dS'(x_0)}{r_0(x_0)} \tag{10.74}$$

The last member here was the form given by Lighthill (1948). Interpreted as a Stieltjes integral, it allows for the possibility of a finite jump in S' such as occurs at the start or the end of a booster transition section.

Equation (10.74) is really all that is needed for numerical calculation of surface pressures—hence forebody pressure drag—in the axisymmetric flow. This is because the only nonzero nonlinear term,

$$\left(\frac{1}{u_0}\frac{\partial\varphi}{\partial r}\right)^2$$

appearing in the Bernoulli equation (10.35) is equal to the squared slope $(dr_0/dx)^2$. Hence we can write

$$C_p[x, r_0(x)] = -\frac{2}{u_0}\frac{\partial\varphi}{\partial x} - \left(\frac{dr_0}{dx}\right)^2 \tag{10.75}$$

and insert (10.74) immediately. Kacprzynski (1965) gives several examples of successful comparisons with test, exact theory for conical noses, and the like.

A similar development, detailed in the cited references, leads to solutions for the body of revolution at angle of attack and for various time-dependent motions. As an example, we reproduce the formula for axial velocity component near the surface of a pointed configuration at arbitrary small α_{ZL}.

$$\frac{1}{u_0}\frac{\partial\varphi_\alpha}{\partial x} = -\frac{\alpha_{ZL}\sin\theta}{\pi r_0(x)}\int_0^x F_{12}\left(\frac{x - x_0}{\sqrt{M^2 - 1}\,r_0(x_0)}\right) dS'(x_0) \tag{10.76a}$$

where

$$F_{12}(x) \equiv \mathscr{L}^{-1}\left\{\frac{K_1(s)}{s^2 K_1'(x)}\right\} \tag{10.76b}$$

K_1 is the modified Bessel function of the second kind and order 1. Numerical integrations are again required to determine F_{12} and, ultimately, the pressures and airloads.

Kacprzynski (1965) makes some imaginative proposals about how certain important nonlinearities can be accounted for during the estimation of these airloads. An instance is his formula for the distribution of lifting pressure at small angle of attack, which allows for the influence of the axial flow due to body thickness on this quantity:

$$\left(\frac{\partial C_p}{\partial \alpha_{ZL}}\right)\bigg|_{\text{Body surface}} = -2\left[\frac{\mathrm{M}_{\text{Surface}}(x)}{\mathrm{M}}\right]^2 \times \left\{1 - \frac{\gamma-1}{2}\left[\frac{2}{u_0}\frac{\partial\varphi}{\partial x} + \left(\frac{1}{u_0}\frac{\partial\varphi}{\partial x}\right)^2 + \left(\frac{dr_0}{dx}\right)^2\right]\right\}\frac{\partial\varphi_\alpha/\partial x}{\alpha_{ZL}u_0} \tag{10.77}$$

Here the first factor, calculable from $\partial\varphi/\partial x$, involves the ratio of local-surface to free-stream Mach numbers. $\partial\varphi/\partial x$ and $\partial\varphi_\alpha/\partial x$ would be taken, respectively, from (10.74) and (10.76).

Figure 10.12 reprints from Kacprzynski (1965) a plot of $C_{N\alpha}$ versus Mach

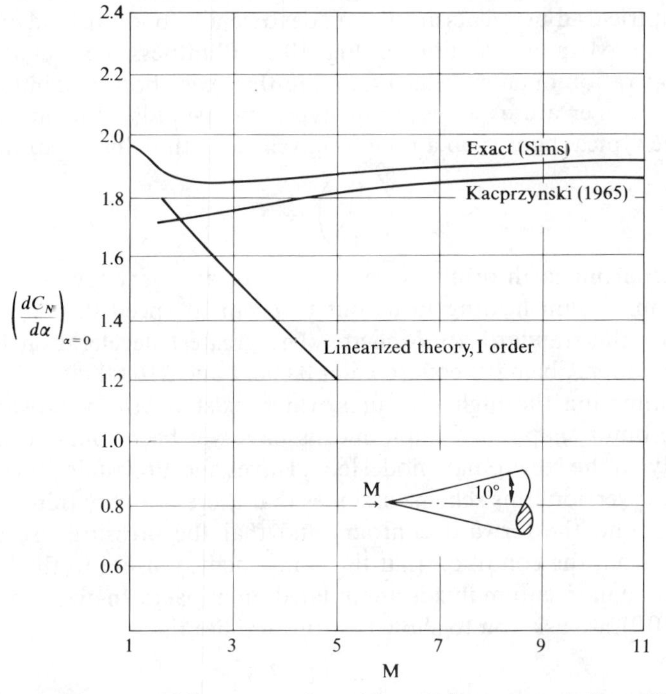

Fig. 10.12. Slope of the normal-force curve for a 10° cone in supersonic flow. The results of Kacprzynski (1965) are compared with exact calculations by Sims (1964) and with the first-order result of linearized theory.

number for a right circular cone of 10° semi-vertex angle. The curve obtained by integrating (10.76, 77) with respect to θ and x agrees better with the exact predictions (Sims, 1964) of inviscid supersonic theory than do any other methods, including slender-body theory and various second-order solutions.

We remark that Lighthill (1960) and Hayes and Probstein (1967, Chapter 2) give critical summaries of theories adaptable to bodies of revolution and other slender three-dimensional shapes, the latter with special reference to the hypersonic regime in which $\text{M}\tau = O(1)$. The *hypersonic small-disturbance theory* of Van Dyke (1954a, 1954b) forms the basis of several practical computation schemes. It has an analogy with the slender-body model, in that the lowest-order flow perturbations are found to be 2-D in crossflow planes. Since the disturbance velocities may be of the same order as a_∞, however, compressibility must be accounted for. Another practically useful approach is the extension (cf. Section 4.6) of shock expansion to axisymmetric shapes at high M, due to Eggers, Savin, and Syvertson (1955).

10.4 NEWTONIAN THEORY AND HEAT TRANSFER

Certain boosters and payload fairings are blunted at the nose, and some sort of semi-empirical adjustments must be made to slender-body airload predictions for the important M-range identified in Fig. 10.1. Bluntness is a much more critical factor to operation of entry vehicles, and in their case both maximum forces and peak heat transfer are experienced at hypersonic speeds. For instance, we can quote some typical figures on a nonlifting vehicle with ballistic parameter

$$\left(\frac{W}{C_D S}\right) \cong 65$$

which enters from earth orbital speed $v_E = \sqrt{gr}$ at a very low initial flight-path angle. Its maximum heating turns out to occur at speed $0.707v_E$ and altitude 226,000 ft in the standard atmosphere. The greatest deceleration is associated with maximum q_∞, being encountered at $0.606v_E$ and 216,000 ft.

For estimating the high pressures which exist on the windward side of a hypersonic blunt shape, no simple means have yet been found which improve significantly on the Newtonian model (e.g., Hayes and Probstein 1967, Chapter 3). In its earliest version, this scheme assumes that there is a very thin "shock layer" wrapped around the forward contours and that the pressure rise above p_∞ is determined from the condition that the component, normal to the local surface, of stream-tube momentum flux is annihilated on impact. In an assigned problem (Problem 10B), we ask you to show that this implies that

$$C_p \cong 2\cos^2\delta \tag{10.78}$$

where δ (see Fig. 10.13) is the angle between the flight direction and the surface normal.

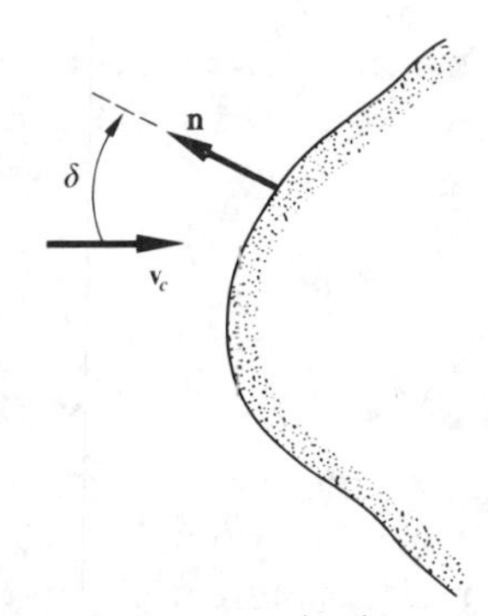

Fig. 10.13. A blunt nose and a hypersonic airstream, illustrating the angle δ.

It has been suggested that the Newtonian C_p on rounded convex bodies should be reduced to account for a normal pressure drop through the shock layer, required to curve the stream tubes in this layer. An equally successful adjustment to (10.78) is based on the observation that the factor 2 is associated with the total rise in impact pressure through a normal shock in the limit $\mathrm{M} \to \infty$, $\gamma \to 1$. Thus an improvement results from writing

$$\boxed{C_p \cong C_{p0} \cos^2 \delta} \tag{10.79}$$

where C_{p0} is the coefficient of stagnation pressure behind a normal shock under the actual conditions of hypersonic flow. For real gases $1 < C_{p0} < 2$, with values close to 2 being characteristic of air at entry speeds.

We close this chapter by furnishing, without derivation, some information on heat transfer, summarized from sources such as Nielsen (1960, Chapter 6) and Loh (1968). A conceptual law governing heat transfer rate $\dot{q}$, per unit area and time, from an airstream to a body surface is

$$\boxed{\dot{q} = C_h \frac{\rho_\infty}{2} v_c^3} \tag{10.80}$$

The dimensions will be found to be correct here, provided thermal energy is measured in mechanical units. The dimensionless C_h coefficient is generally a function of position, body shape, Re, M, and the relationship between specific enthalpy h_w of the gas at the surface temperature and the value h_{w0} which this quantity would have in thermal equilibrium, where $\dot{q} = 0$.

It is customary for preliminary design purposes to predict first the value $\dot{q} = \dot{q}_{\text{Stagnation}}$ at the forward stagnation point on the nose. Then the distribution of $\dot{q}$ over the surface is assumed to be proportional to $\dot{q}_{\text{Stagnation}}$, multiplied by a function of position only. According to Scala (1968), a generally useful relation is

$$\dot{q}_{\text{Stagnation}} = \frac{C\rho_\infty^{1/2} v_c^3}{\sqrt{R_B}} \tag{10.81}$$

where C depends on such information as the properties of the gas. For instance, if the body surface is cold (as it might be initially during entry), the stagnation enthalpy rise is

$$h_{t\,\text{Stagnation}} - h_w \cong \frac{v_c^2}{2} \tag{10.82}$$

and

$$C \cong [9.18 + 0.663\mathscr{M}] \times 10^{-10} \tag{10.83}$$

Here $\mathscr{M}$ is molecular weight of the ambient gas (28.9 for undissociated air), whereas the units of C are [Btu sec^2 ft^{-3} (lbm)$^{-1/2}$].

For some purposes, it is perhaps more useful to have a rough estimate of the total heat load experienced by the vehicle during entry. For instance, Allen and Eggers (1957), in their classic on the subject, made use of the *Reynolds analogy* between skin friction and heat transfer to derive

$$Q_{\text{Total}} = \frac{C_f' \rho_\infty v_E^2 S}{4 \dfrac{W}{C_D S} |\sin \gamma_E|} \tag{10.84}$$

Here γ_E is the initial entry angle corresponding to orbital speed v_E. C_f' is an averaged friction coefficient,

$$C_f' = \frac{1}{S} \int_{\text{Body surface}} C_{f\,\text{Local}} \left(\frac{\rho_{\text{Local}} u_{\text{Local}}}{\rho_\infty v_c} \right) dS \tag{10.85}$$

(10.84) and (10.85) were used to advocate substantial blunting of the nose of the vehicle as a means of reducing Q_{Total}. Forward portions of such a shape cause the speed u_{Local} just outside the boundary layer to be generally much smaller than, say, on a pointed cone or flat plate. The correspondingly reduced C_f' then ameliorates the effects of heating

We emphasize (cf. Scala 1968) that hypersonic heat transfer is actually a complex subject, involving interactions among the various chemical species in the high-temperature boundary layer and, for superorbital entry, even significant radiation. The brief review we offer here is, therefore, mainly to indicate how various parameters enter into the process. To get the full flavor of the field, you must delve deeply into an extensive literature.

10.5 PROBLEMS

10A Determine the forebody pressure drag (subsonic and supersonic) on a slender body of revolution with parabolic shape

$$\frac{r_0}{r_{\max}} = \frac{4x}{l}\left(1 - \frac{x}{l}\right)$$

Explain why slender-body theory for M > 1 shows this to be independent of flight Mach number. Why would you expect such results to be most successful in the higher transonic range?

10B The simplified Newtonian law for airloads on forward portions of blunt bodies in hypersonic flight assumes that the component of fluid momentum (in a stream tube approaching the body) perpendicular to the local body surface is just canceled by impact. The idea is that a shock wave is tightly wrapped around the body, and a high-density "shock layer" of fluid is trapped in between and moves around toward the back. Show that this law implies a local surface pressure [cf. (10.79)]

$$p = p_\infty + \rho_\infty v_c^2 \cos^2 \delta$$

and stagnation pressure $C_{p0} = 2$. On physical grounds, explain why this law could be expected to give good results on a body with constant δ (wedge or cone at zero incidence), but that there is a problem about centrifugal effects in the shock layer on a curved surface.

Determine the forebody drag of a wedge, cone, and hemisphere. Explain why it is usually assumed that the "lee side" of such a body is at a pressure somewhere between p_∞ and vacuum, and why the contribution of this region to forces and moments on blunt bodies is quite unimportant.

10C A supersonic body has the form of a right-circular double cone, which is symmetrical front-to-back (see the figure). Let $(2r_M/l) = 0.1$.

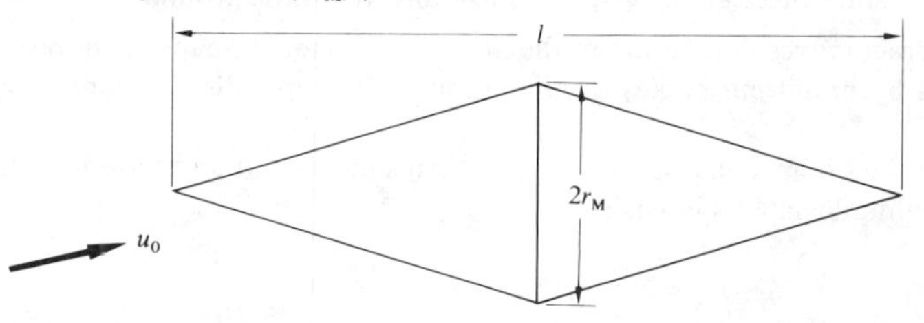

a) Estimate the zero-lift drag by Newtonian theory, assuming that the back half of the body is exposed to vacuum ($p = 0$).
b) Estimate the zero-lift drag by supersonic slender-body theory, including drag due to skin friction at $\text{Re}_l = 10^6$. [Be careful with the singular behavior of $S(x)$ at the middle.]
c) Determine $C_{L\alpha}$ and the location of the aerodynamic center, according to slender-body theory, but with two different hypotheses.
   i) The flow remains attached over the entire length.
   ii) The flow separates at the maximum cross section, so that the body creates the same disturbance as a cone followed by an infinite cylinder of radius r_M.

10D a) Assuming that it is a sphere, what is the size of the smallest reentry device capable of existing at 170,000 ft and 25,000 ft/sec, given that the maximum allowable skin temperature is 3500°R? The heat emitted per unit time due to surface radiation is $\dot{q}_{\text{rad}} = \sigma\epsilon T_{\text{skin}}^4$, where ϵ = skin emissivity = 0.9, and σ = Stefan–Boltzmann constant = 0.482×10^{-12} Btu/ft²-sec-°R.

[*Note:* The temperature at the stagnation point is determined by a balance between convective heat input $\dot{q}$ and output $\dot{q}_{\text{rad}}$.]

b) Which would be more effective in reducing the minimum size necessary for the above reentry device, a 10% decrease in v_c or a 10% increase in the maximum allowable skin temperature?

10E According to the elementary theory of sonic boom in a uniform atmosphere (Whitham 1952, Hayes 1971), the overpressure experienced remotely at the earth's surface due to supersonic overflight by a lifting vehicle is given by

$$\Delta p = \frac{p_\infty K_r \gamma M^2 F(x - \sqrt{M^2 - 1}\,h)}{[2h\sqrt{M^2 - 1}]^{1/2}}$$

h is the vehicle's altitude, and K_r is a reflection factor that would be unity in free air but approaches 2 for a solid plane surface. The function F is defined by the approximate equation

$$F = \frac{1}{2\pi} \int_0^{x - \sqrt{M^2 - 1}\,h} \frac{A''(x_1)\,dx_1}{\sqrt{x - \sqrt{M^2 - 1}\,h - x_1}}$$

where

$$A''(x_1) = S''(x_1) \sin \mu + \frac{[M^2 - 1]^{1/4} n'(x_1)}{\gamma p_\infty M^2}$$

Here μ is the Mach angle, $n(x)$ is the lift force per unit length of the vehicle, and $S(x)$ is the oblique cross-sectional area intercepted by the Mach plane $x + \sqrt{M^2 - 1}\,z = \text{constant}$, along which disturbances are propagated most directly to the ground.

a) Use the foregoing results to examine the effects of volume and shape on the peak overpressure produced by nonlifting bodies of revolution with simple distributions of $r_0(x)$ that you choose.

b) Calculate the boom signature for a typical cruising aircraft and look into design changes which might alleviate its intensity.

CHAPTER 11

DYNAMIC PERFORMANCE: ATMOSPHERIC ENTRY

11.1 INTRODUCTION; EQUATIONS OF MOTION

Our development in Section 9.1 of the laws governing booster operation may readily be adapted to study a vehicle entering a planetary atmosphere from orbital flight. Indeed the latter case is a specialization in which the angle γ is mostly negative, the thrust is zero, and the mass is constant except for possible minor losses due to an ablative thermal protection system. The assumption of planar motion over a nonrotating planet, embodied in (9.7), is more appropriate for entry than for boost. This is because both the peak heat transfer rate and the highest airloads tend to occur near or above one-half the initial speed of entry. Comparing the inertial and relative values of the velocity vector (designated $\mathbf{u}_c$ and $\mathbf{v}_c$, respectively, in Section 9.5), we use earth as an illustration and find from (9.82) that they differ by a westward-pointing vector of magnitude 1519 cos λ ft/sec at sea level. Although this difference may be of order 10% in the range of interest, the errors due to neglecting rotation usually turn out, in practice, to be much smaller. Hence we feel justified in overlooking them for introductory purposes. We warn, however, that all the effects of rotation and nonsphericity treated in Section 9.5 must be considered when constructing trajectories to the precision needed for actual space navigation.

With $T = 0 = \dot{m}$ and (k/r^2) replaced by g, the first three of the equations of motion (9.7) are the important tools needed for the present analysis.

$$m\dot{v}_c = -D - mg \sin \gamma \tag{11.1a}$$

$$mv_c\dot{\gamma} = \frac{mv_c^2 \cos \gamma}{r} + L - mg \cos \gamma \tag{11.1b}$$

$$\dot{r} \equiv \dot{h} = v_c \sin \gamma \tag{11.1c}$$

As noted following (9.10), (9.7d) is also useful for predicting horizontal position over the planet's surface. It may be solved as an afterthought, however, since the angle ν is not coupled into the other members of (9.7).

Now it should be evident that, after we introduce conventional aerodynamic relations and choose initial values for v_c, γ, and h, (11.1) can be numerically integrated, as discussed in Chapter 9. No further approximation would have to be made. A suitable control for the entry maneuver would be angle of attack α_{ZL} or lift coefficient C_L, the latter being a unique function of α_{ZL} and M. Since no particular difficulty nor particular enlightenment results from such an approach, we do not pursue it here. Rather we examine the consequences of a series of further simplifications, starting with an "exponential atmosphere" and constant g in (11.1a) and (11.1b).

11.2 APPROXIMATE ANALYSIS OF GLIDING ENTRY INTO A PLANETARY ATMOSPHERE

Among many schemes that have been published for the approximate solution of (11.1), the one described by Loh (1968) seems to serve best our aims. Not only is it an imaginative example of how to choose properly the variables in a problem, but it is believed to include, as special cases, a majority of the other simplified entry solutions. Loh's article also cites most authors who had given important results prior to 1968—that is, it covers that early period of research in entry and space exploration when interest was at its highest. Three assumptions underlie the entire theory.

- In the range of altitudes in which most of the entry deceleration takes place, the relation between density and altitude can be idealized to

$$\rho \cong \rho_0 e^{-\beta h} \tag{11.2}$$

  Here ρ_0 and β are constant parameters used to obtain a best fit to actual data. For instance, Chapman (1958) proposed $\beta^{-1} = 23{,}500$ ft and $\rho_0 = 0.0034$ slug/ft^3 for the earth's atmosphere (note that ρ_0 does not necessarily correspond to standard sea level). The inverse of β is called *scale height*. By integration of (4.4, 5) under hydrostatic equilibrium with $\mathbf{q} = 0$, we can show that (11.2) is exact for an isothermal layer with constant values of g and gaseous molecular weight.
- An averaged gravitational constant g can be employed. Below $h = 300{,}000$ ft, the error is less than 3% if we adopt the sea-level value g_E, but a much better fit can be made.
- In the term of (11.1b) containing r, the sea level R_E may be substituted. The error in this term alone is of order 1%.

You can imagine the long history of ballistic trajectory analyses that preceded the study of entry, and therefore furnished guidance to this new field. The foregoing assumptions are, for example, quite natural in ballistics. Following several prior investigators, Loh (1968) took over from ballistics the idea of using a quantity related to distance along the flight path as the primary independent variable. Thus (11.1c) enables us to replace time by this distance or by altitude h itself, through

$$(\dot{\ldots}) \equiv \frac{d}{dt}(\ldots) = v_c \sin\gamma \frac{d}{dh}(\ldots) \tag{11.3}$$

When (11.3) is inserted into (11.1a), the longitudinal-acceleration term on the left of (11.1a) becomes

$$m\dot{v}_c = (m \sin \gamma)v_c \frac{dv_c}{dh} = \frac{m \sin \gamma}{2} \frac{d}{dh}(v_c^2) \tag{11.4}$$

We insert (11.4) and the conventional definition of drag coefficient C_D, divide by $mgR_E\beta \sin \gamma$, and manipulate (11.1a) into the form

$$\frac{d}{d(\beta h)}\left[\frac{v_c^2}{gR_E}\right] + \frac{C_D S}{m\beta}\left[\frac{\rho}{\sin \gamma}\right]\frac{v_c^2}{gR_E} = -\frac{2}{\beta R_E} \tag{11.5}$$

We next divide the left side of (11.1b) by $m\beta v_c^2$ and again use (11.3), as follows:

$$\left[\frac{1}{m\beta v_c^2}\right] m v_c \frac{d\gamma}{dt} = \frac{\sin \gamma}{\beta}\frac{d\gamma}{dh} = -\frac{d \cos \gamma}{d(\beta h)} \tag{11.6}$$

The resulting modification of (11.1b) is

$$\frac{d \cos \gamma}{d(\beta h)} - \frac{1}{\beta R_E} \cos \gamma \left[\frac{gR_E}{v_c^2} - 1\right] = -\left(\frac{L}{D}\right)\frac{C_D S \rho}{2m\beta} \tag{11.7}$$

Before proceeding, we comment that (11.5) and (11.7) offer another convenient set of entry equations. With no approximations beyond those already discussed, integrating them numerically yields histories of the angle γ and dimensionless speed $v_c/\sqrt{gR_E}$ as functions of dimensionless altitude βh. Some care is needed with (11.5) on account of division by $\sin \gamma$, which may initially be very small. γ can even pass through zero on a "skip trajectory," in which the vehicle develops excess lift during its dive, executes a pull-up as it encounters higher densities, and is thrown back into space.

Commonly such calculations are carried out at a fixed L/D, since this is the condition that would occur hypersonically if no elevator displacements were applied to change the trimmed α_{ZL}. Alternatively, L/D may be employed as a control, in which case the lift is said to be *modulated*. Figures 11.1 and 11.2, adapted from Loh (1968), are examples of constant-(L/D) trajectories for an earth-entry glider with the rather small *ballistic coefficient* $(mg/C_D S) = 3.2$ lb/ft^2. Note the effects of bringing (L/D) up from -1.0 to the (quite large) hypersonic value of 2.0: The altitudes at which events like peak deceleration occur are increased and the final dive, during which γ approaches large negative values, is delayed to progressively lower speeds. At the two highest lift–drag ratios, we see evidence early during the entry of a phugoid oscillation, whose damping is observed to grow as higher densities are encountered.

As a basis for further approximations, the differential of (11.2)

$$d\rho = -\rho_0 e^{-\beta h}\, d(\beta h) \tag{11.8a}$$

enables us to replace altitude by density as the independent variable through

$$\frac{d}{d(\beta h)}(\ldots) = -\rho_0 e^{-\beta h}\frac{d}{d\rho}(\ldots) = -\rho\frac{d}{d\rho}(\ldots) \tag{11.8b}$$

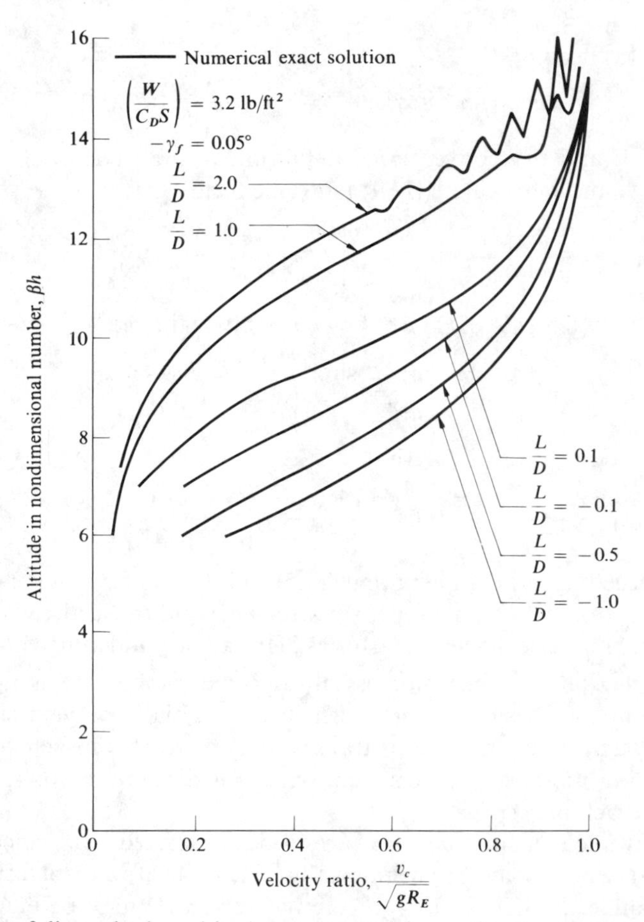

Fig. 11.1. Plot of dimensionless altitude versus speed for earth entry at several values of lift-to-drag ratio and other quantities as indicated. Note the very low ballistic coefficient and the high-altitude oscillations which occur at $(L/D) = 1$ and 2. (Adapted from Loh 1968.)

The corresponding counterparts of (11.5) and (11.7) are

$$\frac{d}{d\rho}\left[\frac{v_c^2}{gR_E}\right] - \frac{C_D S}{m\beta \sin\gamma}\left[\frac{v_c^2}{gR_E}\right] = \frac{2}{\rho\beta R_E} \tag{11.9a}$$

$$\frac{d\cos\gamma}{d\rho} + \frac{\cos\gamma}{\rho}\frac{1}{\beta R_E}\left[\frac{gR_E}{v_c^2} - 1\right] = \left(\frac{L}{D}\right)\frac{C_D S}{2m\beta} \tag{11.9b}$$

Among the dozens of solutions reviewed by Loh (1968) and resulting from various reductions of (11.9), we shall exhibit only three or four. The first is what he calls a "second-order" solution, claiming its validity for preliminary design of almost any entry trajectory. By careful study of exact numerical results, Loh (1968) determined that, for purposes of integration with respect to ρ, the second term on the left of (11.9b) may be treated as a constant. That is, although the

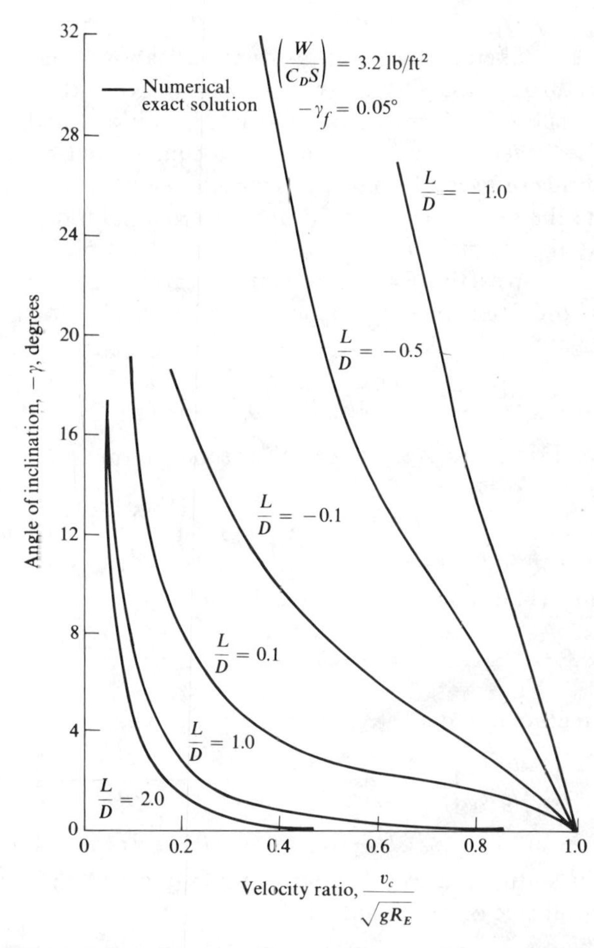

Fig. 11.2. Plot of flight-path angle versus dimensionless speed for earth entry at several values of lift-to-drag ratio and other quantities as indicated. (Adapted from Loh 1968.)

individual quantities ρ, γ, and v_c vary along the trajectory, their combined behavior is such as to render the integral insensitive to changes in this term. Under Loh's assumption, (11.9b) can be solved immediately to give

$$\cos\gamma \cong \cos\gamma_f + \left\{\left(\frac{L}{D}\right)\frac{C_DS}{2m\beta} + \frac{\cos\gamma}{\rho}\frac{1}{\beta R_E}\left[1 - \frac{gR_E}{v_c^2}\right]\right\}[\rho - \rho_f] \quad (11.10)$$

From (11.10) we obtain the following explicit approximation for γ as a function of density and speed:

$$\cos\gamma \cong \frac{\cos\gamma_f + \dfrac{1}{2}\left(\dfrac{L}{D}\right)\dfrac{C_DS}{m\beta}[\rho - \rho_f]}{1 + \dfrac{1}{\beta R_E}\left[\dfrac{gR_E}{v_c^2} - 1\right]\left[1 - \dfrac{\rho_f}{\rho}\right]} \quad (11.11)$$

Here L/D has been taken as constant, as discussed above, and the solution is arranged to pass through the altitude at which $\rho = \rho_f$ when the flight-path angle is γ_f. Although subscript f might be associated with a "final" point on the trajectory (e.g., sea level), this reference state can equally well be chosen at some intermediate altitude or even, in some cases, at initial entry, at which $\rho_f \cong 0$.

We complete the development of the second-order solution with the further assumption that the quantity $2/\beta R_E$ on the right of (11.5) is negligible. The justification is that for earth $\beta R_E \cong 900$, and similarly large values characterize other planets whose atmospheric properties are reasonably well known. This step reduces (11.9a) to

$$\sin\gamma \frac{d}{d\rho}\left[\frac{v_c^2}{gR_E}\right] \cong \frac{C_D S}{m\beta}\left[\frac{v_c^2}{gR_E}\right] \tag{11.12}$$

Again we invoke Loh's discovery about the second term in (11.9b) to rewrite that equation as

$$\sin\gamma \frac{d\gamma}{d\rho} = \frac{\cos\gamma}{\rho}\frac{1}{\beta R_E}\left[\frac{gR_E}{v_c^2} - 1\right] - \left(\frac{L}{D}\right)\frac{C_D S}{2m\beta} \cong \text{constant} \tag{11.13}$$

Dividing (11.12) by (11.13), we get an expression for

$$\frac{d}{d\gamma}\left[\frac{v_c^2}{gR_E}\right]$$

which can be integrated, under the condition $v_c = v_f$ when $\gamma = \gamma_f$, to yield

$$\ln\left[\frac{v_c^2/gR_E}{v_f^2/gR_E}\right] \cong \frac{[C_D S/m\beta][\gamma_f - \gamma]}{\left(\frac{L}{D}\right)\frac{C_D S}{2m\beta} - \frac{\cos\gamma}{\rho}\frac{1}{\beta R_E}\left[\frac{gR_E}{v_c^2} - 1\right]} \tag{11.14}$$

Restoring ρ to its status as a variable, we can finally calculate it explicitly from (11.14) in its dependence on speed and angle,

$$\rho\left[\frac{C_D S}{m\beta}\right] \cong \frac{\frac{\cos\gamma}{\beta R_E}\left[\frac{gR_E}{v_c^2} - 1\right]\ln\left[\frac{v_c^2/gR_E}{v_f^2/gR_E}\right]}{[\gamma - \gamma_f] + \frac{1}{2}\left(\frac{L}{D}\right)\ln\left[\frac{v_c^2/gR_E}{v_f^2/gR_E}\right]} \tag{11.15}$$

[Note that both numerator and denominator on the right of (11.15) are negative during the course of a typical entry, and that increasing density is associated with the increasing magnitude of the numerator as speed v_c drops off.]

The significance of the approximation summarized by (11.11) and (11.15) is that they furnish two transcendental relations among the three principal unknowns γ, ρ, and $v_c/\sqrt{gR_E}$. Therefore, trajectories can be constructed by prescribing γ_f, ρ_f, v_f plus a history for any one of the variables, whereupon the other two may be computed with relative ease. The reference speed $\sqrt{gR_E}$ corresponds, of course, to a hypothetical circular orbit at the earth's surface.

These solution families depend on the following three parameters.

- L/D, which is essentially a vehicle property.
- βR_E, which characterizes the planet and its atmosphere.
- $C_D S/m\beta$, a combined parameter with dimensions of inverse density. We might, for instance, multiply this group by ρ_f and multiply and divide by g. It then becomes the ratio between $\rho_f g/\beta$—the weight of a unit column of air with density ρ_f and height equal to the scale height—and the ballistic parameter $mg/C_D S$.

Loh (1968) furnishes numerous examples of entries from various initial conditions, as well as comparisons either with numerical results from (11.5)–(11.7) or with more restricted approximations. The former seem generally to agree quite successfully, and supercircular entries or even trajectories with multiple skips into the atmosphere can be handled with suitable adjustments. The principal difficulty with the second-order scheme seems to involve matching trajectories with outer-space orbits in which $\rho = 0$. There, of course, the gravitational term on the right side of (11.5) or (11.9a) predominates over the drag term, so that neglecting it cannot be justified.

An important special case of the foregoing analysis involves ballistic entry. The condition $(L/D) \cong 0$ typifies most missiles, and also the recovery vehicles used in early manned space programs. Equations (11.11) and (11.14) simplify to

$$\cos\gamma \cong \frac{\cos\gamma_f}{1 + \dfrac{1}{\beta R_E}\left[\dfrac{gR_E}{v_c^2} - 1\right]\left[1 - \dfrac{\rho_f}{\rho}\right]} \tag{11.16}$$

and

$$\ln\left[\frac{v_f^2/gR_E}{v_c^2/gR_E}\right] \cong \frac{[C_D S/m\beta][\gamma_f - \gamma]}{\dfrac{\cos\gamma}{\rho}\dfrac{1}{\beta R_E}\left[\dfrac{gR_E}{v_c^2} - 1\right]} \tag{11.17}$$

We remark that, if we choose $\rho_f \cong 0$, (11.16) suggests that there is an almost unique relation between speed and flight-path angle along the zero-lift trajectory. Perhaps more interesting, however, is the fact that large βR_E causes the denominator of (11.16) to be close to unity, whence the angle is seen to remain nearly invariant, $\gamma \cong \gamma_f$. In actuality, the assumption of constant-γ ballistic entry proves quite accurate only when $|\gamma_f| > 5°$ or so, because smaller angles lead to long times of flight, which give the neglected gravitational term and other effects more opportunity to curve the path. Several early investigators, including Gazley (1957), Allen and Eggers (1957), and Chapman (1958), arrived at these results by initially setting $\gamma \cong$ constant and noting that drag far outweighs the gravity term in (11.1a) during critical portions of the entry.

Although solutions of this type can be derived by manipulations of (11.11) and (11.14, 15), it is more straightforward to return directly to (11.5) and drop the weight term from the right-hand side. Canceling out the constant factor $(\beta g R_E)^{-1}$ and dividing by v_c^2, we obtain

$$\frac{1}{v_c}\frac{dv_c}{dh} \cong -\frac{\rho C_D S}{2m \sin \gamma} \cong -\frac{\rho_0 C_D S}{2m \sin \gamma_f} e^{-\beta h} \tag{11.18}$$

In the last member here, we have further employed (11.2) and assigned γ its fixed value. It is not difficult to integrate (11.18), with the initial condition $v_c = v_f$ when $h \to \infty$, and arrive at

$$v_c = v_f \exp\left[-\frac{\rho_0 C_D S}{2m \sin |\gamma_f|} e^{-\beta h}\right] \tag{11.19}$$

where the proper algebraic sign has been displayed in the brackets, since $\gamma_f < 0$.

You will find it interesting to examine some of the implications of (11.19). You will discover, for instance, that the peak entry deceleration is

$$\left|\frac{dv_c}{dt}\right|_{\max} = \frac{\beta}{2e} v_f^2 \sin |\gamma_f| \tag{11.20}$$

and that it occurs at an altitude

$$h_{\max} = \frac{1}{\beta} \ln\left[\frac{\rho_0 C_D S}{\beta m \sin |\gamma_f|}\right] \tag{11.21}$$

Using (10.80) or (10.81), we can make somewhat rougher estimates of the most severe conditions of aerodynamic heat transfer.

Among other consequences of (11.20), we can demonstrate why nonlifting manned entry is infeasible from subcircular orbits which extend only part way around the earth. We recall that peak human tolerances demand a $|dv_c/dt|_{\max}$ appreciably less than $20g$, and we observe that (11.20) is independent of the ballistic coefficient. It develops that only certain combinations of initial conditions v_f and γ_f will keep the deceleration within the required bounds. These can be achieved either with nearly tangential orbits close to circular or with short, low-energy trajectories like those used on early Mercury flights. Anything in between, however, proves quite unacceptable.

Somewhat complementary to the steep ballistic solutions are those involving small γ_f and relatively high L/D, typified by the work of Eggers, Allen, and Neice (1957). Such a result may be derived by taking $\cos \gamma \cong \cos \gamma_f \cong 1$ in Loh's equation (11.10), an approximation with less than 1% error for $|\gamma| < 8°$. The quantity in braces is then seen to equal zero, whence follows a simple relation between speed and density or altitude:

$$\frac{v_c^2}{gR_E} \cong \frac{1}{1 + \dfrac{R_E}{2}\dfrac{L}{D}\left[\dfrac{C_D S}{m}\right]\rho} = \left\{1 + \frac{R_E}{2}\frac{L}{D}\left[\frac{\rho_0 C_D S}{m}\right] e^{-\beta h}\right\}^{-1} \tag{11.22}$$

Clearly (11.22) describes circumstances in which v_c approaches $\sqrt{gR_E}$ as $\rho \to 0$, that is, the initial orbit must be low and nearly circular. Hence it is possible to manipulate (11.14) by replacing v_f with $\sqrt{gR_E}$, using (11.22) to eliminate ρ, and evaluating limits associated with the very small γ. One form of the result provides a connection between flight-path angle and speed for the gliding entry,

$$\sin \gamma \cong \frac{-\dfrac{2}{\beta R_E}}{\dfrac{L}{D}\left[\dfrac{v_c^2}{gR_E}\right]} \tag{11.23}$$

The formula

$$\frac{dv_c}{dt} = \frac{dv_c}{dh}\frac{dh}{dt} = v_c \sin \gamma \left[\frac{dv_c}{dh}\right] \tag{11.24}$$

can be combined with (11.22, 23) to study such questions as the location and magnitude of peak deceleration. One vitally important conclusion concerns the effectiveness of even small positive lift–drag ratios in ameliorating the severity of the entry process. Figure 2.29 of Loh (1968) dramatizes this phenomenon, showing that even the large planet Jupiter may be entered at $(L/D) > 1$ with decelerations no greater than four times earth's gravity, despite Jupiter's surface value of $2.64g_E$.

Inasmuch as (11.22, 23) are limited to earth-orbital operations, it seems worthwhile to display another similar approximation, also due to Eggers, Allen, and Neice (1957), which relates to skip from supercircular initial conditions. Let the reference state correspond to the extreme outer atmosphere, in which $\rho_f \cong 0$, and let L/D be 1 or greater. Because of the size of βR_E, it is again logical to neglect the second term in the denominator of (11.11), whereupon we are led to

$$\cos \gamma - \cos \gamma_f \cong \frac{1}{2}\left[\frac{C_L S}{m\beta}\right]\rho = \frac{1}{2}\left[\frac{C_L S}{m\beta}\right]\rho_0 e^{-\beta h} \tag{11.25}$$

The original authors, incidentally, arrived at (11.25) by assuming in (11.1b) that the lift term on the right predominated over the difference between the gravitational and centrifugal terms.

The history of speed along the skip trajectory was first estimated by dropping, on the grounds of large βR_E and v_c^2/gR_E near unity, the second denominator term in (11.14), whence

$$\ln\left[\frac{v_c}{v_f}\right] \cong \frac{\gamma_f - \gamma}{L/D} \tag{11.26}$$

We point out, however, that at very low densities both this omission and the neglect of the gravity term on which (11.14) is based deserve reexamination.

Tentatively accepting (11.25) and (11.26), we see that the former implies a completely symmetrical relationship between angle and density (or altitude) along the flight path. γ starts at its greatest negative value γ_f, at which $\rho \cong 0$, passes through zero, at which

$$\rho = \rho_{\max} \equiv \frac{2(1 - \cos \gamma_f)}{C_L S/m\beta} \tag{11.27}$$

then increases to $(-\gamma_f)$ at exit from the atmosphere. During the skip, (11.26) shows a continuous decay of speed due to dissipation of energy by drag. Starting from the initial v_f, at which $\gamma = \gamma_f$, v_c falls off to

$$v_c = v_f \exp\left(\frac{-|\gamma_f|}{L/D}\right) \tag{11.28a}$$

at the minimum altitude and to

$$v_c = v_{c\,\min} \equiv v_f \exp\left(\frac{-2\,|\gamma_f|}{L/D}\right) \tag{11.28b}$$

at exit.

Furthermore, $v_{c\,\min}$ and $(-\gamma_f)$ may be regarded as initial conditions for the orbit in space which follows the skip. In the absence of other significant gravitational fields, we can decide whether the vehicle will return for another entry or escape by testing whether this orbit is hyperbolic or elliptical. Clearly, if the initial v_f were less than the escape speed, a further return is assured. It is also possible to estimate whether a skip will occur at all: a $\rho_{\max}$ below sea level is one negative indication, and a $v_{c\,\min}$ less than the circular value $\sqrt{gR_E}$ signals almost immediate reentry.

Guidance and control of entry from low, nearly circular planetary orbits is a fairly straightforward matter. Typically, a retro-rocket is fired ahead and upward for the period needed to establish suitable initial values for v_c and γ. With $(L/D) \cong 0$, the instant when this impulse is applied is important for fixing the area in which the trajectory intersects the surface and in which the vehicle is to be recovered. When there are flight controls and a small amount of lifting capability, moreover, maneuvering as a hypersonic glider provides considerable latitude as to the choice of landing site.

On the other hand, guidance of a vehicle during its return from interplanetary flight is likely to be severely constrained by the *entry corridor*. A representative entry corridor is pictured in Fig. 11.3. When it is necessary to complete the maneuver during a single pass into the atmosphere, as with return of the Apollo command module from translunar orbit, corridor widths in the range of 10–20 statute miles are typical. Since v_f is nearly 7 mi/sec, initial conditions of orbit must obviously be established with great precision.

The *overshoot boundary* or upper limit of the corridor corresponds to the initial speed and flight-path angle at which an appreciable skip would first take place. This limit might be estimated, say for preliminary design purposes, from the approximate solution in (11.25, 26) at fixed (L/D). We recommend a more sophisticated analysis, however, because a vehicle which develops an appreciable C_L can extend its overshoot boundary by resort to inverted flight, whereby the lift force is used to "pull" it into the atmosphere from altitudes at which skipping would occur at zero or positive (L/D).

The *undershoot boundary* is usually fixed by the desire to avoid excessively

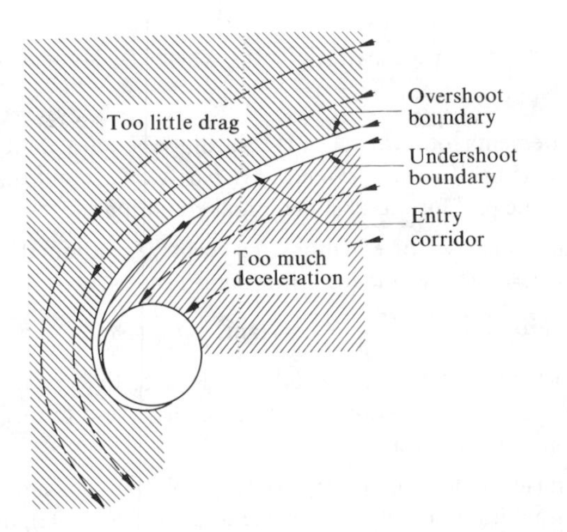

Fig. 11.3. The entry corridor and circumstances which prevent successful entry of vehicles with trajectories outside it.

large peak decelerations and/or unacceptably severe heating. Again these phenomena can be examined by certain of the approximations discussed above. But we reemphasize the need for much more accurate analyses, including such effects as that of planetary rotation, when quantitative trajectories are required for actual space operations. For example, most Apollo flights impacted the ocean within a few thousand yards of a preassigned spot. For this sort of mission, not only must initial conditions be set well within the corridor, but the clock time of entry has to be fixed with similar precision.

11.3 PROBLEMS

11A Specialize (11.1) to describe orbits in the vacuum of space about a single gravitating center. Use the results, together with your knowledge about the universal law of gravitation, to verify the following (polar coordinates have certain advantages).

a) All orbits are conic sections (ellipse, parabola, hyperbola) with the gravitating center at a focus. Discuss how to determine whether a given satellite will "escape."

b) Kepler's laws of planetary motion are as follows
   1) Each planet's orbit is an ellipse, with the sun at one focus.
   2) Area within the ellipse is swept at a constant rate by the radius r connecting the planet to the sun.
   3) The orbital periods squared are directly proportional to the cubes of the semi-major axes of the ellipses.

11B For the nonlifting entry approximations of (11.19), determine the *peak* value of the specific force $f \cong D/m$ and the altitude at which it occurs. This is the inertial deceleration which would be experienced by astronauts in the vehicle. What is the maximum permissible value of γ (in degrees) that will ensure that $f_{\text{peak}} < 5g_E$'s?

Noting the dependence of f on v_i, could you foresee a vehicle such as Apollo, which enters at $v_{esc} = \sqrt{2}v_{sat}$, utilizing a pure ballistic entry scheme as studied here? How would you improve things?

11C Verify the statements following (11.21) regarding the infeasibility of nonlifting manned entry from subcircular orbits which extend only part way around the earth. The results of Problem 11A will be helpful for establishing initial conditions.

11D Develop a simplified set of equations for reentry westbound over the equator on a rotating earth ($\beta = \lambda = 0$). Assume the following:

$$\rho = \rho_0 e^{-\beta h}, \qquad g = \text{constant}, \qquad r \cong R_E, \qquad \gamma \ll 1$$

Evaluate the importance (relative magnitude) of each of the remaining terms, and for an example of your choice, calculate numerical values for these terms. Study the process of solving the equations you develop.

11E Working with (11.9), develop an approximate solution which describes how the nearly circular orbit of a nonlifting satellite decays due to drag. In (11.9a), it would be appropriate to neglect the gravity term on the right because the flight path is nearly horizontal. In (11.9b), $(L/D) = 0$ and $v_c \cong \sqrt{gR_E}$ (what does this say about the flight-path angle?).

Carry out a couple of numerical examples, appropriate to natural and artificial earth satellites, and see what you can learn about the lifetime of such satellites at various altitudes.

CHAPTER 12

THE USES OF OPTIMIZATION

12.1 INTRODUCTION

Every competent engineer is preoccupied with attaining the "best" in his work. In the aerospace field, however, the need to minimize weight for a given performance, to attain the least time of flight between two points, or to find the body of lowest drag is frequently so urgent that it becomes a fetish for the designer. There are many ways, qualitative and quantitative, of measuring what is the best. But since about 1950 the mathematical theory of optimization has removed at least some of the guesswork, in the following sense: If the quantity whose extreme value is sought can be precisely defined, then a unique set of system or operating parameters to achieve this goal can often be calculated. This methodology has become practical through a marriage between variational calculus and the capabilities of the digital computer.

We deem the subject so important for the future that our treatment will go somewhat beyond "trajectory optimization" as an example of dynamic performance. Nevertheless, that example is our tie to the preceding chapters, and it is preeminent among situations in which variational techniques are contributing to aeronautics today. Bryson and Ho (1969) is unquestionably the definitive text in English, with its particular focus on the role of control systems. We have also profited from studying the introductory account in Halfman (1962, Volume II, Chapters 10 and 11) and the elegant presentation of principles of mechanics by Lanczos (1962). Fox (1971) is an excellent recent book with emphasis on minimization of structural weight.

A typical problem in formal optimization is compounded of the following parts.

- Quantitative description of the system whose behavior is to be optimized, in terms of an adequate set of parameters or time functions. These may be thought of as the components of the state vector and control vector that we introduced in Chapter 2; it must be clear which, and to what degree, the designer can select.
- Definition of a single goal,* for which the extremum is to be found. When

* As Bryson has remarked, "You can only make one thing best at a time." The goal has been called by innumerable names, including "performance index," "merit function," "payoff function," "return," and so forth.

the problem involves the course of events over time (e.g., a fastest climb to altitude), the goal is a *functional*, i.e., a function or integrals of functions which are to be determined.

- Quantitative knowledge of all physical laws, human factors, environmental limitations, and the like which govern the system. The resulting collections of equations and inequalities are called the *constraints* of the problem.
- A methodology, embodied in theory and working computer programs, for calculating the best policy which is consistent with the constraints.

For a case of, say, trajectory optimization, you are already in a fair position to carry out the first three of these steps. Accordingly, we devote a section now to the required theoretical background, with special reference to variational techniques.

12.2 THE MATHEMATICS OF OPTIMIZATION

Following Bryson and Ho (1969, Chapter 1), let us first review the classical problem of finding extreme values for a function $L(u_1, \ldots, u_m)$ of m scalars or parameters. If $m = 1$, we have the familiar case (Fig. 12.1) in which a necessary condition is

$$\frac{dL}{du_1} = 0 \tag{12.1}$$

We can further characterize a point on the curve of $L(u_1)$ which satisfies (12.1) as being a minimum, maximum, or saddle point, depending on whether

$$\frac{d^2L}{du_1^2} > 0, \qquad < 0, \qquad \text{or} \qquad = 0 \tag{12.2}$$

respectively, in its vicinity.

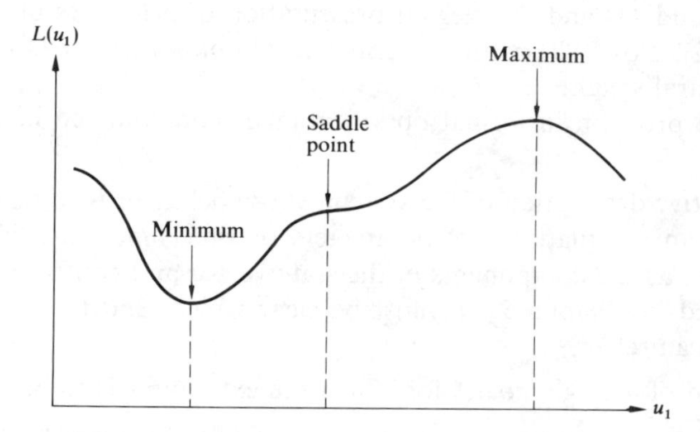

Fig. 12.1. Three circumstances at a point of zero slope on the curve of a function of one variable.

Relations (12.1) and (12.2) can be generalized by examining the first- and second-order terms in the total differential of L.

$$dL = \frac{\partial L}{\partial u_1}\, du_1 + \cdots + \frac{\partial L}{\partial u_m}\, du_m + \frac{1}{2}\left[\frac{\partial^2 L}{\partial u_1^2}(du_1)^2 + 2\frac{\partial^2 L}{\partial u_1\,\partial u_2}\, du_1\, du_2 + \cdots + \frac{\partial^2 L}{\partial u_m^2}(du_m)^2\right] \qquad (12.3)$$

One necessary condition for an extremum, when $dL = 0$, is clearly

$$\frac{\partial L}{\partial u_i} = 0, \qquad \text{for } i = 1, 2, \ldots, m \qquad (12.4)$$

The generalization of (12.2) is more complicated. Thus, to ensure a minimum, the matrix whose components are $\partial^2 L/\partial u_i\, \partial u_j$ must be "positive definite" (Hildebrand 1954, pages 42–47); this test is met if the characteristic roots of the matrix are all positive.

As a simple example, consider $m = 2$, for which dL near the extremum is

$$\begin{aligned} dL &= \frac{1}{2}\left[\frac{\partial^2 L}{\partial u_1^2}(du_1)^2 + 2\frac{\partial^2 L}{\partial u_1\,\partial u_2}\, du_1\, du_2 + \frac{\partial^2 L}{\partial u_2^2}(du_2)^2\right] \\ &= \frac{1}{2\partial^2 L/\partial u_1^2}\left\{\left[\frac{\partial^2 L}{\partial u_1^2}\, du_1 + \frac{\partial^2 L}{\partial u_1\,\partial u_2}\, du_2\right]^2 \right. \\ &\qquad \left. + \left[\frac{\partial^2 L}{\partial u_1^2}\frac{\partial^2 L}{\partial u_2^2} - \left(\frac{\partial^2 L}{\partial u_1\,\partial u_2}\right)^2\right] du_2^2\right\} \end{aligned} \qquad (12.5)$$

For a *minimum*, we require that $dL > 0$ in (12.5) for all possible nonzero du_1 and du_2. Since the first term in braces is obviously positive, sufficient conditions are therefore

$$\frac{\partial^2 L}{\partial u_1^2}\frac{\partial^2 L}{\partial u_2^2} - \left(\frac{\partial^2 L}{\partial u_1\,\partial u_2}\right)^2 > 0 \qquad (12.6a)$$

and simultaneously

$$\frac{\partial^2 L}{\partial u_1^2} > 0 \qquad (12.6b)$$

You can verify that (12.6) also confirm that the two roots λ of

$$\left|\lambda[I] - \begin{bmatrix} \dfrac{\partial^2 L}{\partial u_1^2} & \dfrac{\partial^2 L}{\partial u_1\,\partial u_2} \\ \dfrac{\partial^2 L}{\partial u_1\,\partial u_2} & \dfrac{\partial^2 L}{\partial u_2^2} \end{bmatrix}\right| = 0 \qquad (12.7)$$

are positive. $[I]$ here is the unit matrix.

One way to introduce constraints into algebraic optimization is to make the goal L a function of the m u_i plus n additional "state parameters" x_j. In view of physical laws, etc., which govern it, the system is then required to satisfy m algebraic equations among these scalars, in the form*

$$f_k(u_i, x_j) = 0, \qquad i = 1, \ldots, m; j, k = 1, \ldots, n \tag{12.8}$$

If we may assume that they exist, the straightforward way of calculating the values of u_i which optimize L would seem to consist of using (12.8) to eliminate the x_j from L, after which (12.4) apply. Ordinarily much simpler in practice, however, is to bring in an auxiliary set of n Lagrange multipliers λ_k. One then "adjoins" the constraints (12.8) to L by defining

$$H(u_i, x_j, \lambda_k) = L(u_i, x_j) + \sum_{k=1}^{n} \lambda_k f_k(u_i, x_j) \tag{12.9}$$

The procedure by which H is used to find an extremum of L is as follows: Regarding the u_i and x_j as variables, we write a small change of H in the vicinity of the desired solution as

$$\begin{aligned} dH &= \sum_{i=1}^{m} \frac{\partial H}{\partial u_i} du_i + \sum_{j=1}^{n} \frac{\partial H}{\partial x_j} dx_j \\ &= \sum_{i=1}^{m} \left[\frac{\partial L}{\partial u_i} + \sum_{k=1}^{n} \lambda_k \frac{\partial f_k}{\partial u_i} \right] du_i + \sum_{j=1}^{n} \left[\frac{\partial L}{\partial x_j} + \sum_{k=1}^{n} \lambda_k \frac{\partial f_k}{\partial x_j} \right] dx_j \end{aligned} \tag{12.10}$$

From (12.9), we can also write

$$dH = dL + \sum_{k=1}^{n} \lambda_k \, df_k \tag{12.11}$$

In order to ensure that (12.8) is satisfied, however, we require $df_k = 0$ for all k, whence it follows that

$$dH = dL \tag{12.12}$$

That is, *the extremum of* L ($dL = 0$) *therefore coincides with an extremum of* H ($dH = 0$).

For convenience, we choose the n λ_k as the solutions of the system of n equations

$$\frac{\partial H}{\partial x_j} = \frac{\partial L}{\partial x_j} + \sum_{k=1}^{n} \lambda_k \frac{\partial f_k}{\partial x_j} = 0, \qquad \text{for } j = 1, 2, \ldots, n \tag{12.13}$$

It follows from (12.10), and the condition $dH = 0$ at the extremum, that also

$$\frac{\partial H}{\partial u_i} = 0, \qquad \text{for } i = 1, 2, \ldots, m \tag{12.14}$$

* Note the analogy with (2.45), $\{\mathscr{X}\}$ being replaced by the n-vector of state parameters x_j and $\{U\}$ by the m-vector of parameters u_i. The latter are used to "control" L toward an optimum point.

In summary, (12.8), (12.13), and (12.14) furnish us with $(2n + m)$ equations in $(2n + m)$ u_i, x_j, and λ_k. Although the number of unknowns has been increased by n, we have thus avoided the often difficult step of solving (12.8) explicitly for the x_j; only the derivatives of the function H have to be calculated.

A simple example of how the foregoing is applied, involving only a single constraint and Lagrange multiplier, appeared in Section 9.4 on optimal multistage boosters. The μ_i in that analysis play the role of our controls u_i. The explicit appearance of the system state parameters x_i was suppressed by the way in which the problem was nondimensionalized, but returning to dimensioned quantities will show that the first-stage weight W_1 is a suitable choice for x_1.

We complete our discussion of parameter optimization by remarking that the first-derivative conditions (12.13, 14) are necessary ones and that the establishment of sufficient conditions for a minimum or maximum of L is a more complex question (cf. Bryson and Ho, 1969, Section 1.3). In practical applications, fortunately, it is often possible by inspection, or by resort to common sense, to decide when a solution is the desired one. You should be warned that "optima" are frequently not unique and occasionally nonsense. Nonlinearity is always present, and the questions of how properly to formulate and efficiently to solve such problems are complex and fascinating.

Our introductory treatment of the algebraic case will, we hope, render more plausible the use of Lagrange multipliers in connection with continuous-system optimization, in which the constraints are ordinary differential equations. There are numerous variations on how the latter may be formulated; e.g., the constraints may include inequalities as well as equalities, the initial and final values of the independent variable may be fixed or free, and so forth. In order to be specific, however, we begin by examining equality constraints only and by fixing the range of the solution. For instance, we might think in terms of minimizing the propellants consumed during a boost of the sort discussed in Chapter 9, while also requiring that time t run from 0 to an instant t_f when orbital injection must take place.

In physically meaningful situations, it is almost always possible to find an $n \times 1$ state vector $\{\mathscr{X}(t)\}$ and to write the n constraints, as we have done in (2.45) and elsewhere,

$$\{\dot{\mathscr{X}}(t)\} = f(\{\mathscr{X}(t)\}, \{\mathscr{U}(t)\}, t) \tag{12.15a}$$

with initial conditions

$$\{\mathscr{X}(0)\} = \{\mathscr{X}_0\} \tag{12.15b}$$

Here f is a set of n functions, generally nonlinear in the state and in the $m \times 1$ control vector $\{\mathscr{U}(t)\}$. We seek to optimize a scalar J, which may depend on the final state and/or the course of events between $t = 0$ and t_f:

$$J = \varphi(\{\mathscr{X}(t_f)\}, t_f) + \int_0^{t_f} L(\{\mathscr{X}(t)\}, \{\mathscr{U}(t)\}, t)\, dt \tag{12.16}$$

For instance, in a minimum-propellant boost we could set $L = 0$, let the function φ be just the final booster mass $m(t_f)$, and maximize J. Alternatively, to determine a minimum-distance path between two fixed points on a prescribed surface, curvilinear coordinates r, s could be defined. φ would be equated to zero, and the integral to be minimized in (12.16) would measure arc length along an arbitrary curve $r(s)$ between the points. There are opportunities for ingenuity both in the choice of variables and the manner of expressing the goal.

Our aim is now to find a vector of controls which causes J to have an optimal or extreme value, while (12.15) are simultaneously enforced. To this end, we appeal to the concept of "variation" $\delta(\ldots)$ of a parameter or function. To illustrate, Fig. 12.2 portrays what could be the optimal curve $u_i(t)$ for the ith component

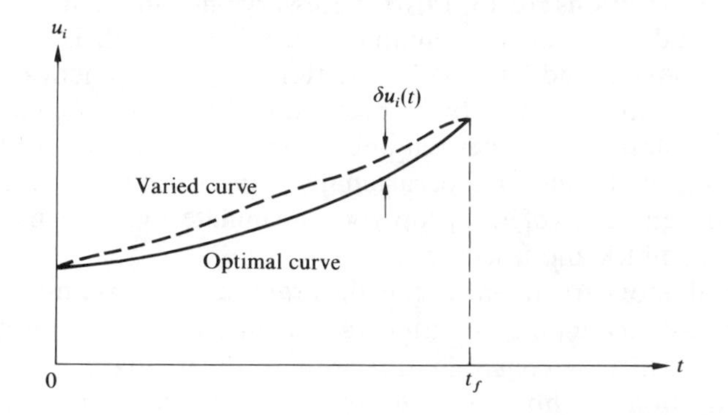

Fig. 12.2. Optimal solution and slightly varied curve of the ith component of a time-dependent control vector.

of $\{\mathscr{U}\}$, together with a typical slightly varied function $[u_i(t) + \delta u_i(t)]$, obtained by making an arbitrary small change to u_i at each value of t in the range. The operator $\delta(\ldots)$ has many of the properties of a differential, but it represents a small change* in a dependent quantity at a particular t. That is, we are studying the effects of varying the different u_i and x_j; the variation of time δt is equated to zero, most particularly at the fixed limits of the range.

We assert that the generalization of the condition $dL = 0$ in algebraic optimization is to require $\delta J = 0$ for all permissible small variations $\delta u_i(t)$ in the m components of $\{\mathscr{U}\}$. In order to apply this observation toward solving the present problem, let us adjoin the constraints to J by introducing a vector or row matrix

* We confine ourselves to working with what are known as "weak" variations, i.e., small changes which are smooth and have sufficient continuous derivatives to make possible the calculus operations that we carry out below. "Strong" variations are jagged, discontinuous modifications to the vector components. They are usually ruled out by the fact that the laws of physics forbid sudden jumps and other singular behavior. It is worth mentioning, however, that "corners" or slope discontinuities in both the $u_i(t)$ and $x_j(t)$ are sometimes acceptable. Special conditions can be derived which govern the system's behavior at such corners.

$\lfloor\Lambda(t)\rfloor$ of m Lagrange multiplier functions $\lambda_k(t)$ and writing [cf. (12.9)]

$$J = \varphi(\{\mathscr{X}(t_f)\}, t_f) + \int_0^{t_f} [L(\{\mathscr{X}(t)\}, \{\mathscr{U}(t)\}, t) + \lfloor\Lambda(t)\rfloor\langle f(\{\mathscr{X}(t)\}, \{\mathscr{U}(t)\}, t) - \{\dot{\mathscr{X}}(t)\}\rangle] \, dt \tag{12.17}$$

(Recall that the product of a row $\lfloor\ldots\rfloor$ by a column $\{\ldots\}$ matrix is a scalar.) In the integrand of (12.17) there appears a function called the *Hamiltonian*,

$$H(\{\mathscr{X}(t)\}, \{\mathscr{U}(t)\}, t) \equiv L(\{\mathscr{X}(t)\}, \{\mathscr{U}(t)\}, t) + \lfloor\Lambda(t)\rfloor f(\{\mathscr{X}(t)\}, \{\mathscr{U}(t)\}, t) \tag{12.18}$$

After inserting (12.18), one finds it convenient to integrate (12.17) by parts, as follows:

$$\begin{aligned} J &= \varphi(\{\mathscr{X}(t_f)\}, t_f) + \int_0^{t_f} [H(\{\mathscr{X}(t)\}, \{\mathscr{U}(t)\}, t) - \lfloor\Lambda(t)\rfloor\{\dot{\mathscr{X}}(t)\}] \, dt \\ &= \varphi(\{\mathscr{X}(t_f)\}, t_f) - \lfloor\Lambda(t_f)\rfloor\{\mathscr{X}(t_f)\} + \lfloor\Lambda(0)\rfloor\{\mathscr{X}(0)\} \\ &\quad + \int_0^{t_f} [H(\{\mathscr{X}(t)\}, \{\mathscr{U}(t)\}, t) + \lfloor\dot{\Lambda}(t)\rfloor\{\mathscr{X}(t)\}] \, dt \end{aligned} \tag{12.19}$$

We next calculate δJ. We recognize that variations $\delta u_i(t)$ automatically bring about changes $\delta x_j(t)$ in the state components, related to the $\delta u_i(t)$ by the constraints (12.15a) which have been adjoined into (12.19). As an example, consider the first term in the final member of (12.19). Since t_f is fixed, the small change in φ is related to the variations δx_j at t_f through the n partial derivatives of φ,

$$\delta\varphi = \sum_{j=1}^{n} \frac{\partial\varphi}{\partial x_j(t_f)} \delta x_j(t_f) \tag{12.20a}$$

We abbreviate (12.20a) by defining a row matrix of these partials,

$$\delta\varphi = \lfloor\partial\varphi/\partial x\rfloor\{\delta\mathscr{X}\}|_{t=t_f} \tag{12.20b}$$

We recall that the integration limits in (12.19) are not to be varied; therefore, expressions similar to (12.20b) may be constructed for the effects of $\delta u_i(t)$ and $\delta x_j(t)$ in changing the H-term of the integrand. When all these steps are completed, we obtain

$$\delta J = [(\lfloor\partial\varphi/\partial x\rfloor - \lfloor\Lambda\rfloor)\{\delta\mathscr{X}\}]|_{t=t_f} + [\lfloor\Lambda\rfloor\{\delta\mathscr{X}\}]|_{t=0} + \int_0^{t_f} [\lfloor\partial H/\partial u\rfloor\{\delta\mathscr{U}(t)\} + (\lfloor\partial H/\partial x\rfloor + \lfloor\dot{\Lambda}(t)\rfloor)\{\delta\mathscr{X}(t)\}] \, dt \tag{12.21}$$

We now put the Lagrange multipliers to work by a step resembling (12.13). It is very inconvenient to find from (12.15a) how the variations in the x_j are related to those of u_i. Accordingly, we *define* the $\lfloor\Lambda(t)\rfloor$ such that they cause the following equations to be satisfied for all t in the range [cf. (12.18)] and thereby eliminate $\{\delta\mathscr{X}(t)\}$ from the integrand of (12.21):

$$\lfloor\dot{\Lambda}(t)\rfloor = -\lfloor\partial H/\partial x\rfloor = -\lfloor\partial L/\partial x\rfloor - \lfloor\Lambda(t)\rfloor[\partial f/\partial x] \tag{12.22}$$

(Note that the square matrix in (12.22) is composed of the n^2 partials of the n functions f with respect to the n x_j.) The seeming artificiality of employing $\lambda_k(t)$ is more than outweighed, in realistic applications, by the complete avoidance of varying the constraint equations.

Equation (12.22) disposes of all but the first term in the (12.21) integral. To ensure that $\delta J = 0$ for arbitrary variations of $\{\mathscr{U}(t)\}$, however, we need to force every other term in (12.21) to equal zero. The considerations are as follows:

- What remains under the integral sign will vanish for any $\delta u_i(t)$ if, and only if,

$$\lfloor\partial H/\partial u\rfloor = 0 \tag{12.23a}$$

or

$$\partial H/\partial u_i = 0, \qquad \text{for } i = 1, 2, \ldots m \tag{12.23b}$$

- At $t = 0$, one or the other factor in each of the n products whose sum is $\lfloor\Lambda\rfloor\{\delta\mathscr{X}\}$ must vanish. When a full set of initial conditions (12.15b) is prescribed, then $\delta x_j(0) = 0$ for all j because such a fixed quantity cannot be varied. In some problems, certain state components (e.g., initial booster mass) might be left free at $t = 0$; the Lagrange multiplier functions connected with the corresponding constant equations are then required to have zero initial values.
- At $t = t_f$, one or the other factor in each of the n products whose sum is $(\lfloor\partial\varphi/\partial x\rfloor - \lfloor\Lambda\rfloor)\{\delta\mathscr{X}\}$ must vanish. If the full set of initial values, and no final values, are given for $\{\mathscr{X}\}$, we discover that all the final λ_k are thus specified through

$$\lambda_k(t_f) = \partial\varphi/\partial x_k(t_f), \qquad \text{for } k = 1, 2, \ldots, n \tag{12.24}$$

Alternatively, elements of $x_j(t_f)$ may be known (e.g., final speed, altitude, flight-path angle during boost). Since $\delta x_j(t_f) = 0$ for these elements, (12.24) does not apply to the corresponding Lagrange multipliers.

The mathematical description of the optimal solution is now complete. Over the range $0 \leq t \leq t_f$, the system of $(2n + m)$ ordinary-differential and algebraic equations (12.15a), (12.22), and (12.23) is to be solved for the $(m + 2n)$ unknowns $u_i(t)$, $x_j(t)$, and $\lambda_k(t)$. The total order of the system is $2n$. The required $2n$ boundary conditions are provided, half at $t = 0$ and half at $t = t_f$, by expressions like (12.15b) and (12.24). We are thus confronted with a two-point boundary-value problem. Since we expect to employ methods of numerical integration starting from $t = 0$, such as those discussed in Section 9.2, it is clear that a trial-and-correction process will be needed to adjust the conditions at $t = t_f$ which are not met by the first tentative solutions. Dealing with this uncertainty poses many of the practical obstacles to successful optimization. Perhaps the greatest theoretical obstacle is again how to know that a particular computed result is indeed the sought-after minimum or maximum.

Although, as we have mentioned, there are numerous variants on the manner in which the goal and constraints are stated, the foregoing development contains the important principles that underly most optimization processes for continuous systems in one independent variable. Again we cite Bryson and Ho (1969, especially Chapters 2 and 3) for details and beautiful illustrative exercises.

12.3 TWO ELEMENTARY EXAMPLES

1. The Brachistochrone

Books on applied variational calculus inevitably devote space to the classical problem of a bead which is permitted to slide on a frictionless wire between two fixed points in a vertical plane (Fig. 12.3). The goal is usually to determine that shape $z(x)$ for the slidewire which minimizes the trip time.

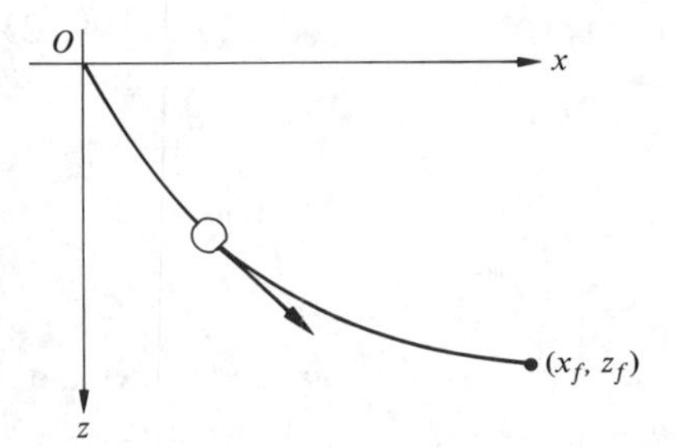

Fig. 12.3. Bead sliding frictionlessly down a wire between the origin and a specified point in the xz-plane.

When stated in terms of $z'(x)$ as the control ($m = 1$), it turns out that the brachistochrone can be brought precisely into the form (12.15, 16). Thus, given that the initial speed is zero, one knows from conservation of energy that the bead's speed at a vertical distance z below the origin is

$$v \equiv \frac{ds}{dt} = \sqrt{1 + \left(\frac{dz}{dx}\right)^2}\,\frac{dx}{dt} = \sqrt{2gz} \tag{12.25}$$

Solving (12.25) for dt, we can integrate to find the time of fall from $x = 0$ to $x = x_f$,

$$t_{\text{Fall}} = \frac{1}{\sqrt{2g}} \int_0^{x_f} \frac{\sqrt{1 + (dz/dx)^2}}{\sqrt{z}}\,dx \tag{12.26}$$

Clearly we have shifted from t to x as the independent variable. You will find it instructive to restate the problem with $\{x(t), z(t)\}$ as the state vector. Then $J = t_f$, and you will find it necessary to deal with a case in which the endpoint is variable. Now suppose that

$$\frac{dz}{dx} = u(x) \tag{12.27a}$$

We already have

$$z(0) = 0 \tag{12.27b}$$

Except for a nonessential constant factor, the goal is to minimize

$$J = \int_0^{x_f} \sqrt{\frac{1 + u^2}{z}}\, dx \tag{12.28}$$

subject to the constraint (12.27a). Equation (12.18) indicates that a single Lagrange multiplier $\lambda(x)$ is needed to construct the Hamiltonian

$$H = \sqrt{\frac{1 + u^2}{z}} + \lambda u \tag{12.29}$$

Since $z(x)$ is here playing the role of $x_1(t)$, (12.22) and (12.23) assume the forms

$$\frac{d\lambda}{dx} = -\frac{\partial H}{\partial z} = \frac{\sqrt{1 + u^2}}{2z^{3/2}} \tag{12.30}$$

and

$$\frac{u}{\sqrt{z(1 + u^2)}} + \lambda = 0 \tag{12.31}$$

(12.27b) furnishes one boundary condition for the system. Since the total order is two, a second is required, and Fig. 12.3 shows it to be

$$z(x_f) = z_f \tag{12.32}$$

In principle, one can eliminate λ between (12.30) and (12.31), whereupon (12.27a) converts the result into a second-order differential equation for $z(x)$. A simpler, but less direct, approach to the solution follows from the observation that the Hamiltonian must here be constant along the trajectory. Indeed, this is a general property of the continuous system optimizations analyzed in Section 12.2, so long as the independent variable does not appear explicitly in H. Returning to the definition (12.18), we obtain for the total rate of change

$$\frac{dH}{dt} = \frac{\partial H}{\partial t} + \lfloor \partial H/\partial x \rfloor \{\dot{\mathscr{X}}\} + \lfloor \partial H/\partial u \rfloor \{\dot{\mathscr{U}}\} + \lfloor \dot{\Lambda} \rfloor f(\{\mathscr{X}(t)\}, \{\mathscr{U}(t)\}, t) \tag{12.33}$$

[The last term in (12.33) must be added to allow for the contributions of explicit variations of $\lfloor \Lambda \rfloor$ with time, which are not otherwise included.] Equation (12.15a) enables us to write

$$\frac{dH}{dt} = \frac{\partial H}{\partial t} + \lfloor \partial H/\partial u \rfloor \{\dot{\mathscr{U}}\} + (\lfloor \partial H/\partial x \rfloor + \lfloor \dot{\Lambda} \rfloor)\{\dot{\mathscr{X}}\} \tag{12.34}$$

Along the optimal trajectory, however, (12.22) and (12.23a) indicate that the last two terms on the right are zero. Hence, if H depends on time only through $\{\mathcal{U}(t)\}$, $\{\mathcal{X}(t)\}$, and $\lfloor\Lambda(t)\rfloor$,

$$\frac{dH}{dt} = \frac{\partial H}{\partial t} = 0 \tag{12.35a}$$

and

$$H = \text{const.} \tag{12.35b}$$

on that trajectory.

Returning to the brachistochrone, we use (12.31) to eliminate $\lambda(x)$ from (12.29). Equation (12.35b) then yields

$$\sqrt{\frac{1 + u^2}{z}} - \frac{u^2}{\sqrt{z(1 + u^2)}} = \frac{1}{\sqrt{z(1 + u^2)}} = \text{const.} \equiv \frac{1}{K} \tag{12.36}$$

Since $u = dz/dx$, (12.36) is equivalent to the differential equation

$$\frac{dz}{dx} = \sqrt{\frac{K^2 - z}{z}} \tag{12.37}$$

the positive radical sign being chosen because positive dz/dx is expected when $x_f, z_f > 0$. The integrals of (12.37) are curves known as *cycloids*. For instance, a solution through (0, 0) may be expressed parametrically by

$$z = \frac{K^2}{2}[1 - \cos\theta] \tag{12.38a}$$

$$x = \frac{K^2}{2}[\theta - \sin\theta] \tag{12.38b}$$

as can be confirmed by substitution into (12.37). $\theta = 0$ is evidently the origin, and boundary condition (12.32) furnishes two relations from which θ_f and K may be calculated corresponding to any given endpoint.

It is of interest that our brachistochrone analysis makes possible quick extension to cover more general problems of the same class. Thus the lower end of the slidewire might be located anywhere on a prescribed curve

$$F(x, z) = 0 \tag{12.39}$$

rather than at a fixed point. An elementary instance in which the question of minimum time is obviously well posed involves fall to any point on the vertical straight line

$$x - x_f = 0 \tag{12.40}$$

The time t_{Final} is then very long either far down to large values of z or to a z near the starting altitude $z = 0$, so that an intermediate minimum must exist.

Since x_f is fixed by (12.40), the only difference from the preceding analysis is that boundary condition (12.32) no longer applies. Since $z(x_f)$ is unknown, we must return to the form (12.24); with $\varphi = 0$, it yields

$$\lambda(x_f) = 0 \tag{12.41}$$

With $u = dz/dx$ given by (12.37), we calculate λ from (12.31) as

$$\lambda(x) = \frac{dz/dx}{\sqrt{z\left[1 + \left(\frac{dz}{dx}\right)^2\right]}} = -\sqrt{\frac{K^2 - z}{K^2 z}} \tag{12.42}$$

In terms of the parameter θ of (12.38a), λ is therefore

$$\lambda = -\frac{1}{K}\sqrt{\frac{1 + \cos\theta}{1 - \cos\theta}} \tag{12.43}$$

(12.43) and (12.41) show that the absolute minimum times to fall to $x = x_f$ is associated with $\theta = \pi$. K is found in terms of x_f from (12.38b),

$$x_f = \frac{K^2}{2}[\pi + 0] \tag{12.44}$$

whence the height of the optimal point is

$$z_f \equiv z(\pi) = K^2 = \frac{2}{\pi} x_f \tag{12.45}$$

2. Minimum Drag in Newtonian Flow

As discussed in Chapter 10, the principal aerodynamic loads on an object in hypersonic flow act over its windward side, and the pressure can be estimated from the Newtonian formula (10.79). In particular, suppose we study a body of revolution, with $\alpha_{ZL} = 0$, radius $r_0(x)$, fixed maximum diameter D, and fixed length from $x = 0$ to $x = l$. The angle δ between the flight direction and surface normal is given by

$$\cos\delta = \frac{dr_0/dx}{\sqrt{1 + (dr_0/dx)^2}} \tag{12.46}$$

Therefore the Newtonian forebody drag is easily seen to be

$$D_B = 2\pi q_\infty C_{po} \int_0^l \frac{r_0(dr_0/dx)^3\, dx}{[1 + (dr_0/dx)^2]} \tag{12.47}$$

where q_∞ is flight dynamic pressure.

We are dealing with fixed endpoints. The obvious application of the Section 12.2 procedure is to use a control $u(x)$ defined by the auxiliary equation

$$\frac{dr_0}{dx} = u(x) \tag{12.48}$$

and set as our goal the minimization of

$$J = \int_0^l \frac{r_0 u^3}{1 + u^2}\, dx \tag{12.49}$$

One boundary condition is already provided by

$$r_0(l) = D/2 \tag{12.50}$$

With the single constraint (12.48), our Hamiltonian is

$$H = \frac{r_0 u^3}{1 + u^2} + \lambda u \tag{12.51}$$

whence the optimizing differential equations (12.22, 23) read

$$\frac{d\lambda}{dx} = -\frac{u^3}{1 + u^2} \tag{12.52}$$

$$\frac{r_0 u^2[3 + u^2]}{(1 + u^2)^2} + \lambda = 0 \tag{12.53}$$

This is another case of constant Hamiltonian, so that we can also use

$$\frac{r_0 u^3}{1 + u^2} + \lambda u = -\frac{2r_0 u^3}{(1 + u^2)^2} = \text{const.} \tag{12.54}$$

where (12.53) has helped by eliminating the Lagrange multiplier.

We have intentionally avoided a second boundary condition to supplement (12.50), because a curious difficulty arises in this problem when we insist, for instance, on a pointed nose $r_0(0) = 0$. This trouble manifests itself if we try to construct a solution by expressing r_0 and x parametrically in terms of u. Thus, replacing the constant in (12.54) by $(-2C)$, we get immediately

$$r_0 = C\frac{(1 + u^2)^2}{u^3} \tag{12.55}$$

Furthermore, from (12.48) and (12.55), we obtain

$$\frac{dx}{du} = \frac{dx}{dr_0}\frac{dr_0}{du} = \frac{1}{u}\frac{dr_0}{du}$$

$$= \frac{C}{u}\frac{d}{du}\left[\frac{(1 + u^2)^2}{u^3}\right] \tag{12.56a}$$

We integrate (12.56a) from $u = u_0$ at the nose $x = 0$ and obtain, by integrating by parts,

$$x = C\int_{u_0}^{u} \frac{d}{du}\left[\frac{(1+u^2)^2}{u^3}\right]\frac{du}{u}$$

$$= C\left\{\left[\frac{1}{u}\cdot\frac{(1+u^2)^2}{u^3}\right]\Bigg|_{u_0}^{u} + \int_{u_0}^{u}\left[\frac{(1+u^2)^2}{u^5}\right]du\right.$$

$$= C\left\{\ln\frac{u}{u_0} + \frac{1}{u^2} - \frac{1}{u_0^2} + \frac{3}{4}\left[\frac{1}{u^4} - \frac{1}{u_0^4}\right]\right\} \tag{12.56b}$$

From (12.55), it is clear that no real u exists which will permit $r_0 = 0$ and, in fact, that r_0 has a minimum value $16C/3\sqrt{3}$ when $u = \sqrt{3}$.

Eggers *et al*. (1957) proposed, as a way out of the dilemma, that meaningful optimal shapes might result if the body were allowed to have a small flat face at the front with radius $r_0(0)$. An additional drag term $r_0^2(0)/2$ is thereby added to the goal formula (12.49). Moreover, if we think of reversing the direction of coordinate x so that $x = 0$ becomes the "final" value of the independent variable, we may associate this new term with the function φ of (12.17) and write

$$\varphi = \frac{r_0^2(0)}{2} \tag{12.57}$$

The new formulation also furnishes us, through (12.24), with a boundary condition on λ. Because of the reversal of endpoints, the sign of the derivative must be changed, and

$$\lambda(0) = -\frac{\partial\varphi}{\partial r_0(0)} = -r_0(0) \tag{12.58}$$

Inasmuch as the steps leading to (12.55, 56) are unaltered, we are now in a position to present the complete solution, at least implicitly. First, u_0 is determined by substituting (12.58) into (12.53) at $x = 0$. Since $r_0(0) \neq 0$ cancels, we get

$$\frac{u_0^2[3 + u_0^2]}{[1 + u_0^2]^2} = 1 \tag{12.59}$$

Equation (12.59) has one physically meaningful root, $u_0 = 1$, which tells us that the slope of the radius just behind the flat face is 45°, irrespective of length and other details of the minimum-drag body. With $u_0 = 1$, (12.55) yields $C = r_0(0)/4$ and

$$\frac{r_0(x)}{r_0(0)} = \frac{(1+u^2)^2}{4u^3} \tag{12.60}$$

The parametric variation of x with u from (12.56b) is

$$x = \frac{r_0(0)}{4}\left\{\ln u + \frac{1}{u^2} + \frac{3}{4u^4} - \frac{7}{4}\right\} \tag{12.61}$$

Finally, boundary condition (12.50) fixes u at $x = l$ in terms of the maximum diameter.

Many examples of these optimal Newtonian shapes will be found illustrated in Eggers *et al.* (1957). Miele (1965), among others, is a good source for many further details regarding the interesting subject of minimization of drag.

12.4 OPTIMAL TRAJECTORIES

We shall close this chapter with a necessarily brief introduction to the ideas involved in determining the best flight paths for interception, climb to altitude, boost, and the like. This is a rich and rewarding subject, which has recently been extended to include such refinements as effects of random external disturbances, incomplete knowledge of the state vector, and optimal interaction between two vehicles ("differential game theory"). The statement of necessary conditions for maximum performance or minimum time turns out to be a fairly straightforward matter. But interesting subtleties then arise in connection with efficiently solving the resulting sets of nonlinear equations under two-point boundary values—not to mention the question of proving that a solution, once obtained, fulfills all the requirements ascribed to it.

For a single vehicle in the atmosphere, our starting point would be constraint equations like (2.39), (2.40), or (9.7). The goal can usually be cast in the form (12.16), except that t_f is often a variable quantity. In fact, when the aim is to minimize time of flight, J is greatly simplified because $\varphi = t_f$ and $L = 0$. A second difference between this and our development in Section 12.2 is that some or all of the x_j may be prescribed at both $t = 0$ and t_f.

Necessary conditions for the desired optimum are fully derived by Bryson and Ho (1969, Sections 2.4 and 2.7). Again it proves convenient to adjoin the constraints (12.15a) as is done in (12.17). We recall that the variation $\delta(\ldots)$ is defined to be carried out at a fixed value of t. Accordingly, it is not $\delta J = 0$ but the total differential $dJ = 0$ that locates our optimum here. Allowing, therefore, for a change in t_f, we write

$$dJ = \left[\frac{\partial\varphi}{\partial t}\,dt_f + \lfloor\partial\varphi/\partial x\rfloor\{d\mathscr{X}\}\right]\bigg|_{t=t_f} + L\,|_{t=t_f}\,dt_f$$
$$+ \int_0^{t_f} \langle[\lfloor\partial L/\partial x\rfloor + \lfloor\Lambda\rfloor[\partial f/\partial x]]\{\delta\mathscr{X}\}$$
$$+ [\lfloor\partial L/\partial u\rfloor + \lfloor\Lambda\rfloor[\partial f/\partial u]]\{\delta\mathscr{U}\} - \lfloor\Lambda\rfloor\{\delta\dot{\mathscr{X}}\}\rangle\,dt \qquad (12.62)$$

Combining terms and integrating the $\{\delta\dot{\mathscr{X}}\}$ by parts, we can obtain a form resembling (12.21). At the same time, let us recognize that $\{d\mathscr{X}(t_f)\}$ is actually made up of a part due to the variation and one due to t_f itself.

$$\{d\mathscr{X}(t_f)\} = \{\delta\mathscr{X}(t_f)\} + \{\dot{\mathscr{X}}(t_f)\}\,dt_f \qquad (12.63)$$

Since the initial time is not to be varied, we get

$$dJ = \left\langle \frac{\partial \varphi}{\partial t} + L + \lfloor \Lambda \rfloor \{\dot{\mathscr{X}}\} \right\rangle \bigg|_{t=t_f} dt_f$$
$$+ \langle \lfloor \partial\varphi/\partial x \rfloor - \lfloor \Lambda \rfloor \rangle \big|_{t=t_f} \{\delta\mathscr{X}(t_f)\} + [\lfloor \Lambda \rfloor \{\delta\mathscr{X}\}] \big|_{t=0}$$
$$+ \int_0^{t_f} \langle (\lfloor \partial L/\partial x \rfloor + \lfloor \Lambda \rfloor [\partial f/\partial x] + \lfloor \dot{\Lambda} \rfloor)\{\delta\mathscr{X}\}$$
$$+ [\lfloor \partial L/\partial u \rfloor + \lfloor \Lambda \rfloor [\partial f/\partial u]]\{\delta\mathscr{U}\}\rangle \, dt \tag{12.64}$$

Aiming to make $dJ = 0$, let us generalize (12.22) and define the $\lambda_k(t)$ so as to eliminate the coefficient of $\{\delta\mathscr{X}\}$ in the integral:

$$\lfloor \dot{\Lambda} \rfloor = -\lfloor \partial L/\partial x \rfloor - \lfloor \Lambda \rfloor [\partial f/\partial x] \tag{12.65}$$

Let us also suppose that the entire state is known at $t = 0$ [cf. (12.15b)], so that $\{\delta\mathscr{X}(0)\} = 0$. At $t = t_f$, we can separate the $x_j(t_f)$ into one group ($j = 1, \ldots, q$) for which boundary conditions are available and a second group ($j = q + 1, \ldots, n$) for which (12.64) leads to

$$\lambda_j(t_f) = \frac{\partial \varphi}{\partial x_j}\bigg|_{t=t_f} \tag{12.66}$$

The straightforward way of suppressing the remaining terms in (12.64) would seem to be to enforce the differential equations [cf. (12.23)]

$$\lfloor \partial L/\partial u \rfloor + \lfloor \Lambda \rfloor [\partial f/\partial u] = 0 \tag{12.67}$$

and the additional boundary condition

$$\left\langle \frac{\partial \varphi}{\partial t} + L + \lfloor \Lambda \rfloor \{\dot{\mathscr{X}}\} \right\rangle \bigg|_{t=t_f} = \left\langle \frac{\partial \varphi}{\partial t} + L + \lfloor \Lambda \rfloor \{f\} \right\rangle \bigg|_{t=t_f} = 0 \tag{12.68}$$

We should call attention, however, to the alternative procedure in Bryson and Ho (1969, Section 2.7), whereby additional constant "control parameters" ν_i are introduced into (12.67) and (12.68).

We illustrate optimization of trajectory by sketching an early result of Bryson and Denham (1962), whose application to the F-4 airplane was a landmark in the history of the subject. Their aim was to discover how the F-4 could be flown in minimum time from low speed (M = 0.38) at sea level to M = 1 at a nominal ceiling of 20 kilometers (about 65,000 ft). They made certain reasonable simplifying assumptions, the most important of which was that the pitching inertia could be neglected. This made it possible for the vehicle to be analyzed as a point mass, with angle of attack $\alpha_{ZL}(t)$ treated as an instantaneously adjustable control variable. The thrust and zero-lift lines were taken to coincide. With constant gravity and a flat earth ($r \to \infty$), we may adapt (9.7) as equations of motion. We replace φ_T by α_{ZL}, use $\dot{x} = v_c \cos \gamma$ for horizontal speed, and regard the fuel consumption $(-\dot{m})$ as a known function of altitude and Mach number. Thus are obtained for (12.15a)

$$\dot{v}_c = \frac{T \cos \alpha_{ZL}}{m} - \frac{D}{m} - g \sin \gamma \tag{12.69a}$$

$$\dot{\gamma} = \frac{T \sin \alpha_{ZL}}{v_c m} + \frac{L}{v_c m} - \frac{g \cos \gamma}{v_c} \tag{12.69b}$$

$$\dot{h} = v_c \sin \gamma \tag{12.69c}$$

$$\dot{x} = v_c \cos \gamma \tag{12.69d}$$

$$\dot{m} = F(h, M) \tag{12.69e}$$

[Note that, unless winds and horizontal variations in atmospheric properties must be accounted for, x may be calculated from (12.69d) as an afterthought. Thus the system basically has state 4.]

The initial conditions selected for (12.69) called for horizontal flight at sea level and weight W_0. They read

$$\mathrm{M}(0) \equiv \frac{v_c(0)}{a_{\infty\,\mathrm{SL}}} = 0.38 \tag{12.70a}$$

$$\gamma(0) = h(0) = x(0) = 0 \tag{12.70b,c,d}$$

$$m(0) = W_0/g \tag{12.70e}$$

Two explicit final conditions are

$$\mathrm{M}(t_f) = 1.0 \tag{12.71a}$$

$$\gamma(t_f) = 0 \tag{12.71b}$$

The further requirement

$$h(t_f) = 20 \text{ km} \tag{12.71c}$$

is actually a *stopping condition* which defines t_f, while $m(t_f)$ and $x(t_f)$ are left free.

For minimum time, we have the goal function

$$J = \varphi = t_f \tag{12.72}$$

so that $L = 0$. It is convenient to identify the five Lagrange-multiplier functions with subscripts related to the corresponding state variables. The explicit forms of the optimizing equations (12.65) and (12.67) work out as follows:

$$\begin{aligned}\dot{\lambda}_v = & -\lambda_v\left[\frac{\cos \alpha_{ZL}}{m a_\infty}\frac{\partial T}{\partial \mathrm{M}} - \frac{1}{m a_\infty}\frac{\partial D}{\partial \mathrm{M}}\right] - \lambda_h \sin \gamma - \lambda_x \cos \gamma - \frac{\lambda_m}{a_\infty}\frac{\partial F}{\partial \mathrm{M}} \\ & - \lambda_\gamma\left[-\frac{T \sin \alpha_{ZL}}{v_c^2 m} + \frac{\sin \alpha_{ZL}}{v_c m a_\infty}\frac{\partial T}{\partial \mathrm{M}} - \frac{L}{v_c^2 m} + \frac{1}{v_c m a_\infty}\frac{\partial L}{\partial \mathrm{M}} + \frac{g \cos \gamma}{v_c^2}\right]\end{aligned} \tag{12.73a}$$

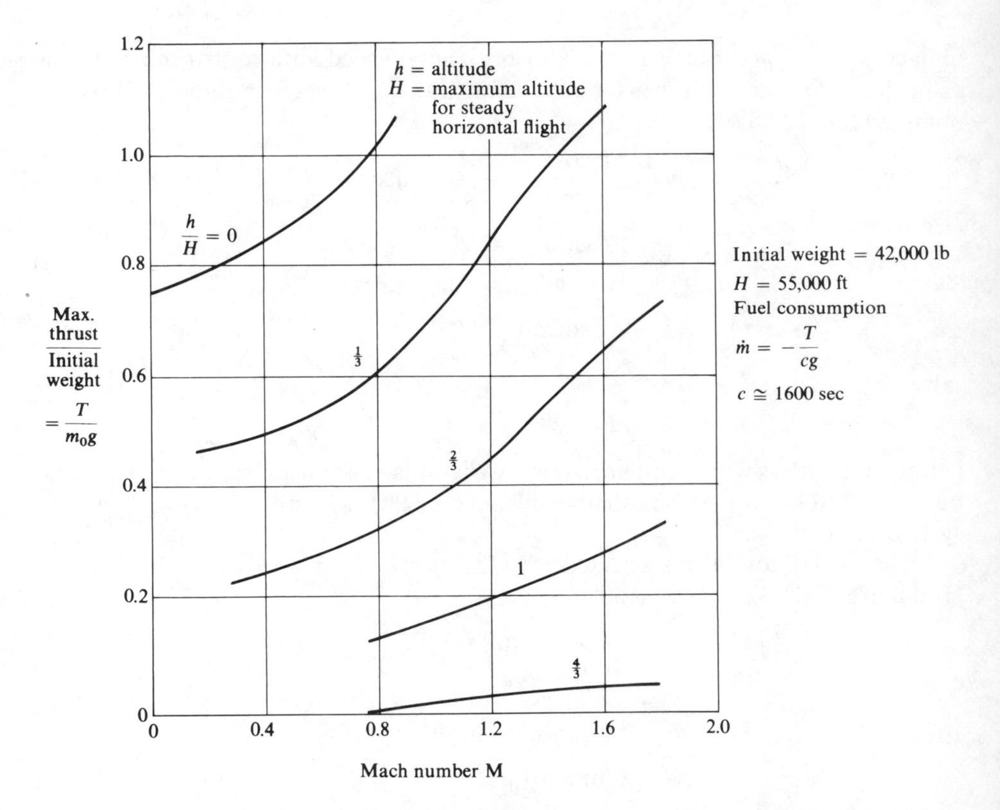

Fig. 12.4. Data on dimensionless maximum thrust for the F-4H airplane, plotted versus flight Mach number. Parameter on these curves is altitude referred to steady-state ceiling. [From Bryson and Denham (1962)]

$$\dot{\lambda}_\gamma = \lambda_v g \cos \gamma + \lambda_\gamma \frac{g \sin \gamma}{v_c} - \lambda_h v_c \cos \gamma + \lambda_x v_c \sin \gamma \tag{12.73b}$$

$$\dot{\lambda}_h = -\lambda_v \left[\frac{\cos \alpha_{ZL}}{m} \frac{\partial T}{\partial h} - \frac{1}{m} \frac{\partial D}{\partial h} \right] - \lambda_\gamma \left[\frac{\sin \alpha_{ZL}}{v_c m} \frac{\partial T}{\partial h} + \frac{1}{v_c m} \frac{\partial L}{\partial h} \right] - \lambda_m \frac{\partial F}{\partial h} \tag{12.73c}$$

$$\dot{\lambda}_x = 0 \tag{12.73d}$$

$$\dot{\lambda}_m = -\lambda_v \left[-\frac{T \cos \alpha_{ZL}}{m^2} + \frac{D}{m^2} \right] + \lambda_\gamma \left[\frac{T \sin \alpha_{ZL}}{v_c m^2} + \frac{L}{v_c m^2} \right] \tag{12.73e}$$

$$-\lambda_v \left[\frac{T \sin \alpha_{ZL}}{m} + \frac{1}{m} \frac{\partial D}{\partial \alpha_{ZL}} \right] + \lambda_\gamma \left[\frac{T \cos \alpha_{ZL}}{v_c m} + \frac{1}{v_c m} \frac{\partial L}{\partial \alpha_{ZL}} \right] = 0 \tag{12.74}$$

In essence, (12.69), (12.73), and (12.74) are 11 ordinary differential equations in the 10 state and Lagrange variables plus the control $\alpha_{ZL}(t)$. The total order is

10, since (12.74) is algebraic. The required 10 boundary conditions are (12.70) and (12.71), together with (12.66) for x and m.

$$\lambda_x(t_f) = 0 \tag{12.75a}$$

$$\lambda_m(t_f) = 0 \tag{12.75b}$$

The unimportance of x for the primary computation is verified by (12.73d) and (12.75a), which show that λ_x is identically zero.

With the system thus reduced to eighth order, we need to describe the data used by Bryson and Denham (1962) in their solution. Figures 12.4 and 12.5 reproduce the raw information with which they worked. The thrust T was assumed always to be at its maximum value corresponding to instantaneous h and M. The derivatives $\partial T/\partial M$ and $\partial T/\partial h$ were also taken numerically from Fig. 12.4. A constant I_{sp} of 1600 "seconds" made it possible for the function $F(h, M) = -T/I_{sp}$ and its derivatives to be computed directly from thrust data. Figure 12.5 furnishes

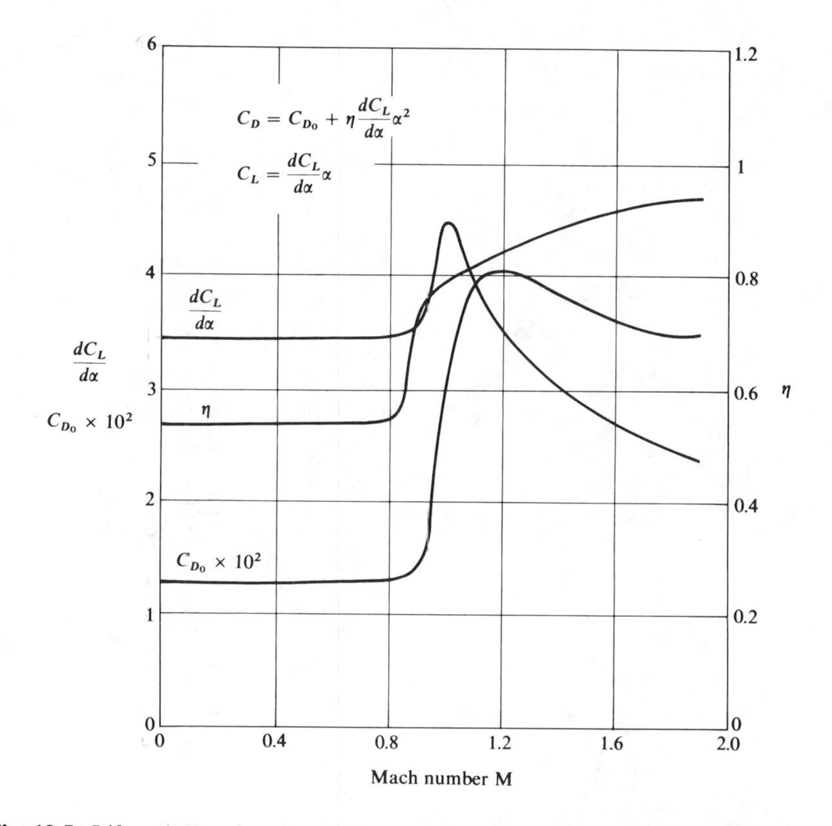

Fig. 12.5. Lift and drag data for the F-4H as functions of Mach number. [From Bryson and Denham (1962)]

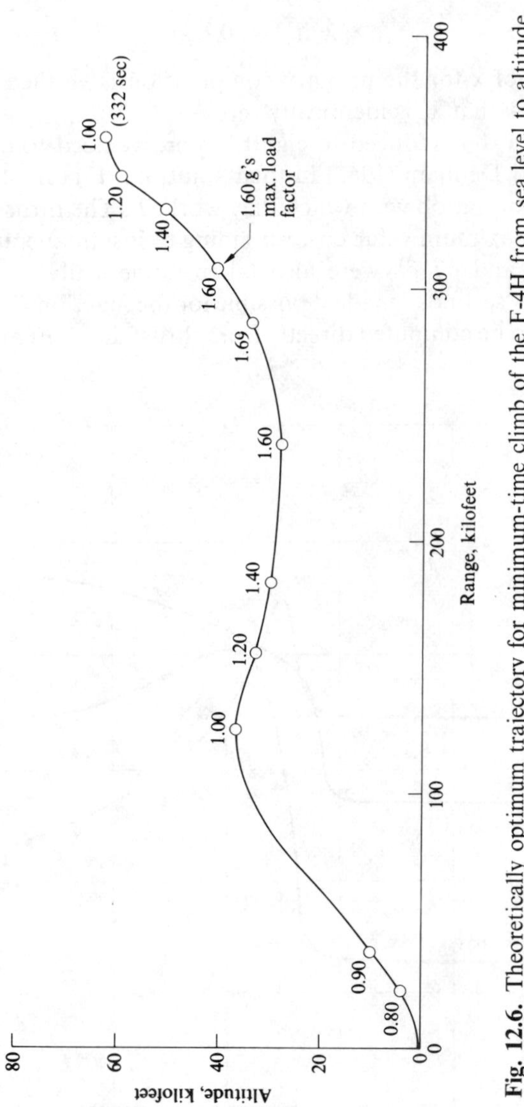

Fig. 12.6. Theoretically optimum trajectory for minimum-time climb of the F-4H from sea level to altitude 20 km. Values of Mach number are marked on the curve. [From Bryson and Denham (1962)]

quantities from which lift and drag can be found through

$$L = \frac{\rho_\infty(h)}{2} v_c^2 S \frac{dC_L}{d\alpha} \alpha_{ZL} \tag{12.76}$$

and

$$D = \frac{\rho_\infty(h)}{2} v_c^2 S C_D \cong \frac{\rho_\infty(h)}{2} v_c^2 S\left[C_{DO} + \eta \frac{dC_L}{d\alpha} \alpha_{ZL}^2\right] \tag{12.77}$$

Reference area $S = 530$ ft^2; density ρ_∞ and the speed of sound a_∞ were taken as functions of h from a standard-atmosphere table. A typical derivative in (12.73) would be

$$\begin{aligned}\frac{\partial D}{\partial \mathrm{M}} &= \frac{\partial}{\partial \mathrm{M}}\left[\frac{\gamma}{2} p_\infty(h)\mathrm{M}^2 S C_D\right] \\ &= \frac{\gamma}{2} p_\infty(h) S\left\{2\mathrm{M}C_D + \mathrm{M}^2\left[\frac{dC_{DO}}{d\mathrm{M}} + \frac{d\eta}{d\mathrm{M}}\frac{dC_L}{d\alpha}\alpha_{ZL}^2 + \eta\alpha_{ZL}^2 \frac{d}{d\mathrm{M}}\left(\frac{dC_L}{d\alpha}\right)\right]\right\}\end{aligned} \tag{12.78}$$

Quantities like $dC_{DO}/d\mathrm{M}$ come from numerical differentiations in Fig. 12.5.

Numerical integrations of the system (12.69)–(12.73)–(12.74) were done by a program similar to those discussed in Section 9.2. A version of the method of steepest ascent was employed to proceed toward a solution of the two-point boundary-value problem. Essentially, this method starts from a trial integral which fails to meet some of the final conditions. By perturbing the equations, an efficient scheme is generated for making adjustments which drive this solution toward the one sought.

Figure 12.6 presents the final minimum-time trajectory on a plot of altitude versus horizontal distance, with M indicated as a parameter. Perhaps its most interesting feature is that a dive is used to assist with acceleration through the high-drag region near $\mathrm{M} = 1$. Thereafter the airplane "zooms," trading kinetic energy for potential energy of altitude in such a way that the Mach number approaches unity despite the low degree of thrust available near the peak at $h =$ 20 km. This result has some unexpected features which show the inefficiency of applying quasi-static performance methods to such a powerful vehicle. If it were flown according to the recipe of tracking maximum static rate of climb in each altitude range, the time to climb would be nearly twice the optimal 332 sec predicted by Bryson and Denham (1962). They were especially rewarded when a Navy test pilot achieved 330 sec the first time he tried the maneuver.

12.5 PROBLEMS

[In view of the extensive collection of relevant problems in the first few chapters of Bryson and Ho (1969), it is recommended that this text be used as a source for such assignments.]

REFERENCES

Abbott, I. H., and A. E. von Doenhoff, 1959, *Theory of Wing Sections*, New York: Dover Publications

Abbott, I. H., A. E. von Doenhoff, and L. S. Stivers, Jr., 1945, Summary of Airfoil Data, *NACA Rept. 824*

Abramowitz, M., and I. A. Stegun, editors, 1965, *Handbook of Mathematical Functions with Formulas, Graphs, and Mathematical Tables*, New York: Dover Publications

Albano, E., and W. P. Rodden, 1969, A Doublet-Lattice Method for Calculating Lift Distributions on Oscillating Surfaces in Subsonic Flows, *AIAA J.* 7:279–285, 292

Allen, H. J., 1945, General Theory of Airfoil Sections Having Arbitrary Shape or Pressure Distribution, *NACA Rept. 833*

Allen, H. J., and A. J. Eggers, Jr., 1957, A Study of the Motion and Aerodynamic Heating of Missiles Entering the Earth's Atmosphere at High Supersonic Speeds, *NACA TN 4047*

Allen, H. J., and E. W. Perkins, 1951, A Study of Effects of Viscosity on Flow over Slender Inclined Bodies of Revolution, *NACA TN 1048*

Ames Research Staff, 1953, Equations, Tables, and Charts for Compressible Flow, *NACA Rept. 1135*

Anonymous, 1969, Flying Qualities of Piloted Airplanes, *Military Specification MIL-F-8785B (ASG)*

Anonymous, *Royal Aeronautical Society Data Sheets, Aero Series*, published by R.A.S. Engineering Sciences Data Unit, London (see yearly Index and Supplements)

Ashley, H., and M. Landahl, 1965, *Aerodynamics of Wings and Bodies*, Reading, Mass.: Addison-Wesley

Ashley, H., and W. P. Rodden, 1972, Wing-Body Aerodynamic Interaction, *Ann. Rev. Fluid Mech.* 4:431–472

Batchelor, G. K., 1967, *An Introduction to Fluid Dynamics*, New York: Cambridge University Press

Battin, R. H., 1964, *Astronautical Guidance*, New York: McGraw-Hill

Berry, D. T., and B. G. Powers, 1970, Handling Qualities of the XB-70 Airplane in the Landing Approach, *NASA TN D-5676*

Bihrle, W., Jr., 1966, A Handling Qualities Theory for Precise Flight Path Control, *USAF Technical Rept. AFFDL-TR-65-198*

Bisplinghoff, R. L., and H. Ashley, 1962, *Principles of Aeroelasticity*, New York: John Wiley (out of print; may be obtained in paperback reprint from Ashley)

Blackwell, J. A., Jr., 1966, Supersonic Investigation of Effects of Configuration Geometry on Pressure-Coefficient and Section Normal-Force-Coefficient Distributions for a Two-Stage Launch Vehicle, *NASA TN D-3408*

Bradley, R. G., and B. D. Miller, 1971, Application of Finite-Element Theory to Airplane Configurations, *J. Aircr.* 8:400–405

Braslow, A. L., and T. G. Ayers, 1971, Application of Advanced Aerodynamics to Future Transport Aircraft, Vol. 1 of *NASA Aircraft Safety and Operation Problems*; paper presented at Conference, NASA Langley Res. Ctr., May 4–6, 1971

Bryan, G. H., 1911, *Stability in Aviation*, New York: Macmillan

Bryson, A. E., Jr., 1972, *Control Theory of Random Systems*, Proceedings of 13th International Congress of Theoretical and Applied Mechanics (to be published)

Bryson, A. E., Jr., and Y. C. Ho, 1969, *Applied Optimal Control*, New York: Blaisdell Publishing Co.

Chapman, D. R., 1958, An Approximate Analytical Method for Studying Entry into Planetary Atmospheres, *NACA TN 4276*

Churchill, R. V., 1944, *Modern Operational Mathematics in Engineering*, New York: McGraw-Hill

Cleveland, F. A., 1970, Size Effects in Conventional Aircraft Design, *J. Aircr.* 7:483–512

Doetsch, G., 1944, *Theorie und Anwendung der Laplace-Transformation*, New York: Dover Publications

Donovan, A. F., and H. R. Lawrence, editors, 1957, Aerodynamic Components of Aircraft at High Speeds, in *High Speed Aerodynamics and Jet Propulsion*, Vol. VII, Princeton, N.J.: Princeton University Press

Eggers, A. J., Jr., H. J. Allen, and S. E. Neice, 1957, A Comparative Analysis of the Performance of Long Range Hypervelocity Vehicles, *NACA TN 4046*

Eggers, A. J., Jr., *et al.*, 1957, Bodies of Revolution Having Minimum Drag at High Supersonic Airspeeds, *NACA Rept. 1306*

Eggers, A. J., Jr., R. C. Savin, and C. A. Syvertson, 1955, The Generalized Shock-Expansion Method and its Application to Bodies Traveling at High Supersonic Air Speeds, *J. Aeronaut. Sci.* 22:231–238, 248

Ellison, D. E., and L. V. Malthan, 1963, *USAF Stability and Control Methods*, handbook issued by USAF Flight Control Division, Flight Dynamics Lab., WPAFB, Dayton, Ohio

Etkin, B., 1959, *Dynamics of Flight*, New York: John Wiley

Etkin, B., 1972, *Dynamics of Atmospheric Flight*, New York: John Wiley

Evvard, J. C., 1950, Use of Source Distributions for Evaluating Theoretical Aerodynamics of Thin Finite Wings at Supersonic Speeds, *NACA Rept. 951*

Falkner, V. M., 1948, The Solution of Lifting Plane Problems by Vortex Lattice Theory, Aeronautical Research Council *R. & M. 2591*

Ferri, A., 1968, Review of SCRAMJET Propulsion Technology, *J. Aircr.* 5:3–10

Fox, R. L., 1971, *Optimization Methods for Engineering Design*, Reading, Mass.: Addison-Wesley

Frazer, R. A., W. J. Duncan, and A. R. Collar, 1938, *Elementary Matrices*, New York: Cambridge University Press

Gardner, M. F., and J. L. Barnes, 1942, Transients in Linear Systems, Vol. 1, *Lumped-Constant Systems*, New York: John Wiley

Gates, S. B., 1942, Proposal for an Elevator Manoeuverability Criterion, Aeronautical Research Council *R. & M. 2677*

Gazley, C., 1957, Deceleration and Heating of a Body Entering a Planetary Atmosphere from Space, *The RAND Corp. Rept. P-955*

Gilruth, R. R., 1943, Requirements for Satisfactory Flying Qualities of Airplanes, *NACA TN 755*

Glauert, H., 1926, *The Elements of Aerofoil and Airscrew Theory*, New York: Cambridge University Press

Goldstein, H., 1950, *Classical Mechanics*, Reading, Mass.: Addison-Wesley

Graham, M. E., and B. M. Ryan, 1960, Trim Drag at Supersonic Speeds of Various Delta Planform Configurations, *NASA TN D-425*

Halfman, R. L., 1962, *Dynamics*, Vols. I and II, Reading, Mass.: Addison-Wesley

Harder, K. C., and E. B. Klunker, 1957, On Slender-Body Theory and the Area Rule at Transonic Speeds, *NACA Rept. 1315*

Hayes, W. D., 1968, Linearized Supersonic Flow, Princeton, N.J.: Princeton University *AMS Rept. No. 852* (reprinted from North Am. Av. Rept. AL-222)

Hayes, W. D., 1971, *Sonic Boom*, edited by M. Van Dyke and W. G. Vincenti, *Ann. Rev. Fluid Mech.*, pages 269–290, Palo Alto, Calif.: Annual Reviews, Inc.

Hayes, W. D., and R. F. Probstein, 1967, *Hypersonic Flow Theory*, second edition, New York: Academic Press

Heaslet, M. A., and H. Lomax, 1954, Supersonic and Transonic Small Perturbation Theory, in *High Speed Aerodynamics and Jet Propulsion*, edited by W. R. Sears, Section D, Vol. VI, Princeton, N.J.: Princeton University Press

Henrici, P., 1962, *Discrete Variable Methods in Ordinary Differential Equations*, New York: John Wiley

Hildebrand, F. B., 1954, *Advanced Calculus for Engineers*, Englewood Cliffs, N.J.: Prentice-Hall, first edition, pages 42–47

Hildebrand, F. B., 1956, *Introduction to Numerical Analysis*, New York: McGraw-Hill

Hill, P. G., and C. R. Peterson, 1965, *Mechanics and Thermodynamics of Propulsion*, Reading, Mass.: Addison-Wesley

Hoerner, S. F., 1958, *Fluid-Dynamic Drag*, published by the author

Hoff, N. J., 1974, *Planes, Plane Makers and Pilots—Introduction to Aeronautics and Astronautics*, New York: Holt, Rinehart and Winston

Holloway, R. B., P. M. Burris, and R. P. Johannes, 1970, Aircraft Performance Benefits from Modern Control Systems Technology, *J. Aircr.* 7:550–553

Jones, R. T., 1946, Properties of Low-Aspect-Ratio Pointed Wings at Speeds Below and Above the Speed of Sound, *NACA Rept. 835*

Jones, R. T., 1972, Reduction of Wave Drag by Antisymmetric Arrangement of Wings and Bodies, *AIAA J.* 10:171–176

Jones, R. T., and D. Cohen, 1960, *High Speed Wing Theory*, Princeton Aeronautical Paperback No. 6, Princeton, N.J.: Princeton University Press

Jones, R. T., and M. D. Van Dyke, 1958, The Compressibility Rule for Drag of Airfoil Noses, *J. Aeronaut. Sci.* 25:171–172, 180

Kacprzynski, J. J., 1965, Supersonic Steady and Unsteady Flows over Slender Axisymmetric Bodies with Continuous or Discontinuous Surface Slopes, Part I, *AFOSR 66-0132*, MIT Fluid Dynamics Research Laboratory

Kacprzynski, J. J., and M. T. Landahl, 1967, Recent Developments in the Supersonic Flow over Axisymmetric Bodies with Continuous or Discontinuous Slope, *AIAA Paper No. 67*-5

Karamcheti, K., 1966, *Principles of Ideal-Fluid Aerodynamics*, New York: John Wiley

Kaula, W. M., 1966, *Theory of Satellite Geodesy*, New York: Blaisdell Publishing Co.

Kevorkian, J., 1966, The Two-Variable Expansion Procedure for the Approximate Solution of Certain Nonlinear Differential Equations, *Lectures in Applied Mathematics*, Vol. 7, *Space Mathematics*, Part III, American Mathematical Society

Kerrebrock, J. L., 1974, *Airbreathing Engines*, to be published, Cambridge, Mass.: MIT Press

Klawans, B. B., 1956, A Simple Method for Calculating the Characteristics of the Dutch Roll Motion of an Airplane, *NACA TN 3754*

Klopp, W. D., W. R. Witzke, and P. L. Raffo, 1966, Mechanical Properties of Dilute Tungsten-Rhenium Alloys, *NASA TN D-3483*

Kolk, W. R., 1961, *Modern Flight Dynamics*, Englewood Cliffs, N.J.: Prentice-Hall

Krasilshchikova, E. A., 1951, Uchenye Zapiski 154, Mekhanika 4:191–239 (translated as NACA TM 1383, 1956)

Kuethe, A. M., and J. D. Schetzer, 1959, *Foundations of Aerodynamics*, second edition, New York: John Wiley

Kutta, W., 1902, Auftriebskräfte in strömenden Flüssigkeiten, *Illustrierte Aeronaut. Mitt.*

Lamar, J. E., 1968, A Modified Multhopp Approach for Predicting Lifting Pressures and Camber Shape for Composite Planforms in Subsonic Flow, *NASA TN D-4427*

Lamb, Sir Horace, 1945, *Hydrodynamics*, sixth edition, New York: Dover Publications

Lanczos, C., 1962, *Variational Principles of Mechanics*, second edition, Toronto: University of Toronto Press

Langewiesche, W., 1944, *Stick and Rudder, An Explanation of the Art of Flying*, New York: McGraw-Hill

Liepmann, H. W., and A. Roshko, 1957, *Elements of Gasdynamics*, New York: John Wiley

Lighthill, M. J., 1948, Supersonic Flow Past Slender Pointed Bodies the Slope of Whose Meridian Section is Discontinuous, *Quart. J. Mech. Appl. Math.* 1:90–102

Lighthill, M. J., 1958, *Introduction to Fourier Analysis and Generalized Functions*, New York: Cambridge University Press

Lighthill, M. J., 1960. *Higher Approximations in Aerodynamic Theory*, Princeton Aeronautical Paperback No. 5, Princeton, N.J.: Princeton University Press

Loh, W. H. T., 1968. *Entry Mechanics*, Chapter 2 of *Re-entry and Planetary Entry Physics and Technology*, edited by W. H. T. Loh, Berlin: Springer-Verlag

Lotka, A. J., 1956, *Elements of Mathematical Biology*, New York: Dover Publications

Malina, F. J., and M. Summerfield, 1947, The Problem of Escape from the Earth by Rocket, *J. Aeronaut. Sci.* 14:471–480

Malthan, L. V., D. E. Hook, and J. W. Carlson, 1961, *U.S. Air Force Stability and Control Handbook*, Wright-Patterson Air Development Division, Flight Control Laboratory

Mangler, K. W., 1951, Improper Integrals in Theoretical Aerodynamics, Royal Aircraft Establishment *Rept. Aero 2424*

McKinney, L. W., and S. M. Dollyhigh, 1971, Some Trim Drag Considerations for Maneuvering Aircraft, *J. Aircr.* 8:623–629

Miele, A., editor, 1965, Theory of Optimum Aerodynamic Shapes, *Appl. Math. and Mech.*, Vol. 9, New York: Academic Press

Milne, R. D., 1964, Dynamics of the Deformable Airplane, Aeronautical Research Council *R. & M. 3345*

Milne-Thomson, L. M., 1962, *Theoretical Hydrodynamics*, fourth edition, New York: Macmillan

Morrisette, E. L., and D. L. Romeo, 1965, Aerodynamic Characteristics of a Family of Multistage Missiles at a Mach Number of 6.0, *NASA TN D-2853*

Morse, P. M., and H. Feshbach, 1953, *Methods of Theoretical Physics*, Parts I and II, New York: McGraw-Hill

Multhopp, H., 1950, Methods for Calculating the Lift Distribution of Wings (Subsonic Lifting Surface Theory), Aeronautical Research Council *R. & M. 2884*

Munk, M. M., 1924, The Aerodynamic Forces on Airship Hulls, *NACA TN 184*

Nielsen, J. N., 1960, *Missile Aerodynamics*, New York: McGraw-Hill

Nielsen, K. L., and J. L. Synge, 1946, On the Motion of a Spinning Shell, *Q. Appl. Math.* IV:201–226

Oswatitsch, K., and F. Keune, 1955, Ein Äquivalenzsatz für nichtangestellte Flügel kleiner Spannweite in Schallnaher Strömung, *Z. Flugwiss.* 3:29–46

Oswald, W. B., 1932, General Formulas and Charts for the Calculation of Airplane Performance, *NACA TN 408*

Phillips, W. H., 1948a, Effect of Steady Rolling on Longitudinal and Directional Stability, *NACA TN 1627*

Phillips, W. H., 1948b, Appreciation and Prediction of Flying Qualities, *NACA Rept. 927*

Pratt and Whitney Aircraft, 1968, *Aeronautical Vest-Pocket Handbook*, eleventh edition

Puckett, A. E., and H. J. Stewart, 1947, Aerodynamic Performance of Delta Wings at Supersonic Speeds, *J. Aeronaut. Sci.* 14:567–578

Ralston, A., and H. Wilf, editors, 1960, *Mathematical Methods for Digital Computers*, New York: John Wiley

Rauscher, M., 1953, *Introduction to Aeronautical Dynamics*, New York: John Wiley

Reynolds, W. C., 1965, *Thermodynamics*, New York: McGraw-Hill

Reynolds, P. A., and A. R. Pruner, 1972, Total In-Flight Simulator (TIFS)—A New Aircraft Design Tool, *J. Aircr.* 9:392–398

Ribner, H. S., 1945, Propellers in Yaw, *NACA TN 820*

Rice, C. M., and W. H. Arnold, 1969, Recent NERVA Technology Development, *J. Spacecraft & Rockets* 6:565–569

Rodden, W. P., and J. P. Giesing, 1970, Application of Oscillatory Aerodynamic Theory to Estimation of Dynamic Stability Derivatives, *J. Aircr.* 7:272–273

Roskam, J. R., *et al.*, 1968–69, An Analysis of Methods for Predicting the Stability Characteristics of an Elastic Airplane, *NASA Contractor Repts. 73274 through 73277*

Scala, S. M., 1968, *Entry Heat Transfer and Material Response*, Chapter 8 of *Re-entry and Planetary Entry Physics and Technology*, edited by W. H. T. Loh, Berlin: Springer-Verlag

Schlichting, H., 1960, *Boundary Layer Theory*, fourth edition, New York: McGraw-Hill

Sears, W. R., 1954, Small Perturbation Theory, in *High Speed Aerodynamics and Jet Propulsion*, edited by W. R. Sears, Section C, Vol. VI, Princeton, N.J.: Princeton University Press

Sears, W. R., editor, 1954, General Theory of High Speed Aerodynamics, in *High Speed Aerodynamics and Jet Propulsion*, Vol. VI, Princeton, N.J.: Princeton University Press

Seckel, E., 1964, *Stability and Control of Airplanes and Helicopters*, New York: Academic Press

Shapiro, A. H., 1953, *The Dynamics and Thermodynamics of Compressible Fluid Flow*, Vols. I and II, New York: Ronald Press

Sims, J. L., 1964, Tables for Supersonic Flow Around Right Circular Cones at Small Angle of Attack, *NASA SP-3007*

Spencer, B., Jr., and C. H. Fox, Jr., 1966, Hypersonic Aerodynamic Performance of Minimum-Wave-Drag Bodies, *NASA TR R-250*

Stein, G., and A. H. Henke, 1971, A Design Procedure and Handling-Quality Criteria for Lateral-Directional Flight Control Systems, *USAF Technical Rept. AFFDL-TR-70-152*

Stinton, D. E., 1966, *The Anatomy of the Airplane*, New York: American Elsevier

Stoney, W. E., Jr., 1958, Collection of Zero-Lift Drag Data on Bodies of Revolution from Free-Flight Investigations, *NACA TN 4201*

Sutton, G. P., 1956, *Rocket Propulsion Elements*, second edition, New York: John Wiley

Taylor, J. W. R., editor, published yearly, *Jane's All the World's Aircraft*, New York: McGraw-Hill

Thwaites, B., editor, 1960, *Incompressible Aerodynamics*, Oxford: Clarendon Press

United States Air Force Handbook of Geophysics, 1960, New York: Macmillan

Ursell, F., and G. N. Ward, 1950, On Some General Theorems in the Linearized Theory of Compressible Flow, *Quart. J. Mech. Appl. Math.* 3:326–348

Van Dyke, M. D., 1954a, A Study of Hypersonic Small-Disturbance Theory, *NACA Rept. 1194*

Van Dyke, M. D., 1954b, Applications of Hypersonic Small-Disturbance Theory, *J. Aeronaut. Sci.* 21:179–186

Van Dyke, M. D. 1964, *Perturbation Methods in Fluid Mechanics*, New York: Academic Press

Wagner, S., 1969, On the Singularity Method of Subsonic Lifting-Surface Theory, *J. Aircr.* 6:549–557

Ward, G. N., 1949, Supersonic Flow Past Slender Pointed Bodies, *Quart. J. Mech. Appl. Math.* 2:75–97

Watkins, C. E., H. L. Runyan, and D. S. Woolston, 1959, A Systematic Kernel Function Procedure for Determining Aerodynamic Forces on Oscillatory or Steady Finite Wings at Subsonic Speeds, *NASA TR R-48*

Whitcomb, R. T., 1956, A Study of the Zero-Lift Drag-Rise Characteristics of Wing-Body Combinations Near the Speed of Sound, *NACA Rept. 1273*

Whitham, G. B., 1952, The Flow Pattern of a Supersonic Projectile, *Commun. Pure and Appl. Math.* 5:301–348

Wolowicz, C. H., L. W. Strutz, G. B. Gilyard, and N. W. Matheny, 1968, Preliminary Flight Evaluation of the Stability and Control Derivatives and Dynamic Characteristics of the Unaugmented XB-70-1 Airplane, Including Comparisons with Predictions, *NASA TN D-4578*

Woodward, F. A., 1968, Analysis and Design of Wing-Body Combinations at Subsonic and Supersonic Speeds, *J. Aircr.* 5:528–534

Wykes, J. H., and R. E. Lawrence, 1971, Estimated Performance and Stability and Control Data for Correlation with XB-70-1 Flight Test Data, *NASA CR 114335*

Zadeh, L. A., and C. A. Desouer, 1963, *Linear System Theory*, New York: McGraw-Hill

Zucrow, M. J., 1958, *Aircraft and Missile Propulsion*, New York: John Wiley

INDEX

Acceleration, 7, 22, 38, 45, 278
longitudinal, 2
linear, 47
inertial, 276
Actuator, hydraulic, 42
Admittances, impulsive, 208
indicial, 208, 216
Aerodynamicist, 4, 34, 67, 81
Aerodynamics, 1, 13, 26, 64, 65, 68, 108, 303
Aeroelasticity, 45
Afterburner, 134
Aileron, 4, 11, 41, 42, 182, 201, 203, 207, 208, 222, 239, 256
Aircraft, 3, 5, 7, 9, 14, 24, 30, 34, 41, 42, 47, 49, 53, 62, 65, 113, 151, 169, 171, 197, 211, 238, 239, 241, 251, 254, 257, 261, 262, 263
conventional, 24
experimental, 5
military, 5
rotary-wing, 42
subsonic, 260
supersonic, 16
supersonic cruising, 16
transport, 5
Airfoil, 58, 74, 75, 76, 78, 79, 83, 84, 89, 90, 91, 92, 93, 94, 95, 96, 97, 98, 102, 110, 121, 125, 126, 127, 128, 130
supersonic, 327
transonic, 12
two-dimensional, 87
Airframe, 42
Airloads, 34, 40, 46, 53, 86, 87 97, 128, 131, 190, 255, 303, 319
vibratory, 19
Airplane, 4, 13, 32, 48, 58, 61, 63, 68, 139, 203, 204, 209, 244, 262
Airspeed, 3, 4, 51, 86, 242, 258, 271
Airstream, 11
Alleviator, gust, 19
Altitude, 2, 33, 39, 40, 63
Analogy, Reynolds, 334
Angle, downwash, 192
Euler, 34, 35, 36, 37, 38, 47, 58, 64, 69, 257
from zero lift, 182
of attack, 3, 13, 26, 33, 34, 38, 40, 46, 51, 52, 58, 62, 63, 65, 73, 74, 76, 80, 85, 88, 94, 102, 108, 109, 120, 125, 126, 129, 131, 132, 137, 182, 183, 188, 189, 191, 195, 200, 201, 231, 237, 243, 248, 249, 252, 278, 295, 296, 303, 310, 313, 328, 330, 364
of sideslip, 182
sidewash, 198
of stall, 67
variable, 79
Anhedral, 15
Approximation, inviscid, 65
Atmosphere, 38, 46
Augmenter, 134
Autopilots, 4

Axis, 28, 32, 50, 76, 78, 83, 84, 109, 137, 190
  vertical, 3
Axes, 32, 34, 35, 43, 47
  aerodynamic, 34
  inertial, 35

Balloons, 1
Barrier, sonic, 14
Beam, structural, 7
Bending, wing, 45
Biplane, wing, 7
  wire-braced, 5
  Wright, 5
Bipropellant, 152
Boeing B-52, 9
Boeing Company, 10
Bombers, V-, 13
Boom, sonic, 17
Boost, 24, 25, 276, 281, 286, 337
  power, 42
Booster, 22, 266, 279, 281, 289, 296, 303, 306, 307, 310, 311, 312, 353
  ballistic, 24
Box, beam, 26
  depth, 8
  structural, 7
Brayton Cycle, 135

Cabins, 8
Cable, elastic, 42
  control, 43
Camber, 85, 94, 102, 108, 116, 120, 125, 126, 129, 192, 195, 247
Canard, 3, 17
Capsules, 23
Ceiling, 7, 25
Chamber, combustion, 2
Chord, 14, 85
Chordwise, 11, 90
Climb, 32
Cockpit, 4, 8
Coefficient, 51
  ballistic, 339
  of thrust, 186
  viscosity, 51, 93
Collocation, 107
Combustion, 2
Component, 5, 41
  airload, 51
  Cartesian, 29, 31, 37, 68
  velocity, 38
Compressibility, 19
Compression, ram, 141
Compressor, 136, 156
Configuration, 5
  aerodynamic, 17
  optimum, 4
  supersonic, 20
  unswept, 19
  vehicle, 2, 27
Control, 1, 3, 16, 17, 42
Core, 11, 17, 76
Coriolis Law, 29, 45
Cornell Aeronautical Laboratory, 239
Corridor, entry, 346
Cowlings, 8
Crossflow, 89
Cruise, 5, 15, 16, 19, 32
  supersonic, 18, 20
Cruisers, supersonic, 16
Cryogenic, 152
Curtiss R-4, 5
Curtiss-Wright Corporation, 6
Cycloids, 359
Cylinders, 8

Damping, structural, 46
Dash, supersonic, 13
Density, 11, 180
  air, 51
  atmospheric, 7
Design, 4, 5, 10, 27
  aerodynamic, 58
Dimensions, cross-sectional, 2
  linear, 26
  spanwise, 85
  streamwise, 14
Direction, 29
  zero lift, 52

Disc, actuator, 139
Displacement, 37, 41, 42, 43
Doublet, 114
  supersonic, 322
Douglas Aircraft Company, 8
  DC-3, 7
Downdrafts, 60
Drag, 2, 4, 7, 16, 17, 22, 23, 24, 25, 34, 48, 52, 53, 58, 62, 63, 66, 67, 84, 93, 94, 95, 99, 102, 110, 126, 131, 133, 138, 139, 188, 189, 191, 237, 246, 248, 259, 260, 279, 289, 293, 303, 305, 306, 307, 308, 319, 321, 334, 341, 348, 360, 362
  induced, 4, 7, 110, 113, 369
  supersonic, 13, 14
  transonic, 14
  vehicle, 139
  wave-making, 19
Dynamics, 34, 41
  structural, 45

Efficiency, adiabatic, 158
  burner, 159
  Carnot, 154
  cruising, 4, 9
  kinetic-energy, 171
  work, 139
Electromagnetics, 31
Elevator, 3, 13, 41, 42, 43, 44
  trailing-edge, 17
Elevons, 13, 17
Empennage, 3, 10, 15, 178
Energy, kinetic, 25, 43, 44, 60, 110, 111, 237, 243, 256
  potential, 25, 43, 60, 243, 256, 280
  specific, 25
  thermodynamic, 97
Engineer, 65, 349
  aeronautical, 256
  aerospace, 5
  structural, 11
Engineering, 5, 134
  control-system, 22
  science, 27
Engine, 41, 134, 137, 148, 151, 156, 163, 166, 168, 170, 276
  cycle, 160
  jet, 134
  masses, 11
  parts, 7
  throttle, 41
  turbojet, 12
Engines, 2, 5, 8, 16, 24
  reciprocating, 2
  turbofan, 19, 41
  turbojet, 10
  wing-mounted, 11
England's National Science Museum, 5
Entry, atmospheric, 2
  ballistic, 25
Equation, Kelvin's, 69
  Laplace's 71, 74, 81, 106
  of motion, 1
  vector, 39
Extrapolation, 276

Fan, 99
Fields, gravitational, 2
  wing flow, 11
Fighters, air-superiority, 15
  jet, 12
  single-engine, 13
  supersonic, 15
Fin, 3, 48, 63, 201
Fingers, 17
Flaps, 3, 41
  wing, 4
Flight, 1, 2, 3, 17, 25, 32, 33, 38, 42, 51, 52, 58, 60, 61, 63, 80, 84, 117, 131, 137, 146, 177, 203, 278
  atmospheric, 24
  dynamics, 58
  gliding, 9
  rectilinear, 33
  subsonic, 26, 102
  supersonic, 13, 16
Flight-path, 61, 62
  angle, 40

determination, 34
modification, 23
Flow, field, 34
subsonic, 85
supersonic, 328, 331
transonic, 11
wing, 84
Flutter, 11, 45
Force, 62, 184, 185
aerodynamic, 2, 23, 43, 256, 318
chord, 187
dynamic, 11
external, 28
inertia, 22
propulsive, 2
"spring," 41
Formulas, aerodynamic, 7
Frame, inertial, 34, 43
Friction, 42, 49
skin, 4, 7
Function, stream, 75
Fuselage, 2, 3, 8, 9, 11, 13, 14, 15, 18, 20, 23, 41, 53, 67, 139, 190, 197, 198, 200
axis, 3
frame, 8

Gases, 51
atmospheric, 2
Gear, retractable landing, 4
General Dynamics, 14, 18
Generator, gas, 148
Gliders, 60, 133
blunt-nosed, 23
hypersonic, 346
Gliding, 2, 9
Grain, 152
Gravity, 40, 48, 68
Guidance, 1

Hammerhead, 249
Helicopter, rotor, 41
Hingeline, 44, 45
control-surface, 42
Horsepower, 7

Inertia, 25, 26, 30, 31, 33, 42, 44, 47, 48
coupling, 42, 45
"elastic, " 46
principal moments of, 31
Inlet, 16, 133, 134, 157
air, 2
arrangements, 15
under-wing, 17
Inputs, 41, 58
Instruments, gyroscopic, 38
Interceptor, aircraft, 12
delta-winged, 13, 14
supersonic, 15

Jet, 133, 169
subsonic swept-winged, 12

Lagrange multiplier, 353, 355, 358, 361
Lagrange polynomial interpolation, 276
Lagrange's technique, 42
Landcraft, 1
Landing, 4, 19, 26, 61
touchdown, 14
Layout, 4
Length, chord, 13
chordwise, 26
Life, structural, 13
Lift, 2, 4, 7, 13, 15, 18, 19, 22, 23, 25, 26, 30, 34, 46, 52, 53, 58, 94, 102, 109, 120, 125, 130, 131, 180, 191, 195, 248, 250, 278, 283, 303, 341, 369
coefficient, 108, 132
loads, 7
shock, 97
wing, 13
Liftoff, 14, 276, 279
Lines, Mach, 96, 119, 122, 202
Lipschitz condition, 270
Loading, wing, 15
Loads, aerodynamic, 63
Lockheed, 20, 28, 52
Long-period, 62

Magnitude, 29
Maneuverability, 5
Maneuvers, 25
Mass, 27, 28, 97, 276
  aircraft center of, 47
  center of, 3
  generalized, 46
Matrix, 36, 38, 39, 47
  notation, 36
Mechanics, 31
  fluid, 1
  solid, 1
Missile, 49, 203
  air-to-air, 47
  cruciform, 32
Mode, *i*th, 46
  short-period, 45
Module, Apollo Command, 23
Mohr circle relations, 33
Moment, 38, 42, 46, 49, 50, 52, 180
  aerodynamic, 22
  bending, 7
  external, 28
  hinge, 43, 49
  pitching, 3, 13, 19, 30, 46, 49, 51, 53, 60, 61, 74, 83, 94, 108, 109, 125, 128, 131, 183, 190, 191, 192, 193, 195, 196, 203, 251, 303, 325
  rolling, 3, 4, 11, 30, 39, 49, 108, 109, 131, 199, 200, 201
  yawing, 3, 30, 39, 49, 201, 202
Momentum, 2, 28, 41, 97
  angular, 32, 41
Monoplane, cantilever, 8
Monopropellant, 152
Morphology, 7
  of flight vehicles, 1
Motion, 38, 39, 41, 48, 52, 62, 63, 65, 207
  harmonic, 45
  inviscid fluid, 65, 84, 110
  longitudinal, 41, 58
  transonic fluid, 88
Motors, rocket, 2

Nacelle, 2, 138, 190
Nap of the earth, 5, 19
NASA, 21, 24, 52, 56
NASA Langley Research Center, 12
Newton's Second Law, 27
North American Rockwell, 12, 16, 219
Nose, fuselage, 13
Nozzle, 134, 136, 143, 153, 156, 160, 169
  exhaust, 2
Number, Mach, 4, 11, 13, 30, 48, 52, 63, 70, 81, 84, 100, 101, 102, 117, 127, 129, 136, 142, 143, 144, 146, 147, 158, 159, 164, 166, 167, 171, 172, 188, 189, 225, 231, 243, 263, 303, 306, 307, 310, 311, 312, 318, 319, 323, 334, 365, 366, 367, 369
  Reynolds, 53, 54, 56, 66, 243, 305

Orbit, inertial, 2
Origin, gravitational, 38
Oscillation, 45
Outlets, 133, 134

Parameters, 2, 8, 51
  hypersonic, 327
Payload, 2, 21, 22, 23, 25, 192
Performance, dynamic, 25, 265
  cruising, 15
Permutation, cyclic, 30
Phugoid, 62, 230, 236
Pilot, automatic, 41
Pitch, 22, 38, 45, 190, 231, 235, 277, 295
  damping, 195
Plane, 3, 44, 51, 53, 93
  horizontal, 3
  of symmetry, 3, 10, 32, 33, 39, 51, 52, 177, 190, 230, 257, 265
  of the wing, 103
  symmetry, 32
  vertical, 40
Planform, 109, 114, 117
  lifting-surface, 13
Pod, jet-engine, 2

Point, design, 4, 5, 16, 20, 150
Pressure, 86, 98
  doublets, 103
  dynamic, 25, 53
  positive, 88
Projectiles, 21
Propeller, 2, 41
  variable-pitch, 7
Properties, ambient air, 39
  stagnation, 134
Propulsion, 1, 38
  higher speed, 12
  reaction, 2
Pylons, 19

Quarter-chordline, 18

Radian, foot-pounds per, 42
Ramjet, 2, 139, 140, 141, 142, 143, 144, 145, 146, 148, 150, 156
  hypersonic, 2
Ramps, 17
Ratio, amplitude, 209
  aspect, 26, 53
  by-pass, 11, 169
  lift-to-drag, 16
  low-bypass, 19
  structural aspect, 13
References, 371
Research, aerodynamic, 8, 17, 19
"Rigid-body," 46
Rocket, 2, 133, 139, 152, 154, 170, 300
  liquid, 2
  propulsion, 152
  solid, 2, 24
  solid-propellant, 24
Rods, push, 42
Roll, 38, 46, 207, 222
  convergence, 225
  Dutch, 211, 243
Rolling, 4, 13, 49, 241
Roots-locus technique, 219, 225
Rotation, 12, 33, 34, 35, 37, 42
  antisymmetrical, 13
  differential, 11
  machinery, 41
  matrix, 36, 37
Rotational displacements, 41
Rudder, 3, 17, 41, 42, 48, 197, 256

Sailplane, 3, 9, 60
  Diamant, 9, 18
Satellite, 50, 300
Scalar, 27, 29, 31
Semispan, 25
  structural, 25
Shears, 7
Ships, submersible, 1
  surface, 1
Shuttle, 24, 25
Sideforce, 39
Sideslip, 3, 34, 108, 197, 199, 217, 239, 241, 256, 257, 295, 296
  angle, 197, 222, 278
Slipstream, 2, 138
Smithsonian Institution, 5
Soaring Society of America, 9
Space, inertial, 38
Spacecraft, 1, 2, 5
Space Shuttle, 156, 281
Span, 4, 10
  aerodynamic, 19
Speed, 33, 86, 90, 190
  of sound, 51
Speeds, flight, 25
  flying, 7
  subsonic, 23
  supersonic, 15, 17, 188
Spoilers, 4, 11, 41
Spring, torsional, 42
Stability, 1, 34, 244
  closed-loop, 22
Stabilizer, canard, 7
  horizontal, 3, 11
  plane of the, 51
  vertical, 3, 17
Stalling, 8
Strips, chine, 20
Structures, 1, 14, 26
  primary wing, 11

truss, 7
wing, 7
Strut, 5
compression, 7
Surface, aerodynamic, 26
canard trimming, 16
lifting, 10, 11, 17
swept lifting, 12
Symmetry, 73, 76
axis of, 33
longitudinal, 32
System, 50
inertial navigational, 38
linear, 26
nonlinear, 62
propulsion, 2, 7, 51
stability augmentation, 4

Tabs, 42
Tail, 3, 190, 248
Takeoff, 4, 5, 19, 26, 61
Tank, gas, 22
wing, 11
Tensors, 31
Cartesian, 47
first-rank, 37
Theory, aerodynamic, 14, 66, 67, 71, 208
hypersonic small-disturbance, 332
Thermodynamics, 68, 136
Throttle, 58, 163, 165
Thrust, 2, 7, 10, 22, 41, 58, 133, 139, 189, 237, 246, 259, 280, 293
gross, 139
horsepower, 139, 138
HP, 7
net, 138, 139
rocket, 22
static, 148
Torque, 7, 134
Tower, launch escape, 23
Tradeoffs, 4
Trajectory, 22, 62
control, 23
in state space, 41
Transformation, fast Fourier, 214
Transport, 19
supersonic, 5, 17
Transportation, 2
system, 24
Trim, 7, 51
Trimming, 3, 4, 16
Triplanes, 7
Tube, Pitot, 38
shock, 97
Turbine, 156, 158, 159, 160, 163
gas, 41
Turbofan, 2, 11, 135, 140, 148
engines, 41
Turbojet, 2, 11, 133, 134, 135, 136, 139, 140, 141, 145, 148, 150, 151, 156, 162, 163, 164, 168, 169, 170
Turboprop, 2
Turbopumps, 152
Turbulence, atmospheric, 19
Turn, helical, 47

Updrafts, 60

Vacuum, 45
Vector, 27, 28, 31, 35, 37, 39, 41, 45, 47, 63, 138
airspeed, 3, 33, 51
angular velocity, 27
axial, 37
component, 29
control, 40
downward force, 38
identity, 30
radius, 50
state, 39, 40
unit, 29
velocity, 23, 27, 32, 44, 48
Vehicle, 2, 5, 8, 10, 25, 27, 28, 29, 32, 33, 34, 37, 38, 40, 42, 43, 45, 48, 50, 51, 52, 58, 61, 62, 174, 192, 197, 203, 208, 219, 230, 237, 238, 239, 251, 267, 276, 279, 281, 283
aerodynamic, 51
atmospheric entry, 23
axes, 32, 34, 35, 39, 41

ballistic, 23
boost, 21
design, 10
dynamics, 26
flight, 1, 4, 5, 27, 42, 47, 65, 67, 133, 173
rigid, 45
transport, 11
Velocity, 12, 27, 28, 29, 38, 40, 41, 44, 47, 49, 97, 138, 207, 280
angular, 17, 27, 37
linear, 47
pitch, 183
potential, 69
rolling, 108
Volume, displacement, 18
Vortex, 15, 17
drag, 123
horseshoe, 105
interaction, 15
Vortices, 17, 76, 79

Warping, antisymmetric, 7
Waves, Mach, 89, 90, 92, 98, 99, 102
shock, 17, 18
Weathercock, 198
Weathervane, 3, 48, 64
Weight, 2, 33
gross vehicle, 8
maximum, 10
tradeoff, 14
Wind-tunnel, 65
two-dimensional, 74
Wing, 3, 8, 9, 10, 13, 14, 15, 17, 19, 26, 53, 58, 65, 66, 67, 71, 72, 73, 85, 87, 88, 93, 95, 96, 108, 110, 118, 128, 131, 132, 139, 190, 191, 192, 195, 197, 201, 202, 241, 247, 249, 251
biplane, 7
cantilever monoplane, 8, 9, 10
flying, 2
large-aspect-ratio, 18
streamlined, 14
swept, 11
swept-back, 11
unswept, 25
variable-sweep, 18, 19
Wing-body blending, 20
Wingspan, 4, 26, 51, 107, 113
Wingtips, 17
Wires, bracing, 7
Wright Flyers, 7
Wright-Patterson Air Force Base, 5

Yaw, 22, 38, 197, 228, 260
Yawing, 49